# EXPLORING Tech CAREERS

## Real People Tell You What You Need to Know



Ferguson Publishing Company • Chicago

# Editorial Staff

***Project Editor:*** Elizabeth Oakes

***Editors:*** Andrew Morkes, Carol Yehling

***Writers:*** Dale Gregory Anderson, Jerry Barker, Marty Breheney, Bill Brymer, Shawna Brynildssen, Kerstan Cohen, Mickey Cohen, Mark Cornell, Anita Coryell, Patty Cronin, Deborah Douglas, Jennifer Elcano, Deidre Elliott, Cathy Foster, Mariko Fujinaka, Nic Gengler, Mike Gosling, Bonnie Anne Griffin, Monique Hamzé, Laura Heller, Mathew T. Hohmann, Louise Howe, Sally Jaskold, Kim Kazemi, Jane Lawrence, Nicholas J. Mangano, Andy Morkes, Andrea Page, Melissa Rigney Baxter, Susan Risland, Alicia Saposnik, Tim Schaffert, Frances Stocker-Burton, Patricia S. Stoeppelwerth, Carol Yehling

***Interior Layout and Graphics:*** Joe Grossmann, Grossmann Design & Consulting

***Cover Design:*** Norman Baugher, Baugher Design

***Bibliographers:*** Lynne Perrigo, Alan Wieder

**Exploring Tech Careers: Real People Tell You What You Need To Know**



ISBN 0-89434-244-4

Printed in the United States of America

Published and distributed by
Ferguson Publishing Company
200 West Madison Street, Suite 300
Chicago, Illinois 60606
800-306-9941
Web site: http://www.fergpubco.com

V-8



Library of Congress Cataloging-in-Publication Data

Exploring tech careers: real people tell you what you need to know, 2nd ed.
Elizabeth Oakes, editor
p. cm.
Includes bibliographical references and index.
ISBN 0-89434-244-4 (set)
1. Technology—Vocational guidance. I. Oakes, Elizabeth
T65.3.E95 1998
602.3' 73—dc21

98-29296
CIP
AC

# contents

# acknowledgements

The editorial staff of Ferguson Publishing Company would like to express its appreciation to all of the individuals profiled here. Your enthusiasm for your work and eagerness to share your experiences with others have made this set possible. Also, many thanks to all of the educators, employers, and associations whose additional comments and all-important factual data help make this reference work as up to date as possible.

# introduction

**Technicians** are popping up all over the place. Your auto mechanic isn't a mechanic anymore, he's an automotive technician. The person at the nursery isn't a gardener anymore, she's a horticultural technician. The person who installs your cable isn't just an installer, she's a cable TV Technician. And air traffic controllers, airplane pilots, drafters? These are technicians too? What gives? Is this just another in a seemingly endless line of changes in the name of political correctness? Who do these people think they are, anyway?

## what is a technician?

Technicians are highly specialized workers who work with scientists, physicians, engineers, and other professionals, as well as with clients and customers. They assist these professionals in many activities, assist clients or customers, and they frequently direct skilled workers. They work in factories, businesses, science labs, hospitals, law offices, clinics, shops, and private homes. Some work for themselves as consultants. They are found in all facets of the work world and are one of the fastest growing job ranks. *Exploring Tech Careers* seeks to motivate readers about this classification of work by profiling individual technicians and technologists who do more than one hundred "technical" jobs.

First and foremost, a technician is a specialist. When looking at the range of job classifications in the traditional, hierarchical sense, the technician is the middleman, between the scientist in the laboratory and the worker on the floor;

between the physician and the patient; between the engineer and the factory worker. In short, the technician's realm is where the scientific meets the practical application, where theory meets product.

Little by little, however, industries and businesses are recasting the traditional paradigm. All workers, be they scientist, manager, technician, or line worker, are coming to be viewed as part of a team. Competency and knowledge are the new standards by which workers are valued, not rank alone. Thus technicians are seen by many as valuable employees in their own right, not simply junior scientists or engineers or wannabe doctors. The work they do is normally supportive, yes, of their more prestigious professional colleagues; in fact basic to the definition of technician is their supportive role—their work adds to and enhances the work of someone else. What is often overlooked, however, is that the work of the professionals frequently could not be done without them. Think about it. Would you really want your radiologist operating the X-ray machine? Or your lawyer spending hours of time in the library researching your case? Or the automotive engineer fixing your car? Of course not. Assuming that the professional could even remember that far back in his or her training to perform the work, it is certain that he or she does not know the technician's specialty as well as the technician. No, the radiologist should be hard at work diagnosing health problems; the lawyer creating brilliant summations to win cases; the engineer designing more efficient, better cars. The technician's job is to make all of this possible. (No matter what your educational background and natural abilities, it's hard to be brilliant, after all, when you have no material to work with.)

One reason for the growing reliance on technicians is, of course, our own growing reliance on technology. As businesses switch to automated systems, as products we buy become more technologically complex, technicians are needed to help design, implement, run, and repair such systems or equipment, be they automobiles, airplanes, X-ray machines, or computer networks. Even in areas that aren't themselves obviously "techno," there exists a certain expectation of speed, efficiency, and quality that is frequently the domain of the technician.

Another reason for the ascension of technicians is their overall cost when compared to that of highly paid professionals such as doctors, lawyers, and engineers and scientists. While in some industries technicians work alongside professionals as valued members of the team, contributing their unique knowledge and skills in areas professionals lack the time to fully develop, technicians in other industries are actually replacing professionals because they are cheaper. The field of nursing, for example, is one area in which the replacement of professionals with technicians is a hotly debated issue. Hospitals, on the one hand, are faced with the need to cut costs, while professional, registered nurses, on the other hand, argue that patient care will be compromised by relying on less highly trained staff.

In most cases the increasing reliance on technology creates more jobs for technicians. In a few areas, however, technology is actually replacing technicians. Meteorological technicians, for example, are being phased out completely from jobs with the federal government (National Weather Service). Their painstaking instrument readings and weather data collection work can now be done entirely with high tech instruments. Telephone and cable TV line installers and repairers,

EKG technicians, and home electronic entertainment equipment repairers are other technician specialties that are in decline because of advances in technology.

Overall, however, the future for technicians looks excellent. The U.S. Department of Labor estimates technician careers to grow at a rate of 32 percent through 2005. The overall growth rate for all jobs combined is estimated at 21 percent. As one writer put it, "As the farm hand was to the agrarian economy of a century ago and the machine operator was to the electromechanical industrial era of recent decades, the technician is becoming the core employee of the digital Information Age" (Louis S. Richman, "The New Worker Elite" *Fortune* 130 (1994): 56).

Technician careers are appealing for another, very practical reason: they are a short track to a good job. Most technician careers initially require two years or less of postsecondary training. A few require a bachelor's degree to be competitive, and a very few do not require a high school diploma. For someone interested in medicine but who views with dread eight or more years of college—along with the tremendous responsibilities of being a doctor—a career as a medical technician may be just the ticket. For the aspiring engineer more interested in the day to day, practical applications of engineering rather than the theoretical side of science, a career as an engineering technician may be perfect. Technicians often work right alongside the professionals. They are there for and often directly involved in the ground-breaking discoveries, long-awaited advances, and cutting-edge leaps. For a comparatively small investment in time and money, a person can emerge with a practical, highly marketable career.

Perhaps most importantly, a technician's career is a respectable one; the niche is not populated by failed doctors and dentists and engineers, but by intelligent, highly skilled, and dynamic people who enjoy being on the cutting edge and lending their special kind of expertise to all areas of the work world. It is our hope by letting real technicians talk about what they do, by allowing readers to experience what it's really like to do the work from the people who do it, that *Exploring Tech Careers* will help readers—primarily high school students, but in reality anyone interested in a new career—explore and evaluate some of the exciting possibilities a technician's career can offer.

## how to use this book

Each chapter of *Exploring Tech Careers* is divided into twelve main sections (and their relevant subsections) that discuss specific aspects of each job. Integral to each chapter is an interview with a person who actually does the job. What better way to learn about a career than through someone who actually does it? Some people we interviewed are just starting out; some have been in the business for several years, or are even nearing the end of their careers. All are fervent about their work and were glad to have the opportunity to share some advice with students and others who are just beginning to consider a career or a career change.

## the opening

The opening page of each chapter gives the reader an at-a-glance overview of the job being discussed: definition, alternative job titles, salary range (beginning, experienced, and very experienced), educational requirements, certification or licensing, and outlook. Also given here are the *Guide to Occupational Exploration* (GOE) subgroup number, the *Dictionary of Occupational Titles* (DOT) group number, and the Occupational Information Network numbers from the *O*NET Dictionary of Occupational Titles*. Most of the jobs described in this work correspond exactly to a particular GOE and DOT category; however, some are combinations of several DOT or GOE listings. When this is the case, the numbers listed in the opening are followed by asterisks. A discussion of GOE and DOT classifications appears later in each chapter under the heading "What Are Some Related Jobs." Also appearing on the first page are lists of high school subjects and personal interests relevant to the specific job.

## the introduction

The opening text section of each chapter is an introductory scene showing the profiled person at work. Some are dramatic, as in "Air Traffic Controller," where the scene begins with a pilot calmly telling the controller that he is going to crash his plane; and some are mundane, as in "Dental Hygienist," where we are introduced to a busy, but organized, Mom as she gets her kids off to the baby-sitter's and gets to work well before her first appointment. By appealing to the reader on an emotional, imaginative level, the introduction involves the reader immediately and gives him or her a feeling about the particular type of work.

## what does a technician do?

This section describes in broad terms the basics of what the technician does. Any variations in job duties and titles are covered as well; actual job duties and titles may vary according to where a technician works, and an effort is made here to discuss some of the more common ones. Primary duties and secondary duties are also covered.

## what is it like to be a technician?

In this section, the reader gets a first-person account of what it's like to do the job. This section prompts the reader to ask himself or herself, "Could I really do this?" and, more importantly, "Do I want to?" Here the reader formally meets the profiled person and sees him or her at work on a typical day or series of days. Typical tasks or duties are covered, and a distinction is made between primary duties and secondary ones. As the technicians describe what their work days are like, readers can begin asking themselves if this is something they really want to do. For example, jobs that sound glamourous on the surface may seem more monotonous after getting down to the details of the day; or, conversely, jobs that don't sound that exciting at first glance can be seen as very rewarding when viewed through the eyes of an enthusiastic technician. Readers should remember, however, that these are personal accounts; not all technicians will have similar situations and experiences.

## have I got what it takes to be a technician?

In this section, the profiled person tells the reader what personal qualities are important for success in the job and discusses what he or she likes and dislikes about the job. Where appropriate, educators and employers of technicians add their own opinions about personal qualities and what they like to see in an employee.

What is the most common first priority skill required (or, more realistically, hoped for) in a technician? Communication skills. This may be surprising at first because, when thinking technicians and technical work, one might consider computer skills to be first. Computer skills *are* critical in most of the jobs covered here, but the so-called soft skills—being able to clearly communicate ideas, problems, and solutions, both verbally and in writing; being able to take direction from superiors or co-workers; being able to get along well with co-workers and superiors (and frequently, clients or customers); as well as being able to think critically, that is, solve problems—are what often make the difference between being qualified for a job and actually getting hired, not to mention being able to advance.

Another common trait important to being a successful technician is the willingness to commit to lifelong learning. Nearly every technician interviewed expressed frustration at the amount of information there is to keep up-to-date with (although most of them liked the challenge). Most of the technicians regularly read trade magazines and books, and attend conferences and seminars to keep up with constantly changing technology and developments. Their dedication is demonstrated by the fact that such self-improvement is often done on the technician's own time.

## how do I become a technician?

This section explores the educational path to the job, frequently giving examples from the profiled person's own experiences. A subsection on high school details important classes and activities to take in preparation for the job. It is important to note here that technicians' jobs are not for academic slouches. Many require complex math skills, as well as skills in a science such as physics or chemistry. Shop classes, such as auto shop, machine shop, and electronics shop, are also frequently very useful, and nearly everyone stressed taking English composition and speech classes to improve communication skills.

A subsection on postsecondary training discusses the options of community college, vocational/technical schools, apprenticeships, armed forces training, and on-the-job training. Typical courses of study are outlined, including specific courses and any practical clinical or internship experience required. On-the-job training is the way most technicians hone their classroom skills, or for some, their primary method of training. Future technicians should be prepared to learn from a variety of sources and methods.

A section on certification or licensing outlines the options and requirements of the profession, whether voluntary or mandatory. Details about tests and information about contacting certifying or licensing agencies are also given. For most technicians' jobs that do not formally require certification, those interviewed recommended it as a way to show commitment and dedication to one's profession and to stand out from other job applicants or as a way to advance.

A section on scholarships and grants highlights any special monies set aside for the particular profession. Readers are reminded to check with their high school guidance counselor and to contact their school's financial aid office for additional information.

A section on internships and volunteerships details options for prejob experience. Many of the jobs covered here, or at least their settings, such as hospitals, can be explored first in volunteer situations. Enterprising students can often create their own "internships" by offering to work for experience alone at places such as churches or charitable organizations that can't afford to hire professional help.

Finally, a section on labor unions (if applicable) identifies any major unions that are affiliated with the profession.

## who will hire me?

This section gives the reader an idea about how to get a job in the particular field, frequently discussing how the profiled person got his or her first job in the field. Identified are the most likely places of employment (hospital, clinic, factory, airline, etc.) and how to contact such places. Sources of employment, such as job lines, resume services, and trade magazine classifieds, are also discussed, as is the importance of making and maintaining contacts in one's field. Many of the technicians interviewed here got their first jobs through a personal contact.

## where can I go from here?

This section tells the reader about advancement possibilities, illustrated with the profiled person's own goals about where he or she eventually sees himself or herself going in the field. Advancement usually connotes two things: an increase in responsibilities and an increase in pay. The same is true for technicians, for whom advancement usually means graduated salary increases up to a certain point, and corresponding changes in one's title, such as "Senior" or "Supervisor" added. In some fields earning potential for technicians is virtually unlimited, and these technicians are able to stay in their field and continue to enjoy respectable salaries (this includes those fields that are able to support "consulting technicians," those who are able to open their own businesses and hire out their services to the highest bidder). In other fields, technicians' salaries eventually top off, so that no matter how many years of experience one has, as long as one remains a technician, one is never going to earn over a certain amount. (This scenario is not, of course, limited to technicians, but an experience of workers in all ranges of jobs.) For those technicians who require greater earning power, there is a choice. They can attempt to increase their salary by changing jobs, going to work for a bigger, more prestigious employer (or a smaller, more specialized employer) who pays more, perhaps even switching to a related technician career that has more earning potential; or they can go back to school to become a technologist or even a professional, such as an engineer or a doctor.

Many technicians are satisfied with being technicians; had they wanted to be engineers or doctors, they would have gone that route in the first place. Many, however, treat their experience as a technician as a sort of internship for a professional position. It *is,* after all, a way for a person to quickly begin working in their chosen field, making a decent income and having the opportunity to make important contacts and continue their education. Roughly 40 percent of licensed practical nurses, for example, go on to become registered nurses; on the other hand, very few dental hygienists would even think of being a dentist.

## what are some related jobs?

This section gives the reader some options in other fields or areas that use similar job skills. As already noted, many technician career tracks are limited in that salaries can eventually top off, or worse, the career track itself might be glutted or become obsolete. It is, therefore, worth considering other areas that might value the skills the technician has worked so hard to attain. This section also identifies the U.S. Department of Labor classifications from the most recent editions of the *Dictionary of Occupational Titles* and the *Guide for Occupational Exploration.*

## what are the salary ranges?

This section gives the reader an idea of the average beginning salary, experienced worker salary, and maximum salary. Where available, salaries in particular geographic regions or cities are indicated. When considering salaries, readers should keep the entire benefits package in mind—insurance, pension, tuition reimbursement, and vacation, sick, and holiday pay.

## what is the job outlook?

The purpose of this section is to get the reader thinking about the big picture and his or her long-term objectives. It relies heavily on the U.S. Department of Labor's *Occupational Outlook Handbook* to indicate the projected job growth (decline, slow, average, faster than average, etc.) for the particular job and explains to the reader how the future of the job is tied to the overall economy or subsections of the economy—in short, how it fits into the scheme of things.

## how do I learn more?

This section includes two subsections: professional organizations, which lists relevant organizations that provide information to students or their members on the profession, scholarships, employers, or education; and a bibliography, which lists a sampling of relevant books and periodicals that students can begin looking through to learn about the jobs. Out of consideration for timeliness, only books published since 1990 are included here. In some cases, such as the computer field, even recent books are not listed because they quickly become outdated.

## additional material

Each chapter of *Exploring Tech Careers* also has several sidebars that are quick and fun to read. Sidebars lighten the look of the text, making it visually more interesting and inviting. They also provide useful "for your information" quick facts, "lingo to learn" (glossary of terms relevant to the profession), advancement possibilities, personal traits necessary for success in the job, related jobs, graphs, and in-depth features on history, famous people, and events.

## indexes

At the end of volume two are three indexes: DOT, GOE, and Job Title. The DOT index lists by DOT group all major jobs discussed in this work. A description of the group is given, followed by specific jobs in alphabetical order. DOT numbers are given for the jobs, as are page numbers so the reader can easily turn to the appropriate page in the set. The GOE index is organized in a similar fashion by GOE area and work group. Again, jobs are listed in alphabetical order under their GOE work groups, followed by a page number for easy reference. The Job Title index lists all major jobs discussed in this work and their page numbers, with appropriate cross references.

# a final note

It should be obvious from reading the interviews on these pages that people love to talk about what they do. We encourage readers to take this set a step further and seek out people who do jobs they are interested in. Talk to them. Find out what they like and don't like, and what they would do differently if they had to do it all over again. There is no better teacher than experience. Students would do well to learn from the experiences of those who have been there.

The editors

# agricultural food and processing engineer

**Definition**

Agricultural food and processing engineers design processes and systems for the efficient manufacturing and processing of food products. They develop new food products, design equipment and machinery, and ensure that plants are running efficiently.

**Alternative job titles**

Food engineers
Food processing engineers
Plant engineers
Process engineers
Research and development engineers

**Salary range**

$49,800 to $83,400 to $117,000

**Educational requirements**

Bachelor's degree

**Certification or licensing**

Voluntary

**Outlook**

About as fast as the average

GOE 02.02.04

DOT 041

O*NET 22123

## High School Subjects

Chemistry
Mathematics
Physics

## Personal Interests

Figuring out how things work
Science

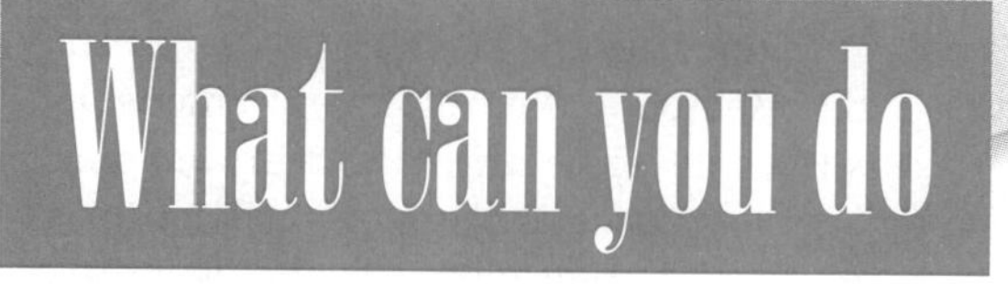


with soybeans? If you're an agricultural food and processing engineer, you can make crayons. At least, that's what Jocelyn Wong and her classmate did. They created color crayons from soybean oil for a college competition seeking innovative uses of soybeans. Using their agricultural food and process engineering knowledge, they created prototypes of the environmentally friendly, nontoxic, biodegradable crayons. They conducted literature and patent searches, ran laboratory tests, completed cost analyses, and used a structured trial and error process to research and test their prototypes. Most problems that could arise did—some crayons were too brittle and broke too easily, some didn't transfer color well, some contained too much pigment and stained their hands, and others were too soft. Little did they know they would go on to win first prize, reap heaps of publicity, and sell their crayons to a major manufacturer. On the fourth of July, 1997, Jocelyn attended a kickoff party at the Indianapolis Zoo, where soybean farmers, children, and the media all gathered to celebrate her crayons. She's not kidding when she says, "Food engineering gives you a lot of possibilities."

## what does an agricultural food and processing engineer do?

*Agricultural food and processing engineers* work in food processing and food manufacturing plants. This can include plants that make products such as potato chips and candy or plants that make intermediate products such as soybean oil. The agricultural food and processing engineer's job is to create new processes for the production of biological products such as food and biochemicals, design and plan how to successfully produce these products, including the layout of the processing lines, and assist in the building and maintenance of equipment and processes. These processes can involve mixing, storing, sterilizing, refrigerating, packaging, extracting, and more. Very simply put, agricultural food and processing engineers figure out how to make food for our consumption.

Agricultural food and process engineering fits under the larger umbrella of agricultural engineering and is closely related to chemical and mechanical engineering. Agricultural engineers may work in natural resources conservation, environmental protection, food processing, and agricultural production. Often, they deal with a single commodity, such as wheat, and design equipment for farm operations and production. But while agricultural engineers may work on farms or in forests, agricultural food and processing engineers work primarily in offices, in laboratories, and on the floors of manufacturing plants. Food process engineering also involves less agricultural science and a lot more chemistry.

An extensive chain of various disciplines is involved with producing food and food products. The chain may start with the *agricultural scientist*, who deals with the science of growing the food. For example, the agricultural scientist may figure out how to improve the soil or control the pests that damage soybean crops. Next in line is the *agricultural engineer*, who may design the machinery and equipment needed to harvest those soybeans. The *food science technologist* or *food scientist* is involved with microbiology, biochemistry, and food chemistry and may develop ideas about what to do with the soybeans. The food scientist hands off to the agricultural food and processing engineer, who designs the equipment so the plant can produce the finished product. These areas often overlap and are dependent on one another.

Agricultural food and processing engineers usually fall into three primary categories: research and development (R&D) engineer, project engineer, and plant or process engineer.

The research and development engineer's task is to come up with new or improved food products and processes. Working in laboratories and offices with food scientists and chemists, R&D engineers are responsible for giving us baked potato chips, sugar substitutes, and fat-free salad dressings. They may also focus on improving existing products as well, such as studying and designing the shapes and sizes of macaroni noodles so they will cook and boil to the correct texture and consistency. Combining market research, engineering, and science, R&D engineers study and develop new processes and determine their feasibility and usefulness.

The project engineer takes what the R&D engineer has developed in the laboratory and designs the equipment and processes to make it happen on a larger scale. Relying upon chemical

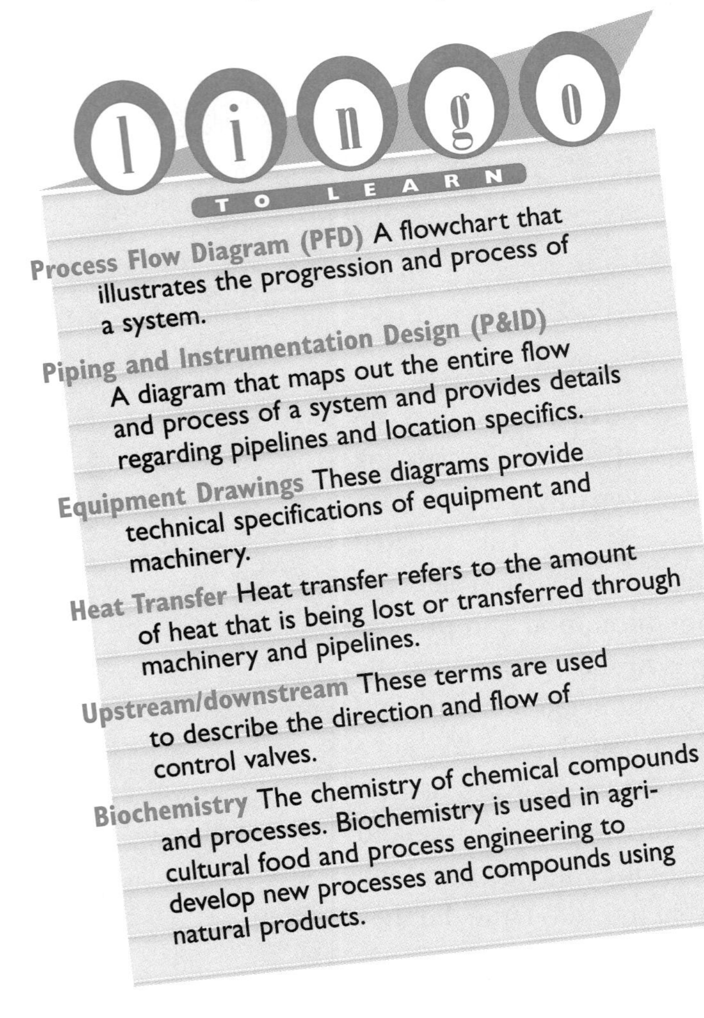


**Process Flow Diagram (PFD)** A flowchart that illustrates the progression and process of a system.

**Piping and Instrumentation Design (P&ID)** A diagram that maps out the entire flow and process of a system and provides details regarding pipelines and location specifics.

**Equipment Drawings** These diagrams provide technical specifications of equipment and machinery.

**Heat Transfer** Heat transfer refers to the amount of heat that is being lost or transferred through machinery and pipelines.

**Upstream/downstream** These terms are used to describe the direction and flow of control valves.

**Biochemistry** The chemistry of chemical compounds and processes. Biochemistry is used in agricultural food and process engineering to develop new processes and compounds using natural products.

and hydraulic engineering principles, the project engineer has to consider every aspect of the process, such as the number of tanks that will be needed, the number of pumps that will be required, and so on.

The plant or process engineer is responsible for the actual process of producing the final product. After the project engineer has designed the processes and equipment and the installation is complete, the plant or process engineer handles the day-to-day supervision of the plant. This entails making sure the plant is running the way it is designed to and is manufacturing product every day like it is supposed to. The plant or process engineer may supervise the technicians who run the plant, deal with personnel issues, and troubleshoot mechanical problems and equipment failures.

Closely related to the plant or process engineer is the reliability engineer, whose task is to ensure that the plant is running efficiently and producing a maximum percentage of product. Training employees, running reports, and setting up systems for measuring reliability are all responsibilities of the reliability engineer.

## what is it like to be an agricultural food and processing engineer?

Jocelyn Wong is a project engineer/reliability engineer for Procter & Gamble's (P&G) Olean ™ plant in Cincinnati, Ohio. She typically works a forty-hour week, but sometimes the hours can grow longer, depending on project deadlines. Her project engineer duties include designing equipment and processes and getting them into place. The start of her workday is devoted to checking E-mail and messages, making phone calls, and planning the day's activities. She reviews equipment drawings and piping and instrumentation designs (P&IDs), which are like blueprints or maps of the plant from a technical standpoint. If, for instance, they are installing a heat exchanger and a pump, Jocelyn reviews the appropriate P&IDs and drawings to determine how large the condenser is, the size of the motor, where the equipment needs to go, and what other pipelines must go in and out of it.

Jocelyn also meets with vendors or contractors, communicating with them about what needs to be accomplished, including how and when. Jocelyn then finds out about their needs—they may have to obtain specific permits to perform the job, or they may require that the plant be shut down for a certain number of hours in order to complete their tasks.

> "I will not lie—college and engineering are very, very hard. You have to be determined and hard working."

It's Jocelyn's job to follow through on these requests and disseminate the appropriate information to all parties.

Sitting through meetings is another part of an agricultural food and processing engineer's job. Jocelyn usually attends several meetings each day, receiving direction on projects, taking on new assignments, or working on design issues. She always works as a team member, and each project may have about eight members. Sometimes she might lead the effort on a project, while other times she works on one portion.

Jocelyn's reliability engineer tasks include making sure the plant is running the way it was designed 85 percent of the time. She must make sure that work systems are in place to measure the plant's reliability, and this can include writing reports, training employees, and receiving updates and progress reports from her staff.

Agricultural food and processing engineers work on many projects at once, so it's important to be able to prioritize. "You have to be smart about what you think the priorities are," emphasizes Jocelyn. "Sure I'll have six or seven different things to work on, but if something is more urgent at the time, then that becomes my first priority. It's not like in school where you have one project at a time." It all boils down to the same goal, however: make sure the plant is running well and producing product. Working on different projects means Jocelyn is always learning something new and that there's plenty of variety in her job.

Jocelyn enjoys most every aspect of being an agricultural food and processing engineer. She likes seeing the fruits of her labor, stating, "You get to use your creativity and see it through to the end, which is very nice. You're working on things that you've designed and you've built." Jocelyn works with people from many different industries to accomplish her tasks, which she finds interesting. "I could work with construction workers, architects, business and accounting people, and cost engineers. It gives you an opportunity to work with many different facets." Of course, being an agricultural food and processing engineer can be stressful. Jocelyn works with deadlines all the time, and if she is unable to deliver on a project, everyone else hits a snag as well and the entire project may be held up. And engineering itself is no day at the park—it's very technical and difficult. Having this technical knowledge, however, is a big plus that can open a lot of doors. "A pro of being an engineer is you have a deep technical understanding of how things work," says Jocelyn, "which is wonderful because you can leverage that, and in ten years I could go down a completely different path than engineering. I could go into advertising or sales, but I'll always have that technical foundation."

Food process engineers work in offices and laboratories, as well as in food processing and manufacturing plants, so you must be willing to get your hands dirty and work with loud machinery in occasionally adverse environments or conditions. The plant Jocelyn works in is located outdoors, which means she must sometimes brave the rain and the cold to accomplish her work.

The technical aspect of engineering threw Jocelyn the biggest curve. "I was never very technical," she recalls. "I was never one of those people who was born to be an engineer. I had to work twice as hard as other engineers." Besides tenacity, Jocelyn possesses other qualities that have helped her succeed as an agricultural food and processing engineer, including strong initiative and leadership skills. Jocelyn adds, "I'm very quick to say what I don't know, and I'm good at finding the resources to get what I don't know. I'm good at asking questions and being aware of my faults and being able to work on them, thinking of them as challenges rather than faults." Believing in your capabilities, as Jocelyn has proved, can carry you far.

## have I got what it takes to be an agricultural food and processing engineer?

Agricultural food and processing engineers must have problem-solving skills to analyze difficult concepts and communication skills to express their ideas clearly. Creativity is also critical in order to come up with design ideas and possible solutions to problems. It helps to have technical aptitude, and if you don't have it, you must work to get it. Jocelyn recalls, "I will not lie—college and engineering are very, very hard. You have to be determined and hard working."

An interest in mathematics and the sciences, particularly chemistry and physics, is also important if you are thinking about a career as an agricultural food and processing engineer. You need to be able to work well with others because all projects are team based, and if you are easily rattled by deadlines, then engineering may not be the right choice. "Engineers work under pressure a lot, because you're always schedule driven," says Jocelyn.

**To be a successful agricultural food and processing engineer, you should**

- Have written and verbal communication skills
- Enjoy working with others on team-oriented projects
- Have problem-solving and analytical thinking skills
- Be detail oriented and well organized
- Be resourceful and creative

## how do I become an agricultural food and processing engineer?

Jocelyn hadn't planned to study engineering in college until her father suggested it. As Jocelyn recalls, "I was more interested in business and management, and it turns out that if you have an engineering degree, you're a much stronger manager." The idea of job security and the flexibility engineering can offer in career choices down the road appealed to her. Jocelyn chose food engineering as her major because "there was a lot of room for creativity," she says.

## education

### High School

To prepare for life as an agricultural food and processing engineer, focus on math and sciences, such as calculus, chemistry, physics, algebra, and geometry. All of these subjects will come into play as an engineer, and it's important to build up the knowledge base and analytical thinking skills. Jocelyn remembers being inspired by physics projects that required students to create solutions to problems, like how to drop an egg off of a building without having the egg smash on the ground.

Jocelyn suggests getting involved in science fairs and clubs, which will provide you with the opportunity to design and create science projects and compete against other schools. Activities such as these will also help you build up communication skills and learn the art of working as a team member.

### Postsecondary Training

A bachelor's degree is required for agricultural food and processing engineers. Jocelyn attended Purdue University in Indiana, majoring in both biochemistry and food process engineering in the school of agricultural and biological engineering. Food engineering programs are often offered jointly by the departments of agricultural engineering, chemical engineering, and food science and technology.

Jocelyn enjoyed her upper division classes more than her core engineering courses. "I hated my beginning college classes," she recalls, "like the basics, like physics and chemistry. But later on, in your junior and senior year, you start to take classes like plant engineering." These project-based courses gave Jocelyn the opportunity to work with fellow students, creating and designing plants. "We had to come up with an idea of a plant we wanted to build, and we came up with a winery. So we had to do lab scale tests to make wine, and then we had to get into the nitty gritty of actually building a plant that makes wine. So we had to get into what equipment you would use, and so on." Seeing a project through from start to finish was a rewarding learning experience for Jocelyn.

Joining a student chapter of a professional organization such as the American Society of Agricultural Engineers or the American Institute of Chemical Engineers is a good idea while you're in college. You are able to meet engineers working in the field, thus gaining the opportunity to establish contacts, build relationships with those in the industry, and learn more about the field. "I tried to be as involved as I could in the organizations so I could learn more about the job," says Jocelyn, "and what I was getting myself into." Jocelyn and a friend founded the Society of Biochemical and Food Process Engineers at Purdue University, which is still going strong.

## certification or licensing

There is no certification or licensing specifically geared toward agricultural food and processing engineers. Most employers require only that their engineers have graduated from an engineering program accredited by the Accreditation Board for Engineering and Technology (ABET). Engineers may, if they choose, earn a professional engineering license from the National Society of Professional Engineers. To begin the process, college students take a written examination during their senior year on fundamental engineering principles. If they pass and graduate from school, they gain the distinction of being an "engineer in training" (EIT). After four years in the workforce as an engineer, EITs take another written examination. If they pass, they become professional engineers (PEs). PEs can sign off on engineering plans and drawings or start their own businesses.

## scholarships and grants

Scholarship and grant opportunities are plentiful for students in engineering disciplines. It would be a good idea to contact various associations and professional organizations with engineers as members, such as the American Society of Agricultural Engineers, the National Society of Professional Engineers, or the American Chemical Society (*see* "Sources of Additional Information"). Universities offer a wide variety of scholarships in various departments and with assorted eligibility requirements. Contact the financial aid office or a guidance counselor for further information.

Large companies are another source of financial aid. Companies that hire and employ engineers are likely to offer scholarship programs for aspiring engineers. Some may also provide summer work opportunities as part of a scholarship award. Your school advisor or guidance counselor should have access to this information, but you can also find this information at the library or on the Internet. In fact, most large companies have web sites that list scholarship and grant information.

## internships and volunteerships

Internships are common for agricultural food and processing engineers, and Jocelyn recommends them highly. "I would say that 75 percent of the people I graduated with had an internship or coop experience." Jocelyn participated in a cooperative work-study program in which she alternated semesters and worked in the industry. She also had two summer internships where she worked for different companies and gained hands-on experience. Engineering internships usually pay well, which is another bonus.

Internships are an excellent way to learn what the job of an agricultural food and processing engineer may be like after graduation. Jocelyn enjoyed her internship experiences immensely, and they reminded her why she was in school studying engineering. "The classes are frustrating at first, but after you actually get a taste of what the industry is like, it is extremely satisfying."

## who will hire me?

Agricultural food and processing engineers are needed in almost all facets of the food industry, and food manufacturing plants and food processing companies are the primary employers. Entry-level agricultural food and processing engineers often begin as plant or project engineers, and some may find positions in research and development. Jocelyn's fellow graduates landed jobs with companies such as General Mills, Kraft, Calistoga, Frito-Lay, Campbell Soup Company, and other large companies.

Consulting agencies also hire agricultural food and processing engineers. These agencies consult with a variety of food processing plants and offer solutions to manufacturing problems, suggest ways to make the plants more efficient and more productive, and design equipment and processes. Consulting jobs often involve travel. Some jobs may be available with companies that manufacture machinery or food-related equipment as well.

The best way to find a job if you are in college is to go through your school's career planning and placement center. Jocelyn found her job thanks to Purdue University's placement center, which, she says, offers a great placement program. Recruiters often visit campuses to meet with and interview prospective employees. Purdue, for example, hosts an industry roundtable every year, with hundreds of companies in attendance. Engineering students can go from table to table, passing out resumes, learning about the companies and what they can offer, networking, interviewing, and finding entry-level jobs. Many of Jocelyn's classmates landed jobs through the industry roundtable.

Networking and contacts are another avenue for finding jobs in the industry, so it's important to be involved with professional organizations and clubs. These organizations frequently publish trade magazines, and it's possible to find job openings listed. Many professional associations also host annual conferences or national meetings, which is another opportunity to find out about jobs. The American Chemical Society (ACS), for example, provides various career services to its members. ACS has a national database of job openings that matches members' qualifications with appropriate jobs, offers employment clearinghouses where members can interview with hiring companies at ACS meetings, and has career consultants on staff who can help you with your resume, provide interview tips, and get you on a career path.

Some agricultural food and processing engineers have found jobs through headhunters and placement agencies, while others have discovered job opportunities through classified advertisements in newspapers. Almost all professional organizations and major newspapers can be found on the World Wide Web, which is an excellent tool for finding job leads or openings and for researching companies you may be interested in working for. Large companies often list job opportunities on their web sites, along with the option to apply directly on-line.

## where can I go from here?

Jocelyn has been working for Procter & Gamble for less than two years, and she plans to continue working there, gaining more experience and working on as many projects as possible. Concurrently, she would like to attend school and earn a master of business administration. Because Jocelyn enjoys interacting with others, she hopes to shift away from project engineering

### Advancement Possibilities

**Vice presidents** direct the operations of various company units, such as engineering or manufacturing.

**Section heads** or **department managers** supervise and manage divisions of manufacturing plants and oversee technical leaders.

**Technical leaders** or **supervisors** oversee teams of engineers and coordinate large-scale projects.

**Consultants** work on a contract basis for different companies, undertaking a variety of projects at some point: "I would like to go into manufacturing where I work more with people and see the day-to-day business."

Agricultural food and processing engineers typically advance within their companies as they gain more experience. Advancement usually entails additional responsibility, such as supervising more employees, including engineers, and overseeing larger-scale projects. A vice president of engineering may be in charge of an entire company function, such as engineering for all salted snacks, for instance. The job of a vice president tends to focus on the business management end more heavily than the actual engineering end.

Section heads or department managers of companies manage entire divisions or sections of a plant. A section head, for example, may be in charge of all engineering projects for a specific product or oversee a plant start-up, which includes the design and building of a new plant. Section heads must plan, direct, and coordinate all activities and departments to accomplish the overall goals. Section heads report to vice presidents.

Also within a company, technical leaders or supervisors oversee large projects and delegate work to other engineers. Technical leaders also tend to focus on the business management and finance end of projects, and they leave much of the engineering work to the engineers. Technical leaders report to section heads and may be in charge of a subsection, such as the controls portion of a project, the packing area, or specific unit operations.

Independent consultants visit different food companies and work on various projects on a contract basis. For example, a smaller food processing plant may not require a full-time agricultural food and processing engineer, or they may have a unique project or problem they are having trouble resolving.

Agricultural food and processing engineers can also go into teaching. Teaching at a two-year college may be possible with a master's degree, but teaching at a four-year school requires a doctorate degree.

## what are some related jobs?

Agricultural food and processing engineers can shift to other engineering careers, such as chemical or mechanical engineering. They can also go into sales of food-related equipment and machinery. The U.S. Department of Labor classifies agricultural food and processing engineers under the headings *Occupations in Architecture, Engineering, and Surveying* (DOT) and *Engineering: Design* (GOE), which also includes people who work in aerospace engineering, chemical engineering, ceramics engineering, landscape architecture, architecture, environmental engineering, civil engineering, and optical engineering.

| **Related Jobs** |
|---|
| Landscape architects |
| Architects |
| Aerospace engineers |
| Chemical engineers |
| Ceramics engineers |
| Environmental engineers |
| Civil engineers |
| Optical engineers |

## what are the salary ranges?

Agricultural food and processing engineers earn salaries comparable to those of chemical engineers. According to the 1998–99 Occupational Outlook Handbook, starting salaries for chemical engineers averaged $42,817 a year. Engineers with bachelor's degrees in all disciplines earned a median annual salary of $49,800 in 1996. The median annual salary in 1996 of chemical engineers was $52,600. Engineers working at the highest senior managerial level earned a median income of $117,000 a year in 1996.

According to the Institute of Food Technologists, the median annual salary of a food engineer in research and development was $58,950 in 1995. Process engineers and directors earned an average of $65,000, and plant managers and supervisors made $60,000 a year.

Agricultural food and processing engineers usually receive comprehensive benefits packages. Salaries do not vary much by region and are not consistently higher in urban areas, perhaps because rural locations are ideal for large food manufacturing plants.

## what is the job outlook?

Agricultural food and process engineering is a relatively young field, and specific outlook information is scarce. According to Jocelyn, however, the job outlook is extremely positive. She says, "If you think about it, food is a necessity. In today's age, people want more convenient foods and healthier foods. The need for food engineers and the ability to do that kind of work is increasing tremendously." As entire households enter the workforce, packaged and prepared foods are growing in popularity. In addition, the increasingly health-conscious public demands foods lower in fat and calories. The shelves of every grocery store are lined with fat-free and low-fat products, and new products appear regularly. Jocelyn also points to the worldwide population boom, noting that food processing engineers are needed to find ways to feed all the people.

According to the 1998–99 Occupational Outlook Handbook, as technology advances and competition increases, companies will be compelled to come up with new ideas and improve on old products and designs more rapidly. This, in turn, means companies must maximize production and make sure manufacturing processes are as efficient as possible, and this is where agricultural food and processing engineers will enter the picture.

Biochemistry is a growing field within agricultural food and process engineering, which is lucky for Jocelyn, who majored in it along with food process engineering. "Biochemistry," explains Jocelyn, "is using natural products, like soybeans, and using it for a variety of different purposes that we never have even dreamed of. The biochemistry field is going to be really hot." Olestra, the fat substitute used in potato chips and other snack foods, is an example of what biochemistry can provide the public.

Although there are few statistics concerning agricultural food and process engineering, Jocelyn feels her job is secure and that the industry she selected is growing. Jocelyn and her fellow graduates all work for well-established companies and are in demand. "We're contacted by headhunters all the time," she declares.

# how do I learn more?

## sources of additional information

The following organizations provide information on careers in agricultural food and process engineering, accredited schools and educational programs, scholarship opportunities, and employers.

**American Society of Agricultural Engineers**
2950 Niles Road
St. Joseph, MI 49085
616-429-0300
http://www.asae.org

**Biochemical and Food Process Engineering Club**
Purdue University
1146 ABE Building
West Lafayette, IN 47907
765-494-1172
http://pasture.ecn.purdue.edu/~bfpe/

**Institute of Food Technologists**
221 North LaSalle Street, Suite 300
Chicago, IL 60601
312-782-8424
http://www.ift.org

**National Society of Professional Engineers**
1420 King Street
Alexandria, VA 22314
714-684-2800
http://www.nspe.org

## bibliography

Following is a sampling of materials relating to the professional concerns and development of agricultural food and processing engineers.

### Books

Connor, John M.,and William A. Schiek. *Food Processing: An Industrial Powerhouse in Transition.* 2nd ed. Covers the growth, economic development, and business management of the U.S. commercial food processing industry .New York: John Wiley & Sons, 1997.

Heldman, Dennis R., and Richard W. Hartel. *Principles of Food Processing*. Food Science Text Series. 3rd ed. An introduction to the food industry, food trade, food engineering, and food processing. New York: Chapman & Hall, 1997.

Mather, Robin. *A Garden of Unearthly Delights: Bioengineering and the Future of Food.* A consumer guide to bioengineering practices. New York: Plume Press, 1996.

### Periodicals

*Feedstuffs: The Weekly Newspaper for Agribusiness.* Weekly. An agribusiness newspaper that covers food processing, feed and grain, environment, and more. The Miller Publishing Company, 12400 Whitewater Drive, Suite 160, Minnetonka, MN, 55343.

*Institute of Food Technologists On-line*. Daily. Website contains news, surveys of recent research, conference details, and links to food industry trade shows. http://www.ift.com.

*World of Food & Beverage Newsletter.* Monthly. Official newsletter of the FPM & SA, which represents 500 companies that provide processing and packaging equipment, supplies and services to the food and beverage industries domestically and internationally. FPM & SA, 200 Daingerfield Road, Alexandria, VA, 22314-2800.

# agricultural technician

**Definition**

Agricultural technicians assist engineers, scientists and/or conservationists in food, plant, soil, and animal research, production, and processing.

**Alternative job titles**

Agricultural engineering technicians

Conservation technicians

Lab technicians

**Salary range**

$15,500 to $27,000 to $49,500

**Educational requirements**

High school diploma; associate's degree in agricultural technology highly recommended

**Certification or licensing**

Voluntary

**Outlook**

About as fast as the average

GOE 02.02.03

DOT 040

O*NET 225990

## High School Subjects

Agriculture
Biology
Chemistry
Mathematics

## Personal Interests

Animals
Plants/Gardening

**Lloyd Nelson** walks across the farmer's field with notebook in hand. He stops to dig the toe of his boot into the ground, checking to see how deeply the rain water has soaked into the soil. He glances out across the wide-open field; the only building in sight is a farmhouse alongside the winding gravel road. He catches the scent of sage in the wind and regrets having to go back to the office, where he'll have to sit behind a desk and report today's work.

Lloyd hasn't been to this spot of land for three months. He's here now to check that his conservation planning has worked for the land. He looks out across the dam he designed. Because of the recent rain, the water sits trapped in the dam, soaking into the soil. Pleased to see all his plans in place, and successful, Lloyd marks down the results of another completed job.

# what does an agricultural technician do?

If you've spent even one day on a farm, you've seen the variety of hard work that goes into keeping the farm productive. Raising livestock or crops requires a careful understanding of natural resources and animal or insect biosystems. A farmer must have knowledge of complex machinery, structures, and irrigation systems. And different animals and crops have different demands: someone with a pecan orchard in New Mexico must have a different understanding of the soil and of irrigation systems from someone growing corn in Nebraska or tobacco in Virginia; a beekeeper in Oregon must understand different structures and controlled environments from a rancher raising longhorn cattle in Texas. The area of agriculture offers diverse job opportunities for researchers, engineers, scientists, and technicians all across the country.

*Agricultural technicians* work with farmers and ranchers on many different levels—helping them use their resources and equipment for maximum efficiency and keeping up on the latest and most economical tools and designs in farming. With their knowledge of biology, the environment, engineering, and design, technicians bring together the work of many professionals. They study animals and their environments and help develop healthy facilities for raising these animals. They design electrical, irrigation, and food processing systems. They make improvements on farm machinery, operate the equipment, and lay out plans for the conservation of natural resources.

Agricultural technicians can choose to specialize in four areas: food science, plant science, soil science, and animal science. The type of work performed varies according to the specialization.

Fruits, vegetables, grains, baked goods, candy, beverages, meat and dairy products, and other foods all require processing and preparation. Food and processing engineering involves using agricultural materials for food, feed, fiber, and industrial products. Technicians focusing on food science aid in every step of processing food for human consumption, devising ways to preserve the food's taste and nutrition and designing systems for the proper shipment and storage of food. They study the physical properties of food materials and determine the best ways to heat, refrigerate, and dry the materials. They are also involved in handling, packaging, and product development. Agricultural technicians involved in food science may also enforce government regulations and inspect food processing plants to uphold the standards of sanitation, safety, quality, and waste management.

Plant science involves agronomy, crop science, entomology, and plant breeding. Agricultural technicians may conduct tests and experiments to improve the quality and yield of crops or study various ways to increase the resistance of plants to insects and disease. They may also work to improve seed quality and the nutritional value of crops. Agricultural technicians in plant science help farmers and scientists produce the largest, healthiest crops possible, factoring in variables such as pest control, disease, crop management, and genetic engineering to breed stronger plants.

Agricultural technicians who specialize in soil science study soil composition and its relationship to crop and plant growth. This includes the effects of various fertilizers, crop rotation, and plowing and sowing practices. Technicians may also be involved with erosion studies and diagnosing and solving soil problems. Conservation of natural resources is also the task of an agricultural technician. Technicians design and install irrigation systems and dams and lay out conservation plans.

Animal science involves the nutrition, genetics, growth, reproduction, and development of domestic farm animals. Agricultural technicians working in animal science study animals in relation to their environments and develop adequate ventilation, management, and sanitation systems so the animals will be healthy

**Aquaculture** Working with the quality, use, and discharge of water.

**Biotechnology** Working with living systems to aid in the development of commercial processes and products.

**Cryopreservation** The storage of cells in suspended animation at -196 degrees Celsius; used for the storage of plant species that produce seeds.

**Field mapping** Profiling a field on the basis of yield performance.

**Microirrigation** Irrigation distributing a small amount of water; irrigation by drip or trickle.

**Watershed** The region or area drained by a river or stream.

enough to produce high-quality, plentiful products, including milk and eggs. They may assist farmers in designing or upgrading animal housing structures, and they devise nutrition and feeding plans for livestock.

Agricultural technicians often work with power and machinery, designing equipment for the efficient harvesting of crops, the baling of hay, and other field operations. They help manufacturers and engineers develop and research the products that will make farm work easier and more economical, which may involve acquiring new machinery or upgrading existing products. Agricultural technicians also assist in the design of buildings, such as storage structures for grains or waste.

Electrical and electronic systems are also part of an agricultural technician's area of expertise. Farms and other agricultural producers require electrical systems to operate the variety of machines used in their daily work for storage, refrigeration, heating, irrigation, and grain and feed handling. Technicians provide farms with efficient electrical distribution systems, taking these systems from the design stages to installation.

**To be a successful agricultural technician, you should**

- Be able to communicate and work well with people
- Enjoy working outdoors
- Be willing to accept responsibility
- Have an understanding of math and science
- Have a curiosity about plants, animals, or insects

## what is it like to be an agricultural technician?

Lloyd Nelson has been working for the U.S. Department of Agriculture for twenty-three years. As a technician for the Soil Conservation Service, he works in many different areas of agriculture, from engineering and conservation planning to pipeline design and concrete testing. "I'm a jack-of-all-trades," he says, referring to the diversity of projects he heads. His job involves soil and range conservation, work with irrigated crop land, water management, and dairy and waste management. And this list continues to grow as he learns more about agriculture and conservation.

Most of Lloyd's work hours are spent in the field, and his remaining hours are devoted to administrative duties, engineering work, and educating the public about conservation. "I spend about 70 percent of my time in the field, surveying and installing conservation systems for agricultural producers and landowners," Lloyd says. Twenty percent of his time is devoted to paperwork and entering data in the computer. He spends the remaining 10 percent on design and engineering work.

A typical day begins early for Lloyd; he usually checks into the office at about 7:00 AM to read his mail and retrieve messages. The calls start coming in shortly after he arrives—calls from farmers and ranchers requesting conservation planning. Lloyd schedules meetings with these landowners, giving priority to those with land in specific areas. The Soil and Water Conservation District Board determines these priority areas based on conservation needs. "But I hate to have to keep somebody waiting a month," Lloyd says. "My job is to get out there and get conservation on the land."

During his initial meeting with a farmer, Lloyd will gather information, getting the farmer's thoughts, plans, and conservation objectives. "I have to be pretty thorough," Lloyd says. "I have to get involved in solving a problem. I approach the problem like it's my own, so I can provide the best solution." He not only tries to find the best solution for the farm, but the most cost-efficient one as well. "And I look for shortcuts," he says, "ways to get the job done as quickly as I can." He later visits the actual job site for layout, design, and surveying. Once the project is complete, and conservation practices are in place, Lloyd schedules follow-up visits to make sure everything is going as planned. Lloyd must also document all his work, providing full reports of his goals and achievements.

Though most jobs for agricultural technicians don't require certification, Lloyd is certified by the American Concrete Institute in field testing. With this certification, he has worked with the design and management of pipelines, and has served as concrete inspector and concrete surveyor on various projects.

Lloyd generally works a forty-hour workweek either in the field or in the office. Some time is also spent traveling from farm to farm and from farm to office. The most active months are between April and October, when farmers use their land for summer crops.

## have I got what it takes to be an agricultural technician?

Both in speech and writing, agricultural technicians must have the skills necessary to communicate ideas clearly and completely. Farmers and agricultural professionals work closely with technicians and must be able to rely on the technician's perceptions, planning, and reports. A technician is also relied upon for creative solutions to complex problems in agricultural production.

Lloyd credits much of his success to working well alongside others. "I can get along with just about anybody I come into contact with," he says. This rapport makes it easier to gather information from the farmers, and to get the answers to his many questions. With this clear line of communication, he can offer good solutions to a farmer's problems.

The problem-solving itself can be difficult and requires a lot of creativity, but Lloyd enjoys these challenges. "I like solving other people's problems," he says, "and I like being responsible for getting the work done." Sometimes the challenges for an agricultural technician can involve design and engineering practices, so a background in math, mechanics, and the sciences is valuable.

A familiarity with and interest in farming can give a technician an understanding of the terminology and practices of farm work. Lloyd was hired at the USDA after having grown up on a farm, and he continues to pursue his agricultural interests at home as well as at work. "I love growing plants," he says. "I have a backyard that looks like a jungle." Although you don't have to have lived on a farm to get a job as an agricultural technician, you should have an interest in nature and the outdoors. You may be spending many hours in the field and focusing on natural resources, plant growth or animal life. A curiosity in one or all of these areas will direct you in your research and design.

Lloyd enjoys the variety of work available in agriculture, and he welcomes the challenges, but he wishes there were more opportunities for technicians to advance. The USDA offers many opportunities for advancement but has a basic education requirement for professional positions. Although Lloyd does work similar to that of professional farm planners, he can't move into that position without the minimum number of college credit hours.

## how do I become an agricultural technician?

Before going to work for the USDA, Lloyd took some community college courses specific to his interests in agriculture. Having lived on a farm, he was familiar with farm practices and terminology, and this helped him with his course work and his on-the-job training. Lloyd fondly remembers his on-the-job training, made easier by the experienced and helpful engineers and employees he worked with.

## education

### High School

Courses in the agricultural sciences will give you a good background in farm study and might also provide you with field experience. The National FFA (Future Farmers of America) Organization has chapters all over the nation and offers many programs for those of you interested in agriculture. You can learn about farm practices, agricultural issues, and business management and acquire hands-on experience operating farm machinery and equipment, raising livestock, and identifying plants and trees. Your school should be able to provide you with information about your local chapter.

If your high school doesn't offer courses specifically in agricultural science, any science course can prepare you for agricultural work. Math courses will prepare you for design and engineering. Also, register for any course that will give you an understanding of electrical systems and electronics. Courses that require reading and composition can help you to develop the communication skills you'll need in working with farmers and agricultural professionals. And don't ignore computer classes—more and more farms are relying upon sophisticated computer systems for farm management.

In addition to this, your state may also have an education commission that can provide information about high school work experience programs. These programs combine classroom study with on-the-job experience.

### Postsecondary Training

While formal postsecondary training is not required to become an agricultural technician, more two-year schools are offering associate's degrees in agricultural technology and related fields. This schooling can provide you with a solid foundation and technical training that will

Advancement Possibilities

**Agricultural equipment design engineers** design agricultural machinery and equipment.

**Agricultural equipment test engineers** conduct tests on agricultural machinery and equipment.

**Agricultural research engineers** conduct research to develop agricultural machinery and equipment.

**Agricultural scientists** study farm crops and animals, devising ways to improve quality and increase quantity through research and development.

be helpful in the field. Coursework completed in an agricultural technology program can also be transferred and applied toward a four-year degree if you choose to continue your education. In fact, many of the two-year schools are involved in cooperative programs with four-year colleges or universities, which means you can take classes at the four-year college while enrolled in the two-year school. Some of the courses you may take include science and management of range lands, forests and watersheds, fish and wildlife, local and regional landscapes, resource planning and design, agricultural mechanics, and computers.

Other postsecondary opportunities for agricultural technicians are emerging as well and can be attributed to the increase in technological demands in farming and the proliferation of larger, commercial farms. For example, Hopkinsville Community College in Kentucky offers a one-year certificate program for agricultural technicians. Credits can be used toward a two-year degree as well. The certificate program was developed out of a need for technicians trained in modern agricultural techniques. The program focuses on farm operations and management, emphasizing hands-on learning.

Many universities and colleges offer cooperative education programs that can give you full- or part-time work experience as you pursue a degree. This can sometimes lead directly to a professional position. Check with your advisor or career placement office. You may also want to look into continuing education. The National FFA Organization offers a variety of programs for those working in the agriculture industry.

## certification or licensing

Certification is generally not required for agricultural technicians. Technicians who apply herbicides or pesticides must be licensed, but most agricultural technicians need not seek licensing.

The American Society of Agronomy (ASA) offers a Certified Crop Advisor (CCA) Program that requires passing international and local examinations. Applicants must also have up to four years of education and experience, although a college degree is not mandatory.

The ASA, along with the Crop Science Society of America and the Soil Science Society of America, offers other certification programs. The programs require bachelor's degrees and professional work experience.

## scholarships and grants

The National FFA Organization offers scholarships to students in agriculture, as well as contests and monetary awards. Colleges offering instruction in agricultural technology and related fields may also offer scholarships, and you should also research large agricultural companies and farms for financial aid opportunities.

If you choose to pursue a degree in agricultural engineering or engineering technology, you can contact the American Society of Agricultural Engineers for scholarship information. The American Chemical Society and the American Institute of Biological Sciences can also direct you toward fellowships available in agricultural study (*see* "Sources of Additional Information"). Be sure to allow plenty of time for the application process.

## internships and volunteerships

There are many opportunities for internships and volunteerships. The National FFA Organization and your local FFA chapter can provide you with information and resources. Some schools collaborate with farmers and provide summer field work opportunities for high school students. You may wish to contact local farms or agricultural organizations to find out about internships or to volunteer your time.

Some two-year college programs require students to complete internships before granting degrees. Contact your advisor or placement office for information. You may also request information about internships in your area by writing to the USDA.

## who will hire me?

Lloyd was recruited by the USDA just a few weeks out of high school. The Soil Conservation Service was looking for young workers with a farm background. Lloyd had already become familiar with the USDA, having met some of the engineers and conservationists through past farm business and also through the FFA.

Federal, state, and local governments are actively involved in protecting natural resources and regularly employ technicians. The USDA also hires technicians for work with the Agricultural Research Service, the Bureau of Land Management, the Fish and Wildlife Service, the Forest Service, and the National Park Service. Write the branch offices of USDA agencies for career information (*see* "Sources of Additional Information").

Technicians work with manufacturing companies, testing and selling tractors, farm implements, and other equipment. Agricultural supply organizations, grain and feed handling companies, and electrical utilities also hire technicians.

Agricultural workers are needed in most stages of food processing, and are employed by meat packers, distribution companies, dairy companies, vegetable canning companies, and other food product and processing businesses. Large farms also employ agricultural technicians to help operate and manage all stages of farm production.

## where can I go from here?

Lloyd has received some credit from the local community college, but he needs to take nine more credit hours to move into a professional position with the USDA. "There are good opportunities for advancement," he says, "if you've got the education." By taking community college or university courses specific to agriculture, technicians can meet the basic education requirements for advancement and better pay. Lloyd has plans to get the credit hours necessary to move into a farm planner position.

Agricultural technicians can continue with schooling and earn higher degrees, such as bachelor's and master's degrees. Specialization usually factors in, and technicians may become agricultural research engineers, agricultural scientists, agricultural equipment design engineers, or agricultural equipment test engineers, just to name a few. Technicians who earn bachelor's degrees often shift to more research-oriented positions.

Technicians can also move into farm management and administration, supervising personnel and overall operation of farms.

## what are some related jobs?

The U.S. Department of Labor classifies agricultural technicians under a variety of headings, such as *Agricultural Engineering Occupations, Occupations in Life Sciences, Not Elsewhere Classified, Farm Mechanics and Repairers,* and *Home Economists and Farm Advisers* (DOT) and *Engineering: Design, Laboratory Technology: Life Sciences, Craft Technology: Mechanical Work,* and *Educational and Library Services: Teaching, Home Economics, Agriculture, and Related* (GOE).

| **Related Jobs** |
| --- |
| Agriculture equipment design engineers |
| Architects |
| Ceramic design engineers |
| Chemical engineers |
| Civil engineers |
| Landscape architects |
| Mechanical design engineers, facilities |
| Tool designers |

## what are the salary ranges?

According to the 1998–99 *Occupational Outlook Handbook,* science technicians earned a median income of $27,000 a year in 1996. The lowest 10 percent made less than $15,500, while the highest 10 percent earned more than $49,500.

Wages vary depending on the region and the technician's educational background and experience. According to the 1998 General Schedule produced by the U.S. Office of Personnel Management, beginning technicians at grade level four earn $17,848 a year. Technicians working for the USDA are eligible for benefits, including health, life, and retirement.

## what is the job outlook?

Technicians can find work in many areas of agricultural production. As more people become concerned about the conservation of natural resources, agricultural technicians are needed to put these conservation practices in place. Management of crops and pests without the use of harmful chemicals will create challenges for agricultural technicians. Environmentally sound production systems are also needed, meaning more jobs in service, sales, development, and application of mechanical systems.

Food product processing is another area expected to expand as engineers and scientists devise new ways to ship and preserve more kinds of foods. By keeping up on the latest in agricultural equipment and products, you can get a sense of which areas of agriculture are growing and developing.

If you are thinking of going to work for the USDA, keep track of which areas receive the best funding. When a USDA program must cut back spending, it will sometimes phase out technician positions and rely only on the work of professionals. Also, because of the competitive job pool, the USDA's educational requirements are becoming more demanding, and you may have a difficult time securing a federal job without at least a bachelor's degree.

**Agriculture for Today and Tomorrow**

Precision Agriculture is the up and coming trend in agriculture and crop management. Precision Agriculture, also known as Precision Farming, Prescription Farming, and Site-Specific Farming, utilizes computers and technology to increase the efficiency and yield of crops and farms. Making use of satellites and global positioning systems (GPS), farmers can map fields, detect crop yields, study and assess the pH levels in the soil, and tailor the soil content and fertilizer in fields. For instance, a farmer drives an all-terrain vehicle equipped with a computerized card over a field. The vehicle is hooked into a satellite system that maps out the field as the farmer progresses across the field. This card can then be downloaded onto a computer system that will map out the field into grids. Soil samples can then be taken from each of the grids and tested for mineral content. The farmer or agricultural technician can then make fertilizer recommendations and program in prescriptions for each grid, place the prescriptions on a computer card, insert the card into a spreader truck, and drive the truck over the field. The card will tell the truck how much lime or other minerals to spread into each grid, making the farmer's job much easier and more manageable.

## how do I learn more?

### sources of additional information

The following organizations can provide information regarding careers in agriculture, scholarship opportunities, accredited schools, and more.

**American Chemical Society**
1155 16th Street, NW
Washington, DC 20036
800-227-5558
http://www.acs.org

**American Institute of Biological Sciences**
107 Carpenter Drive, Suite 100
Sterling, VA 20164
800-992-2427
http://www.aibs.org

**American Society of Agricultural Engineers**
2950 Niles Road
St. Joseph, MI 49085
616-429-0300
http://asae.org

**American Society of Agronomy**
677 South Segoe Road
Madison, WI 53711
608-273-8080
http://www.agronomy.org/asa.html

**Crop Science Society of America**
677 South Segoe Road
Madison, WI 53711
608-273-8080
http://www.crops.org/cssa.html

**National FFA Organization**
5632 Mt. Vernon Memorial Highway
PO Box 15160
Alexandria, VA 22309
703-360-3600
http://www.ffa.org

**Soil Science Society of America**
677 South Segoe Road
Madison, WI 53711
608-273-8080
http://www.soils.org/sssa.html

**U.S. Department of Agriculture**
14th Street and Independence Avenue, SW
Washington, DC 20250
202-720-2791
http://www.usda.gov

## bibliography

Following is a sampling of materials relating to the professional concerns and development of agricultural technicians.

### Periodicals

*Resource.* (St.Joseph) Bimonthly. Covers all aspects of agriculture, including buildings, soil conservation and irrigation, and equipment. American Society of Agricultural Engineers, 2950 Niles Road, St. Joseph, MI 49085-9659, 616-429-0300.

*American Journal of Alternative Agriculture.* Quarterly. Discusses low-input farming systems. Institute for Alternative Agriculture, Inc., 9200 Edmonton Road, Suite 117, Greenbelt, MD 20770-1551, 301-441-8777.

*Communications in Soil Science and Plant Analysis.* Twenty times per year. Reviews all aspects of soil and crops. Marcel Dekker Journals, 270 Madison Avenue, New York, NY 10016, 212-696-9000.

*FFA New Horizons.* Bimonthly. Focuses on general topics such as careers and recreation in the agricultural field. (The National Future Farmer) National Future Farmers of America Organization, 5632 Mt. Vernon Memorial Highway, Box 15160, Alexandria, VA 22309, 703-360-3600.

*Outlook on Agriculture.* Quarterly. Discusses current research and development. CAB International, North American Office, 198 Madison Avenue, New York, NY 10016, 212-726-6490.

# air traffic controller

**Definition**

Air traffic controllers organize and direct the movement of aircraft into, out of, and between airports.

**Alternative job title**

Air traffic control specialists

**Salary range**

$29,500 to $46,000 to $100,000

**Educational requirements**

Some postsecondary training

**Certification or licensing**

Required

**Outlook**

Little change or more slowly than the average

GOE 05.03.03

DOT 193

O*NET 39002

## High School Subjects

Computer science
Mathematics
Speech
Physics

## Personal Interests

Airplanes
Computers

**The sky** was clear over Palwaukee Municipal Airport. Traffic was light on the radar. Fifteen minutes ago, air traffic controller Tom Hjelmgren had cleared a Cessna for takeoff. Now the Cessna radioed in through Tom's headset. Tom's eyes widened.

"Cessna 3. This is Tower. Repeat your last message. Over."

"This is Cessna 3." The pilot's voice was calm. "We're going to crash. Over."

Tom signaled the others in the tower. The data technician radioed Chicago's O'Hare International to clear the airways. Traffic was diverted from the runway. Emergency vehicles raced past the tower.

On radar, the plane held altitude. Through Tom's headset, the pilot was joined by the tower's chief controller, the data technician, and the ground controller, each of whom signalled Tom with instructions. He patched the pilot's transmission into the speakers. The others listened, stunned.

"I'm going down." The pilot's voice was as calm as before. "I'm crashing this bird at the intersection of Lawrence and Cumberland. In Chicago."

The radar showed the Cessna at a steady altitude. Chicago was still fifty minutes away. Why wasn't the pilot returning to the airport?

"Cessna 3, this is Tower. I'm sorry, I seem to be a little slow today. Do you need help? Have you lost your engines?"

"No." The pilot was less calm now. "I'm crashing this plane. There's a Pot 'n Skillet at Cumberland and Lawrence. I'm crashing into that."

There were people down there. A busy intersection. A restaurant brewing with the lunchtime crowd. Behind him, Tom heard someone speaking to the manager of the Pot 'n Skillet, trying to persuade her that a plane was about to crash into her restaurant. The other technicians monitored the approaching traffic, the throughways, and coordinated the response from Chicago's fire department.

*Pilots don't do this.* Tom thought. *Pilots were the most dependable, responsible people he knew. They don't just decide to crash their planes.*

It was up to Tom to talk to the pilot of the Cessna, talk him through this, turn him back safely to Palwaukee. There was still time . . . .

## what does an air traffic controller do?

Like traffic cops at any busy intersection, air traffic controllers (ATCs) regulate the flow of traffic into, out of, and around the airport. With thousands of flights operating every day, safety is a primary concern, and the ATC makes certain that airplanes follow their designated flight paths and maintain safe distances from one another. The ATC keeps flights operating on schedule, helping to minimize delays. The ATC also provides pilots with information about weather conditions that will affect their landing or takeoff or even their ability to handle their planes. ATCs alert pilots to other factors, such as geographical terrain features, the presence and movement of other aircraft in the area, ground taxi instructions, and which air routes—the highways in the sky—the pilot should take. Other ATCs guide aircraft through emergency situations, or conduct searches for late or lost planes.

Using radar, radio, computer automation, and his or her own eyes, an ATC usually coordinates a number of planes at once, preparing one plane for takeoff while advising another plane on its approach, and at the same time directing a third plane safely through the airport's airspace. As an airplane leaves the ATC's airspace—which can reach up to forty miles—the ATC passes control of the plane to an *enroute air traffic controller* further along the plane's flight path. Or an ATC, such as a *radar controller*, will accept control from an enroute controller for an airplane entering his or her airspace.

Air traffic controllers must be able to recognize, remember, and act upon a great deal of information, often coming to them simultaneously, such as an airplane's registration number and flight plans and those of any other aircraft nearby (coming to them as blips on a radar screen), the airplane's speed and altitude, the type of craft, and the instructions they have already given to the pilots. They must be able to respond quickly to every situation and to issue clear and decisive instructions.

### types of controllers

**Arrival Controller** Regulates the flow of traffic entering the airspace, establishes holding patterns, and clears planes for landing.

**Departure Controller** Coordinates the flow of traffic leaving the airport's airspace.

**Enroute Air Traffic Control Specialist** Coordinates the flow of traffic between airports' airspace.

**Flight Data and Clearance Delivery Specialist** Receives and communicates flight plans, airport weather conditions, and other data between pilots and other air traffic control towers along the pilot's route.

**Flight Service Station Air Traffic Control Specialist** Links the many air traffic control facilities, providing data on flight plans, type of plane, weather and terrain conditions, and other essential information.

**Ground Controller** Responsible for maintaining the smooth flow of traffic along the airport's taxiways leading to the runways.

**Local Controller** Speaks directly to pilots, prepares them for takeoff, and guides them on the final landing approach.

An ATC is also responsible for relaying a variety of information to other air traffic control positions and centers. And, because it can often take an hour or more for an aircraft to pass through his or her airspace, an ATC must be able to maintain intense concentration over a long period of time.

The duties of an air traffic controller can vary from airport to airport and among the several types of traffic control centers. In a large, or class V, facility, such as an international airport where traffic is particularly heavy, an ATC may receive a specific assignment, for example, as a *ground controller.* At a combined facility, typically where traffic is less heavy, an ATC can be charged with all the various air traffic control functions. Controllers at an enroute facility, on the other hand, usually work in teams of three or more, with each team responsible for conducting traffic within a specific area of the airway. And flight service stations link all the air traffic towers together, coordinating flight data and weather and terrain conditions. In all cases, however, an ATC never works alone, but in conjunction with other air traffic controllers. Teamwork is an important part of maintaining the safety of our skies.

## what is it like to be an air traffic controller?

Tom Hjelmgren has been an air traffic controller for thirteen years. He's worked at Palwaukee Municipal Airport in Wheeling, Illinois (where, after many tense moments, Tom was able to talk the pilot of the Cessna safely back to the airport), the Greater Rockford Airport in Rockford, Illinois, Chicago O'Hare International Airport, and, now, Mitchell Airport in Milwaukee, Wisconsin. "Each airport is different, depending on what kind of facility it is," he says, "and each day is different, depending on the conditions, the traffic, the weather. All these variables affect what we do. So the job is always changing."

There are forty to forty-three controllers at Mitchell Airport, a twenty-four-hour facility, working rotating shifts to cover the entire day. Controllers generally work a forty-hour week, with four days on the job, and two days off, although their shifts may change from week to week. A shift can last eight to ten hours, depending on the facility, the expected traffic at the airport, and the number of ATCs available. Overtime is always available, and can be paid out as extra days off. Controllers rotate through the day, evening, night, and weekend shifts,

> "I talk fast. And I'm hyper. You have to be, with everything we do during the shift."

which can be difficult getting used to, according to Tom. "But," he says, laughing, "every six weeks I get a whole weekend off."

Tom works in the air traffic control tower, where he tracks airplanes on radar, receives and transmits data (including weather condition reports, flight plans, and traffic patterns), and communicates with the pilots and the ground crews via radio. "Basically, a plane is my responsibility from the moment it enters our airspace until the time it is safely parked at the gate," Tom says. "Usually, I'm in charge of several planes at once. Luckily, I don't work alone."

Approaching aircraft radio into the tower before they arrive in the airport's airspace. The controllers (in a large facility, the *radar controller*) will already know the plane's flight plan and will be monitoring the plane on radar. The aircraft is directed to its designated runway, if that is clear, or guided into a holding pattern with other planes above the airport until the runway is clear for landing. Contact is maintained with the pilot as he makes his approach; other flights are alerted to the plane's presence, and the runway is kept clear until the plane has landed. Once the plane is on the ground, a controller will guide it visually (he'll use radar, too, when visibility is reduced) as it taxis to its gate.

When a plane is departing the airport, the controllers first signal the pilot about the visibility, the direction and speed of the wind, and weather and other conditions that will affect the flight. The plane is then guided along the taxiways to the runway, where it is cleared for takeoff, and then directed through the airport's airspace. From there, enroute controllers are notified and direct the pilot along the flight's designated air route. At all times during the flight, the plane is kept under the guidance and direction of air traffic controllers. Through their teamwork, air traffic controllers keep the airways safe and make sure a plane and its passengers will reach their destination safely.

"Sure, you're always aware that there are people on that plane, there are people on the ground, that lives are at stake," Tom admits, "but most controllers don't dwell on that. You'd burn out quickly if you do." Generally, however, an airport's traffic schedule remains constant. The same flights arrive and depart on a schedule that doesn't vary. This helps the controller "know" what other airplanes are out there and allows him or her to act quickly when anything goes wrong. Because ATCs are highly trained, not only in the FAA training academy but on the job as well, they become instinctually aware of the skies around them. "We're paid for what we know," Tom says. "We have to be able to respond fast, without thinking about it. And the training never stops."

## have I got what it takes to be an air traffic controller?

"I talk fast," Tom says. "And I'm hyper. You have to be, with everything we do during the shift. You're constantly busy. You have to be focused all the time."

Tom believes that his sense of self-discipline and diligence have been key factors throughout his career. Guiding several planes at once and for extended periods of time requires complete concentration, and because an ATC's decisions affect so many lives, they must not only be made correctly but communicated correctly as well. The great amount of teamwork, with other controllers, and with the pilots and the ground personnel make strong communication skills—including listening—essential.

An ATC must be able to express himself clearly, and to recall instantly and act on large amounts of rapidly changing data. An ATC must be able to perform many rapid calculations, and he must be able to "see" the relationships among the many and constantly changing variables involved in safely conducting an airport's traffic. Because of the stress of maintaining such high levels of concentration, an air traffic controller needs to possess a highly developed emotional self-control. This allows him to perform his duties, while giving and receiving instructions calmly, clearly, and decisively.

Tom also counts his willingness to ask for and to give help as an important quality on the job. "Even though I know the others know their jobs as well as I do, I'll still call their attention to, say, a plane just entering the airspace, just to be certain that they've seen it too. And I expect them to do the same for me. You're always aware that you're part of a team here." Being part of a team is one of the things Tom likes most about being an air traffic controller. "I love this job. I couldn't imagine being anything else."

**To be a successful air traffic controller, you should:**

- Be articulate, self-controlled, and patient
- Have a good memory
- Be able to think, act, and speak quickly and decisively
- Be able to work independently and as part of a team
- Be able to visualize spatial relationships
- Have an aptitude for abstract reasoning

## how do I become an air traffic controller?

Competition for air traffic controller training is fierce, in part because of the high pay, generous benefits, and job security an ATC enjoys, but also because of the stringent requirements for this career. All candidates must have three to four years of progressive work experience (that is, work in which their responsibilities steadily increase) or four years of college, or a combination of both, and must score highly on the federal civil service examination. Many candidates come from a military background, where they may even have functioned as an *operations specialist,* the equivalent of a civilian air traffic controller. Others enter with experience in a related field, including pilots and navigators. And some ATCs come into the job with no experience at all. Tom was a supervisor of a graphic design department. "I took the exam on a lark," Tom says. "I'd been out of college and working for a couple of years. I just wanted to see how I'd do."

### education

#### High School

A high school diploma is an essential first step toward becoming an air traffic controller. Candidates at the FAA Academy should have a solid understanding of mathematics; some knowledge of aviation, meteorology, and computers is also helpful. A high school student can thus prepare for a career in air traffic control by developing these skills. Yet, says Rob Reedy, an air traffic controller at DuPage Airport in Illinois and the National Training Coordinator for the National Air Traffic Controllers Association, "the FAA will teach you everything you need to know at the academy and on the job. In fact, they almost prefer an empty sponge, someone who will do what they are told, when they are told. The *ability* to learn is an essential skill. "Other than that, I would say that a high school student should concentrate on his communication skills," Rob adds, "because that is really the whole job."

### Postsecondary Training

Three to four years of college or progressive work or military experience are required before being accepted into training at the FAA Academy. A college degree is also desirable. Equivalent work experience includes administrative, supervisory, professional, or technical positions that have prepared the candidate for the great responsibilities of air traffic control.

Candidates must be at least eighteen years of age (but not older than thirty-one) and articulate, with eyesight correctable to 20/20. Only candidates scoring high on the civil service exam are considered for training. This exam measures a candidate's aptitude for—that is, his ability to learn—the skills needed to become a qualified air traffic controller. Candidates are then subjected to a four-day computer screening to determine if they have the alertness, decisiveness, motivation, emotional self-control, and ability to work under intense pressure that are essential to the work of an air traffic controller.

Candidates must also pass a physical exam and must routinely submit to drug testing (as must controllers). The training program that follows is an intensive eleven- to seventeen-week course at the FAA Academy in Oklahoma City. At this point, trainees are considered employees of the FAA and are paid as such. Training consists of the fundamentals of the airway systems, civil air regulations, radar, and aircraft performance characteristics. Emergency simulations, designed to test the candidates' emotional stability under pressure, contribute to the high (50 percent) failure rate among all candidates.

"The Academy was probably the hardest thing I've ever done," Tom says. "When it came time for training, I was what they called a 'no-knowledge' because I literally knew nothing. I had to learn everything—meteorology, vectors, FAA flight rules—from scratch. It was constant study. But what got me through it was knowing I could always count on one of the others to help me understand something I was missing. There's a great sense of camaraderie among ATCs, and I think that begins at the Academy."

**FYI**

Air traffic controllers who successfully complete the Federal Aviation Administration's (FAA) training are guaranteed FAA jobs.



In recent years, efforts have been made to reduce the stress of the training period. "In the old days," Rob Reedy says, "you had to undergo a sixteen-week screening before you could even begin your actual training. You never knew if you'd wash out, up to the day of graduation. Now, with the four-day computer screening period, you know quickly if you'll be accepted into training. And the atmosphere at the academy has become more like a college campus, where people want you to succeed."

ATC candidates may join academic programs offered in the army, navy, and air force, or at one of several private institutions. However, only candidates trained at the FAA/DOT or military academies are guaranteed jobs with the FAA, where they will receive their crucial on-the-job training.

## certification

On-the-job training (with increasing levels of responsibility) and continued study are required before a candidate can become a fully certified air traffic controller. This training period can range between eighteen months and six years, depending on the facility and the amount of traffic. Senior ATCs supervise and grade trainees' performance each day. Rob Reedy explains: "A new hire at an enroute center can train from three to six years before he is fully certified there; here at DuPage, where we direct mostly private and pilot training flights, someone beginning their training in the summer, when there is much more traffic, can expect to finish their on-the-job training much sooner than someone who begins in the winter, when traffic is light. Here, the training range is from eighteen to twenty-eight months." Candidates must be

### Advancement Possibilities

**District supervisors** coordinate and supervise the responsibilities of all air traffic control facilities in a given district.

**FAA Administrators** work in a variety of administrative posts within the FAA, from policy development to statistical research to candidate recruitment.

**Facility Supervisors** supervise and grade the performances of non-certified air traffic controllers and coordinate the schedules and responsibilities of controllers and other personnel working in air traffic control.

certified at each level of air traffic control, and failure to certify within a specific period of time results in dismissal. Yearly physical exams and twice-yearly job performance reviews are also part of the job.

A candidate fresh from the academy will typically be assigned to an available position in a specific region of the country. Persons with the highest grades will have the widest choice of assignments. This is important because everyone has his or her own career goals and preferences. "Personally, I like it here at DuPage Airport," Rob Reedy says. "At a place like O'Hare, you deal only with professional pilots. But here, where our pilots are mainly leisure flyers and student pilots, things don't always happen the way they're supposed to. That keeps you on your toes."

On-the-job training begins at the flight data and clearance delivery position. The next step is ground controller. "It's only on the job that a controller is really tested," Rob Reedy says. "They may know everything they need to know, but if they have what we call mike fright, that is, if they freeze when speaking to the pilots, they'll never be a good controller." From ground controller, the controller advances through the local controller and departure controller grades, before reaching the grade of arrival controller. The tower supervisor, finally, functions as an extra pair of eyes and ears, overseeing all of the activities of the tower, and is ready to offer assistance whenever the need arises. At an enroute facility, a candidate begins by supporting the teams with printed flight data, then advances to the teams themselves as a radar associate and finally radar controller.

"Of course, as a beginner, you don't work alone," Tom adds. "You're always under the supervision of another controller, who can quickly override your instructions if you make a mistake."

## who will hire me?

There are more than 32,000 air traffic controllers employed by the federal government, mostly in the FAA, while a few civilian controllers work for the Department of Defense. In addition to the nearly 450 airport control towers across the United States, there are also twenty-four enroute air control centers, each with 300 to 700 controllers, and more than 275 flight service stations, which provide pilots with preflight or inflight assistance, linking the airways above the United States. A small number of ATCs are employed at privately owned, non-FAA towers.

## where can I go from here?

Promotion to each successive grade of ATC certification increases the complexity of the controller's responsibilities. At the higher grades, a controller may coordinate all of the traffic control activities at his or her facility, supervise and train controllers at the lower grades, or manage the various aeronautical agencies involved in air traffic control.

A controller can reach supervisory and managerial positions, and assume responsibilities for wider and wider areas. Some controllers continue on to administrative positions within the Federal Aviation Administration. A controller is granted civil service status upon completion of his or her first year on the job, and career status at the end of his or her third year.

Advancement moves from assistant to fully-qualified controller. When fully qualified, a person may move into a chief controller position, air traffic control management, or FAA administration. There is a rapid turnover due to early retirement, yet job competition is stiff since there are more qualified personnel than there are jobs.

## what are some related jobs?

Related jobs include many careers in aviation and aeronautics, military service, and transportation. The U.S. Department of Labor classifies air traffic controllers under the headings *Radio Operators* (DOT) and *Expediting and Coordinating Workers* (GOE). Occupations that involve directing air traffic include airline radio operator, airplane dispatcher, navigator, and data technician. Other careers in aviation include pilot, aviation safety inspector, and technical and engineering careers in the design, maintenance, and repair of aviation systems. Civil, electrical, electronic, and mechanical engineering, as well as electronic and engineering technology are all vital components of maintaining the safety of the skies. In addition, air traffic controllers have successfully entered careers in automation, computer technology, transportation, and teaching.

| Related Jobs |
|---|
| Dispatchers |
| Field engineers |
| Material schedulers |
| Production schedulers |
| Railroad tower operators |
| Transmitter operators |

## what are the salary ranges?

The average salary among air traffic controllers is $46,000 a year. Trainees start at $21,400, while an experienced controller can earn $65,000 a year and more. Salaries vary widely with level of seniority, degree of responsibility, and type of facility. For example, an experienced air traffic controller at a level 5 airport, such as O'Hare, can earn close to $100,000 a year including bonuses and overtime. Generally, the higher the level of the air traffic facility, the higher the salary will be. Air traffic controllers also receive special benefits, such as a higher pay scale than other civil service employees, more vacation days, and a liberal retirement program. Air traffic controllers receive life insurance and health benefits, thirteen paid sick days, and, depending on their length of service, thirteen to twenty-six paid vacation days. Air traffic controllers also enjoy job security; even during times of recession, they are rarely laid off.

## what is the job outlook?

The FAA is currently not hiring air traffic controllers and few new air traffic control positions are expected to be created through the year 2000, despite the increasing demands for air travel. This is because technological advancements in automated equipment will make it possible for controllers to perform their work more efficiently than ever before. Most position openings will come from retiring or leaving controllers. "However," Rob Reedy says, "by the year 2000, 50 percent of all air traffic controllers will be eligible for retirement. Which means that, like it or not, the FAA might find itself hiring a lot of new ATCs." Competition for any openings will remain intense, because the supply of qualified applicants far exceeds the demand. The high pay, benefits, and job security attract many highly qualified people to this field.

Employment opportunities will be particularly strong for candidates with college degrees, a strong background of progressive work experience, or related civil or military experience as controllers, navigators, or pilots. Recent years have seen the active recruitment of women and minorities into this career. Top scores on the civil service exam, excellent performance on the four-day computer screening, and high grades in training will improve a candidate's chances of becoming an ATC and insure that the candidate enjoys a wide range of options when choosing the facility in which to begin his or her career.

## how do I learn more?

### sources of additional information

Following are organizations that provide information on aviation air traffic control careers and employers.

**Air Traffic Control Association**
2300 Clarendon Boulevard, Suite 711
Arlington, VA 22201
703-522-5717

**Federal Aviation Administration**
Office of Personnel and Training
800 Independence Avenue, SW
Washington, DC 20591
202-267-3229
202-366-4000

**National Association of Air Traffic Specialists**
11303 Amherst Avenue
Wheaton, MD 20902
301-933-6228

**Canadian Air Traffic Control Association**
#1100, 400 Cumberland Street
Ottawa ON K1N 8X3
Canada
613-225-3553

**National Air Traffic Controller Association**
1150 17th Street, NW, Suite 701
Washington, DC 20036
202-223-2900
202-659-3991

## bibliography

Following is a sampling of materials related to the professional concerns and development of air traffic controllers.

### Books

Brenlove, Milovan S. *Vectors to Spare: The Life of an Air Traffic Controller.* Nontechnical account of the author's experiences on the job. Ames, Iowa: Iowa State University Press, 1993.

Nolan, Michael S. *Fundamentals of Air Traffic Control.* 2nd ed. Belmont, CA: Wadsworth Publishing Company, 1994.

### Periodicals:

*Air Traffic Control Association Bulletin.* Monthly. Discusses procedures involved in forming and maintaining a safe and efficient air traffic control system. Air Traffic Control Association, 2300 Clarendon Boulevard, Suite 711, Arlington, VA 22201, 703-522-5717.

*Air Traffic Research Quarterly.* Quarterly. International journal delving into traffic and systems operations. John Wiley & Sons, Inc., 605 Third Avenue, New York, NY 10158, 212-850-6000.

*Aviation Safety Institute Monitor.* Monthly. Aviation Safety Institute, 6797 North High Street, Worthington, OH 43085, 614-885-4242.

# aircraft technician

**Definition**

Aircraft technicians service, repair, overhaul, and inspect aircraft and aircraft engines. They also repair, replace, and assemble parts of the airframe.

**Alternative job titles**

Airframe and power plant technicians

Airplane mechanics

Aviation maintenance mechanics

Avionics technicians

Instrument repair workers

**Salary range**

$16,000 to $35,000 to $65,000

**Educational requirements**

High school diploma; 18 to 30 months of specialized training

**Certification or licensing**

Recommended

**Outlook**

About as fast as the average

GOE
05.05.09

DOT
621

O*NET
22599C

## High School Subjects

Chemistry
Mathematics
Physics
Shop (Trade/Vo-tech education)

## Personal Interests

Airplanes
Cars
Computers
Fixing things

A large airline passenger jet appears overhead, gently makes its descent, and rockets across the runway. Moments later, technicians swarm over it. Their job? Get it ready to fly again in thirty minutes. "There's a slight glitch in the INS," one technician is told. "Can you take a look?" This is important: the plane's inertial navigation system (INS), an integral part of its avionics system, provides steering commands to the aircraft's computerized autopilot system. The autopilot system in turn can operate the controls automatically. Climbing into the cockpit, the technician begins to work.

# what does an aircraft technician do?

Just like cars, airplanes and other aircraft need specially trained people to fix them quickly when something is wrong, or to do a major repair or overhaul if the problem is more serious; to do regular maintenance to keep them in good working order; and to inspect them to be sure they are safe to operate. In many ways, what auto mechanics are to cars, *aircraft technicians* are to aircraft.

Different technicians do different types of work on an aircraft, but all of it is critical to the safety of the millions of people who fly each day. Technicians keep thousands of main structural parts (as many as 18,000) and components (50,000 or more) in an aircraft in good working order.

The work a technician does depends on his or her training and certification. *Airframe technicians* work on all aircraft parts except the instruments, power plants, and propellers. "Airframe" refers to the wings, body, tail assembly, and fuel and oil tanks of the aircraft. Work includes checking the sheet metal surfaces of the aircraft for corrosion or other problems, measuring the tension of control cables, and other jobs.

*Power plant technicians* work on engines (and, if the aircraft has them, may do some work on propellers). Power plant technicians may stand on ladders or scaffolds to examine the engine or use a hoist to take the engine out of the plane and work on it.

*Airframe and power plant (A&P) technicians* can work on any part of the plane. Most aircraft technicians are A&P technicians. A&Ps with an *inspector's authorization* can authorize the inspection work done by others.

These four positions are certified by the Federal Aviation Administration (FAA), the government agency entrusted with ensuring aviation safety. Noncertified technicians must work under the supervision of certified technicians. Other technician positions include *instrument repair workers*, who focus on the aircraft's instrument systems, and *avionics technicians* (or, in some airlines or other employers, A&P technicians trained in avionics), who work on an aircraft's avionics systems. *Avionics* is everything having to do with the aircraft's electronic systems. Advanced aircraft increasingly are making use of computer-controlled electronic systems for the plane's primary functions, including flight, navigation, communication, radar, and other operations.

Technicians may perform line maintenance or heavy (base) maintenance. *Line maintenance* is routine upkeep, repairs, and checks on planes currently in service and scheduled to fly. For an airline, this work usually is done at the airline's terminals at the airport. A plane may land with a minor problem, such as a stuck door or a broken seat, and a technician performing line maintenance will have to fix it in the short time before the plane takes off again. Technicians may be told by the pilot, flight engineer, or head mechanic what repairs need to be made, or they may inspect the plane themselves for any malfunctions or defects. In addition, routine maintenance, such as replenishing the hydraulic and oxygen systems, must be performed before the plane can take off again. Technicians must work quickly so the plane can return to service without delay.

*Heavy (base) maintenance*, for planes not in service, may be done at the airline's domicile, or base facility. Work includes major overhauls and repairs, as well as inspections required after the plane has flown a certain number of hours, after a certain number of calendar days have passed, or at some other milestone. For these inspections, the technicians may check the plane's

## lingo TO LEARN

**A&P** Airframe and power plant.

**Airframe** The wings, fuselage, tail assembly, and the fuel and oil tanks of the airplane.

**Domicile, or base** Airline hangar space used to repair that airline's aircraft.

**Fixed base operations** Service facilities offering maintenance services as well as fuel and parking.

**Fuselage** The body of the airplane.

**General aviation** Segment of the aviation industry that includes air taxi and fixed based operations, FAA-certified repair stations, aerial applicator companies, flight training schools, private aircraft, corporate fleets, and aircraft manufacturers.

**Power plant** The aircraft's engines and propellers.

**Terminal** An airline's place of business at an airport where planes dock, passengers and cargo are loaded and unloaded, and basic repairs are made.

**Walkaround** A quick inspection of the plane before a scheduled flight.

engines, landing gear, instruments, pressurized sections, brakes, valves, pumps, air-conditioning, and other parts and systems. They also may repair the plane's sheet metal surfaces. Then, tests are run to make sure everything is working right.

Many technicians work for the airlines, but others work in general aviation or for the military. *General aviation* is a broad category that includes air taxi (commuter plane) and fixed-based operators (service facilities offering maintenance, plus fuel and parking), FAA-certified repair stations, aerial applicator companies (companies that use planes to do crop dusting or fight fires, for example), flight training schools, corporations that have their own planes, and aircraft manufacturers. A general aviation technician might work on any of a variety of aircraft, including small piston-engine aircraft, larger turbine-engine aircraft, propeller-driven aircraft, and helicopters.

In addition, other technicians work for a third segment of the aviation industry, the government. All branches of the military—air force, army, navy, marines—train and use technicians. In fact, the military is a top producer of avionics technicians because it generally has the most advanced facilities and equipment and offers highly specialized training.

Throughout the industry, the number of technicians required to work on an aircraft depends on the type of aircraft and the employer. A large jet air carrier, for example, requires about fourteen technicians per airframe.

## what is it like to be an aircraft technician?

Ralph Ortiz has been an aircraft technician for American Airlines for three years. He is a licensed airframe and power plant (A&P) technician. "I've always been fascinated with planes," he says. "I just think they're really cool."

Like most technicians, Ralph typically works at least a five-day, forty-hour week. In addition, workers often are asked to put in overtime. The three regular shifts at American are 6:00 AM to 2:00 PM, 2:00 PM to 10:30 PM, and 10:30 PM to 6:00 AM. Ralph gets paid double time for overtime and double-time-and-a-half if he works on a holiday. Extra work is usually needed around major holidays. "Those are really the craziest times," Ralph says.

When doing overhauling and major inspection work, aircraft technicians generally work in heated and well-lit hangars. If the hangars are full, however, or if repairs must be made quickly, they may work outdoors, sometimes in bad weather.

The object is to keep a plane flying all day, from 6:00 AM to 10:00 PM.

Outdoor work is the norm for Ralph, who focuses on line maintenance. He works at the American Airlines terminal at Chicago's O'Hare Airport, the busiest airport in the world. His job is to quickly make minor repairs and do preflight checks. With planes and passengers waiting, he often has to work under time pressure. "The object is to keep a plane flying all day, from 6:00 AM to 10:00 PM," Ralph says.

However, he adds, "everything is redundant on a plane," so there are other systems that can take over if something is not working right. As long as the plane is still safe to fly, the mechanics make only necessary repairs during the day and wait until the night shift to do any major work.

Often, the work is physically hard and demanding. Technicians may have to lift or pull up to fifty pounds in weight. They may stand, lie, or kneel in awkward positions as they work and sometimes use a scaffold or ladder to reach the part of the aircraft that needs work. Working in tight spaces is common, and the noise and vibration can be very strong. Regardless of the stresses and strains, aircraft technicians are expected to work quickly and with great precision.

Although the power tools and test equipment are provided by the employer, technicians may be expected to furnish their own hand tools. These may include common tools like screwdrivers, pliers, hammers, and drills. Special X-ray and magnetic inspection instruments allow the technician to check for extremely tiny cracks in the aircraft's fuselage, wings, and tail.

Ralph's normal work day begins at 2:00 PM and ends eight-and-half-hours later, at 10:30 PM. A typical day's work, he says, might include anything from a "walkaround" to changing a part. A 'walkaround' is a quick, routine inspection in which you basically walk around the plane and eyeball it to make sure everything's as it should be. It takes about fifteen minutes. Something like changing a part takes longer. For example, it might take me about an hour to change a generator."

What are his favorite jobs? "Probably the jobs in the cockpit," he says. "Things involving avionics, like working with the flying instruments." Ralph's company does not have separate avionics technicians, although it does have an avionics department.

Sometimes Ralph's day begins when he's not expecting it to. "You could be home sleeping and they call you and say they have a plane" in another city that needs repairs, he says. "If you're up for overtime, you could get a call at any time. You don't have to take it; I've known guys who turn it down every time, but the money's great."

Each time he goes to make a repair, Ralph begins by consulting the manual. This is standard procedure for technicians. "They never want you to do anything from memory," he says. His company has a special room filled with manuals and computer programs to help keep its technicians up to date, and there is frequently time during the shift to do extra studying. Because of the importance of the work and the fact that specifications are constantly changing, even the most experienced technicians must have a manual at their side whenever they are doing any repair work.

## have I got what it takes to be an aircraft technician?

Aircraft technicians make decisions every day that could affect the lives of many people. They take direct responsibility for each task they do, signing off on their work in the FAA log book. "Putting your name in the log book," or signing off on your work, says Ralph, makes this a very stressful job. A technician could actually "be charged with manslaughter" if something goes wrong with a plane and he or she is found at fault. "I've known guys who got fined by the FAA for not writing a paragraph correctly in the FAA log book," he says. They are very strict about how things have to be done.

"Someone from the FAA doesn't personally come and inspect your work," Ralph says. "They can tell if you did the job right by reading the logs. An airline quality control (QC) inspector may come around to check some types of work, though, like major repairs."

Ralph sees the heavy responsibility of his job as one of the good parts, something that gives him a tremendous feeling of importance and pride. Along with that is the satisfaction he gets from fixing something that was broken. He also likes the money he makes, especially the overtime pay, and the flexible hours. "You also get to meet a variety of people," he says. "I know guys at airports all over the country."

Something else he enjoys is taxiing the planes around the airport. What? "You didn't think the pilots moved the planes all around, did you? No, the mechanics move them to where they're supposed to be, and then the pilots fly them."

All in all, Ralph says, you have to be a very patient, meticulous, and careful person to be a successful aircraft technician. "I've known guys who just come in and put in their time. They just want to get the work done." They don't last very long, he says.

**To be a successful aircraft technician, you should**

- Have manual dexterity and mechanical aptitude
- Be able to work with precision and meet rigid standards
- Have more than average strength for lifting heavy parts and tools
- Be agile for reaching and climbing
- Not be afraid of heights

## how do I become an aircraft technician?

After he expressed an interest in getting certified to be a licensed A&P technician, Ralph's company sent him to its own training academy for twenty-four months. He remembers being an average student in high school, "but I always liked working on cars," he says. This proved to be valuable experience for his future career choice. "Cars have pretty much the same systems," he says, and anybody with any kind of mechanical background should do fine.

## education

### High School

Only about half of the global/major airlines, and about a quarter of the rest, require that their new hires have a high school diploma. They do, however, want one or more years' experience. A high school diploma is required, however, for entry into an A&P school. Thus, while there continue to be opportunities for those who have not finished high school, their career paths will be significantly more limited.

High school courses in mathematics, physics, chemistry, and mechanical drawing are important to learn the basic principles involved in how an aircraft works, knowledge that is often necessary in making repairs. Machine shop, auto mechanics, or electrical shop courses are important in order to help develop a student's mechanical aptitude.

### Postsecondary Training

There are a variety of roads to becoming an aircraft technician, including private training schools, some of which are owned by the airlines; community college programs; two- and four-year degree institutions; or simply work experience.

The two most common ways of becoming an aircraft technician are either to join the military and acquire the experience there or to attend one of the more than 220 FAA-certified schools in the United States. Military training usually is highly specialized, and candidates with military experience are much sought after in the industry. Those who trained in the service, however, may need to supplement their military training with additional courses in order to be able to do a wider range of civil aviation work and to qualify for their civilian ratings or certificates.

There also are programs at two- and four-year colleges that permit you to earn maintenance technician ratings and certificates, while simultaneously earning an associate's or bachelor's degree.

Schools that are FAA-certified meet Federal Aviation Regulation Part 147 standards for curriculum and structure. Trade school programs generally take twenty-four to thirty months. Courses cover airframe structures; electrical, hydraulic, environmental, and pneumatic systems; fiberglass, sheet metal, and welding work; physics; chemistry; mathematics; and reciprocating and turbine engines. Courses in composite structures, turbine engine technology, and avionics are especially valuable for the future of the industry. Course work includes both classroom studies as well as hands-on experience in the shop.

Ralph found his electricity and hydraulics classes particularly interesting. "There's a lot of physics in hydraulics," he says. "A lot of it is just common sense." One class project involved taking apart an entire engine and putting it back together again. In another class, students had to make their own airfoils (wings).

Another route to becoming a technician is through an apprenticeship. Apprentices work under the guidance of certified technicians. In order to qualify for certification, an apprentice's work must be both consecutive and concurrent on airframes, power plants, or both.

**FYI**

A&P certification is valid outside the United States only if the technician is working on U.S.-registered aircraft.

## certification or licensing

To advance beyond the entry level, technicians must obtain FAA certifications. Certification is granted for airframe technicians, power plant technicians, airframe and power plant technicians, and aircraft inspectors.

Technicians seeking FAA certification must either have graduated from an FAA-certified school or be able to prove that they have at least eighteen months of practical experience as an airframe or power plant technician, or if they wish to become certified in both, at least thirty months of practical experience as an airframe and power plant technician.

Aircraft technician experience gained in the military may be considered for credit toward the work experience qualification. Those with military experience must present documents attesting to the type and amount of training they have received, their length of service, their occupational specialty codes, and any personal evaluation records.

Once the work experience or school requirement is satisfied, applicants for FAA certification must pass written and oral/practical tests. The written tests for the A&P certificate, for example, cover fifty general questions, one hundred airframe questions, and one hundred power plant questions. Exams cost about $50 each.

After passing the written tests, applicants have up to twenty-four months to pass the oral and practical tests. Tests are given one-on-one or in small groups and are administered by an FAA-designated mechanic examiner (DME) at A&P schools or at the DME's shop. These tests cover the applicant's basic skill in performing the mechanical duties of the job.

Options for preparing to take the exam include attending special seminars or other structured test-taking classes. For self-study, FAA publishes books of questions found on each type of certification test. The books are available from the Government Printing Office (GPO) and other sources.

A&P technicians who have been certified for at least three years may pursue an inspector's certificate, which qualifies them to perform annual inspections, supervise inspections done by other technicians, and authorize the return to service of aircraft, parts, or components after major repairs.

Finally, a license from the Federal Communications Commission (FCC) also is required for some types of work. Avionics technicians may be required to have an FCC operator's license, which covers operation, repair, or maintenance of broadcast equipment. Technicians who maintain or repair FCC-licensed radiotelephone transmitters must have an FCC general radiotelephone operator's license. FCC licenses are obtained by taking a written test. Check with your local FCC office about test locations, samples, and license information.

## scholarships and grants

Several organizations offer scholarships for students pursuing a career in aircraft maintenance (*see* "Sources of Additional Information"). Your chances are much better if you contact them well in advance of your studies and carefully follow all application instructions.

Employment breakdown for aircraft technicians

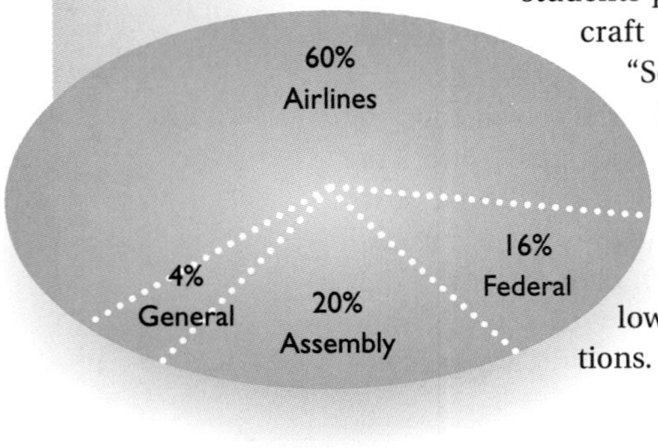

60% Airlines
16% Federal
20% Assembly
4% General

## labor unions

In addition to education and certification, union membership may be a requirement for some jobs, particularly for technicians employed by the major airlines with a mechanics union on the property. The principal unions organizing aircraft technicians are the International Association of Machinists and Aerospace Workers and the Transport Workers Union of America. In addition, some technicians are represented by the International Brotherhood of Teamsters, Chauffeurs, Warehousemen and Helpers of America. Membership in a union usually starts after an initial probationary period, often three to six months. Unionized technicians pay weekly or monthly dues to the union and work under a contract that determines their pay, benefits, and work rules.

## who will hire me?

*Airlines* More than three-fifths of all aircraft technicians currently employed in the United States work for airlines. Each airline usually has one main overhaul base, where most of its technicians are employed. These bases are found along the main airline routes or near large cities, including New York, Chicago, Los Angeles, Atlanta, San Francisco, and Miami.

*Federal government* About one-sixth of aircraft technicians work for the federal government. This includes military service people and civilians employed at military aviation installations, as well as technicians employed by the FAA, mainly at its headquarters in Oklahoma City.

*General aviation* Most other technicians work in the general aviation segment of the industry. About one-fifth of all technicians work for aircraft assembly firms. Aerospace manufacturers and research and development firms also hire aircraft technicians.

Future Aviation Professionals of America assists its members by supplying job reports, an employment guide, a directory of employers, and a salary survey. It also provides toll-free interview briefings and a resume service. Other sources include trade magazines, such as *PAMA News Magazine,* which includes company profiles in each issue.

Also, the Federal Aviation Administration hires technicians, mainly to work at the training center (*see* "Sources of Additional Information" and "Bibliography"). For information about working as a civilian for one of the branches of

**Advancement Possibilities**

**Airplane inspectors,** also known as **airplane and engine inspectors,** examine airframe, engines, and operating equipment to ensure that repairs are made according to specifications, and certify airworthiness of aircraft.

**Field service representatives,** also known as **technical specialists, aircraft systems,** advise and train customers in operation, overhaul, and maintenance of company aircraft, applying knowledge of aircraft systems, company specifications, and FAA inspections procedures and specifications.

**Flight engineers,** also known as **flight mechanics,** make preflight, inflight, and postflight inspections, adjustments, and minor repairs to ensure safe and efficient operation of aircraft.

the U.S. Armed Forces, you should contact the local recruiter of the appropriate branch.

## where can I go from here?

"I could see myself after several years going for my inspector's license," Ralph says, "but what I really want to be is a tech crew chief. If you've got a problem on a plane and can't figure it out, they're the guys you go to. They just know everything. That's what I want to be."

An aircraft technician's advancement depends partly on the size of the organization he or she works for. Ralph may have to wait five to ten years before he realizes his goal. "But that's okay," he says. "The more experience you have, the more you can see and do, the better." Like most technicians, his first promotion was merely a salary increase, not a change in title or job responsibilities. To advance further, many companies require the technician to have a combined airframe and power plant (A&P) certificate, as Ralph does, or perhaps an aircraft inspector's certificate.

Because seniority plays an important part in work assignments, including overtime, moving to another company or airline is not necessarily a way to advance. Although a technician might elect to change jobs for better pay, he or she would begin the new job with no seniority, no matter how many previous years' experience he or she might have.

Advancement possibilities include head mechanic or crew chief, inspector, head inspector, and shop supervisor. With additional training, a mechanic may advance to engineering, administrative, or executive positions. With business training, some technicians open their own repair shops.

## what are some related jobs?

The U.S. Department of Labor classifies aircraft technicians under the headings *Mechanics and Machinery Repairers* (DOT) and *Craft Technology: Mechanical Work* (GOE). Also under these headings are people who service, maintain, and repair motorcycles, automobiles, construction equipment, railroad cars and locomotives, boats, farm machinery, trucks, and industrial machinery. An aircraft mechanic, says Ralph, could work anywhere there are turbine engines, such as utility companies or cruise ships, or anywhere there are hydraulic machines, such as in factories, or even in nuclear plants.

| Related Jobs |
|---|
| Automobile mechanics |
| Diesel mechanics |
| Farm equipment mechanics |
| Heating and air-conditioning technicians |
| Millwrights |
| Powerhouse mechanics |
| Robotics technicians |

## what are the salary ranges?

Aircraft technicians earn an average of about $35,000 a year. Some skilled, experienced technicians make more than $48,000. Beginning technicians can expect to earn from $16,000 to $23,000 a year.

Salaries vary greatly depending on whether you work in general aviation, for an airline, or for the government. In general aviation, for example, the average starting wage for A&P technicians varies from $8.50 an hour to $18 an hour. Maximum earnings for general aviation A&Ps range from $35,000 to $60,000 annually. In contrast, maximum salaries for airline A&Ps range from $38,000 to $65,000 annually. Most major airlines are covered by union agreements.

Their technicians generally earn more than those working for other employers.

An attractive fringe benefit for airline technicians and their immediate families is free or reduced fares on their own and most other airlines.

## what is the job outlook?

Job opportunities for aircraft technicians are expected to increase as fast as the average for all occupations through 2006. Demand for technicians is tied directly to the number and types of aircraft in service, which in turn is tied to the overall economic climate. During periods of recession, for example, airlines may resort to downsizing because there tend to be fewer passengers. As the economy expands, however, demand tends to increase and airlines put more planes into service.

Employment growth will be restricted by a greater use of automated equipment that speeds repairs and maintenance. This will be largely offset by the smaller numbers of younger workers in the labor force, coupled with fewer recruits from the military and a larger number of retirements. Opportunities are on the rise for those with advanced training in electronics, avionics, modern turbine engines, and composites.

In general aviation, qualified technicians, particularly if they are willing to relocate, should find plenty of opportunities. Most of these opportunities will be with small companies. Because the wages paid by small companies are usually lower, there will not be as much competition for these jobs as for jobs with airlines and large private companies.

## how do I learn more?

### sources of additional information

Following are organizations that provide information on aviation maintenance careers, accredited schools and scholarships, and employers.

**Aircraft Electronics Association**
AEA Educational Foundation, Inc.
13700 East 42nd Terrace South
Independence, MO 64055-4748
816-373-6565

**Aviation Technician Education Council**
2090 Wexford Court
Harrisburg, PA 17112-1579
717-540-7121

**Future Aviation Professionals of America**
4959 Massachusetts Boulevard
Atlanta, GA 30337
1-800-JET-JOBS

**National Air Transportation Association**
4226 King Street
Alexandria, VA 22302
703-845-9000

**Professional Aviation Maintenance Association**
1200 18th Street, NW, Suite 401
Washington, DC 20036
202-296-0545

## bibliography

Following is a sampling of materials relating to the professional concerns and development of aircraft technicians.

### Periodicals

*Aircraft Maintenance Technology.* Bimonthly. Focuses on technical and professional topics of interest to the aviation maintenance professional. Johnson Hill Press, Inc., 1233 Janesville Avenue, Fort Atkinson, WI 53538, 920-563-6388.

*AMFI Industry News.* Bimonthly. Newsletter featuring industry information. Aviation Maintenance Foundation International, Box 2826, Redmond, WA 98073, 360-658-8980.

*Aviation Mechanics Bulletin.* Bimonthly. Newsletter providing ground crew personnel with information on new products and methods. Flight Safety Foundation, 601 Madison Street, Suite 300, Alexandria, VA 22314-1756, 703-739-6700

*FAA Aviation News.* Bimonthly. Covers FAA regulations and directives designed to increase aviation safety and includes information on aircraft maintenance, avionics, and accident analysis and prevention. U.S. Federal Aviation Administration, 800 Independence Avenue, SW, Washington, DC 20591, 202-267-8017.

# airplane and helicopter pilot

**Definition**

Airplane and helicopter pilots perform many different kinds of flying jobs, including transporting people and cargo, such as freight or mail.

**Alternative job titles**

Aircraft pilots
Airline pilots
Charter pilots
Private pilots

**Salary range**

$15,000 to $54,000 to $200,000+

**Educational requirements**

Some postsecondary training

**Certification or licensing**

Required

**Outlook**

About as fast as the average

GOE
05.04.01

DOT
196

O*NET
97702B*

## High School Subjects

Mathematics
Physics

## Personal Interests

Airplanes
Fixing things
Travel

**Brett Anker** and his boss are en route to Miami, Florida, flying at a steady 22,000 feet, when Brett checks the moving map and notices some echoes forming a pattern about 200 miles away. He suspects the echoes may indicate a thunderstorm, so he contacts the flight service station, which provides weather updates to pilots, and receives word of a severe weather system over Houston, Texas. He continues to monitor the map, and as they approach the storm, they see a solid black mass hovering in the air at 10,000 feet. Soon they are directly over the storm. Flying in perfectly blue sky, Brett looks straight down over clouds so black they look purple. Lightning zigzags from cloud to cloud, illuminating the whole cloud with a blinding flash. Brett can see the lightning hitting the ground, and the thunder rumbles louder than the engines. Brett watches in amazement and feels lucky—would anyone other than a pilot be able to view something so magnificent?

# what does an airplane and helicopter pilot do?

The best-known pilots are the *commercial airline pilots* who fly for the major airlines, such as United and American. Although these individuals comprise the majority of employed pilots, there are a variety of flying jobs that pilots hold. Some of these jobs include *charter pilots,* who transport passengers on a smaller scale than the large airlines; *agricultural pilots,* who are involved in things such as crop dusting and seed reforestation; and *helicopter pilots,* who transport people and work for law enforcement agencies, television stations, and hospitals, doing such things as monitoring traffic, tracking criminals, searching for missing persons, and transporting patients in emergency situations. Pilots may also be employed as *flight instructors.*

The are three main designations of commercial airline pilots: *captain, copilot,* and *flight engineer.* The captain is usually the pilot with the most seniority. He or she is in charge of the plane, and the copilot is second in command. The flight engineer makes preflight, inflight, and postflight inspections, adjustments, and minor repairs to the aircraft, as well as monitors the plane's instruments during flight to make sure the plane is flying safely.

Except for the takeoff and landing, most of the time a large commercial jet is in the air it is actually being flown on autopilot, a device that controls the plane's flight, and keeps it on course. Planes today may even land on autopilot. That does not mean, however, that pilots can sit back and relax. Pilots must constantly monitor the aircraft's systems, the weather, and be in constant communication with air traffic controllers. Flying for a large commercial airline carries much responsibility. Aircraft flown by airline pilots normally cost millions of dollars; the safety and welfare of perhaps hundreds of passengers are on the line each time a plane makes a flight.

All commercial pilots must undergo continual testing to make sure their skills are in top shape. Each major airline has its own testing requirements, but most of them involve annual or semiannual testing of each pilot's abilities. Flying and navigating are considered primary flying responsibilities. Secondary flying responsibilities include filing flight plans and listing flight reports for the Federal Aviation Agency (FAA), the agency that oversees all flying in the United States.

Charter pilots do essentially what the large commercial pilots do, except on a smaller scale. As with the commercial pilots, charter pilots are involved in continual testing and licensing requirements as they try and gain higher-paying flying positions. Charter pilots are often more involved with secondary flying responsibilities than the commercial pilots.

Charter pilots may also be involved with such tasks as the loading and unloading of baggage and freight. Closely related to the job of charter pilots is the *executive pilot.* Executive pilots are employed by companies to transport employees and company products. The National Business Aircraft Association estimates that more than eighteen thousand firms either own or lease their own planes. In some instances, executive pilots are used to inspect large company holdings or determine the condition of such items and facilities as pipelines and electric power lines.

Agricultural pilots are involved in such activities as seed reforesting, insect control, crop dusting, and also checking the condition of crops and livestock. Pilots who work in this capacity often combine interests in both flying and farming. These pilots also may have specialized training in agriculture that allows them to better understand the duties they are performing. Agriculture pilots may also be involved in fighting forest fires, which at times may be hazardous work, especially when they are forced to fly low over ground fires to drop water or chemicals.

Pilots may also make a living flying helicopters. Although the number of pilots doing this is small, their numbers are expected to

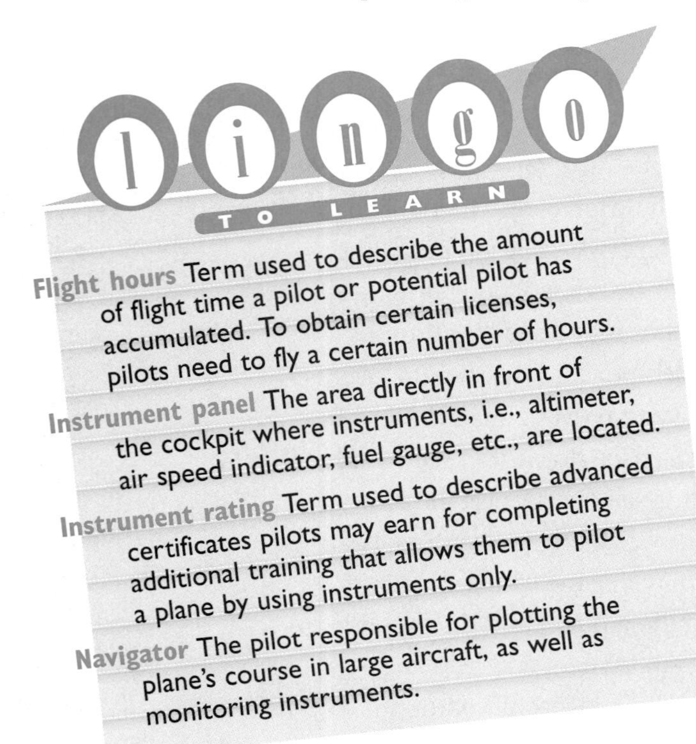

lingo TO LEARN

**Flight hours** Term used to describe the amount of flight time a pilot or potential pilot has accumulated. To obtain certain licenses, pilots need to fly a certain number of hours.

**Instrument panel** The area directly in front of the cockpit where instruments, i.e., altimeter, air speed indicator, fuel gauge, etc., are located.

**Instrument rating** Term used to describe advanced certificates pilots may earn for completing additional training that allows them to pilot a plane by using instruments only.

**Navigator** The pilot responsible for plotting the plane's course in large aircraft, as well as monitoring instruments.

increase in the future. Helicopters have an advantage over planes in that they are able to land in places planes cannot, such as the tops of buildings, or on flat surfaces in open fields where a landing strip may not exist. Helicopters are also used for short trips, such as transporting mail within a limited region, providing air-taxi services, or monitoring traffic patterns for the police or news stations. Helicopter pilots may also be involved with rescue services, sightseeing trips, aerial photography or advertising, and agricultural activities, including fighting forest fires and spraying fields for insect control. Because helicopters often fly at low altitudes, helicopter pilots must be acutely aware of obstacles, such as trees, power lines, and bridges.

Flight instructors teach students how to fly and often help them prepare for license exams. Classes include classroom study of flight principles, safety rules, and federal regulations and in-flight instruction in dual-controlled planes and helicopters. Some pilots work as examiners or check pilots. They may give examinations to pilot's license applicants or fly with experienced pilots as part of a periodic review.

Finally, although few in number, there are test pilots, who are employed by aircraft companies to perform flight tests on new aircraft. These tests measure a new plane's performance and the accuracy of the plane's instruments. Test flying is often hazardous since pilots are flying in aircraft where such things as the plane's maximum airspeed and altitude have yet to be determined. It is the test pilot's job to determine these functions, as well as other factors of the plane's performance.

## what is it like to be an airplane and helicopter pilot?

Brett Anker, a commercial pilot for a private air cargo company in San Diego, California, flies two to four days per week, transporting cargo all over California, Arizona, and Nevada in twin-engine airplanes. Every aspect of each day is different—some days Brett flies the King Air B200, while other days he flies the Navajo or the Aerostar. He picks up and delivers everything from automotive parts to canceled bank checks to human blood. The duration of his workday depends on the number of stops he is scheduled to make; Brett begins work at 10:30 AM and heads home as early as 1 PM or as late as 8 PM.

When Brett arrives at work, one of his first duties is to complete his manifests, which includes getting information about where he has to go, what he must pick up, fuel requirements, and weather conditions. He then takes a walk around the plane, inspecting the tires, the electrical systems, and the fuel to make sure everything is in place and ready. During his runs, Brett is responsible for making sure weight is distributed correctly, picking up receipts for fuel loads, and completing all the required paperwork, such as securing signatures from appropriate parties. Before he goes home, Brett makes sure the fuel is topped off and the plane ready for the next pilot. Brett believes each step is important, whether it's checking the fuel or landing the plane. "I consider everything primary with flying because there are so many variables that I just put everything in that critical stage of checking," he says.

While Brett maintains a fairly regular work schedule, this may change as he prepares to fly the company's Learjet more frequently. Currently, he flies the jet once every few months when he fills in for another pilot. With the jet, Brett is on call and must stay within two hours of the airport at all times. He notes that "it's sort of hard to have a life sometimes," but emphasizes, "I absolutely love flying. It's a lot of fun learning all the systems and flying different planes, and I really enjoy the responsibility."

For Mike Eckstein, being a pilot for Executive Air Services means being on call twenty-four hours per day. "No week is typical. You can work six days per week, or just three or four," he says. There are FAA requirements on how many hours per week a pilot may fly, depending on the type of license he or she holds. With Mike's charter license, he is able to fly up to eight hours per day, six days in a row. After this he must take twenty-four hours off before he can fly again.

To improve his flying skills, Mike will sometimes spend time in flight simulators. These devices are built to simulate actual flight in a variety of aircraft, from simple two-engine prop planes to complex jets. Most flight schools have simulators on hand for students and instructors alike.

"No week is typical. You can work six days per week, or just three or four."

Closely related to Mike's job of charter pilot is that of a flight instructor. Thirty-four-year-old Gary Holom-Bertelsen is an assistant flight instructor at Lewis University in Romeoville, Illinois. It's Gary's job to instruct young pilots on the basics of flight and prepare them for their careers.

A typical day for Gary may include taking three to seven students up for flight instruction. Sometimes, he may take one student up at a time, or even several at once. While in the air, he instructs the students on the basics of flight while sitting in the copilot seat. Once on the ground, he writes performance reports which he later goes over with the students to monitor their progress.

Gary has been flying since 1988 and has seen many changes within the aviation industry. "Aviation takes a dive when the economy goes bad," he says.

Gary enjoys his job and would like to advance to higher-paying positions within the aviation industry, such as that of commercial airline pilot. He is realistic, however, about his chances: he likens the aviation industry to getting acting jobs, where there are only a few top jobs.

## have I got what it takes to be an airplane and helicopter pilot?

Airplane and helicopter pilots are typically people who have strong mechanical skills, as well as very decisive personalities. It's important for pilots to be able to make decisions quickly and accurately, and sometimes under the pressure of having to be a certain place at a specific time. Pilots also need to be able to accept responsibility for others, especially in the case of the passenger pilot, who may be transporting more than two hundred people.

Persistence is key for airplane and helicopter pilots—it's hard to break into the business, and it's not easy accumulating the knowledge and flight hours required to gain a commercial pilot's license. You must have a cool head to make it as a pilot, because "it's very dangerous, obviously," says Brett, "and stressful in a lot of ways with weather and things like that. It can take its toll on people."

Flexibility is important as well, since pilots often spend nights away from home in different cities or countries. And due to the increased competition for flying jobs, pilots must be mobile in their employment, moving from job to job. Flexibility is also important on the job, where pilots are "dealing with so many variables, in terms of weather, in terms of the conditions of the plane, the regulations, the controllers," Brett mentions. Pilots who are on call must be ready to fly at the drop of a hat, and all airplane and helicopter pilots are responsible for keeping abreast of federal regulations and current information.

**To be a successful airplane and helicopter pilot, you should**

- Have a decisive personality
- Be able to accept responsibility for others
- Be willing to be on call twenty-four hours per day
- Be interested in continuing your education your entire career

## how do I become an airplane and helicopter pilot?

Most people aren't as lucky as Brett, who received his pilot training at a small FBO (fixed-base operation) in Florida. Basically, FBOs are small airports that often have flight schools with a few airplanes. When Brett was nearing college graduation, his best friend from prep school, who happened to be a flight instructor, called and offered him free lessons. The decision wasn't difficult. Brett recalls, "It just sort of fell into my lap, and it's one thing that I've always wanted to do, so I just went for it."

Training to become an airplane or commercial pilot isn't easy, though. Brett progressed at his own pace and studied independently, which worked well for him. "I did it very fast. I just wanted to immerse myself in it totally. I was flying at least five days a week, and I also worked at the airport so I was always there studying. I did it as fast as I could afford it," says Brett.

## education

### High School

Completing high school is a must for anyone interested in becoming a pilot. Students should take high school classes in mathematics, particularly algebra and geometry. Physics, meteorology, and shop classes are also helpful. Good activities in which to participate include sports that may improve hand-eye coordination, a local ham radio club, as well as organizations such as local Air Scout troops, part of the Boy Scouts of America. At sixteen, interested students can start taking flying lessons.

### Postsecondary Training

There are two main routes for gaining flight experience: military training and civilian training.

Military pilot training is a two-year program for which a college degree is normally required. The first year is spent on flight basics, including classroom and simulator instruction, as well as officer training. The second year is spent training in a specific type of aircraft. Following completion of this training, pilots are expected to serve at least four years before they may leave the service and pursue a civilian flying career. On average, the military needs about four hundred new pilots per year.

Outside of the military, there are nearly 600 flight schools certified by the FAA. This includes some colleges and universities that award college degrees (in majors such as aeronautical engineering or airport operations) and pilot's licenses simultaneously, as well as small, FBO-based flight schools such as the one Brett attended. Cost is a consideration—both Gary and Mike confirm that a flight education can cost more than $10,000 by the time classroom and flight time are paid for. Brett also advises thinking about the best learning environment for you. He suggests, "If you learn best with a whole bunch of people around, at a given pace, and with lesson plans given to you, then I think it's better to go to school or learn through the military. But if you're a self-starter and you can go at your own pace and you're very motivated, I think the route I took is best, because you have the freedom to explore so many avenues."

The airline industry's primary requirement is that pilots must have a commercial pilot's license issued by the FAA. To obtain this license, pilots must pass simulator and written tests and have accumulated 1,500 hours of flying time. Most airlines also require that applicants have a bachelor's degree, and more flying companies are requiring at least an associate's degree. In fact, according to the *Occupational Outlook Handbook,* nearly 90 percent of all pilots have completed some college. Otherwise, the industry depends mostly on market needs. When there are many military pilots coming out of the service, competition for airline jobs is fierce; when there are fewer pilots, jobs are easier to find. Today, competition for the top jobs is very tight—only the best and most qualified pilots are looked at. Continuing certification and the accumulation of many flight hours are strongly recommended for airplane and helicopter pilots. Flying hours are recorded in a log book either by a flying instructor or by the pilots themselves. Pilots must also complete in-flight tests. These tests are called *check rides* and are done by flying instructors. During check rides, the pilot's flying performance is rated, and a pass or fail is given.

Flying through a variety of conditions in many types of aircraft is crucial for pilots in training. Pilots can test their skills in flight simulators, which simulate such flying scenarios as night flying, thunderstorm flying, and landing without the use of an engine. High scores in flight simulators can translate into better job opportunities.

In addition to accumulating flight hours and passing tests, all pilots must maintain good health and have excellent vision. They are constantly evaluated and re-evaluated on the job. Continual testing is an integral part of being a pilot.

## certification or licensing

Licensing of all pilots is governed by the FAA. To obtain different licenses through the FAA, flying hours must be accumulated and pilots must pass in-flight and written tests. Licenses include the *private pilot certificate,* the *commercial pilot's license,* and a beginning *recreational pilot's permit* (25 hours). Helicopter pilots must have a commercial pilot's license with a helicopter rating.

Closely related to getting different flying licenses is obtaining *instrument ratings.* Instrument ratings show that a pilot is able to fly based on reading instruments alone, without the help of landmarks, clouds, or other visuals. These ratings change as pilots progress from flying single-engine to multiengine planes—all the way to jets.

## scholarships and grants

Several organizations offer scholarships for students pursuing a career in airplane piloting (*see* "Sources of Additional Information"). Your chances are much better if you contact them well in advance of your studies and carefully follow all application instructions.

## labor unions

The Air Line Pilots Association, International represents more than 49,000 pilots at approximately forty-eight airlines in the United States and Canada. Many airlines have their own pilot unions.

## who will hire me?

According to the 1998-99 *Occupational Outlook Handbook,* about 60 percent of the 110,000 civilian pilot jobs in 1996 were held by pilots working for airlines. The remainder of the civilian pilots worked for the government, for private corporations (usually transporting people or cargo), or for small businesses.

Many pilots entering the aviation field want to work toward flying for one of the major airlines. In order to accomplish this, a college degree, along with the necessary flight requirements (i.e., the pilot's license and flight hours) is necessary. Once you log enough realistic flight hours, Brett advises that you print up your resume and cover letter and send it to the companies you are interested in working for, whether they are the major airlines, commuter airlines, or cargo companies. Follow up by sending monthly updates. That's right, monthly. "It's very important," Brett emphasizes. "I've been told that companies like Delta and Sky West get 10,000 applications a day. And they keep files on you. Some companies do it for six months, and some do it for a year, and they really want to see that you're very interested in the company."

After completing the military or civilian route, prospective pilots often start out as flight instructors because it's a practical way to accumulate flight hours—you get paid to fly—and your chances of landing a flying job without those hours are poor. Brett indicates that most students, by the time they exit school, will only have around 350 hours; as a flight instructor, it's possible to acquire enough hours to surpass the 1,000 mark.

Other employment possibilities include sightseeing companies, government opportunities, private industry, and agricultural flying. Aviation Information Resources, Inc. can assist members by supplying job reports, an employment guide, a directory of employers, and a salary survey. It also provides interview briefings and a resume service (*see* "Sources of Additional Information").

## where can I go from here

With Brett's commercial pilot's license, he could work for one of the major airlines, but he chooses not to. He prefers flying faster jets and has no desire to fly larger, slower airplanes. Brett hopes to land the position of chief pilot in his company, shifting his focus to the business aspect of flying. Chief pilots oversee the company fleet and are responsible for hiring, acquiring new accounts, sales, and upkeep of the business.

In the airlines, seniority is everything. The flight engineer, or navigator, may spend one to five years before being promoted to first officer, then spend another five to fifteen years before becoming captain. When going to another airline, pilots must start over at the bottom and build up their seniority all over again.

There are a number of other positions a pilot might hold before working for the airlines, including owning and operating a flying business and working as a flight instructor, charter pilot, or agriculture pilot. Pilots may also work as check pilots, testing other pilots for advanced ratings.

| Related Jobs |
| --- |
| Aerial-applicator pilots |
| Air traffic controllers |
| Facilities flight check pilots |
| Flight instructors |
| Flight operations inspectors |
| Navigators |
| Remotely piloted vehicle controllers |
| Test pilots |

**Advancement Possibilities**

**Captains** of large airlines are high in seniority and have years of experience. Captains are highly paid and assume responsibility of large aircraft and crews.

**Check Pilots** test the flying skills of pilots and issue advanced ratings.

**Chief Pilots** supervise the work of other pilots and are responsible for general upkeep of the business.

## what are some related jobs?

The U.S. Department of Labor classifies airplane and helicopter pilots under the headings *Airline Pilots and Navigators* (DOT) and *Air and Water Vehicle Operation: Air* (GOE). Also included under these headings are test pilots, check pilots, flying instructors, helicopter pilots, navigators, flight operations inspectors, and remotely piloted vehicle controllers.

FYI

**Orville and Wilbur**

Today, in the skies above Kitty Hawk, North Carolina, private planes circle the famous field where the first official airplane flight took place on December 17, 1903. The entire flight was a scant 120 feet, but proved beyond a reasonable doubt that people possessed the ability to build a machine that could fly.

Near the site where the Wright brothers made their first flight stands a museum where some of the remnants of their career are housed. The Wright brothers were actually bicycle makers. The intricate tools they used on their bikes are displayed, showing that even the first pilots had a precision and mechanical aptitude that has been one of the mainstays of the pilot personality ever since.

## what are the salary ranges?

Salary ranges for airplane and helicopter pilots vary greatly, from lows of around $15,000 per year for beginning flight instructors to more than $200,000 for senior captains of large airlines. In 1996, experienced pilots earned an average of $76,800 at the major airlines, according to a salary survey conducted by the Future Aviation Professionals of America, a now defunct organization whose services have largely been taken over by Aviation Information Resources, Inc. Salaries vary by airline, experience, and by the size and type of plane flown.

According to the same survey, commercial helicopter pilots earned anywhere from $33,700 to $59,900 in 1996, while corporate helicopter pilot salaries ranged from $47,900 to $72,500. Brett believes commercial cargo pilots frequently start at $15,000 and top out at more than $70,000.

Brett feels jobs are more prevalent and the wages higher in big cities and the east and west coasts. According to the 1998-99 *Occupational Outlook Handbook,* pilot distribution is concentrated in areas with higher amounts of flying activity, such as California, Nevada, Hawaii, Texas, Georgia, Washington, and Alaska.

Most airplane and helicopter pilots receive benefits that include life and health insurance, retirement plans, and disability payments. Pilots with major airlines often receive travel benefits and discounts at hotels and rental car companies.

## what is the job outlook?

Brett feels that the job outlook is currently very strong. "The Vietnam and Korea era pilots are retiring now," Brett notes. "As soon as the people at the top leave and retire, everybody from the bottom goes up to the top, and then they need to hire all new people on the bottom." Brett also points out that new routes and countries are opening up, thereby increasing flight demand. The growth, however, is tempered by the demise of many major and commuter airlines and the high cost of flying. Many commercial flights, if not at full capacity, are often canceled, inconveniencing both passengers and pilots.

According to the 1998–99 *Occupational Outlook Handbook,* pilots may face stiff competition in coming years. There may be a glut of prospective hires due to airline mergers and bankruptcies resulting in layoffs, as well as federal budget cuts that caused an outpouring of pilots from the military. The plus side, however, is that demand may grow because cargo and passenger traffic are expected to increase. The services that a helicopter pilot can provide are also forecast to expand. You're better off if you choose not to become a flight engineer, though, because engineers are being phased out with the dominance of computers and automation.

When it comes right down to it, the more experience you have, and the better pilot you are, the better your prospects. Continuing education and acquiring as many FAA licenses and ratings as possible will be to your benefit. It may also be helpful to identify your niche, whether it's flying a 747 for a major airline or piloting a Learjet for a corporate client, and concentrate on gaining as much experience and knowledge in that niche as possible.

# how do I learn more?

## sources of additional information

The following organizations provide information on airplane and helicopter pilot careers, schools, mentor programs, and employers.

**Aircraft Owners and Pilots Association**
421 Aviation Way
Frederick, MD 21701
301-695-2000
http://www.aopa.org

**Air Line Pilots Association, International**
1625 Massachusetts Avenue, NW
Washington, DC 20036
703-689-2270
http://www.alpa.org/

**Air Transport Association of America**
1301 Pennsylvania Avenue, NW, Suite 1100
Washington, DC 20004-1707
202-626-4038
http://www.air-transport.org/

**Aviation Information Resources, Inc.**
1001 Riverdale Court
Atlanta, GA 30337-6018
800-AIR-APPS
http://www.airapps.com/

**Experimental Aircraft Association**
PO Box 3086
Oshkosh, WI 54903-3086
800-564-6322
http://www.eaa.org/

**Helicopter Association International**
1635 Prince Street
Alexandria, VA 22314
703-683-4646
http://www.rotor.com

**International Organization of Women Pilots, the Ninety-Nines**
PO Box 965, 7100 Terminal Drive
Oklahoma City, OK 73159
800-994-1929
http://www.ninety-nines.org/

**National Agricultural Aviation Association**
1005 E Street, SE
Washington, DC 20003
202-546-5722
naaa@aol.com

**National Air Transportation Association**
4226 King Street
Alexandria, VA 22302
800-808-6282
http://www.nata-online.org

**National Business Aircraft Association**
1200 18th Street, NW, Suite 400
Washington, DC 20036
202-783-9000
http://www.nbaa.org

**Society of Expert Test Pilots**
PO Box 986
Lancaster, CA 93584
805-942-9574
http://www.netport.com/setp/

## bibliography

Following is a sampling of materials relating to the professional concerns and development of airplane and helicopter pilots.

### Books

Griffin, Jeff. *Becoming an Airline Pilot.* Blue Ridge Summit, PA: TAB Books, 1990.

### Periodicals

*Aerolog.* Bimonthly. Newsletter published for club members. National Aviation Club, 1500 North Beauregard Street, Suite 104, Alexandria, VA 22311-1715, 703-379-1506.

*Air Line Pilot.* Monthly. Features coverage of aviation history as well as developments in air safety and aviation technology. Air Line Pilots Association, 535 Herndon Parkway, Box 1169, Herndon, VA 20172, 703-481-4460.

*Career Pilot.* Monthly. Provides prospective commercial pilots with information on career, financial, and legal topics. Future Aviation Professionals of America, 4002 Riverdale Court, Atlanta, GA 30337-6018, 404-997-8097.

*FAA Aviation News.* Bimonthly. Covers FAA regulations and directives designed to increase aviation safety and includes information regarding rule changes for student and professional pilots. U.S. Federal Aviation Administration, 800 Independence Avenue, SW, Washington, DC 20591, 202-267-8017.

*Flight Training.* Monthly. Discusses information useful for pilots working toward certification. Smooth Propeller Co., 201 Main Street, Parkville, MO 64152, 816-741-5151.

# animal breeder

**Definition**

Animal breeders breed and raise animals to improve traits, to develop new breeds, to maintain standards of existing breeds, or to preserve threatened or endangered species.

**Alternative job titles**

Animal husbandry technicians

Artificial Inseminators

Livestock production technicians

**Salary range**

$16,000 to $25,000 to $50,000

**Educational requirements**

High school diploma

**Certification or licensing**

None

**Outlook**

About as fast as the average

GOE
03.01.02

DOT
410

O*NET
79015

## High School Subjects

Agriculture
Anatomy and Physiology
Biology
Computer science

## Personal Interests

Animals
Science

**Kyle Alexander** moves quickly in the dark, stowing his breeding equipment in the back seat of his car, setting his coffee in the cup holder, and tuning the radio to an early morning talk show. Ten minutes later, the sky nearly light with the first rays of the sun, he pulls up in front of a barn. Daniel Evans, a local farmer, greets Kyle cheerfully, "Cold enough for ya?" His frosty breath hangs in small clouds in the air.

He leads Kyle into the barn, to the dam, or cow, ready to be inseminated. "Hey there, girl," Kyle says, and runs a hand along her back. The dam is tied in her stall to discourage resistance. Kyle and Bob talk for a few minutes about the bull who provided the semen and what they hope the calves will be like.

All day long, Kyle will talk to farmers and soothe cows ready for breeding. Then he will take a semen straw from the temperature-controlled case in the back of his car and use a special "gun" to efficiently shoot the semen into the cervix of the cow. Although the procedure is quick and painless, Daniel talks to the cow to keep her from fidgeting while Kyle performs the procedure. Kyle puts the gun back in its case and looks at his patient, who is now standing calmly, tail swinging, large brown eyes fixed in front of her. He chats with Daniel for a moment, arranging a follow-up visit, then gets back into his car. The radio announcers are reviewing a new movie as Kyle drives down the road to his next appointment.

## what does an animal breeder do?

Selective breeding is usually intended to improve the genetic makeup of common animals such as cattle, horses, sheep, poultry, dogs, and cats, along with more exotic species such as llamas and monkeys. Some breeding programs help preserve threatened or endangered species; these efforts often take place at zoos or facilities that raise animals to be released into the wild.

By mating males and females with preferred traits, breeders encourage the production of young with the best traits of both. For example, a cattle breeder might mate a bull whose offspring have lean meat with a cow who is uncommonly large. The calf would most likely be a large animal with more than the average amount of lean meat. Cattle breeders typically try to develop animals that are large, have less fat than in the past, give birth easily, eat less, and are not very susceptible to diseases. Strong, healthy, meaty cattle bring higher profits for their owners.

Horse breeders also try to develop strong, healthy animals, but for a different reason. Doug Lindsay, dean of animal sciences at Lakeshore Technical College in Cleveland, Wisconsin, explains, "The aim is to breed horses that perform well for their riders, horses more physically perfect, as far as judges at horse shows are concerned." Horse breeders are always trying to find ways to make horses jump higher, run faster, or ride more smoothly.

Some dog or cat breeders, on the other hand, want animals that meet the standards of a breed association. They might aim for a collie with a perfect, long nose and small, dainty feet or a Persian cat with eyes of a certain shape and fur of a certain color. Other breeders want working dogs with a natural ability and desire to perform a certain job, such as herding sheep.

Domestic animals are usually bred to please humans, but animals in zoos are bred mainly to maintain their populations, both in zoos and in the wild. Pandas, gorillas, Chinese alligators, and the Arabian oryx (a graceful antelope) are just a few of the species that have been bred in captivity in an attempt to save them from extinction. The breeding of exotic animals requires special skills and training, but it's done in basically the same way as livestock breeding.

Like other animal breeding endeavors, zoo breeding now relies heavily on artificial insemination, although it's also still common to bring the animals together and let nature take its course. Artificial insemination is often easier, cheaper, and more predictable than the old-fashioned method. In-vitro fertilization (which produces "test-tube babies") is also becoming more common, particularly for zoo breeding and other work with endangered species.

Modern technology has also affected the way breeders decide which animals to match. Computers are used to keep track of schedules and to enter and analyze data about each animal. A computer can generate pictures that show how the offspring of any two animals would be apt to look. Computers are essential tools for laboratory specialists who do genetic work.

Artificial-breeding laboratory technicians usually focus on laboratory tests, measuring, and other procedures to improve the quality of

**Dam** A female animal that is or will be a mother.

**Daughter** The female offspring of bulls whose semen is sold for breeding purposes.

**Linear traits** Confirmation characteristics of bull's daughters, such as how tall they are or what mammary structure they have.

**Calving ease** The ease or difficulty with which a cow delivers her calves; it depends to some degree on the calf's size.

**Semen straws** The container in which semen is stored in preparation for inseminating dams. The straws are approximately four inches long and three-sixteenths of an inch in diameter and are inserted into the dam using an insemination gun.

stored semen. Artificial-breeding technicians collect and package semen for artificial insemination, but they rarely perform the actual insemination. Artificial insemination technicians inseminate female animals and sometimes collect semen from males. Poultry inseminators collect semen from roosters and use it to fertilize the eggs of chickens and other fowl.

Animal breeders often spend time on preparatory tasks and follow-up visits to animals that have been inseminated. Before they begin work on an animal, they make sure the equipment is sterile and functioning properly and that the semen is stored at the optimum temperature. They work closely with the owners and handlers of the animals, sometimes visiting the site ahead of time to make sure the procedure will go as planned. When it's time for the animals to give birth, breeders frequently assist with the delivery and make sure the babies get a healthy start in life.

**To be a successful animal breeder, you should**

- Love working with animals
- Be able to withstand extreme conditions
- Be patient and careful
- Be able to work independently
- Be confident in making decisions

## what is it like to be an animal breeder?

Kyle Alexander is a livestock breeder for 21st Century Genetics in Shawano, Wisconsin. Like most professionals in this field, he has an erratic schedule. "During heavy breeding season, you work a lot of hours," he says. "You might start at six or seven and go until seven in the evening or later. By the same token, when times are slower, you put in less time. It balances out."

Many breeders are employed by companies that sell semen, but the breeders operate as though they work for themselves; they're responsible for managing their own time, and their schedules are flexible. They tend to average about forty hours a week, but they receive commissions and have other incentives to work long hours. They drive their own cars and are reimbursed for their mileage.

In addition to inseminating animals, they make sales calls. For that reason, they're sometimes called company representatives. "There are a number of farmers who inseminate their own cows, who've taken courses and know what they're doing," Kyle says. "Obviously, they save a lot of money that way, but they still need to get the semen from somewhere. We call on them, explain to them how we can help them, mainly by describing the quality and variety of bulls we use to provide the semen, and showing them that we have a good product."

When he visits a farm, Kyle might find himself working in a barn, a field, or a feedlot. Barns are not always the most pleasant places; sometimes they're clean, and sometimes they're filthy. In feedlots animal breeders work in tight quarters with large animals. The job can be physically demanding. Breeders have to struggle with heavy animals sometimes, and they often spend long periods of time on their feet or kneeling while they wait for insemination to take hold or a baby to be born. There is a risk of being injured by uncooperative or clumsy animals. In the winter technicians spend long hours outdoors in bitterly cold weather.

"Animals don't always have the best timing," Kyle notes. "You have to be adaptable. Births needing assistance often occur in the middle of the night. On the coldest days, cows take the longest to breed." It's all part of the job, he remarks, adding that he doesn't know an animal breeder who doesn't enjoy the trade.

Kyle's work could involve as many as twenty appointments to inseminate cows, which might not leave much time for sales calls. At noon and in the evening he checks his message centers to find out who needs his services that afternoon or the next morning. Since cows, horses, dogs, and some other animals can be bred only at certain times in their hormonal cycles, technicians must be available to perform their work when the animals are ready, even on Sundays and holidays. The technician must be able to interpret the animal's body language and other clues that indicate whether the time is right. Sometimes the technician will encourage a mare, for example, to give him these clues by "teasing" her (bringing her near a stallion).

Technicians also perform physical examinations, either manually or by ultrasound, to determine when an animal is ready for breeding and whether she is pregnant afterward. Sometimes animal breeders administer hormones to manipulate the animal's natural cycles so the baby will be born at the most desirable season of the year. To stay informed about the techniques and other developments of their profession, animal breeders read trade magazines and books, and they practice using new computer programs.

## have I got what it takes to be an animal breeder?

Animal breeders need to enjoy animals and feel comfortable and confident working with them. For a livestock breeder, Kyle says it's also essential to enjoy meeting and visiting with new people. "You're out there every day," he says, "and it's your job to listen and try to help the farmers with their needs, to ascertain how best the company can help, not just in the moment of breeding, but how to best improve the genetics of the daughters these cows will bear. You meet all kinds of people, people who are looking to you for advice and help."

In addition, self-discipline is important. Animal breeders need to be good self-starters, because the way they manage their time is more or less left up to them. It's up to them to complete their breeding calls and then make sales calls. Livestock breeders need to be able to handle a regular work day (visiting farms, meeting with farmers, inseminating cows, calling into the center for assignments, making sales calls), as well as any emergencies that come their way.

A background in farming or other breeding operation is not required, but some experience with animals is a big plus. Great respect for and interest in animals are essential. "You need to be a patient person," says Kyle, "as well as someone who likes the outdoors. Animals don't always cooperate. You need to be confident at handling the animals, as well as willing to wait sometimes. You have to work through all the seasons, and that can be rough. It's important to be a person who enjoys the seasons."

Doug Lindsay, who runs a program that teaches students to breed horses agrees that a level head and the ability to work independently are important qualities for an animal breeder. "Breeders need to be well-rounded. We look for intelligent, self-starters with a lot of self-confidence. You want someone who will be able to think in emergencies, as well as being able to direct and supervise employees, someday," he says.

## how do I become an animal breeder?

Between semesters at college, Kyle interned with several animal breeders, getting a feel for what they did. When he graduated and got a job with 21st Century Genetics, the company put him through its own two-week course on livestock breeding and the art of artificial insemination. It's common for companies to require new employees to complete such a program, whether they're college graduates or have learned on the job. "It's a combination of classroom training and hands-on work," he says. "You learn insemination techniques, the reproductive aspects of cattle, their hormones, et cetera. Then they take you out in the field where you actually practice inseminating the animals. You go either to a stockyard or slaughter plant to learn to do it."

Was he nervous the first time? "Sure," Kyle says. "But it's not tremendously hard. It's more a matter of patience than anything. Even people with no experience around cattle learn quickly. I grew up on a dairy farm, so it was probably less nerve-wracking for me."

### education

#### High School

Nearly all animal breeding organizations require that their new hires have a high school diploma, and a growing number require some post-secondary training. Without a high school diploma, it is possible to start out as a stable hand and move up through experience, although at some point along the way you won't be able to advance further without some formal training. Stable hands work for an hourly wage that is not much higher than the minimum wage, but with time and effort it is possible to earn much more.

Students who have spent time around animals have an advantage over those with no experience. Future Farmers of America and 4-H offer useful opportunities to learn about animals and the future of the industry. You can also work a summer job at a horse or cattle farm or ranch, an animal shelter, or a zoo to gain experience. Raising a pet is a good way to become familiar with animals and their needs. It's also advisable to read books about animals.

In school, the more classes you have with an agricultural base, the better prepared you will be. Courses in genetics, biology, health, mathematics, chemistry, communications, mechanics, natural science, and the environment are helpful. In case you end up owning a business or moving into management of a company some day, it's also wise to take some business classes.

### Advancement Possibilities

**Sales managers** supervise technicians and direct herd sales representatives for specific geographic areas.

**Production managers** are responsible for overall sire management and semen production.

**Herd evaluators** offer advice on sires to help producers improve their herds.

**Zoo keepers/aquarists** care for animals and fish; they prepare food, clean animals and their quarters, maintain exhibits, and keep records.

**Conservation biologists/zoologists** provide scientific expertise in the management of zoo animals and participate in research and conservation projects.

### Postsecondary Training

"It used to be that many people got into breeding straight from their farms, through hands-on experience," Kyle says. "But now a lot of people have some postsecondary training. I'd say at least 40 to 50 percent."

Nine months to two years of technical schooling provides the basic training for an animal breeder. A four-year degree in animal science or animal husbandry is recommended if you plan to become a farm or ranch manager eventually. For zoo careers, on-the-job experience is sometimes enough for an entry-level position, but usually a four-year degree in animal science, zoology, marine biology, conservation biology, wildlife management, or animal behavior is necessary. Significant experience in the care of exotic animals is also required for many zoo positions.

College will help you gain a background in animal sciences, and you'll also learn business skills that will be helpful if you need to sell a product or if you move into management. Courses in animal science tend to cover a range of subjects. For example, at the University of California's Animal Science Horse Facility, students work with a herd of about thirty mares, four stallions, and five to ten foals. Classes cover feeding, breeding, unsoundness, health management, physiology, sports medicine, biology, pharmacology, equine nutrition, equine exercise, law, and marketing. The program also features internships that include experience with studs, broodmares, and foals.

Doug Lindsay explains how, at Lakeshore Technical College, the curriculum is varied and prepares students to do almost anything they want. "There are no limits," he says. "Our students are required to take certain standard classes in animal science and business. Once they've done this, the rest is up to them."

## internships and volunteerships

At some junior colleges you can participate in a sort of apprenticeship program that combines course work and on-the-job training. The college provides classroom studies, and an employer pays you for part-time work at a facility that produces livestock or other animals.

Many artificial insemination organizations offer internships to help prospective animal breeders learn about the industry before they commit to it as a career. Most are summer positions, although some are offered during the school year. You might also find work in veterinary clinics, animal hospitals, and animal shelters. Although artificial insemination internships will give you the most valuable insights into a career as an animal breeder, any job that involves working with animals will provide relevant experience (*see* "Sources of Additional Information").

Zoos sometimes offer internships, and most accept volunteers. If you'd like to volunteer at a zoo or aquarium, inquire at a local facility or contact:

**Association of Zoo and Aquarium Docents**
Columbus Zoological Gardens
Attn: Sue Kiebler
PO Box 400
Powell, OH 43065-0400
614-645-3400

## who will hire me?

Kyle Alexander began work as an animal breeder immediately after college; his current employer approached him, not vice-versa. Half of the company's four hundred employees work as breeders. "I knew enough about the company to know that I'd like working there," he says.

Animal breeders work in various settings. Some travel to various farms and ranches, as Kyle does. Others work for large operations, such as a zoo or the hatchery of a poultry farm. Some work in laboratories, where they develop better products and techniques for the industry. Some people with degrees in animal science use their training as a springboard into other animal-related careers, such as livestock sales, selling horses on video or over the telephone, professional horse photography, or working in advertising that features animals.

Finding a job is sometimes easier for college students, who can take advantage of the placement offices at universities and technical colleges. Many of these offices place at least 85 percent of their graduates in their fields of expertise; some companies recruit directly from the campuses. Companies also sometimes offer permanent employment to outstanding technicians who have completed internships there. Additional information on the trade and possible employment is often available in *Hoardes Dairyman,* a trade magazine that covers all aspects of farming.

## where can I go from here?

Kyle has been in the business ever since he graduated from college. He wouldn't dream of doing anything else. "It's an exciting time," he says. "There are a lot of places to go with this, many areas to explore."

Most breeders, like Kyle, begin their careers by working for established companies. Some go into business for themselves after they gain enough experience; you could operate a freelance, animal breeding venture, or if you could raise the money, you could own a farm or ranch. With sufficient education, you could advance to manager of a stud farm or other livestock operation. You could also become a feedlot manager, supervisor, or distributor of artificial breeding products.

## what are some related jobs?

The U.S. Department of Labor classifies animal breeders under the headings *Managerial Work: Plants and Animals* (GOE). Also under this heading are people who work in farming, specialty breeding, specialty cropping, forestry, and logging.

| **Related Jobs** |
| --- |
| Livestock production technicians |
| Poultry production technicians |
| Feed buyers and feedlot managers |
| Veterinarians' assistants |
| Farm managers |
| Flock supervisors |
| Egg processing technicians |
| Poultry graders |
| Field contact technicians |
| Fur farmers |
| Ranchers |
| Rare animal breeders |
| Fish farmers |

## what are the salary ranges?

According to the *Encyclopedia of Careers and Vocational Guidance,* animal breeders generally earn $16,000 to $50,000 a year, depending on their level of education, the employer, and geographic location. According to the U.S. Bureau of Labor Statistics, employees with animal science degrees made about $65,500 a year in 1997, and agronomists made $52,000. Beginning salaries for animal scientists averaged about $24,900 a year in 1997, according to the National Association of Colleges and Employers. According to the American Zoo and Aquarium Association, animal keepers at zoos can earn from the minimum wage up to more than $30,000 annually.

Earnings for animal breeders tend to be higher in California and some Midwest states, including Iowa and Minnesota, and lowest in the Northeast. Most animal breeders work as representatives for companies that sell semen; commissions from sales of semen can increase a technician's earnings significantly, depending on the individual's drive and sales skills. Some companies offer food and housing allowances that can add up to a value of several thousand dollars annually. Other benefits vary, but often they include health insurance, paid vacation time, and pension plans.

## what is the job outlook?

The animal production industries have been going through a period of change recently, and the way animals are raised and marketed is expected to continue changing rapidly in coming years. There is great emphasis on uniform products. In addition, animal husbandry operations must cope with small profit margins

## in-depth

### Pedigreed Animals

Raising pedigreed animals, such as dogs and cats, differs from the multi-million-dollar breeding industries of horse and livestock breeding. State agencies regulate the care and breeding of animals to some extent, but because of the large number of breeders and the wide variety of pedigreed animals, this is often difficult. Clubs organize and set standards for each type and breed of animal. They also officially document and confirm each animal's lineage, or pedigree. Pedigreed animals, because of the purity of their breeding, are more expensive than nonpedigreed animals, and breeders often can make several hundred dollars per animal. Unfortunately, this has led some individuals to enter the field purely for monetary reasons. The result is what Ed Kilby of Daytona Beach, Florida, calls "puppy mills," which generate large numbers of unwanted dogs.

Ed has been breeding, showing, and judging bloodhounds for thirty-five years, and today he serves as secretary of the American Bloodhound Club. "Someone interested in dog breeding, or any other kind of breeding in which the animals are pedigreed, should know it's serious business, not fun and games," he says. "Too many folks already get into it for all the wrong reasons. You don't just slap two animals together. You have to know what you're doing. You have to care about the breed."

Ed says dog breeders should be familiar with genetics and how to research a pedigree, have a basic knowledge of veterinary procedures, and understand the anatomy of dogs. "If you lack any of these, you're just wasting your time and perhaps helping ruin the breed," he adds.

The goal of a reputable dog breeder is always to breed the perfect dog and to improve the breed by taking the best female, or dam, and matching her with the best male, or sire. Each club defines the ideal dog, establishing a standard by which every generation of the breed is judged. "If you can look at a dog and he looks exactly like the standard, you know you've not only bred a great dog, but you've helped maintain the breed," Ed says. For example, weight and color are among the standard requirements for bloodhounds. Bloodhounds should weigh, on the average, from 90 to 110 pounds and be black and tan, red, or liver and tan—the only accepted colors. Ed explains the purpose of the standard: "Over one thousand years ago, the bloodhound was bred to find things and people. Of all the canines in the world, this dog has the best nose. It's the only dog whose scent trail to a criminal can be used as evidence in a court of law. Now, if today he can't find anything, he's not a bloodhound."

Researching a dog's lineage is crucial to improving the breed. "You want to go back at least three generations and know each individual relative of both the sire and the dam. You don't want to pass along any medical or behavioral problems to the next generation, so you have to be extremely careful," Ed says. The pedigree is important because, although the breeder is not registered with the state, the puppies he raises are. The American Kennel Club can refuse to register a dog if the breeder has not tried to follow the standard or, in trying to cut costs, has actually harmed or mistreated the animal. Without registration, the breeder cannot sell the animal as an officially pedigreed dog.

"Breeding is hard work. It takes a lot of research and study before the dog is born and then more hard work to train it. You've got to love the breed and respect the standard," Ed says.

and competition from foreign markets. In agricultural endeavors, such as the breeding of sheep and cattle, it's now almost impossible for a one-person business to remain solvent; large corporations and cooperatives of consultants are becoming the norm.

Professionals who can offer an area of specialization will be in demand, however. Technicians who have completed animal production technology programs should have good prospects, particularly those who hold degrees in animal science from universities or technical institutes. Employment for agricultural scientists is expected to increase about as fast as the average through the year 2006. Opportunities for breeders who inseminate animals should be relatively stable.

As people pay more and more attention to what they eat, there will continue to be an increasing emphasis on the improved genetic make-up of animals to ensure the healthiest possible food. Along with continued progress in technology, people will have an interest in experimenting with and manipulating the genetics of animals. Specialists will be in demand. "There are a lot of jobs and not so many qualified people," says Doug Lindsey. "Trained workers are in demand. Jobs are plentiful, if graduates are willing to be mobile."

Kyle Alexander also predicts that there will be plenty of jobs in the future. "Farmers are interested in bettering their animals. The better the animals, the more money they can make. As long as this attitude is around, there will be plenty of work for breeders. Of course, it's better in some states than others," he says, pointing out that Wisconsin, for example, isn't known as the Dairy State for nothing, which means there is plenty of demand for breeders there.

# how do I learn more?

## sources of additional information

Following are organizations that provide information on animal breeder careers, accredited schools, and possible sources of employment.

**American Dog Breeders Association**
PO Box 1771
Salt Lake City, UT 84110
801-298-7513

**American Society of Animal Science**
111 North Dunlap Avenue
Savoy, IL 61874
217-356-3182
http://www.asas.uiuc.edu

**Humane Society of the United States**
2100 L Street, NW
Washington, DC 20037
202-452-1100
http://www.hsus.org

**National Pedigreed Livestock Council**
272 Meeting House Lane
Brattleboro, VT 05301
802-257-9396
http://www.sover.net/nplczva

**United States Animal Health Association**
PO Box K227
8100 Three Chopt Road
Richmond, VA 23229
804-285-3210
http://www.usaha.org

## bibliography

Following is a sampling of materials relating to the professional concerns and development of animal breeders.

### Periodicals

*AKC Gazette*. Monthly. Features information about dog breeding and showing, health and grooming, and legal considerations, among other topics. American Kennel Club, 51 Madison Avenue, New York, NY 10010, 212-696-8333.

*Bird Breeder.* Monthly. Dedicated to the care, breeding, and nutrition of caged birds. Fancy Publications, Inc., 3 Burroughs Street, Irvine, CA 92618-2804, 714-855-8822.

*Bloodlines*. Bimonthly. Deals with topics pertaining to different U.K.C. registered dogs. United Kennel Club, Inc., 100 East Kilgore Road, Kalamazoo, MI 4902-5584, 616-343-9020.

*Livestock, Meat, and Wool Market News*. Weekly. General news and statistics covering the livestock industry. Agricultural Marketing Service, U.S. Dept. of Agriculture, South Building, Room 2623, Box 96456, Washington, DC 20090-6456, 202-720-6231.

*Ranch & Rural Living Magazine*. Monthly. General articles discussing sheep, angora goats, and cattle. Ranch & Rural Living Magazine, Box 2678, San Angelo, TX 76902, 915-655-4434.

*Sheep Breeder & Sheepman*. Ten times per year. Includes purebred and commercial sheep. Mead Livestock Services, Box 796, Columbia, MO 65205, 314-442-8257.

# animal caretaker

**Definition**

Animal caretakers feed, water, exercise, and monitor the general health of animals. They also clean and repair cages and provide companionship for animals. Those involved with veterinary medicine or research may assist during animal surgery or medical testing procedures.

**Salary range**

$10,000 to $14,000 to $30,000+

**Educational requirements**

Some postsecondary training

**Certification or licensing**

Required by certain states

**Outlook**

Faster than the average

GOE
03.03.02

DOT
410

O*NET
79017A

## High School Subjects

Anatomy and Physiology
Biology
English (writing/literature)
Mathematics

## Personal Interests

Animals
Fixing things
Science

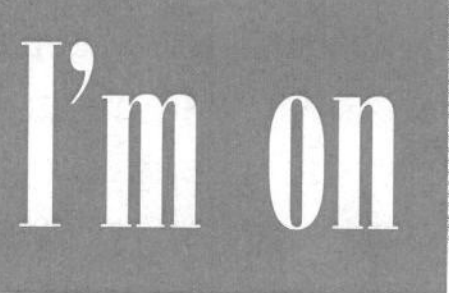


the cutting edge of scientific breakthroughs and health care," says Deborah Donohoe. "I gown up, assist in the operating room, and look out for the welfare of the patient before and after surgery." Then she smiles and adds, "But none of our patients are human. Instead, as an animal laboratory technologist, I work with pigs, monkeys, rabbits, cats, dogs, mice, turtles, rats, and chinchillas." Donohoe explains that research physicians, M.D.s with experience caring for people, often are not familiar with the medical needs of animals. It's the animal lab technologist's job to read up on the biology of other species, inform the research team, and then make sure that each different animal is cared for properly during the procedure.

"This morning," she says, "we're working with rabbits in order to test a new technology involving lasers. Everything that's done for humans in a hospital setting is done for these animals. For instance, I make sure that the rabbits fast properly before the procedure, inject them with antibiotics, test the equipment we use, and after advising the physicians about rabbit behavior and physiology, assist in surgery. That's a lot of responsibility, but I thrive on it. I have to use my brain every day and no two days are ever the same."

## what does an animal caretaker do?

People and animals have coexisted for millions of years. In exchange for a warm place by the fire and a few bones to chew on, dogs helped prehistoric people hunt. Later, cats kept grain storehouses free of rodents and pet birds cheered people with their beautiful songs. Throughout history, humans have used and enjoyed animals and throughout history, someone took responsibility for the health and welfare of the animals that lived side-by-side with people.

Today animal caretakers work with animals in a variety of settings. They are responsible for the basic welfare of other creatures. They feed, water, groom, and exercise animals. They clean the animals' living quarters and provide companionship. They watch for signs of ill health and inform medical staff when necessary.

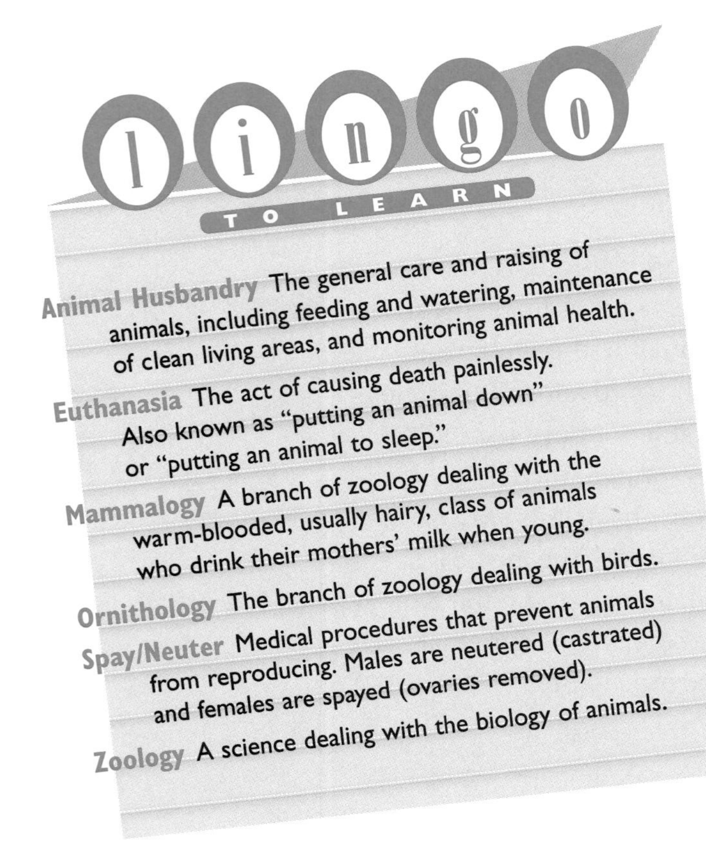


**Animal Husbandry** The general care and raising of animals, including feeding and watering, maintenance of clean living areas, and monitoring animal health.

**Euthanasia** The act of causing death painlessly. Also known as "putting an animal down" or "putting an animal to sleep."

**Mammalogy** A branch of zoology dealing with the warm-blooded, usually hairy, class of animals who drink their mothers' milk when young.

**Ornithology** The branch of zoology dealing with birds.

**Spay/Neuter** Medical procedures that prevent animals from reproducing. Males are neutered (castrated) and females are spayed (ovaries removed).

**Zoology** A science dealing with the biology of animals.

Animal caretaking is a broad field that may be divided into six major categories: animal laboratory workers, veterinary hospital employees, animal shelter workers, zookeepers, stable and farm workers, and groomers and kennel employees.

*Animal laboratory workers* are employed by institutions that use live animals for research, testing, or educational purposes. Laboratory workers are classified according to their education levels and job duties. *Assistant laboratory animal technicians*, the entry-level position for this branch of animal caretakers, provide the most basic animal care. They clean cages and feed animals. They help with general animal husbandry and handling. In the second tier position, laboratory animal technicians take on additional responsibilities. They provide more direct care. They may give prescribed medications, take medical specimens, perform lab tests, record daily scientific observations, and help with minor surgeries.

*Laboratory animal technologists* work at the highest level and supervise the work of lab assistants and technicians. They oversee the advanced care of animals and assist in surgery and other complicated medical procedures.

*Veterinary hospital employees* also may be classified according to job duties and educational level. *Veterinary technicians* have the most responsibility. While they do not diagnose illness or prescribe medicines, they prepare animals for surgery, assist during medical procedures and examinations, perform lab tests, dress wounds, take specimens, and keep records. (See the chapter on veterinary technicians for a complete description of this job.) *Veterinary assistants* feed and bathe animals, administer medications as directed, and assist veterinarians and veterinary technicians. *Animal attendants* (often part-time, entry-level workers) clean cages, exercise animals, and observe sick and recovering animals for signs of illness.

Another type of animal caretaker is employed by private or city-run shelters. These people are in charge of general husbandry, vaccinating new arrivals, arranging for adoptions or animal foster care, and when necessary, euthanizing sick, injured, or unwanted animals.

*Zookeepers* work in zoological parks or aquariums. Every day they clean and maintain the animal enclosures. They prepare the animals' food and observe their behavior in order to detect signs of illness. They are often concerned with conservation issues and may be involved in breeding loans with other zoos, with public educational projects, and with other programs dedicated to preserving threatened and endangered species.

Some animal caretakers work in stables or on farms. *Stable workers* may saddle and unsaddle horses, feed and groom them, muck out stalls, and organize tack rooms, supplies, and feed. Others exercise or help train race-horses. Stable workers may also assist other professionals who care for horses, such as farriers and veterinarians. *Farm workers* perform similar animal husbandry functions on a variety of domestic animals. They may feed and water chickens, sheep, pigs, or goats. They also clean and maintain their coops, stalls, corrals, and pens.

*Kennel workers* care for pet animals such as dogs or cats. They may bathe, groom, and exercise animals while their owners are out of town or while the pets are under the care of veterinarians. Kennel workers also feed and water the animals and clean their cages or enclosed runs. Some kennel workers move up to overseeing other employees while others become small business owners and manage their own kennels.

*Groomers* specialize in maintaining the appearance of animals, usually dogs. They bathe and clip the dogs according to standards for each specific breed. Groomers may own their own businesses or work for kennels or pet supply firms. Other animal caretakers may work for pet stores or teach animal obedience classes.

## what is it like to be an animal caretaker?

Deborah Donohoe works for the Medical College of Wisconsin as an animal lab technologist. One of her projects involved testing the effectiveness of using shock waves to break apart gallstones. Twenty million people in the United States have problems with gallstones. "Because surgery can be dangerous," she says, "we wanted to investigate a new, less invasive procedure." Her research team used two hundred-pound domestic pigs as substitutes for human subjects.

"But pigs can be peculiar," she notes, "because they have an enormous amount of body fat. Consequently, anesthesia may collect in the fat and the animal can be overdosed very easily." Her job involved providing researchers with up-to-date information on pig biology as well as watching over the pigs before, during, and after laboratory tests. Thanks to her efforts, four hours after the first surgery, the pigs were on their feet and eating and drinking—well, like pigs.

> "I have a chance to be with animals that many people never get close to... I've touched tigers."

But not all of Deborah's work occurs in the operating room, and it involves much more than just feeding and cleaning animals. She spends a lot of time sterilizing, preparing, and testing equipment. She also contacts other departments within her institution to set up collaborative research projects in order to decrease the total number of experimental animals used. In addition, Deborah has the opportunity to write professional papers and train other animal caretakers. "All that I do keeps me on my toes," she says. "Plus, I'm the first person to see the animals in the morning. I know my job is important. Animal care technologists are what make the research go."

Jewel Waldrip works in another branch of animal caretaking, and her job is quite different from laboratory research. She works at the Tucson Humane Society in Tucson, Arizona. She meets anxious pet owners searching for lost cats and dogs, receives pets from people who can no longer keep them, and arranges for animal adoptions.

On a typical day, she may also vaccinate new arrivals, clean kennels, feed and water the animals, train co-workers, or write press releases, newsletter articles, or informational handouts.

Jewel loves animals. "I've worked with them for twenty-three years," she says. "I started as a kennel worker, then moved on to grooming. I've shown dogs and done a bit of desktop publishing about handling horses and dogs. For a while I volunteered at the Humane Society and then when a job opened up, I applied for it and got it. Now I can use all my skills and experience in one place."

Another Tucson resident, Taylor Edwards, works as a zookeeper at the Arizona-Sonora Desert Museum. Taylor works in mammalogy and ornithology. His duties include cleaning, feeding, and servicing the bird and mammal enclosures. For instance, he puts out fresh drinking water and makes sure that all the bird perches are stable in the museum's walk-through aviary. If any of the birds' "furniture" needs replacing, he fixes it. He also feeds the birds, but not by simply pouring pellets into a trough. "Keepers are also in charge of providing enrichment activities for the animals," he explains. "So I scatter the food around the aviary so that the birds will have to search for it. Sometimes I stack logs or move rocks into interesting positions and place the food there. This keeps the birds mentally healthy and active."

"I also write daily reports on the behavior of the animals," Taylor says. "But one of my very favorite things to do as a keeper is to talk to museum visitors, especially children. There's a connection people can make with animals whether it's at home with their dog or with animals in the zoo. I help people understand the purpose of zoos and teach them a little bit about each animal."

Because they are dealing with live animals, caretakers often work flexible hours. Some shelter employees, for instance, must arrive early to feed and water the animals and clean their cages. Others must stay late or work weekends to accommodate extended adoption and receiving hours. Research laboratory technicians and technologists monitoring ongoing research projects also put in some night and weekend hours. Dog groomers, kennel workers, veterinary assistants, stable workers, and farm workers also work evenings and weekends according to the needs of their employers. Much of their work may also be seasonal and tied to the horseracing calendar or to the vacations of clients.

## have I got what it takes to be an animal caretaker?

"I've touched tigers," says Taylor Edwards. "I have a chance to be with animals that many people never get close to." In addition, by being involved with captive breeding and reintroduction programs like the Mexican wolf study at the Arizona-Sonora Desert Museum, Taylor has found a way to make sure that those animals can continue to have a place in the wild. He says that, "Compared to Ph.D. scientists who work on population genetics but mostly sit before computer terminals, zookeepers are on the forefront of each project."

Jewel Waldrip agrees. "It's so gratifying to see that what you're doing makes a difference. One time some people brought a standard poodle into the Humane Society. He'd been a cute puppy once, but as he grew up they ignored and neglected him. By the time we saw him his fur was matted to the skin and his ears were all infected and full of foxtails. I knew the shelter would have to put him down if I couldn't find a foster family to care for him." Jewel smiles as she relates the story's happy ending. "The foster family groomed the poodle, took him to obedience school, and last week brought him in to sign his adoption papers. That dog is now one of the happiest, most well-trained animals I've ever seen."

However, working with animals and being involved with the general public does have its downside. Animal shelter workers must deal with distraught people who've lost their pets. Also they must try not to be judgmental when people deliver unwanted pets to the shelter for what may seem like selfish reasons. It can be easy to get depressed when confronted with so many unwanted pets. Shelter workers try to combat the pet overpopulation problem through educational campaigns and by providing low-cost spay and neuter services.

Dog groomers should be aware that theirs is a physically demanding job. Occasionally, they may be bitten or scratched by frightened dogs. Pet hair and dander along with the chemicals used in flea dips, sprays, and shampoos can irritate humans. Some groomers also suffer from carpal tunnel syndrome due to the repetitive nature of their work.

The hardest aspect of animal caretaking is one that most workers must deal with at some point—euthanasia. Animal caretakers are usually the ones to put animals to death humanely. Shelters can't keep every animal forever, hoping for adoptions that never come. Besides, some animals are simply too sick or too young or too old to be adopted successfully. In addition, the end point for all animal research is also euthanasia. "Intellectually you know that there's a termination plan," says Deborah Donohoe. "You put it to the back of your mind and yet you can't help but form an attachment. Even at the completion of short studies, I've cried." Maintaining compassion while doing what's best—or necessary—is often quite difficult.

**To be a successful animal caretaker, you should:**

- Enjoy working with animals
- Have at least a high school diploma for basic animal caretaking
- Have a bachelor's degree for zookeeping
- Be able to take direction and work independently
- Be able to lift heavy animals, supplies, or equipment

# how do I become an animal caretaker?

Taylor Edwards began his career at his hometown zoo in Albuquerque, New Mexico, as a high school volunteer. He worked in the petting zoo, showing baby animals to school children, and then moved up to junior zookeeper. "Ever since I was ten or eleven years old," he says, "I knew I liked animals. Volunteering at the Rio Grande Zoo helped me to see different job possibilities: zookeeper, veterinarian, research scientist." After high school, he attended college and received a bachelor's degree in zoology. His hands-on experience combined with a college education earned him a job at the Arizona-Sonora Desert Museum.

Less than 40 years ago, polio was one of the most feared diseases. Today, because of animal research, vaccines exist for polio, typhus, diphtheria, whooping cough, smallpox, and tetanus.

## education

### High School

Entry-level animal caretakers often begin their careers with only a high school diploma. Classes in biology, math, and English, along with additional courses in other sciences, are helpful. Computer literacy, business skills, and verbal and written proficiency are also recommended. In addition, those animal caretakers who plan to work as zookeepers or work toward professional certification need to fulfill the entrance requirements of the college or university they plan to attend.

### Postsecondary Training

On-the-job training is a large part of the education of all animal caretakers. People working in kennels, animal shelters, stables, farms, pet stores, and as dog groomers are almost exclusively trained on the job. Most dog groomers, for instance, begin by practicing on their own pets. Then they groom dogs belonging to their friends and neighbors. They may also attend a dog grooming school. Finally they apprentice to a professional groomer and learn more about the business while on the job. In addition, the American Boarding Kennel Association and the National Dog Groomers Association provide workshops and educational videos for their members (*see* "Sources of Additional Information").

Several colleges offer course work in animal health, usually leading to an associate's degree after two years of study. Future animal caretakers at Harnell College in Salinas, California, for instance, take classes in anatomy, physiology, biology, chemistry, mathematics, clinical procedures, animal care and handling, clinical pharmacology, infectious diseases, pathology, radiology, surgical assisting, anesthesia, and current veterinary therapy. These students go on to work in veterinary practices, zoos and marine mammal parks, and for wildlife rehabilitators, humane societies, pharmaceutical companies, and laboratory animal research institutions.

Zookeepers need a college degree in zoology, biology, or an animal-related field. Experience as a zoo volunteer, animal intern, or veterinary assistant is also very beneficial.

## certification or licensing

Licensing or certification for animal caretakers varies according to the job performed and the state of residence. For example, some states require veterinary technicians to be licensed while others do not, and no state regulates the dog-grooming industry. Animal caretakers should check with their municipalities to determine local governmental regulations.

While many states do not regulate animal caretakers, several professional organizations offer certification. Professional certification, combined with work experience, helps animal caretakers earn higher pay and obtain better jobs.

Through correspondence courses, the American Boarding Kennel Association offers three levels of educational programs for its members: pet care technicians, advanced pet care technicians, and certified kennel operators. The National Dog Groomers Association offers a two-tiered program for its members as well as periodic, one-day, intensive workshops. Certification at these workshops is based on hands-on performance. The applicant's work is critiqued by dog grooming professionals.

To determine certification for its members, the American Association for Laboratory Animal Science combines on-the-job work experience with college courses and rigorous AALAS examinations. AALAS certifies three categories: assistant laboratory animal technician, laboratory animal technician, and laboratory animal technologist. (*See* "Sources of Additional Information.")

## internships and volunteerships

Colleges and universities, along with professional organizations, are sources of information on work-study projects and student internships. Any type of volunteer work with animals is beneficial for future animal caretakers. Often, it is the only way for entry level personnel to get important hands-on experience.

## labor unions

Most animal caretaker positions are not unionized. The exceptions are zookeepers who work for city or state zoos or aquariums that are under union contract. In these places, zookeepers may become union members.

# who will hire me?

Deborah Donohoe began her career in laboratory animal science with a job layoff. While she enjoyed her work as a receptionist, it seemed a logical time to make a change. She began talking to lots of people about career alternatives. She already had an associate's degree in business when she decided to take the Purina Laboratory Animal Care Course. Her first job as a laboratory technician involved caring for seventy old world macaques and twenty squirrel monkeys at the Yerkes Primate Center in Atlanta.

Deborah believes in setting goals. When she started in laboratory animal science, she was determined to learn as much as she could about her new career and to advance as far as possible. "After my regular job duties were done, I asked if I could watch animal autopsies," she says. "I knew I didn't want to stay on the lowest rung of the career ladder, so I talked to pathologists, veterinarians, microbiologists . . . anyone who would listen to my questions! Everyone was very receptive and I learned a lot." She also credits her American Association of Laboratory Animal Science certifications with helping her move forward. Now, fourteen years after her first job with laboratory animals, she has reached the highest rung on the professional ladder and is a certified laboratory animal technologist.

Animal caretaking is a very broad field. Technical personnel work for research laboratories. Others may be employed by pharmaceutical and chemical companies, food production companies, medical schools, teaching hospitals, federal and state governments, or a branch of the armed forces. The majority of animal caretakers work in veterinary offices or boarding kennels. Zookeepers are employed by both private and public zoos. Some animal caretakers, especially those who are dog groomers or pet store owners, may operate their own businesses.

Business directories listing professional organizations and major employers may be obtained from local libraries. Federal job centers and state employment services may have information on jobs in animal caretaking.

### Advancement Possibilities

**Animal shelter directors** run animal shelters and, if the shelter is a private organization, determine and enforce policy regarding adoptions and euthanasia; they also hire and supervise other workers.

**Laboratory animal technologists** supervise lab assistants and technicians and oversee the advanced care of animals, assisting in surgery and other complicated medical procedures.

**Senior animal keepers** oversee the entire routine and care of a particular zoo animal.

## where can I go from here?

Because the field is so varied, there is no typical career path for animal caretakers. Dog groomers or pet store workers may become store managers. Kennel workers may move up to kennel supervisor, assistant manager, and manager. With enough capital, groomers, kennel workers and pet store caretakers may open their own businesses. Animal shelter workers may may be promoted to animal control officers, assistant shelter managers, or shelter directors. Laboratory staff may advance from assistant laboratory animal technician to laboratory animal technician to technologist.

Beginning zookeepers start as animal attendants and may move to lead keeper in one area of the zoo. After that, they become senior keepers and oversee the entire routine for an animal. They also supervise the work of other employees. From there, zookeepers can move to administrative positions. These positions, though salaried, are less hands-on and require additional education. Curators, zoo registrars (people who keep track of breeding loans and animal stud books), and reproductive and behavior physiologists are examples of other salaried positions at zoos.

## what are some related jobs?

The U.S. Department of Labor classifies animal caretakers under the headings Domestic Animal Farming Occupations, Game Farming Occupations, and Animal Service Occupations (DOT), and Animal Training and Service (GOE). Also under these headings are stable attendants who care for horses and mules, clean the animals' quarters, and treat injured livestock according to instructions; game farm supervisors who oversee the breeding, raising, and protection of wild animals and fowl on private or state game farms; fur workers who kill and skin animals and pack pelts in crates; and cowhands, sheep shearers, and dairy farm workers.

| Related Jobs |
|---|
| Animal-nursery workers |
| Animal-ride managers |
| Animal trainers |
| Aquarists |
| Beekeepers |
| Dog bathers |
| Horse exercisers |
| Livestock-yard attendants |

## what are the salary ranges?

Experience, level of education, employer, and work performed determine the salary ranges for animal caretakers. Most entry-level personnel are hourly employees who receive minimum wage and make around $10,000 a year. However, many entry-level personnel are hired part-time and make far less.

Dog groomers earn between $12,000 and $30,000 a year. Some work on a commission basis and some own their own shops. One advantage to starting a dog grooming operation is that it does not take a lot of capital to open a shop.

Veterinarian technicians, laboratory animal technologists, and zookeepers earn more than other animal caretakers. Zookeepers may start work at around $5.00 an hour and move up to $12 to $15 an hour. Curators receive salaries of $20,000 to $40,000 or more a year, depending on the size and location of the zoo. A membership survey done by the American Association of Zookeepers found that southern states (except Florida) generally pay their zookeepers less. States in the Pacific Northwest offer the best pay. However, this difference may be offset somewhat by a higher cost of living in the Northwest.

## what is the job outlook?

Positions as animal caretakers in zoos, aquariums, and wildlife rehabilitation centers are scarce. Far more people want to work with wild and exotic animals than there are available opportunities. Zoos usually have small budgets, and are not expected to grow through the year 2006.

Funding for both private and public research with animals is less now than what it has been. Fewer research dollars means fewer animal caretaker jobs. Nevertheless, there remains a shortage of qualified laboratory animal research technicians and technologists. Veterinary practices and animal shelters often can't find enough staff. And certainly every area of the country has pet stores, dog groomers, and kennels. Turnover, especially among part-time workers in these businesses, is high. Dedicated and qualified animal caretakers, particularly those with lots of hands-on experience, are in high demand.

# how do I learn more?

## sources of additional information

Following are organizations that provide information on animal caretaker careers, accredited schools, and employers.

**American Association for Laboratory Animal Science**
70 Timber Creek Drive
Cordova, TN 38018-4233
901-754-8620
info@aalas.org

**American Boarding Kennels Association**
4575 Galley Road, Suite 400-A
Colorado Springs, CO 80915-2799
719-591-1113
http://www.abka.com

**Humane Society of the United States**
2100 L Street, NW
Washington, DC 20037-1525
202-452-1100

**National Dog Groomers Association of America**
1750 Kenray Drive
Hermitage, PA 16148-3055
412-962-2711

## bibliography

Following is a sampling of materials relating to the professional concerns and development of animal caretakers.

### Books

Miller, Louise. *Careers for Animal Lovers & Other Zoological Types.* Includes sections on zoo keepers, researchers, shelter personnel, etc. Lincolnwood, IL, VGM Career Horizons, 1991.

### Periodicals

*Animal Behaviour.* Monthly. Association for the Study of Animal Behavior, Harcourt-Brace and Co., 24-28 Oval Road, London NW1 7DX, England, 44-171-267-4466.

*Animal Sheltering.* Ten times per year. Newsletter for personnel working in humane societies. Humane Society of the United States, 2100 L Street, NW, Washington, DC 20037, 202-452-1100.

*Animals.* Bimonthly. General magazine covering domestic and wild animals, including such topics as pet care, humane issues, and the environment. Massachusetts Society for the Prevention of Cruelty to Animals, 350 South Huntington Avenue, Boston, MA 02130, 617-541-5065.

*Groom & Board.* Nine times/year. Directed toward the grooming and kennel industry. H. H. Backer Associates, Inc., 20 East Jackson Boulevard, Suite 200, Chicago, IL 60604, 312-663-4040.

*Our Animals.* Quarterly. Free, award-winning magazine bringing upbeat news about pets and people to the general public. San Francisco Society for the Prevention of Cruelty to Animals, 2500 16th Street, San Francisco, CA 94103, 415-554-3009.

*Shoptalk.* Bimonthly. Free newsletter with information for animal control professionals. American Humane Association, 63 Inverness Drive East, Englewood, CO 80112-5117, 303-792-9900.

*Veterinary Technician.* Monthly. Journal with issues handy for animal health care professionals. Veterinary Learning Systems Co., Inc., 425 Phillips Boulevard, Suite 100, Trenton, NJ 08618, 609-882-5600.

*Zoo Biology.* Bimonthly. Discusses topics related to the care and maintenance of wild animals in wildlife institutions. John Wiley & Sons, Inc., 605 Third Avenue, New York, NY 10158, 212-850-6645.

# audiometric technician

**Definition**

Audiometric technicians conduct hearing screenings, test and clean hearing aids, and make ear mold impressions. They often also sell hearing aids and accessories.

**Alternative job titles**

Audiology assistants
Audiometrists
Auditory prosthologists
Hearing aid dispensers
Hearing aid specialists
Licensed hearing specialists

**Salary range**

$20,000 to $40,000 to $50,000+

**Educational requirements**

Some postsecondary training

**Certification or licensing**

Required by certain states

**Outlook**

Faster than the average

GOE
02.03.04

DOT
076

O*NET
32314*

## High School Subjects

Anatomy and Physiology
Biology
Chemistry
Health
Speech

## Personal Interests

Fixing things
Helping people: physical health/medicine
Science

I can hear!” the client keeps repeating as he walks around the office, marveling at the effectiveness of his new hearing aid. For ten years he has been missing out on many of the sounds in the world. Finally he has come to Don Sutton to purchase an assistive device.

The client was concerned about the price, but Don has helped him select a reliable, inexpensive model that fits behind his ear. The customer is overjoyed, and his wife is also relieved that he can hear again.

She is perhaps not as relieved as another client who comes to Don because she has accidentally broken off the head of a Q-Tip cotton swab inside her ear canal. As he examines her ear, Don can see the Q-Tip through his video otoscope, an instrument with a camera so tiny that it can be placed inside the patient's ear canal. He is advising the woman to consult a medical doctor when the instrument happens to touch the Q-Tip, sticks to it, and pulls it out.

Although the remedy was unexpected, Don is glad to have solved the woman's problem. Helping people is a big part of an audiometric technician's job.

## what does an audiometric technician do?

A hearing disorder can result from trauma at birth or other injury, an infection, the overuse of antibiotics or aspirin, smoking, a genetic defect, exposure to loud noises, unhealthy teeth and other causes. People of all ages can suffer from hearing impairment, but it is most common among the elderly. To help people cope with hearing problems, *audiometric technicians* test clients' hearing, fit and clean hearing aids, and teach clients how to use assistive listening devices. They make ear mold impressions, prepare charts and graphs to track the patient's progress, teach clients about hearing and hearing disorders, and sell assistive listening devices and accessories.

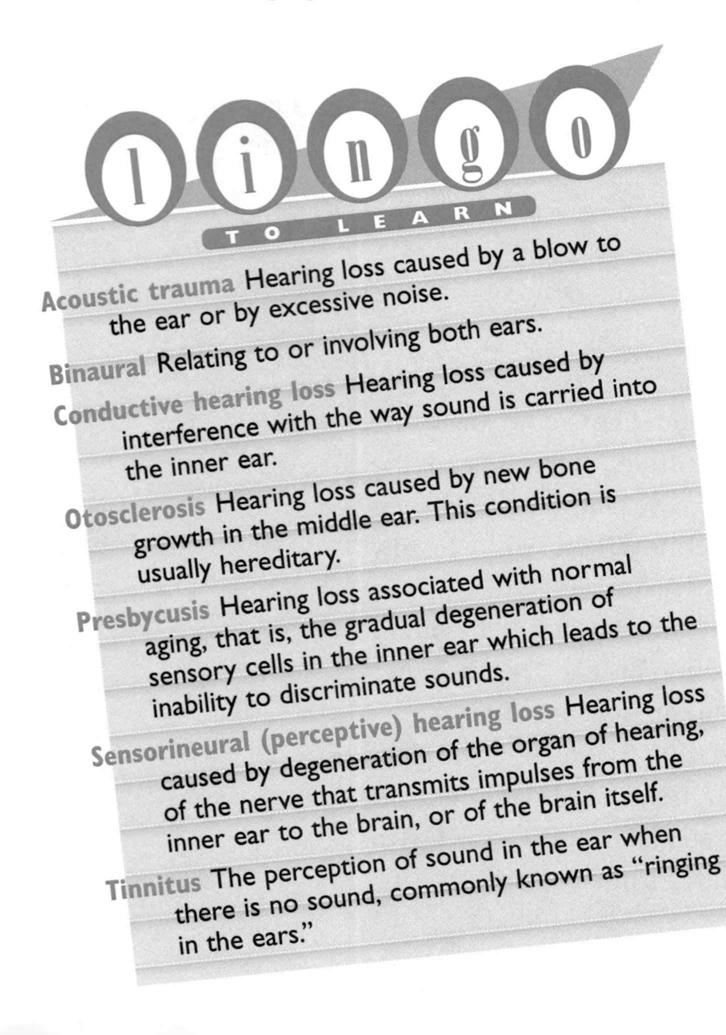


**Acoustic trauma** Hearing loss caused by a blow to the ear or by excessive noise.

**Binaural** Relating to or involving both ears.

**Conductive hearing loss** Hearing loss caused by interference with the way sound is carried into the inner ear.

**Otosclerosis** Hearing loss caused by new bone growth in the middle ear. This condition is usually hereditary.

**Presbycusis** Hearing loss associated with normal aging, that is, the gradual degeneration of sensory cells in the inner ear which leads to the inability to discriminate sounds.

**Sensorineural (perceptive) hearing loss** Hearing loss caused by degeneration of the organ of hearing, of the nerve that transmits impulses from the inner ear to the brain, or of the brain itself.

**Tinnitus** The perception of sound in the ear when there is no sound, commonly known as "ringing in the ears."

To test hearing, the technician might have clients sit in a soundproof booth, where they listen through earphones to sounds such as human voices, automobiles, birds, and rain. The technician pronounces various words and uses an audiometer to determine how well the customer can discern them. Audiometric technicians also use other types of sophisticated equipment to measure a person's sensitivity to pitch, intensity, and loudness.

To see inside a client's ears, the technician may use a video otoscope. The camera creates a magnified picture that the patient and technician can observe on a video screen. If a medical problem or significant blockage from ear wax is detected, the technician refers the client to a physician.

If a patient simply needs a hearing aid, the technician makes an ear mold impression by combining a special oil and powder to make liquid wax. The technician inserts cotton and a string in the ear canal and pours the liquid wax over the cotton. The wax instantly hardens into a perfect duplicate of the ear canal. It is pulled out with the string and sent to the hearing aid manufacturer to be used as a model for making an assistive listening device that will fit comfortably inside the client's ear.

After fitting a hearing aid, the technician usually helps the client become oriented to the often overwhelming array of new sounds. The technician typically performs various follow-up services, such as changing batteries, monitoring for infection, adjusting the fit to eliminate any soreness, and checking periodically to ensure that the device is functioning properly.

Audiometric technicians who perform hearing examinations and who fit, dispense, clean, repair, and sell hearing aids and accessories are called *hearing instrument specialists, hearing aid dispensers*, or *licensed hearing specialists*. Often, they work in retail hearing aid stores. Sometimes they operate in conjunction with *medical doctors*. Requirements for this specialization vary from state to state, but in general a hearing instrument specialist has earned a license by passing a rigorous examination. An *auditory prosthologist* is a hearing instrument specialist with advanced certification.

Other audiometric technicians are employed as *audiology assistants*. They perform basic duties under the direction of certified *audiologists*, who hold master's or doctoral degrees. Audiologists have completed years of graduate-level studies in subjects such as anatomy, neuroanatomy, human perception of language, and the medical causes of hearing loss. They scientifically measure hearing ability, test for diseases in the middle ear, fit hearing aids, teach lip reading, and perform other rehabilitative services. Audiology assistants typically gather information that will be assessed by the audiologist, but they do not interpret test results or discuss evaluation and treatment with patients. The American Speech-Language-Hearing Association (ASHA) standardized the requirements for audiology assistants only about thirty years ago, in 1969.

To be a successful audiometric technician, you should

- Have excellent communication skills
- Enjoy working with people
- Have patience
- Be compassionate
- Be tactful
- Have some aptitude for working with electronics

## what is it like to be an audiometric technician?

Don Sutton is a hearing instrument specialist for Hearing Aid Counselors, a company that operates a chain of hearing aid outlets in Oregon. He is the office manager and the only audiometric technician in the store at Albany, where he works with one assistant and two telemarketers. He sees about fifteen clients each day, mostly elderly people. The customers are generally friendly and often entertaining, but they can sometimes be difficult. Don is patient, chatting with them and learning each person's unique story.

Some customers visit only briefly to have their hearing aids cleaned, adjusted, and repaired. Don or his assistant scrub dust and wax out of the openings on the hearing aid with a small brush, then coat the device with a lubricating lotion that makes it easier to place in the ear. If the hearing aid is still not working properly, Don cleans it with a special vacuum or partially dismantles it and cleans the inside. The battery usually needs to be changed, and sometimes Don performs minor repairs on the hearing aid or ships it back to the manufacturer for more extensive repairs.

Don also tests people's hearing, using a video otoscope and an audiometer. As he performs the tests, he explains how the ear functions, what the test results reveal about the individual's hearing, how a hearing aid could help, and what types are available. Most new clients ask numerous questions, and Don can spend anywhere from thirty minutes to two hours talking with each of them.

In addition to his work at the store, Don sometimes makes house calls, an unusual service for an audiometric technician. In addition to his regular visits to six retirement homes in the community, he occasionally performs hearing tests at the homes of shut-in clients who cannot come to the store.

Another hearing aid specialist, Grant Gording, manages the Hearing Aid Counselors office at a shopping mall in Eugene, Oregon. He and his secretary open the store at about ten in the morning, take an hour for lunch, and go home at five or five-thirty. Working six- or seven-hour days and not having to work outside in bad weather are two things he says he loves about his job.

In a typical day Grant sees about four to six clients with appointments and perhaps a few walk-ins. He spends about an hour with each one. New customers fill out paperwork that details the history of their hearing, and then they are escorted to Grant's office for a hearing examination. If a client needs a hearing aid, Grant discusses the options and helps them select an appropriate model. About 60 percent of his customers already have hearing aids that must be examined, cleaned, adjusted for a proper fit, repaired in the office, or shipped away for repair.

Grant's does much more than merely sell hearing aids. For instance, he explains, "Sometimes someone will say, 'My ear hurts,' and I have to find out where the problem is." With experience, he has become more adept at handling that type of challenge.

Grant talks to the owner of the Hearing Aid Consultants chain of stores every few days but is basically in charge of the store in Eugene. It's almost like being self-employed, except that the home office pays the overhead.

Unlike Don and Grant, Fran Cosgrove is a licensed hearing aid specialist who operates her own store and sets her own working hours. She agrees that, although retail sales are important for her business, customer service is her primary responsibility: "You can't just sell hearing aids. You need to offer service, too. A person can make a lot of money in this field, but we're here to help people hear better, not to sell someone something they don't want or need." Since manufacturers recommend replacing a hearing aid every four to six years, Fran knows that she will have continuing business as satisfied customers return to purchase new models.

Fran offers the same basic services as Don and Grant. She begins her examination by using a video otoscope to check for blockage from ear wax or other physical problems that should be referred to a medical doctor. She checks each ear separately, since hearing loss is not necessarily the same in both ears. All the while, she explains what tests she is conducting and why. "You need to gain the confidence of your client, so it's important to keep them informed about each procedure," she says.

## have I got what it takes to be an audiometric technician?

Audiometric technicians need to be great communicators who are tactful and sensitive to the needs and feelings of their customers. They should also be patient, because some clients will probably be children or senior citizens who have special needs and comfort levels. A genuine desire to help people is perhaps a technician's most important quality.

Grant Gording says he entered the field "to be able to help people hear better. I knew a few people that needed help with their hearing." Most of his customers are elderly, and he notes, "You have to be good with people, with seniors particularly."

It's also important to be thorough and pay attention to detail. Many tests need to be repeated time and again, and some patients progress slowly. The technician must demonstrate perseverance and self-confidence. "It takes time for people to get used to wearing a hearing aid," Fran says. "I like working with people, and that's what you have to do. Find the right aid and make a good fit for the patient."

An aptitude for math and science is valuable in this field. Some knowledge of electronics, anatomy, physiology, linguistics, and psychology is also helpful.

## how do I become an audiometric technician?

Grant Gording had two other careers before he became an audiometric technician about six years ago. He began investigating the profession by talking to hearing instrument specialists in his community. When he knew that he wanted to obtain a license in the field, he began studying to prepare for the examination.

There is no standardized program of study for audiometric technicians, and requirements for licensure vary from state to state. In many states hearing instrument specialists need only learn enough to pass the licensing examination. In other states some experience, such as an apprenticeship, is required. In contrast, to become an audiology assistant, you must be sponsored by a certified audiologist.

### Advancement Possibilities

**Audiologists** determine the type and degree of hearing impairment and provide a range of services to help clients cope with impaired hearing. They also test noise levels where people work; and they conduct hearing protection programs for businesses, schools, and communities.

**Speech pathologists**, also known as **speech therapists**, work with people who cannot speak; those who cannot speak clearly, smoothly, and at the proper pitch; and those who cannot understand language. For example, they work with patients who stutter or have had a stroke. Sometimes they also diagnose and treat patients who have difficulty in swallowing and eating.

**Speech, language, and hearing scientists** study the complexities of human communication; investigate the way social and psychological factors influence communication; and help develop new ways to treat speech, language, and hearing disorders.

**Otolaryngologists**, also known as **otorhinolaryngologists**, are medical doctors who specialize in treating diseases of the ear, nose, and throat.

## in-depth

### Parts of the Ear

The **outer ear** includes cartilage from which earrings are hung and the external *auditory canal,* a tunnel leading from the ear's opening to the *tympanic membrane* or *eardrum.*

The **middle ear** includes the inner surface of the eardrum and the three tiny, bony *ossicles.* They are named for their shape, the *hammer, anvil,* and the *stirrup.* In Latin, they are called the *malleus, incus,* and the *stapes.* When a sound causes the air to vibrate, these bones vibrate in response, and their vibrations are transmitted to the inner ear.

The **inner ear** includes chambers that are completely filled with fluid, which is jostled by the ossicles vibrating against a thin membrane called the *oval window.* This membrane separates the middle ear from the inner ear. Another flexible membrane, *the round window,* prevents the motion of the inner ear fluid from becoming too violent. The *cochlea* (Latin for snail or snail shell, which the organ resembles) is a bony structure about the size of a pea. Behind it is the *organ of Corti,* thousands of specialized nerve endings that are the individual sense receptors for sound. These are in the form of tiny hairs projecting from the membrane lining the cochlea; they wave like stalks of underwater plants in response to the oscillating currents of the inner ear fluid. There are some twenty thousand of these hairs within the cochlea. They merge at the core of the cochlea and exit from its floor as the *auditory nerve.*

## education

### High School

You will probably not be allowed to apply for a license as an audiometric technician without a high school diploma or GED. To begin preparing for a career in the hearing sciences, you should take high school classes in biology, mathematics, electronics, psychology, speech, linguistics, and perhaps music. Some knowledge of computers will be helpful, since computer interactive programs are often part of the patient's course of treatment. Some high schools also offer classes that give potential teachers an opportunity to interact with children in a classroom; experience with children would be helpful if you became an audiometric technician who works at a school.

### Postsecondary Training

Usually some training beyond high school is required to become an audiometric technician, but preparation for the career can be as simple as reading library books independently to prepare for the often rigorous licensing tests. This is what Grant Gording did, studying at home for about four months. He also completed an optional, two-day, preparatory course offered by a private company in Portland.

Don Sutton enrolled in a training program sponsored by a company that manufactures hearing aids. These programs can last from several months to a year and include instruction in human anatomy, with emphasis on topics such as the nerves within and around the ears.

The National Board for Certification and Hearing Instrument Sciences offers a home-study program for people interested in the hearing health care field (*see* "Sources of Additional Information"). The course covers the human ear, audiometric testing, hearing instruments, and fitting.

You could also prepare by studying communication sciences and disorders, audiology, or speech pathology at a college or university. A general background in liberal arts would serve you well in various speech and language fields. After earning a bachelor's degree, you could work as an audiometric technician, return to school at some later time, and obtain the master's degree required to become an audiologist or speech pathologist.

Regardless of how you trained to enter the profession, you might be required to complete continuing education to retain your license. For example, in North Carolina audiometric technicians must complete eight to twelve hours of continuing education annually.

## certification or licensing

A license or certification is required to practice in this field, but licensing requirements vary from state to state. Grant Gording earned his certification by passing a written and practical test administered by the state. He and Fran Cosgrove both found that their licensing examinations were demanding and required intensive preparation.

Certification and licensing is typically handled through state departments of regulation and licensing. Those departments often provide candidates with study guides to help them prepare for the examinations. The exams typically consist of a written test and a hands-on demonstration of your skills in the trade. For example, in Wisconsin you would have to perform hearing tests for all types and degrees of hearing loss during the examination. You would also be required to demonstrate your skills in audiometry and in making ear molds.

Your state's licensing and regulatory board can provide specific information about obtaining credentials in your region. If you expect to relocate to another state some day, it might be wise to investigate reciprocity agreements; some states do not accept credentials granted in other states.

## scholarships and grants

Some grants are available through hearing aid manufacturers, such as Beltone, Starkeys, and Miracle Ear, but most scholarships and grants are given to students pursuing a master's degree in audiology or speech pathology. Most graduate programs administer their own financial aid programs. The American Speech-Language-Hearing Association offers a few scholarships and grants for college and university students at the undergraduate, master's, and doctoral levels (*see* "Sources of Additional Information").

## internships and volunteerships

Volunteering at a nursing home, community center, hospital, or public agency can help you decide whether you enjoy working with the public and, in particular, the elderly people who make up the largest percentage of an audiometric technician's clientele. In some states you will need more structured on-the-job experience to pass the licensing examinations, because candidates must demonstrate their ability to perform the work of an audiometric technician. This gives you insights into the career and helps you prepare for the practical and written licensing examinations.

In some states you can learn this profession through apprenticeship. For example, in Alabama an apprentice permit is granted in certain circumstances, but apprenticeship training is not required. In North Carolina candidates are required to complete 750 hours of on-the-job training under the supervision of a licensed hearing aid dealer and fitter before applying for a license through the state Hearing Aid Dealers and Fitters Board. In Wisconsin candidates can obtain trainee permits and practice the trade for one year under the supervision of a licensed hearing instrument specialist before taking the examination to obtain a permanent license.

Fran Cosgrove learned the trade by assisting another audiometric technician on the job. "He really took me under his wing and taught me the things I needed to know," she says. She says a formal training program, such as the one offered by the National Board for Certification and Hearing Instrument Sciences, is useful, but "there is nothing that helps as much as practical experience. It certainly helped me."

## who will hire me?

Grant Gording found a job by calling hearing aid dealers on the telephone and explaining that he was a licensed hearing aid specialist in search of employment. Some of them invited him to be interviewed, and Hearing Aid Counselors offered him the position he now holds. That is an acceptable and fairly common way to find employment in this field. "Once you have a license, a lot of companies will have room for you," Grant notes.

As an audiometric technician, you would have a wide range of potential employers. You might work for the federal government, perhaps with the Department of Veterans Affairs, the National Institutes of Health, the Department of Health and Human Services, or the Department of Education.

You could work at a city or county public health department, community clinic, psychiatric institution, retirement home, nursing home, university, research laboratory, or private office. Many audiometric technicians work in hospitals or rehabilitation centers. Others work for public schools and in offices where hearing aids are sold. Manufacturers of hearing aids, including the Beltone, Starkey, and Danavox companies, also employ audiometric technicians.

To find job openings in the field, watch the classified advertising sections of newspapers. This is how Fran Cosgrove found her first position and the mentor who taught her the basics of the profession. Some associations in the field maintain job banks on the Internet and publish newsletters that list employment opportunities (*see* "Sources of Additional Information").

You can also send resumes to audiologists; they are listed in the yellow pages of local telephone books. Many technicians learn of job openings through friends in the field, professors, and other personal contacts.

## where can I go from here?

Grant Gording likes his job and plans to stay where he is indefinitely. He doesn't expect to be promoted, since he works in a small office with no positions to which he could advance, but he will be paid more as the office does more business. He could operate his own store, but he says he has no desire to do that.

In contrast, Fran Cosgrove learned the trade by working under the tutelage of another audiometric technician, then went into business for herself. She could expand on her success as an entrepreneur and eventually own a chain of hearing aid stores.

On the other hand, for an audiology assistant, the probability of advancement is small. Slight raises in pay would come with experience, but the American Speech-Language-Hearing Association strictly limits the scope of an assistant's responsibilities. The assistant would always have to work under the supervision of a certified audiologist. Any audiometric technician has the option of advancement, however, by obtaining a master's degree and becoming an audiologist; speech pathologist; or speech, language, and hearing scientist.

## what are some related jobs?

Audiometric technicians are skilled at working with people who have hearing impairments. With additional training, you could translate spoken words into sign language for the deaf. You could also enter the field of speech and language disorders, perhaps becoming a speech-language pathologist.

The U.S. Department of Labor classifies audiometric technicians under the heading *Medical Sciences: Health Specialties* (GOE). Also included under this heading are people who work as speech pathologists, audiologists, chiropractors, and optometrists.

| Related Jobs |
|---|
| Acoustical engineers |
| Industrial hygienists |
| Music therapists |
| Occupational therapists |
| Orientation therapists (for the blind) |
| Physical therapists |
| Radiation therapy technologists |
| Recreational therapists |
| Rehabilitation counselors |
| Speech pathology assistants |

## what are the salary ranges?

An audiometric technician can invest a relatively small amount of money for training and a license, then earn a substantial income. For example, in Wisconsin a candidate can obtain a temporary permit for $10, work under the supervision of a licensed hearing instrument specialist for a year, pay $285 to take the licensure examination, and pay another $41 upon obtaining the license. The license must be renewed every other year at a cost of $200. In Alabama the examination fee is $125, and initial and renewal license fees range from $100 to $150; a $50 apprentice permit is also available but not required.

According to the Alabama Board of Hearing Instrument Dealers, the average hearing instrument dealer in that state is paid minimum wage plus commission. Nationwide, many hearing instrument specialists work entirely on commission, and their earnings depend on how many sales they make. Grant Gording works on commission and is paid no benefits. He estimates that typical annual earnings in this field would be $20,000 to $30,000 to start and $35,000 to $40,000 for the average worker. He knows of some hearing instrument specialists who earn $150,000 to $200,000 annually.

Audiologists, who usually hold advanced college degrees, had somewhat higher average earnings than did hearing instrument specialists in 1997. According to a survey by the American Speech-Language-Hearing Association, the median annual salary was $43,000 for full-time, certified audiologists. The median annual salary for licensed audiologists with one to three years of experience was $32,000. With about twenty years of experience or a doctoral degree, audiologists earned median salaries of about $55,000. Earnings were highest in the West and Northeast and lowest in the Midwest.

A survey by *Medical Device and Diagnostic Industry Magazine* found that the average salary for various professionals involved in the manufacture and sale of medical devices was about

$66,500 in 1997. The average salary was about $7,000 less in the South. The highest salary was $225,000; the lowest was $30,000; and the median bonus (paid in addition to the salary) was $12,000.

## what is the job outlook?

The general outlook in any field that deals with hearing impairments is better than the average for all occupations through the year 2006, largely because the number of elderly people will increase during the next few decades. There is also a growing awareness of childhood hearing problems and of the need to help people who have suffered a hearing loss on the job. *Money* magazine listed audiology and speech-language pathology in eleventh place among its "Fifty Hottest Jobs" in 1995.

Currently, the number of audiologists and speech-language pathologists in private practice is small, but it is expected to increase rapidly as schools, hospitals, nursing homes, and managed care practices use more contract services. Audiometric technicians will be hired as assistants at many of those private offices.

There will be continued demand for hearing aids, but Grant Gording remarks, "With all the competition out there, it's getting harder and harder in this field." In addition to competing with numerous people entering the profession, he has seen profits drop as hearing aids are discounted. Third, he says, some hearing aids being built now are priced lower because their quality is not as high as it could be. "Unfortunately, a lot of consumers are buying into that," he notes.

Still, a large number of people will purchase hearing aids, particularly as technology improves the looks, quality, reliability, and comfort of assistive devices. Audiometric technicians will be needed to sell and service hearing aids until cures are found for the many causes of hearing impairment.

## how do I learn more?

### sources of additional information

Following are organizations that provide information on audiometric technician careers, accredited schools, and employers.

**American Academy of Audiology**
8201 Greensboro Drive, Suite 300
McLean, VA 22102
703-610-9022
http://www.audiology.org/aaahome.htm

**American Speech-Language-Hearing Association**
10801 Rockville Pike
Rockville, Maryland 20852
301-897-5700
http://www.asha.org/
careers@asha.org

**International Hearing Society**
**National Board for Certification and Hearing Instrument Sciences**
20361 Middlebelt Road
Livonia, MI 48152
734-522-2900

## bibliography

Following is a sampling of materials relating to the professional concerns and development of audiometric technicians.

### Periodicals

*Acoustical Society of America Journal.* Monthly. Directed toward professionals involved with speech and hearing problems. American Institute of Physics, One Physics Ellipse, College Park, MD 20740-3843, 301-209-3100.

*ASHA.* Eleven times per year. General speech, language, and hearing discussions. American Speech-Language-Hearing Association, 10801 Rockville Pike, Rockville, MD 20852, 301-897-5700.

*Audiology and Neuro-Otology.* Bimonthly. S. Karger Publishers, Inc., 26 West Avon Road, PO Box 529, Farmington, CT 06085, 203-675-7834.

*Ear and Hearing.* Bimonthly. Covers topics related to hearing conservation, rehabilitation, research, etc. Williams & Wilkins, 351 West Camden Street, Baltimore, MD 21201-2436, 410-528-4068.

*Hearing Health.* Bimonthly. focuses on all topics of hearing and hearing loss including prevention, research, and legislation. Voice International Publications, Inc., Box V, Ingleside, TX 78362-0500, 512-776-7240.

# automobile mechanic

**Definition**

Automobile mechanics repair, service, and overhaul the mechanical, electrical, hydraulic, and electronic parts of automobiles, light trucks, vans, and other gasoline-powered vehicles.

**Alternative job titles**

Automotive service technicians

Bus mechanics

Truck mechanics

**Salary range**

$13,000 to $24,856 to $43,200+

**Educational requirements**

Some postsecondary training

**Certification or licensing**

Voluntary

**Outlook**

About as fast as the average

GOE
05.05.09

DOT
620

O*NET
85302A

## High School Subjects

English
(writing/literature)
Mathematics
Physics
Shop (Trade/Vo-tech education)

## Personal Interests

Cars
Computers
Figuring out how things work
Fixing things

"Mechanical aptitude and good training are important," says Steve Heying as he raises a late model Honda on the alignment rack, "but the only way you're going to learn to do wheel alignments is to do wheel alignments."

He enters the automobile make, model, and year into the computer, which moments later produces a specification sheet of clearances and tolerances particular to this type of car. In performing a wheel alignment, Steve adjusts all parts of the steering system so that they are in correct relation to one another. A properly aligned car will not immediately wander off the road should the driver let go of the wheel. Proper alignment also prevents the tires from wearing unevenly. As Steve calibrates each wheel, the rack monitors his efforts with infrared beams. The computer screen displays an automobile schematic, marking red the areas that are out of specification. Steve adjusts the camber, or tilt, of the front wheels, and caster, or tilt, of the steering knuckle pivots. The task requires great patience.

"Being a mechanic means that you're a troubleshooter," he says, standing up from the left front wheel. "You try one thing and if that doesn't work you try another. You ask a lot of questions. You're always learning."

## what does an automobile mechanic do?

Americans have become dependent upon their automotive vehicles. Each year we buy more cars and trucks and vans, which in time require routine services and repairs. In addition, almost everyone has experienced the frustration and inconvenience of an automobile breakdown. Whether we are dealing with a complete breakdown or a routine oil change, we bring our vehicles to *automobile mechanics,* who service and repair automobiles and other gasoline-powered automotive vehicles.

When mechanical or electrical troubles occur, mechanics either discuss the nature of the trouble with the vehicle owner or, if they work in a dealership or large repair shop, they may consult a repair service estimator or access the information on a computer system. Relying on solid analytical and troubleshooting skills and a thorough knowledge of automobiles, mechanics diagnose the source of the problem with the aid of test equipment such as engine analyzers, scanners, spark plug testers, and compression gauges. Most mechanics value the challenge of making an accurate diagnosis, counting it among their favorite duties.

After the cause of the problem has been established, mechanics make adjustments or repairs. If the part in question is damaged or worn beyond repair, it is ordered, often with the use of a computer system, and replaced, usually with the authorization of the vehicle owner.

During routine service and preventive maintenance—an oil change, for instance—mechanics consult a checklist to ensure they service all important items. They inspect a vehicle's belts, hoses, steering system, fuel injectors, air filter, battery, spark plugs, transmission fluid, and brake systems. They lubricate and adjust engine components and repair or replace parts to avoid future breakdowns.

In their work, mechanics use a variety of power tools, such as pneumatic wrenches, machine tools, such as lathes and grinders to rebuild brakes, and hand tools. Most mechanics furnish their own hand tools, acquired gradually over the first several years of their careers. They may spend between $2,000 and $15,000 for a complete set of tools. In addition, they use welding equipment to repair exhaust systems, jacks and hoists to lift vehicles, and a growing variety of sophisticated electronic equipment, such as infrared analyzers and computerized diagnostic devices.

While most mechanics are able to perform a wide array of repairs, as automobiles have become increasingly sophisticated, mechanics have become specialized.

*Electrical systems mechanics* service and repair batteries, starting and charging systems, lighting and signaling systems, electrical instruments, and accessories.

*Automotive radiator mechanics* flush radiators with chemical solutions, locate and solder leaks, install new radiator cores, and replace old radiators with new ones.

*Automotive exhaust technicians* conduct tests on vehicles to ensure that exhaust emission levels comply with state regulations.

*Engine performance mechanics* perform general and specific engine diagnosis and service and repair ignition systems, fuel and exhaust systems, emission control systems, and cooling systems.

*Automotive field-test technicians* work in manufacturing, research, and development, preparing vehicles for road tests in field proving grounds.

**Carburetor** The device that mixes fuel and air and delivers the combustible mixture to the engine.

**Chassis** The supporting framework of an automobile.

**Cylinder** The tubular opening in which the piston moves up and down.

**Motor** A device that converts electrical energy into mechanical energy.

**Muffler** The noise-absorbing device through which exhaust gases pass.

**Starter** The motor that cranks the engine to start it.

**Transmission** The device that provides different gear ratios between the engine and wheels.

**V-type engine** An engine with two banks of cylinders set at an angle to one another in the shape of a V.

*Brake mechanics* inspect and repair drum, disc, and combination brake systems, parking brake systems, power assist units, and hydraulic application systems. Some perform front-end work as well.

**To be a successful automobile mechanic, you should:**

- Be patient and have excellent troubleshooting skills
- Have a mechanical aptitude
- Be comfortable talking to strangers
- Have excellent customer service and communication skills
- Have a working knowledge of computers

## what is it like to be an automobile mechanic?

Steve Heying has been an automobile mechanic with Sears Auto and Tire Center for almost nine years. "I've always liked to work with my hands," he says. "If you want to be a mechanic, you have to be curious, you have to want to learn how things work." Typically, Steve is on the job Monday through Friday, 12:30 PM to 9:00 PM, with some overtime, though that, he says, is largely dependent upon the weather. Repair shops can be slow in the winter, which results in layoffs of many part-time, unskilled workers. At Sears, Steve works with between twenty-four and thirty other technicians and installers, under the supervision of a lead technician.

"The best part of my job is working with the other technicians as a team," he says. Each technician in Steve's repair shop specializes in one or more aspects of automotive technology. Steve works on brakes and alignments, inspecting all brake components, in both drum and disk systems. A *front end mechanic,* Steve uses a micrometer to check the depth of the shoe pads and the width of the rotors, and performs wheel alignments, using a computerized rack. He also balances wheels and repairs steering and suspension systems.

Automobile mechanics do most of their work indoors, in well-ventilated, lighted repair shops, though some make outdoor service calls to perform emergency repairs on vehicles that have broken down on the road. Mechanics often work in awkward, cramped positions, beneath vehicles or under the hoods, handling greasy and dirty parts. Design, test, research, and development technicians often work in engineering departments or in laboratories which are cleaner than repair shops.

While mechanics must be able to work independently, they must also have the ability to function as members of a team. Above all, technicians must understand the basic scientific and mathematical theories that support their work. As the field changes and automobiles become more complex, technicians need a clear understanding of the underlying science in order to keep abreast of the industry.

"Computers are everywhere," Steve says. "They're in the cars and in the tools and machines I use. I get my work orders on a computer. And I use a computer to look up parts everyday." His employer, Sears, recently installed a multimedia computer system on which technicians are being trained. "If you want to be a mechanic, you're going to have to know how to use a computer. You just can't be afraid of change."

## have I got what it takes to be an automobile mechanic?

In addition to having mechanical aptitude, mechanics need strong interpersonal and customer service skills. They are often required to listen to customers explain automotive problems and, in turn, they must be able to explain diagnostic results and required repairs, translating technical language into lay terms.

"Customers don't always want to talk to a salesperson," Steve says. "They insist on talking face to face with the mechanic who's going to fix their car." Admittedly, dealing with customers who do not understand the basic functions of an automobile can be a trying experience. Steve says this is the hardest part of his job. "They should require a general course in auto mechanics when you get your driver's license," he says. "That would help."

Over the years, Steve has watched a number of unskilled mechanic trainees try to break into the industry fresh out of high school. The problem, he says, is that without some technical training these individuals lack the fundamental mechanical know-how to get the job done right. He recommends that those who want to become mechanics attend a two-year program in order to learn general automotive theory. "Anyone with mechanical aptitude can change oil and mount tires," he says, "but no one wants to do that forever." As automobiles have become more complex, the problems have become harder to diagnose. "Troubleshooting skills are a big plus," he says. "Some mechanics won't be able to pinpoint exactly what's wrong and will give up and start throwing parts on the car, but that's really the last resort. It's not a very effective way to fix a car."

Individuals with a desire to learn new service and repair techniques are excellent candidate for a postsecondary automotive training program. Mechanics must be disciplined, systematic, and analytical, and have careful work habits. They must understand every detail of every part, how the parts work together, and what would happen if all parts were not assembled or working properly. Technicians today also must have a working knowledge of electronics, because a growing variety of automotive components are using electronic devices. In automotive vehicles, electrical systems crank the engine for starting, furnish high voltage sparks to the cylinders to fire compressed fuel-air charges, and operate the lights, the heater motor, and other accessories. Mechanics must be able to determine when an electronic malfunction is responsible for a problem.

## how do I become an automobile mechanic?

Although Steve did not attend a formal automotive training program, after graduating from high school he attended an electronics institute and earned a diploma in 1986. As automotive technology rapidly increases in sophistication, most training authorities strongly recommend that individuals seeking a career in the industry complete a formal postsecondary training program.

Some mechanics train on the job through three- or four-year apprenticeships offered by independent repair shops and automobile manufacturers, though such programs have declined sharply in recent years. In addition, certain branches of the U.S. Army, Navy, and Marines offer this type of training.

FYI

In 1992, the U.S. Department of Transportation found that, on average, Americans had 746 registered motor vehicles per 1,000 residents.

## education

### High School

Chuck Bowen heads the automotive technician and collision program at the Dunwoody Institute in Minneapolis, Minnesota. He recommends that high school students interested in pursuing a diploma in automotive technology take mathematics, physics, and English courses. "Applied math and applied science courses are ideal," he says. In order to keep up with new technology and understand technical manuals, high school students are told to take courses in algebra, plane geometry, chemistry or another laboratory science, basic drafting, automotive technology, electronics, and metals. Strong reading, grammar, and writing skills are essential.

### Postsecondary Training

Chuck Bowen's postsecondary program is typical of those offered by vocational-technical schools and community colleges. While there are some four-year programs, most institutions spread the training over two years, after which students earn a diploma. Some community colleges supplement the technical training with courses in English, speech, composition, and social science, awarding an associate's degree at the end of two years. High school programs in automotive technology offer an introduction to the field, but they vary in quality, and it is difficult to get the necessary science background and the hands-on experience that a formal program offers.

In order to be admitted to the Dunwoody Institute, students must have graduated in the top half of their high school classes; otherwise they must pass an entrance exam. Training is broken into eight units: suspension and steering, automatic transmission/transaxle, manual drive train and axles, heating and air conditioning, engine repair, engine performance, brakes, and electrical systems. Students spend two hours each day, Monday through Friday, in the classroom exploring automotive theories, followed by three hours of practical, hands-on application. Students also take general education courses in computers, speech, basic mathematics, and English. The institute offers some partial scholarships, though many students help finance their education through part-time jobs mounting tires and performing routine automotive services.

The Dunwoody Institute is certified by the National Automotive Technicians Education Foundation (NATEF), an affiliate of the National Institute for Automotive Service Excellence (ASE). While certification is voluntary for training institutions, it ensures that a program meets uniform standards for instructional facilities, equipment, staff credentials, and curriculum. In 1993, more than 650 high school and postsecondary automobile mechanic training programs had been NATEF-certified.

Several automobile manufacturers sponsor two-year training programs in automotive service technology. For more information contact the following organizations:

**ASSET Program**
Ford Motor Company
Ford Parts and Service Division
Training Department
Dearborn, MI 48121

**Chrysler Dealer Apprenticeship Program**
National CAP Coordinator
26001 Lawrence Avenue
Center Line, MI 48015

**General Motors Automotive Service Educational Program**
National College Coordinator
General Motors Technical Service
30501 Van Dyke Avenue
Warren, MI 48090

## certification or licensing

Voluntary certification by ASE is recognized as a standard of achievement for automotive mechanics, who become certified in one or more of eight different areas—the same eight units by which the Dunwoody Institute organizes its educational program.

Steve is ASE-certified in suspension and steering and in brakes. "I went down one Thursday night and took a couple of multiple choice tests," he says. "It only took a few hours." Certification requires at least two years of experience and the passing of the examination. Completing an automotive mechanics program in high school or community or junior college may be substituted for one year of experience. Individuals who pass all eight tests earn the title General Automobile Mechanic. Certified mechanics must retake the examinations at least every five years.

"For me the tests were relatively easy," Steve says; he already had a great deal of hands-on experience at the time he took them. Voluntary ASE certification is a good way to demonstrate initiative and often, Steve says, plays a key role in pay increases.

Mechanics must continually update their skills. Sears, which has an internal training program, offers clinics and seminars for its technicians. Steve recently attended a seminar on anti-lock brake systems. Many manufacturers of automotive parts, equipment, and vehicles, as well as trade associations, offer seminars and host meetings and trade shows.

**Service shops use different methods to compute pay;** This graph shows the percentage of shops that use the following methods to pay technicians

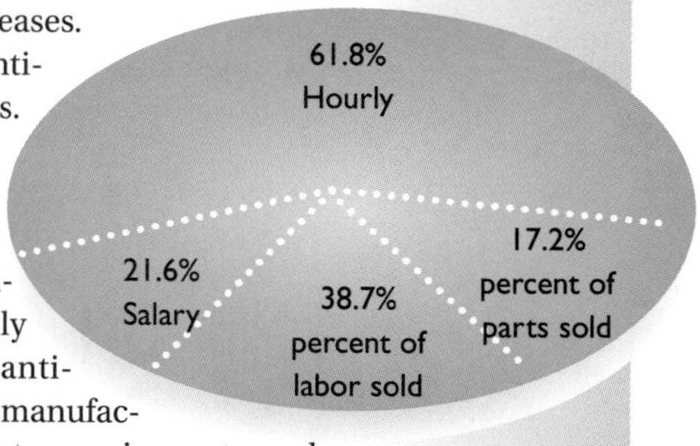

Note: Percentages may add up to more than 100 percent because shops use more than one method to pay their technicians (NATEF, 1995).

## labor unions

While Steve doesn't belong to a union, some automobile mechanics belong to the International Association of Machinists and Aerospace Workers or the International Brotherhood of Teamsters, Chauffeurs, Warehousemen and Helpers of America. Another union is the United Automobile, Aerospace, and Agricultural Implement Workers of America.

## who will hire me?

Steve was first hired at Sears as a tire installer, performing many routine services. Though after only a few months of experience he was able to make simple repairs, he worked three years before being promoted to technician. Typically, trainee mechanics work one or two years, acquiring the proficiency to perform the more difficult types of diagnostic testing, service, and repairs. Some specialists, such as automotive radiator and brake mechanics, learn the skills quickly, while the more difficult specialties, such as automatic transmission repair, take a great deal of time to learn.

The majority of mechanics work in franchised dealerships, such as Chrysler, Honda, GM, and Saturn. Other jobs exist in service stations, auto body repair shops, tire stores, parts supply stores, and department store automotive service centers, such as Sears. Federal, state, and local governments hire technicians to service their vehicles, as do taxi cab companies, trucking firms, and bus lines. Technical program students can learn more about job openings from their department heads, teachers, or placement officers. Graduating students may also apply directly to the personnel office of any auto, truck, or tractor dealership or to the many related industries.

According to Chuck Bowen, the Dunwoody Institute places approximately 97 percent of its graduates in employment positions. A school's placement rate is one factor that may help students decide which school to attend.

## where can I go from here?

In the winter of 1993, Steve enrolled in a two-year diesel technology program. *Diesel mechanics* repair and maintain diesel engines that power heavy transportation equipment, such as trucks, buses, locomotives, construction equipment, and farm equipment. Steve is interested in pursuing a career in diesel technology because wages are relatively high and the specialized repair work is challenging and varied.

As mechanics gain practical experience and become ASE-certified in one or more specialties, they become more valuable to their employers. Experienced mechanics with demonstrated leadership skills may advance to garage supervisor positions or service management. Mechanics adept at working with customers may become automotive repair estimators or sales persons. Many mechanics, in time, open their own independent shops.

## what are some related jobs?

The Employment and Training Administration of the U.S. Department of Labor classifies automobile mechanics, or technicians, under the headings *Engineering, Craft Technology: Mechanical Work* (GOE) and *Motorized Vehicle and Engineering Equipment Mechanics and Repairers* (DOT). Among these workers are individuals who skillfully use hand tools, precision-measuring instruments, and machine tools to repair engines, engineering equipment, generators, alternators, and motors. Careers include air conditioning and refrigeration mechanics, aircraft technicians, industrial mechanics, lawn and garden equipment technicians, locksmiths, motorboat mechanics, and solar energy system installers.

| **Related Jobs** |
|---|
| Automobile salespersons |
| Automotive body repairers |
| Diesel mechanics |
| Endless track vehicle mechanics |
| Farm equipment mechanics |
| Garage supervisors |
| Gas engine repairers |
| Maintenance mechanics |
| Motorcycle mechanics |
| Tune–up mechanics |

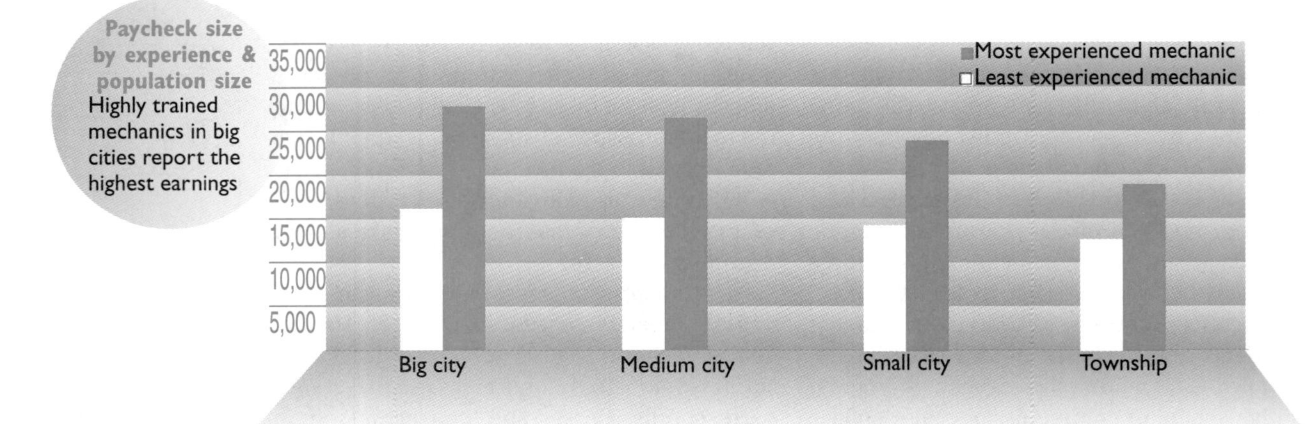

**Advancement Possibilities**

**Automatic transmission technicians** service power trains, couplings, hydraulic pumps, mechanical systems, and other parts of automatic transmissions.

**High–performance engine technicians** analyze, test, and maintain high–performance engines in the auto racing industry. They also redesign parts to enhance engine performance.

**Sales representatives** advise customers and recommend the parts or vehicles best suited to their needs.

**Service managers** manage automobile dealership service departments, hire and train employees, select equipment, and assist in diagnosing customer automotive problems.

## what are the salary ranges?

Chuck Bowen estimates that Dunwoody Institute automotive technician graduates start out earning between $8 and $11 an hour. Some earn commission as well. According to the U.S. Department of Labor, the median weekly salary for automobile mechanics in 1996 was $478, or $24,856 a year. The lowest ten percent earned $250 or less a week, while the highest ten percent earned more than $850 a week. .

While salaries are dependent upon the geographic location of the shop, the main factors affecting earnings are experience and training. The key to getting a good-paying job is good training.

## what is the job outlook?

In general, it is important to remember the simple fact that machinery and equipment eventually break down, and thus skilled repair technicians will always be in demand. The employment outlook for automobile mechanics is expected to be very good into the next century, provided individuals have the specialized technical skills that are in demand. The government projects job growth to be a strong 22 percent in most areas. Openings are expected to increase along with the rise in the driving age population and the increasing number of multicar families.

Federal, state, and local regulations governing safety and pollution control, along with more extensive warranties on new cars and trucks, will create jobs for skilled technicians. In addition, the overall economic climate has little effect on the automobile repair business. Vehicles break down irrespective of an increase in interest rates or a sharp dive by the Dow Jones industrial average.

## how do I learn more?

### sources of additional information

Following are organizations that provide information on automobile mechanic careers, accredited schools and scholarships, and employers:

**Automotive Service Association (ASA)**
1901 Airport Freeway, Suite 100
PO Box 929
Bedford, TX 76021-5732
817-283-6205

**Automotive Service Industry Association (ASIA)**
25 Northwest Point
Elk Grove Village, IL 60007-1035
312-836-1300

**Motor and Equipment Manufacturers Association (MEMA)**
300 Sylvan Avenue
PO Box 1638
Englewood Cliffs, NJ 07632-0638
201-569-8500

**National Automotive Technicians Education Foundation (NATEF)**
13505 Dulles Technology Drive
Herndon, VA 22071-3415
703-904-0100

**National Institute for Automotive Service Excellence (ASE)**
13505 Dulles Technology Drive
Herndon, VA 22071-3415
703-742-3800

# bibliography

Following is a sampling of materials relating to the professional concerns and development of automobile mechanics.

## Periodicals

*AutoInc.* Monthly. Addresses technical and business concerns and provides repair shop profiles and legislative information. Automotive Service Association, 1901 Airport Freeway, Bedford, TX 76095-0929, 817-283-6205.

*The Blue Seal.* Semiannual. Newsletter covering training and new technology for certified mechanics and employers. National Institute for Automotive Service Excellence, 13505 Dulles Technology Drive, Herndon, Virginia 22071-0001, 610-964-4410.

*Chilton's Motor Age.* Monthly. Presents technical developments effecting members of the automotive service industry. Chilton Publishing, One Chilton Way, Radnor, PA 19089, 215-964-4341.

*Motor.* Monthly. Addresses students and technicians in the field of automotive repair and service. Hearst Business Publishing, 645 Stewart Avenue, Garden City, NY 11530, 516-227-1370.

*Motor Service.* Monthly. Directed at auto repair shops and service departments. Hunter Publishing Ltd. Partnership, 2101 South Arlington Heights Road, Suite 150, Arlington Heights, IL 60005, 847-427-9512.

# automotive body repairer

**Definition**

Automotive body repairers straighten bent automobile bodies, remove dents, and replace the crumbled parts of cars that have been damaged in traffic accidents. They repair all types of vehicles to look and drive like new.

**Alternative job titles**

Automobile collision repairers

Collision-repair technicians

**Salary range**

$12,480 to $24,500 to $32,300

**Educational requirements**

High school diploma

**Certification or licensing**

None

**Outlook**

About as fast as the average

GOE
05.05.06

DOT
807

O*NET
85305B

## High School Subjects

Computer science
English
Mathematics
Shop (Trade/Vo-tech education)

## Personal Interests

Building things
Cars
Figuring out how things work
Fixing things
Reading/books

**Shawn Basolo** looks over the white Toyota pickup truck and rereads the estimate. The truck had hit a deer, the most common accident on the highways in and around Missoula, Montana, where Shawn works as an automotive body repairer. Yesterday he finished up a white Chevrolet pickup that had also hit a deer, but he's glad this one is a Toyota. He needs to take most of the truck apart to get at the repairs, and he could probably do it all with a ten and twelve millimeter socket. "Working on domestic cars takes five times the amount of tools needed to work on foreign cars," Shawn says. "And the tools change every year because the cars change every year."

After Shawn gets the hood and radiator off the car, he'll need to put the entire body up on the frame rack to straighten it out. The right front corner was seriously squashed, and the whole front end bent sideways as a result. "Must have been a buck," Shawn muses. Then he gets to work.

## what does an automotive body repairer do?

*Automotive body repairers* fix all kinds of vehicles, but cars and small trucks make up the bulk of their work load. Some repair buses, large trucks, and tractor trailers. Thousands of motor vehicles are damaged in traffic everyday, and although some are scrapped for salvage, most can be made to look and drive like new again.

Automotive body repairers use alignment machines to straighten out bent vehicle bodies and frames. They chain or clamp frames and sections to the machines, which use hydraulic pressure to straighten the damaged parts. "Unibody vehicles," where the body and frames are welded together, must be restored precisely to their original specifications, just as they left the factory, in order to run properly. To do this, repairers use bench systems to measure exactly how much each section is out of alignment, then use hydraulic machines to return it to its original shape. Most cars are unibodies. Trucks, on the other hand, are built with separate frames and bodies.

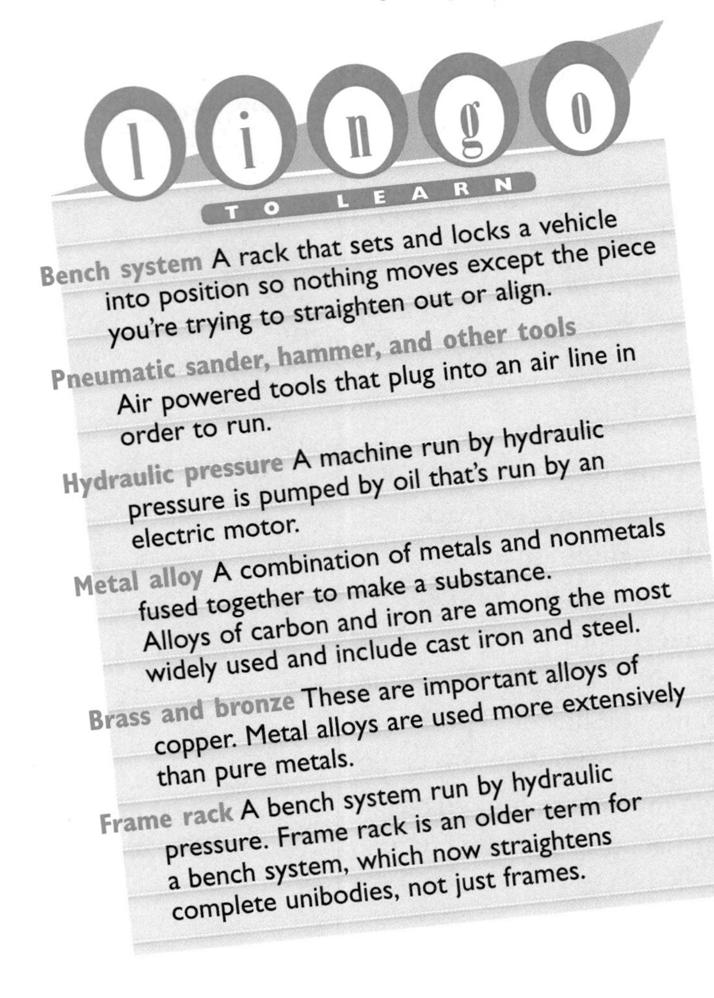


**Bench system** A rack that sets and locks a vehicle into position so nothing moves except the piece you're trying to straighten out or align.

**Pneumatic sander, hammer, and other tools** Air powered tools that plug into an air line in order to run.

**Hydraulic pressure** A machine run by hydraulic pressure is pumped by oil that's run by an electric motor.

**Metal alloy** A combination of metals and nonmetals fused together to make a substance. Alloys of carbon and iron are among the most widely used and include cast iron and steel.

**Brass and bronze** These are important alloys of copper. Metal alloys are used more extensively than pure metals.

**Frame rack** A bench system run by hydraulic pressure. Frame rack is an older term for a bench system, which now straightens complete unibodies, not just frames.

Automotive body repairers remove the badly damaged sections of vehicles, like hoods, grills, and bumpers, and weld on or install new replacement parts. Less serious dents are banged or hammered out, usually with a hand-prying bar, a pneumatic hammer, a hydraulic jack, or some other hand tool. Small dents and creases in the metal are repaired by holding a small anvil on one side of the dent and hammering from the opposite side. Very small pits and dimples are removed with pick hammers and punches by a process called metal finishing.

Usually, a dent is never completely knocked out so it's flush with the undamaged area, and the technician must fill in the gaps with special fillers that bond to the car. Called "bondo," these fillers come in different consistencies and makes, depending on what material they're being used with. Some bondos are made for metal, some for fiberglass and plastic, and some for sheet-metal compound. Some are harder than others, and some need a sealer on the metal before the bondo is applied. After the bondo goes on and dries, the technician sands down the repaired area till it's flush and smooth with the undamaged area. The repaired dent is usually sent over to the painter for the final touches, who does his or her best to match the car's original color. "I know how to do painting and could if I had to," says Shawn, "but we like to have our own painter who specializes in that field." In small shops, repair technicians often do the painting as well as the repairing.

Technicians also fix the electrical wiring in headlamps, blinkers, and taillights if it gets damaged. After they find the break in the current, they can usually splice the wires together to make the repair. In large shops, body repairers sometimes specialize in a certain type of repair, such as frame straightening or door repair. Some exclusively replace glass windshields and windows.

Automotive body repairers face challenges with each new car. Every job is unique and requires a different solution. Usually, the repairer is given the estimate of damage and must go from there to make sure the actual damage matches the appraised damage. "Sometimes it's very different," says Shawn, "because you never know what's going on till you get some pieces off and take a look."

## what is it like to be an automotive body repairer?

Shawn arrives at his job at 8:00 AM every morning, Monday through Friday. He usually knocks off around 5:00 PM, then returns home to do a little automotive body repair on the side. His day begins with inspections—of the work he did yesterday and the work he needs to do today. New stuff that comes in? "Well," says Shawn, "that'll come to me eventually."

Shawn first takes a look at the new door skin he put on a Volvo the day before and at a few other repairs he made on the car. "When you put a door skin on, you have to bend it around the old door frame with a hammer. I looked to make sure I didn't put any dents in the door when I fixed it. Sometimes that happens because the door skin is so flexible." Shawn, in fact, had put a few minor dents in the new door, so he gently pulled them out. He inspected the other damage he repaired the day before: three dents in the front fender and a broken blinker and park light. "I replaced the lenses," says Shawn. "It was not an electrical job at all." When Shawn fixes a dent, he tries to use as little bondo as possible, and that means getting the dent almost completely out. "A car is made of metal, not bondo," says Shawn. "You ruin the structural integrity of the car if you use too much filler."

After inspecting his work, Shawn writes up the order for the painter. The door is ready to be edged, and the painter will have to blend the inside edge of the old door with the new door skin, which came already painted from the factory.

By the time Shawn gets the Volvo door over to the painter, the white Toyota pickup he finished the day before, the one that hit the deer, is back from the painter. Its owner is coming this afternoon to pick up the car, and Shawn needs to get a new radiator in and install the bumper and grill that just arrived from the factory. This takes him till lunch.

"I work on about fifty cars or pickups a month," says Shawn, "and sometimes up to five a day, depending on what needs to be done to them." The hardest part of Shawn's job is working on dirty, trashed vehicles. "You can tell when the owners don't take care of their rig," says Shawn. "If they don't care, why should I?"

"When a new car comes in, I'm the first person to deal with fixing it," says Shawn. "I look at the estimate and evaluate the damage and how to go about the repair. There's never one way to make a repair. Every repairer develops their own technique." Shawn finds it incredibly rewarding to fix a damaged car and make it so new looking you can't even tell where the repair was made. "That's always my goal," Shawn says, "and I take pride in my work."

**To be a successful automotive body repairer, you should**

- Be able to read and follow complex diagrams and instructions
- Have a good grasp of basic math and computer skills
- Be organized, neat, and detail-oriented
- Take pride in your work and be able to work without supervision

## have I got what it takes to be an automotive body repairer?

Today's advanced technology has greatly changed the structure, materials, and parts used in automobiles. Because of this, repair technicians needs to know the newest techniques in repairs and be skilled at implementing them. For example, the bodies of cars are now made of a combination of metals, alloys, and plastics, each one requiring different techniques and materials to smooth out dents and pits and repair damaged parts. Restoring unibody cars to their original state requires a precise measuring and synchronization of body parts. Therefore, being able to read and follow complex instructions, diagrams, and three-dimensional measurements in technical manuals is essential for an automotive repair technician. Basic math and computer skills are also needed to use the manuals that come with today's sophisticated vehicles. It's been estimated that technicians must be able to interpret 500,000 pages of technical text to repair any car on the road, and new information is being produced at a rate of 100,000 pages a year. Today's automotive technicians, besides being good with their hands, must also have a solid understanding of math, science, and electronics. These skills are necessary because 80 percent of a vehicle's functions are now controlled by computer—compared to about 18 percent five years ago.

Repair technicians need to be organized, detail-oriented, and neat. Sloppy work is unacceptable. Shawn believes taking pride in one's work is the most important trait a repair technician can have. The technician usually works alone, unsupervised, and must manage his or her time well to be productive on the job. Obviously, liking cars and understanding cars are primary traits to have. Along with that comes the dirt. "You can't mind getting dirty," says Shawn.

Some drawbacks to being an automotive body repairer include working with toxic chemicals. Repairers are exposed to materials that contain hazardous fumes and substances, and Shawn often wears a face mask and gloves for protection when he works. "That's the one part of the job I can do without," says Shawn. "The mask filters out about 99 percent of the fumes, but health-wise, there aren't many pros. I'm on my knees a lot, too. This job is just hard on the body."

## how do I become an automotive body repairer?

Automotive body repairers do not need formal training to become technicians. Most employers prefer to hire repairers who have completed a training program, however, because of the highly sophisticated advances that have been made in automotive technology. Having an education in automotive body repair could help your chances to get a job and speed up promotion and salary increases.

Some employers, though, favor hiring helpers, or people with little or no experience, and training them on the job. Most employers look for helpers who are high school graduates, know how to use hand tools, and are quick to learn. Shawn worked in the restaurant business before he got hired on as a helper at Jax Auto Body & Paint Inc. "My first question," says Shawn, "was are you willing to hire someone who knows very little?" Shawn's supervisor, and the owner of Jax, was willing.

"I prefer to hire people with no formal training. They don't come in with set ideas about how to do something," says Keith Koch, who has owned the business for thirteen years. "But there are requirements. I look for people who have a good mechanical sense and a genuine desire to learn. In today's technical world, you also need a high school diploma."

Helpers begin by assisting body repairers with small dents and the removal and installation of body parts. They observe technicians doing the bigger jobs, such as repairing a car on a frame rack. Then they begin their own projects, starting with small dents and moving into more complicated repairs. It usually takes three to four years to become a skilled automotive repair technician. Shawn has been learning for three years now. "I'm amazed at what I've learned on the job," says Shawn. "And I picked it up really fast."

## education

### High School

In high school, you should pay attention in math, English, and computer science to get the skills you need to read technical car manuals and work on computerized cars. Shawn took wood shop in high school, a class that taught him to appreciate the finished product and take pride in his work. Shawn wishes he had taken welding and metals class in high school, and recommends that would-be repair technicians select both these classes if available. Auto shop is also a must, and many high schools offer vocational training programs in automotive repair.

When Shawn was little, he remembers watching and helping his older brother renovate a classic Chevrolet. In high school, Shawn continued to tinker around with cars. "All my friends came to me when they had car problems. They still do," says Shawn. Shawn believes any work you can do on cars is good experience. "Most of what I've learned," says Shawn, "I couldn't get from a book."

### Postsecondary Training

Most vocational and technical schools, and community colleges, offer training programs in automotive body repair. Nationwide, there are 119 programs certified by Automotive Service Excellence (ASE), an organization that sets standards for collision-repair training programs. If you attend a vocational school or community college to learn the automotive body repairer trade, you can expect to take courses in structural analysis and damage repair, mechanical and electrical components, plastics and adhesives, and painting and refinishing. Many schools offer a separate training program in painting and refinishing for those interested in that part of the trade.

Because the trade is becoming so technical, schools are now starting to recommend or require academic course work along with vocational training. Classes in computer science, communication, mathematics, and English are becoming more common requirements for those who want to graduate from a collision-repair technician program.

## FYI

On January 14, 1914, Henry Ford did the inconceivable. He raised the wages of his Ford Motor Company employees to an unheard of amount of five dollars a day. Then he decided to earmark ten million of his company's twenty-five-million-dollar earnings to an employee profit-sharing plan. His critics went wild. This will ruin the industry, they cried! We'll all have to pay workers more!

But Ford remained undaunted "I like to see folks who work hard get their fair share," he said in response. To remain a Ford employee, however, the workers had certain standards they had to abide by. For one, they had to live in a comfortable house and were forbidden to take in a large number of boarders, a common practice for making additional income in Ford's day. Workers had to prove they were saving their money and often had to report how they were spending it! Ford became so adamant about how he wanted his employees to live their lives, he sent inspectors out to their homes. If you didn't follow the rules, you were fired.

## certification

Certification for automotive body repairers is offered by the National Institute for Automotive Service Excellence (*see* "Sources of Additional Information"). Although voluntary, it is the recognized standard of achievement for repair technicians. To be certified, a repairer must pass a written exam and have at least two years of experience working as a technician. You can take exams in a variety of subject areas to become certified, for example, in painting and refinishing. If you complete a postsecondary training program, it may be substituted for one year of experience. Every five years, you must take a recertification exam.

I-CAR, a nonprofit training organization, offers educational classes nationwide for automotive body repairers in such subjects as electronics collision repairs, electrical repairs, and structural work (*see* "Sources of Additional Information"). They also do qualification testing in welding. Shawn's employer has paid for him and other employees to take several courses, for which they receive certificates of completion. I-CAR offers a toll-free technical assistance number to anyone who has taken their courses.

## scholarships and grants

Sometimes large companies such as NAPA Auto Parts offer scholarships or apprenticeships to vocational schools and training programs in their area for students learning the trade. Specific automobile dealerships also offer opportunities for students to learn how to work on Ford Assets, for example, or Toyota T-10s. Usually these scholarships are more for collision than repair. For information about scholarship and opportunities for automotive body repairers, contact large companies in the industry directly or your school's financial aid office.

## internships and volunteerships

Automotive repair technicians who begin as helpers are paid to train on the job. Shawn began his career as a helper at $6.00 an hour. It is not unusual for repair shops to accept apprentices, or helpers, even though no formal apprenticeship exists for the automotive repair technician field.

School programs often have apprenticeships and internships as part of their course requirements. The school usually sets up internships with area dealers and shops so students have the chance to practice on-the-job skills.

## labor unions

Some automotive body repairers are members of unions, including the International Association of Machinists and Aerospace Workers; the International Union, United Automobile, Aerospace and Agricultural Implement Workers of America; the Sheet Metal Workers' International Association; and the International Brotherhood of Teamsters. Most auto body technicians in a union work for large automobile manufacturers, trucking companies, and bus lines.

## who will hire me?

Automotive body repairers held about 225,000 jobs in 1996, and most repairers worked for automobile and truck dealers who specialize in body repairs and painting or privately owned auto repair shops. Some body workers are employed by bus lines, trucking companies, and other organizations like car rental companies that have a frequent need for body work done on their vehicles. A few worked for motor vehicle manufacturers. About 1 percent of automotive body repairers is self-employed.

According to the Automotive Service Association (AAS), 36 percent of all employees in the automotive industry are in production positions and 15 percent are in a combination of production and management positions. Twelve percent work in management alone, and less than 7 percent work in offices.

A recent survey done by the AAS found that auto body repairer shops find the majority of their employees through referrals made by other employers. Thirty percent find new technicians through vocational and technical schools, and another 39 percent advertise and hire new positions through newspaper want ads.

Shawn knew the owner of Jax from the restaurant where he worked. "He knew I was good at woodworking, and I knew he had an automotive repair shop. He needed some carpentry work done, and I needed some repair work done, so we traded," Shawn remembers. "He liked my work so much he offered to train me on the job."

## where can I go from here?

Shawn likes working on cars so much, he would hesitate to do anything that would take him away from that. He believes he could supervise someday, but for right now he wouldn't want to stop repairing cars. "Why stop doing something if you're good at it?" says Shawn. Someday, though, he would like to operate his own automotive body repair shop.

Most automotive repair technicians have the opportunity to continue to learn and train on the job, both by taking education courses offered by I-CAR and working with experienced people. By doing so, an automotive body repairer can earn journeyman status and eventually work into production management and supervisory positions in their shop. They can also become estimators, estimating car repair costs, or automobile damage appraisers for insurance companies. Owning your own repair shop business, as Shawn would like to do someday, is another option.

## what are some related jobs?

"Automotive body repairers know a lot about paint and paint compounds," says Shawn, "and could easily move into a retail position at a paint store, especially an automotive paint store." Welding is also something repair technicians learn, as well as how to install electrical equipment.

The U.S. Department of Labor classifies automobile collision repairers under the heading *Metal Fabrication and Repair* (GOE) and *Structural Work Occupations* (DOT). Also under these headings are people who repair and make boilers, do welding, and work with sheet metal and iron. A further classification lists *Electrical Assembling, Installing, and Repairing Occupations*. Under this heading are people who work as cable television technicians, line installers and cable splicers, and electricians.

| Related jobs |
|---|
| Ironworker |
| Welder |
| Sheet metal worker |
| Boilermaker and mechanic |
| Communications equipment technician |
| Signal mechanic |
| Line installer and cable splicer |
| Elevator installer and repairer |

## in-depth

According to the Automobile Service Association, more and more employers are sending their automotive body repairers to training sessions so they can learn job skills. Past surveys conducted by the association showed the industry spent about 1 percent of their revenues for employee training. In a recent survey, however, 81.8 percent of all respondents said they've sent at least one employee to a training session in the year surveyed. When asked for the number of days the shop has spent altogether in training for that year, 46.5 percent said one to five days, 23.9 percent said six to ten days, 7.6 percent said eleven to fifteen days, and 22.1 percent said more than fifteen days (up from 18.7 percent).

Almost 58 percent of those surveyed said they had at least one ASE-certified employee, and of those 58, an average of 3.79 techs are ASE-certified. The ASA believes the more training an employee has, the greater the overall productivity of the trade. For every 10 percent increase in educational level, they state, productivity jumps 8.6 percent. What's more, educational levels raise salaries for employees: an educated technician makes 47 percent more than a technician who has received no training. Why such a pay increase for trained techs? Because the industry is becoming more technical and complex everyday.

## what are the salary ranges?

According to the U.S. Department of Labor, in 1996 the average annual salary of automotive body repairers was $24,076. The middle 50 percent earned between $16,500 and $35,670 a year. The lowest-paid 10 percent made less than $12,900 a year while the highest-paid 10 percent earned a yearly wage of $44,460.

The ASA reports that a collision-repair technician with journeyman or production manager status earned an average of $32,300 a year, while an entry-level helper with no education could expect to earn less than half a journeyman's wage.

Many collision-repair technicians are paid on an incentive basis for the work they finish. Under this method, body repairers are paid a predetermined amount for various tasks, and what they earn depends on the amount of work assigned to them and how fast they get it done. Some shops guarantee their repairers a minimum amount of work or weekly salary. Helpers and trainees usually receive an hourly wage until they gain the skills and experience to take on their own projects. Body repairers who work for bus and trucking companies usually receive an hourly wage.

The ASA reports that 41.3 percent of all repair shops pay their technicians an hourly wage, while 21.4 pay a flat rate, 15.5 percent pay a percentage-of-labor rate, and 12.6 percent pay a combination of hourly wage plus commission. Shawn is paid an hourly wage, and feels commission pay inspires quantity instead of quality work.

Most automotive body repair shops do not offer benefits. According to U.S. Department of Labor statistics on health and retirement benefits for full-time workers, the collision-repair industry offers fewer benefits than most other types of businesses. The ASA believes the lack of benefits in the industry accounts for a huge turnover of workers: 22 percent annually. For shops with no benefits at all, the turnover rate is 45 percent annually. Shawn receives no benefits at all, but he is allowed to work on his own cars at the shop, something which he considers a benefit.

## what is the job outlook?

Employment opportunities for automotive body repairers is expected to increase about as fast as the average through the year 2,006. Opportunities for those with formal training will probably be better than for those with no training or experience. This is mainly due to the sophisticated technology of cars now being produced, and the numerous changes made to cars each year. With automotive technology changing at such a rapid pace, the most knowledgeable and up-to-date repair technician is going to get the job.

The U.S. Department of Labor cites the nation's population growth as the main factor for such a healthy job outlook for automotive body repairers. As the number of people increases, the number of vehicles is expected to grow as well; and with that growth will come a greater number of cars in accidents, getting damaged and needing repairs. Another important factor is the automobile itself. The new, lighter automobile materials now being used, such as plastics, aluminum, and metal alloys, are more prone to damage than the heavier steel car parts of yesteryear. These materials are also harder to repair and take more time than those of older cars. The majority of jobs will be replacement jobs as automotive body repairers retire, move on to management positions, or go on to other occupations.

Economic conditions also account for the job market in the collision-repair industry. Most repairs are paid by car insurance, which is mandatory in a majority of states. And since most cars need to be restored in order to operate safely, people usually bring their cars in to be fixed. Minor dents and fender benders, however, are more subject to the economy. If the economy is slow, people tend to defer these repairs, reasoning that they can live with the dent rather than pay a hefty insurance deductible .

# how do I learn more?

## sources of additional information

The following associations and organizations offer information on automotive body repairer schools, training sessions, and ASE-accredited programs as well as opportunities for a career as a collision-repair technician.

**Accrediting Commission of Career Schools and Colleges of Technology**
2101 Wilson Boulevard, Suite 302
Arlington, VA 22201
703-247-4212

**Automotive Service Association (ASA)**
1901 Airport Freeway
Bedford, TX 76021
817-267-1376
817-885-0225

**Automotive Service Excellence (ASE)**
13505 Dulles Technology Drive
Herndon, VA 20171
703-713-3800
703-713-0727
http://www.asecert.org

**Automotive Service Industry Association**
25 Northwest Point
Elk Grove, IL 60007
847-228-1310

**I-CAR**
3701 Algonquin Road, Suite 400
Rolling Meadows, IL 60008
800-422-7872
800-590-1215
http://www.icar.com

**National Automobile Dealers Association**
8400 Westpark Drive
McLean, VA 22102
703-821-7000

**National Automotive Technician Education Foundation**
13505 Dulles Technology Drive
Herndon, VA 20171
703-713-0100
http://www.natef.org

## bibliography

Following is a sampling of materials relating to the professional concerns and development of automotive body repairers.

### Books

Society of Automotive Engineers. *Analysis of Autobody Stamping Technology.* Covers the process of autobody stamping, from the use of computer simulation to study formability, springback, and wrinkles to the development of new steels. Warrendale: Society of Automotive Engineers, 1994.

Toboldt, William K. and Terry Richardson. *Auto Body: Repairing and Refinishing.* Introduction to the maintenance and repair of car bodies, including welding, paint, and aerodynamics. South Holland: Goodheart-Willcox Co., 1993.

Witzel, Michael. *The American Gas Station.* History of gas and service stations. Osceola: Motorbooks International, 1992.

### Periodicals

*Autobody Online.* Daily. Resource for collision repair specialists. http://www.autobodyonline.com.

*Autobody Pro.* Daily. Product and service listings, discussion boards, and industry news. http://www.autobodypro.com.

*Car and Driver.* Monthly. Magazine for auto lovers and auto professionals. Hachette Filipacchi, 2002 Hogback Road, Ann Arbor, MI, 48105.

*Collision Repair Industry INSIGHT.* Daily. Insider's guide to collision repair, including a shop finder index with reviews. http://www.collision-insight.com.

# avionics technician

**Definition**

Avionics technicians install, repair, test, and service electronic equipment used in the aviation and aerospace industry.

**Alternative job titles**

Avionics engineering technicians

Avionics mechanics

Avionics repair specialists

**Salary range**

$18,600 to $25,000 to $50,000

**Educational requirements**

Some postsecondary training

**Certification or licensing**

Voluntary

**Outlook**

About as fast as the average

GOE 05.05.10

DOT 823

O*NET 22599C

## High School Subjects

Mathematics
Shop (Trade/Vo-tech education)
Physics

## Personal Interests

Airplanes
Figuring out how things work
Fixing things

**The autopilot** has been repaired and reinstalled, but a test flight is needed to make sure it is functioning correctly. An extension cable is attached to the autopilot control, and Craig Johnson sits with his head down in the cockpit, unable to see out of the plane but focused on making adjustments to the unit. The pilot sits behind the wheel as they fly with the autopilot engaged. Craig notices that the plane seems to have picked up some speed, but he continues to work. The pilot asks, "Well, should I disconnect?" Craig finally looks up and sees the ground when he should be seeing the sky. He tells the pilot to disconnect the autopilot, and the plane lifts up. "The autopilot was taking it someplace we didn't really want it to go," Craig says matter-of-factly, "and I didn't realize the plane was pointed in that direction just because the way the autopilot put it there was so smooth I didn't notice it." Craig has plenty of similar stories, and the occasional unusual flight situation is not so unusual for an avionics technician.

## what does an avionics technician do?

Pilots do a lot more than steer when they're flying a plane. They are responsible for navigating the plane, monitoring the instruments during the flight, communicating with air traffic controllers, and more. Pilots rely heavily on the electronic equipment that allows them to carry out these duties—without radios, navigational equipment, autopilots, flight recorders, and so on, pilots would be unable to fly safely. *Avionics technicians* make sure this equipment is in top working condition.

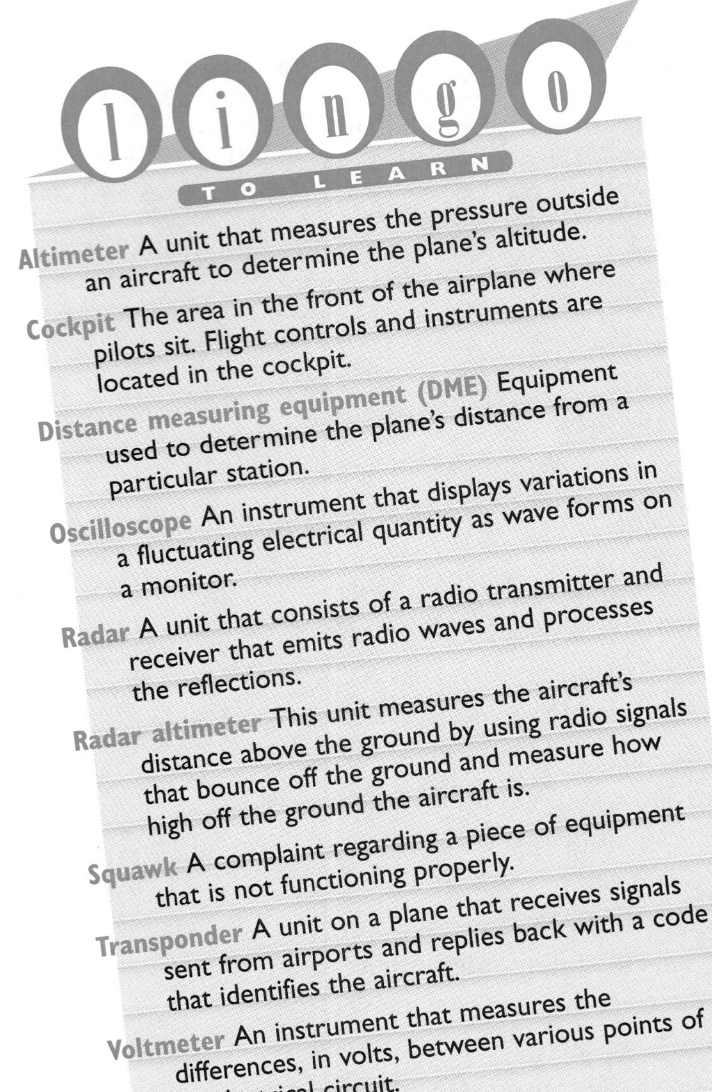


Avionics technicians use their knowledge of electronics to install, repair, test, and service electronic equipment used for navigation, communications, flight control, and other functions in aircraft and spacecraft. After installing new systems, avionics technicians test and calibrate the equipment to ensure that it meets specifications set by the manufacturer and the Federal Aviation Administration (FAA). They adjust the frequencies of radio units and other communications equipment by signaling ground stations and making adjustments until the desired frequency is set. Avionics technicians also perform preventive maintenance checks so equipment will perform effectively.

Avionics technicians may work in a shop atmosphere on individual pieces of equipment or outdoors on large aircraft. To comply with FAA rules and regulations, avionics technicians must keep detailed records and log all repairs and parts replaced. They use equipment and tools such as oscilloscopes, voltmeters, circuit analyzers, and signal generators to test and evaluate units that are in need of repair or maintenance. Signal generators are designed specifically for the avionics industry and are used to simulate and generate the signals used by equipment such as transponders and distance measuring equipment. Transponders receive signals from the airport and reply back with a code that identifies the aircraft. Air traffic controllers are thus able to recognize the plane.

Technicians may be involved in the design and development of new electronic equipment. They must consider operating conditions, including weight limitations, resistance to physical shock, atmospheric conditions the device will have to withstand, magnetic field interference, and other crucial factors. Avionics technicians in research and development must anticipate potential problems and rigorously test the new components. For some projects, technicians may have to design and manufacture their tools before they can begin to construct and test new electronic equipment.

*Installation technicians* are responsible for installing avionics equipment in aircraft. They must make sure the equipment works in conjunction with other components and test the equipment thoroughly. *Bench technicians* work in shops and repair and service defective units. They can specialize in communications equipment, which generally uses analog signals, or pulse equipment, such as radar. Bench technicians test, troubleshoot, and analyze faulty units to determine the problems. They may test down to the component level and replace defective components on the circuit board. Some technicians work down to the board level and replace entire boards instead of dealing with circuits. Most avionics technicians spend time on both installations and bench work, especially if they are employed in small repair shops.

Because the range of equipment in the avionics field is so broad, avionics technicians may specialize even further, especially if they are employed by large manufacturing companies or repair facilities. Technicians can focus on computerized guidance equipment, flight-control systems, or radio equipment, to name a few possibilities. New specialty areas arise frequently due to constant innovations in technology, and avionics technicians must keep informed by reading trade magazines, equipment manuals, and technical articles. Even if they do not plan to specialize in these new areas, avionics technicians must learn about new equipment and have a good understanding of all innovations in the field.

## what is it like to be an avionics technician?

Craig Johnson works for the Avionics Shop, Inc., a small, FAA-certified repair shop in Fresno, California, owned and operated by his family. Craig's workday generally alternates between installations and repairs, and some days he may be out in the field installing parts, while other days he is in the shop working on the bench. "To be a good overall technician," Craig says, "you need to have some experience in installation to understand how the equipment is installed and planned and what can fail. A lot of times, you'll have a failure in a unit on an airplane, but when you put it on the bench it works great. You need to have a good understanding of both installation and the bench."

One of the Avionics Shop's contracts is with a helicopter company, and when a unit is squawked, or reported to be experiencing problems, Craig drives from the shop, which is located on the Fresno Air Terminal, to the other side of the airfield where the helicopter company is located. The problem, for instance, may be a communication unit that doesn't transmit on one of the frequencies. Craig tests the unit in the aircraft using a portable service monitor to verify the problem and begins evaluating the cause. "The most important thing is to get a good description of the problem from the pilot or whoever squawked it," explains Craig. He checks the wires and connectors to make sure nothing is loose, and if the problem still exists, Craig removes all the suspected units and returns to the shop.

> "I think analytically and always want to figure out what makes things work."

When Craig is back in the shop, he first tests the unit on the bench before disassembling it. He hooks the unit up to a test panel and reverifies the problem, making sure the defect is in the unit itself. He then removes the cover and performs a visual check, looking for moisture, corrosion, or leaking parts. He then refers to the appropriate manual for descriptions of the circuits and a troubleshooting guide. Attaching oscilloscope probes to different points on the circuit board, Craig discovers the culprit—a bad integrated circuit in the control head. The shop has a spare in stock, so Craig replaces the circuit, makes sure everything is working, puts the unit back together, and tests it again. Craig cleans the unit, fills out the repair sheet and other paperwork, then heads back out across the field to install the unit back into the helicopter. Craig checks and tests the unit when it's reinstalled in the aircraft and makes entries in the aircraft's logbook.

Craig starts on a repair that has been sitting on the shelf when he returns to the shop, but he doesn't get very far when he receives a call to fix a radio on a helicopter used for medical emergencies. The radio is used to communicate with dispatch and all emergency units, including the police department and fire department, so its repair is urgent. Craig heads back across the field and verifies the squawk. He removes the unit and searches for a spare to insert temporarily. "Usually we'll take a spare unit and put it in there," Craig says, "and that puts them back on call again without much of a delay." None of the spares are in working order, so the helicopter must go off service while Craig repairs the radio. The problem turns out to be a loose solder connection, discovered by tapping on the unit and components to cause some vibration. Craig resolders the connection, checks the entire unit for other potential problems, cleans it, reassembles it, completes the paperwork, then reinstalls the unit in the helicopter, checking and testing it again. Craig finally heads home at 9 PM, the end of a long day. "It's supposed to be a forty-hour week," Craig admits, but it usually doesn't work out that way.

## have I got what it takes to be an avionics technician

Avionics technicians must have an aptitude for this type of work, so an interest in electronics and aviation, as well as mechanical skills, are crucial. You must have excellent analytical and problem-solving skills—you will be relying on those skills every day. Self-motivation, persistence, and the ability to follow through on projects are important as well, since many problems are elusive and you must continue to work to determine the causes.

People's lives may depend on whether the avionics equipment is working or not, so you must be thorough and detail oriented. There are many procedures that must be followed, such as checking the equipment in the aircraft, repairing all existing and potential problems on the bench, and completing all the paperwork. Units must be tested repeatedly in the aircraft and on the bench, so you need to be able to follow steps.

Craig has always been around electronics and avionics. "I think analytically and always want to figure out what makes things work," he explains, and this makes him suited for the avionics field. He admits to being a bit of a perfectionist, which assists him in his work because he will do whatever it takes to get to the root of a problem and solve it. Craig believes initiative and the desire to learn are important to be a successful avionics technician. "There's always something new coming out and something more to learn," Craig asserts.

Craig finds himself working in all kinds of weather, and the hours can be quite long. "Everyone always wants everything done yesterday," Craig notes, and this can cause pressure. And because airplane cockpits aren't very spacious, equipment may be located in hard-to-reach spots. "You have to put yourself in pretty awkward positions," admits Craig. Being an avionics technician carries a lot of responsibility, but Craig willingly takes it on—he enjoys his work and finds the avionics industry is exciting. "A lot of people are in this industry because they like being around planes and love aviation. That's a good motivator."

**To be a successful avionics technician, you should**

- Be meticulous, thorough, and detail oriented
- Enjoy troubleshooting and solving complex problems
- Have analytical skills and the ability to follow through on projects
- Possess strong initiative and a desire to learn
- Have technical aptitude and a love for aviation

## how do I become an avionics technician?

Craig's father started the Avionics Shop in the 1960s, and Craig grew up surrounded by airplanes and electronics. "I've been doing this since I was in high school," says Craig. "I've always been around electronics." Because Craig was exposed to electronics at a young age, he gained a solid fundamental understanding of electronics, which he has continued to build upon.

### education

#### High School

A solid educational background in mathematics and electronics is the most crucial element if you want to become an avionics technician. "Electronics is the most important if you want to go into this field," asserts Craig. Shop classes in electronics or related fields may be available at your school, and any class where you must use tools such as soldering irons or voltmeters will help you gain some manual skills and dexterity. As an avionics technician, you will read schematics and diagrams, so a blueprint reading or drawing class will come in handy. Science courses, especially physics, and English classes are also helpful. English is important for being able to write and read technical reports and to communicate clearly with others both verbally and on paper.

Opportunities to join clubs or organizations involved with electronics should be plentiful in high school. An amateur radio club will familiarize you with radio operation, frequencies, and some federal regulations. Many high schools often participate in science fairs or contests that require competitors to use their knowledge of electronics, problem-solving skills, and ability to build and fix things.

### Advancement Possibilities

**Shop managers** oversee the work of avionics technicians and take care of the daily operation of the repair shop. They may also work as the chief inspector and sign off on the work performed by the technicians.

**Regional managers** of manufacturing companies visit clients to install and service equipment. They may also train clients and solicit new business.

**Avionics engineers** apply engineering and electrical principles to design avionics equipment and systems.

High school is an excellent time to learn about aviation as well. Many organizations offer programs designed to teach youths about flying. You may wish to contact a small regional airport to ask about any such opportunities. A class in ground school can teach you about avionics equipment and familiarize you with aviation terms, but the cost of such a class may be prohibitive. If you can land a part-time or summer job working in an airport or on an airfield, even if it's in the coffee shop, you will learn many of these terms as well.

### Postsecondary Training

Some postsecondary training is necessary to obtain the basic skills you will need to be an avionics technician. Some community colleges and technical schools offer one-year or two-year programs in avionics that can lead to associate's degrees or certificates of completion. Some of the FAA-certified trade schools may also have four-year programs in avionics or aviation technology. If an avionics program or course is not available in your area, you should take classes in electronics. "I would say that a community college course or associate's degree in electronics would be a very good step," Craig advises. Some large corporations, especially those in the aerospace industry, have their own schools and training facilities. The U.S. Armed Forces also provide training in avionics and electronics.

Along with some education in avionics and electronics, knowledge of aviation and the aviation field is important as well. Flying lessons may be a good idea, as you will gain more understanding of how planes function and how the equipment affects the planes. "If you're going to work in the industry," asserts Craig, "you have to understand the industry." Craig also indicates that flying lessons will demonstrate to a potential employer that you have initiative and an interest in the industry overall.

In an avionics program, some of the courses you may take include analog electronics, airframe, avionics line maintenance, radio fundamentals and FCC license preparation, digital circuits, and microprocessor fundamentals.

## certification

Two types of certification are available for avionics technicians, and generally speaking, neither is mandatory. Repair shops can be certified by the FAA as certified repair stations. These shops have a chief inspector who is authorized to sign off on other technicians' repairs and work. The individual technicians may receive FAA certification through their employer. For instance, Craig is the chief inspector at the Avionics Shop. If a technician were interested in obtaining certification as a repair person, Craig would write a letter of recommendation to the FAA indicating that the technician possesses the skills and qualifications necessary to work as an avionics technician for that particular shop. If the certified technician left the Avionics Shop to work for another employer, he or she would need to seek certification from the new employer.

Avionics technicians who work with radios, transmitters, or other communications equipment should obtain a license from the Federal Communications Commission (FCC). Applicants must pass a detailed written examination to obtain licensing. In recent years, the FCC has adjusted regulations in such a way that technicians who work for licensed avionics technicians do not need individual FCC licenses. It is a good idea, however, to seek licensing. As Craig notes, "It proves that you have the incentive to do this type of work."

## scholarships and grants

Technical schools and community colleges with avionics or electronics programs are good sources for scholarships. You should contact the financial aid office or your departmental advisor for information. Professional associations often provide scholarship opportunities as well. The Aircraft Electronics Association (AEA) awards twenty-one scholarships a year, ranging anywhere from $1,000 up to $16,000 (*see* "Sources of Additional Information"). Large manufacturers or companies in the aviation or aerospace industries may also provide financial awards. You should contact these companies directly or search for company home pages on the World Wide Web. These companies may also list their scholarships with relevant associations such as the AEA.

## internships and volunteerships

An internship may be required for students in avionics technology programs. The departmental advisor or student placement office will generally assist in locating internship opportunities and setting them up.

If you are not enrolled in school, however, internship opportunities may be a challenge to locate. Some large manufacturers or companies may have internship programs, and if they do not, you might want to contact them and inquire about volunteer possibilities. Small repair shops with fewer than ten employees generally do not have the resources to offer internships. Usually these shops are operated by the owner and may employ only one or two other technicians. It never hurts, however, to ask about volunteer opportunities. As Craig has indicated, initiative is the key in the avionics field.

## labor unions

There is no union specifically designed to represent avionics technicians. Union membership will depend on the employer. Small repair shops are generally not unionized, but avionics technicians who work for major commercial airlines may be required to join a union, such as the International Association of Machinists and Aerospace Workers.

## who will hire me?

Craig didn't have to look far for his first job in the avionics field. Although he worked for the family shop during his high school years, it wasn't until after college that he decided he wanted to pursue a career in avionics. "I went to college so I could do something else," he recalls, "but I came back to it."

The main employers of avionics technicians are commercial airlines, avionics equipment or aircraft manufacturers, and repair shops, both small, independent stations and large shops with several locations. The federal government may also have job opportunities. Technicians who work for manufacturers will not only perform repairs and maintenance, but they will also be responsible for assembling units. If you have a particular fondness for autopilot equipment or transponders, for example, you may wish to narrow your job search to companies that specialize in producing or repairing these units. Because avionics is a highly specialized field, however, avionics technician jobs are not particularly plentiful, and you shouldn't narrow your search too severely.

If you are in school, the placement office should be able to help you locate prospective employers. Schools frequently host job fairs where students can meet prospective employers and undergo job interviews. If you cannot take advantage of a placement program, however, many repair stations and manufacturers have web sites with job openings. The Internet is also an excellent tool for researching companies or airlines you may be interested in working for.

Craig believes that word of mouth is a good way to learn about job opportunities, since avionics is a close-knit industry. Maintaining membership in aviation clubs and professional organizations and attending meetings is one method for keeping in touch with others in the industry. You can also visit small shops to inquire about job opportunities. "If you went to one shop and they didn't have a position, they'd probably tell you who did," Craig notes. Craig also suggests sending resumes to all the shops in the area where you live. You can look through the classified ads of newspapers, but Craig indicates that small shops generally don't advertise because most are always seeking qualified, experienced technicians. If you are just starting out, relocation may be unavoidable.

Professional associations usually publish magazines or newsletters with classified ads. "The best place to look is probably in the AEA magazine *Avionics News*," Craig says. "They have more job opportunities in there than any other publication I can think of." Associations representing the general aviation industry may also list avionics technician opportunities (*see* chapter "Airplane and Helicopter Pilot").

## where can I go from here?

Craig plans to continue working in his family's repair shop, building the reputation of the shop. He hopes the Avionics Shop will someday be known as "the best shop in California." Craig is interested in doing some design work and looks forward to working with new, high-end equipment. "That's fun to work on because it's very complicated equipment," Craig says.

As avionics technicians gain experience and skills, they begin to work more independently, with only minimal supervision. Because most shops have fewer than ten employees and may be operated by the owner, there aren't many opportunities for advancement. Technicians can advance to *shop manager* and supervise the staff of avionics technicians and oversee the daily operation of the shop. Shop managers usually act as the chief inspector and sign off on technicians' paperwork. Many technicians who reach this level may leave the shop to open their own avionics repair businesses.

Avionics technicians who work for manufacturing companies may move into supervision or administrative positions, such as *regional manager*. The regional manager may travel and visit customers, installing and servicing equipment and providing training sessions. They may also solicit new business or oversee a team of sales representatives.

With additional education, avionics technicians can become *avionics engineers*. This position generally requires a bachelor's degree in engineering or electrical engineering. Avionics engineers plan and design equipment and systems, figuring out how units will best work together.

## what are some related jobs?

With their extensive background in electronics, avionics technicians can enter any number of related occupations. "If you weren't an avionics technician, you could fall back on just about anything that involves electronics," Craig believes. Examples he provides include electronics technicians, radio repair technicians, and surveillance equipment technicians.

The U.S. Department of Labor classifies avionics technicians under the headings *Occupations in Assembly, Installation, and Repair of Electronic Communication, Detection, and Signaling Equipment* (DOT) and *Craft Technology: Electrical-Electronic Equipment Repair* (GOE). Also under the heading *Craft Technology* are people who perform mechanical work, electrical-electronic systems installation and repair, and scientific, medical, and technical equipment fabrication and repair.

| **Related Jobs** |
|---|
| Communications equipment technicians |
| Line installers |
| Systems set up specialists |
| Electricians |
| Signal mechanics |
| Cable splicers |
| Telephone and PBX installers and repairers |

## what are the salary ranges?

According to a 1996 salary survey published in *Avionics News*, wages for entry-level avionics technicians working in the East averaged $10.35 an hour. Technicians working on installations made the least at $9.69 an hour, and those working on radar equipment earned the highest hourly wage at $10.78 for entry-level employees. Avionics technicians at the top of the pay scales earned an average of $17.41 an hour in the East. A salary survey by the magazine *Aircraft Maintenance Technology* indicates that avionics technicians earned the highest wages in the West with an average of $25.00 an hour in 1997. By contrast, technicians in the East earned $16.00 an hour.

According to the 1998–99 *Occupational Outlook Handbook*, the median annual salary of aircraft mechanics, which includes avionics technicians, was about $35,000 in 1996. The top 10 percent earned more than $48,000 a year, and the lowest 10 percent earned less than $23,200.

Craig believes the best pay for avionics technicians is in California, with an hourly wage of $24.00 as the highest a technician can earn. He feels the average pay is probably about $15.00 to $17.00 an hour, depending on the area. Small shops may not be able to provide medical benefits.

## what is the job outlook?

There will always be a need for skilled avionics technicians. Even though there are relatively few avionics jobs, Craig believes that if you are an experienced technician with plenty of initiative, you could walk into any repair shop in California and land a job. "It's hard to find somebody who is qualified," he indicates. Craig also feels this is a good time to become an avionics technician. Most of the skilled technicians are reaching retirement age, and there aren't enough younger avionics technicians to fill the void. "If someone got into this area," he believes, "they'd always have a job."

According to the 1998–99 *Occupational Outlook Handbook*, the job outlook is positive for avionics technicians and aircraft mechanics, with opportunities likely to be best at FAA-certified repair stations and smaller airlines. The commercial airline industry is usually affected by the state of the national economy. When the economy is strong, more people take advantage of air travel, both for business and pleasure. With more people flying, the demand for airplanes increases as well, creating job opportunities for avionics technicians and other employees in the aviation industry. Innovations and advances in technology will affect the avionics field positively as well. Avionics technicians are needed to assemble and test the new equipment, and they must also install and service these new units.

Many of the customers of small avionics repair shops are private citizens who own planes. Because flying is expensive, these customers generally have incomes that can support their flying activities. The avionics industry, therefore, is not as heavily tied to the national economy as other industries. Furthermore, many businesses depend on airplanes and helicopters to perform their work. For example, firefighters often rely on helicopters to help put out forest fires; search and rescue teams must rely on aircraft to locate and rescue injured individuals; and the police department and highway patrol use aircraft to track suspected criminals. Employees of these businesses rely on avionics technicians to keep their electronic equipment in top working order so they can communicate with headquarters, navigate the aircraft, and fly safely and accurately; without avionics technicians, these businesses would be out of work.

# how do I learn more?

## sources of additional information

The following organizations provide information on avionics technician careers, schools, and scholarship opportunities.

**Aircraft Electronics Association (AEA)**
4217 South Hocker
Independence, MO 64055
816-373-6565
http://www.aeaavnews.org

**Professional Aviation Maintenance Association (PAMA)**
636 Eye Street, NW, Suite 300
Washington, DC 20001
202-216-9220
http://www.pama.org

## bibliography

Following is a sampling of materials relating to the professional concerns and development of avionics technicians.

### Books

Collinson, R.P.G. *Introduction to Avionics.* Explains the basic principles underlying the theory of modern avionics systems and how they are implemented with current technology for both civil and military aircraft. Piscataway: IEEE Press, 1998.

Helfrick, Albert D. *Modern Aviation Electronics.* Paramus: Covers programs in both the civil and military sectors of aviation communication, navigation, landing, surveillance, and flight control systems. Prentice Hall, 1994.

Jones, Grady R. *How Airplanes Fly: A Flight Science Primer.* An introduction to avionics and aerospace technology. Dubuque: Kendall/Hunt Publishing Company, 1994.

Kayton, Myron ed. and Walter Fried. *Avionics Navigation Systems.* 2nd ed. A standard introduction to avionics and aerial navigation. New York: John Wiley & Sons, 1997.

### Periodicals

*AV Web.* Daily. Aviation news, reviews, training links, up-to-the-minute articles on new aircraft, and more. http:www/avweb.com.

*Overhaul & Maintenance.* Monthly. A leading maintenance-specific magazine in the aviational aerospace field. Aviation Week Group, McGraw-Hill Companies, 11 West 19th St., New York, NY 10011.

# barber/cosmetologist

**Definition**

Cosmetologists perform a wide range of personal services to improve the look of their clients' hair, skin, nails, and makeup. Barbers cut, shape, wash, color, and bleach hair, beards, and mustaches.

**Alternative job titles**

Beauticians

Beauty operators

Hairstylists

**Salary range**

$12,000 to $20,000 to $30,000

**Educational requirements**

Some postsecondary training

**Certification or licensing**

Required

**Outlook**

About as fast as the average

GOE 09.02

DOT 330 & 332

O*NET 68005A*

## High School Subjects

Anatomy and Physiology
Business
Health

## Personal Interests

Business management
Exercise/Personal fitness
Helping people: personal service
Selling/Making a deal

**It was** an awkward situation, to say the least. Colleen was visiting pleasantly with a client in her beauty salon when a second client arrived, and tension suddenly filled the room. "I had no idea they were both divorced from the same guy. And it wasn't like twenty years apart. It was like five years. That was kind of scary," she remembers with a rueful smile. "They both just sat there and got real quiet. I was almost done with one of them, and she got out of there, and the other one told me, 'Don't ever book us at the same time again.'"

One of the women was not angry about the encounter, although the other one was. Colleen kept her poise, handled the situation with tact, and managed to avoid losing either client. "I just talked about different things, like the weather," she recalls. "You have to be diplomatic and keep them both happy. In this business you have to be a doctor, divorce lawyer, counselor, psychologist."

Cosmetologists, like bartenders, are expected to be good conversationalists who can be trusted with confidences. "When one lady leaves," Colleen notes, "you don't want to talk about her to the next lady in the chair, or she'll think you'll talk about her. They told us in beauty school, don't talk about politics, religion, or the other customer."

## what does a barber or cosmetologist do?

An attractive appearance increases a person's self-esteem and can contribute to professional success. In the United States beauty salons and barber shops are big business, comprising one of the largest personal service industries.

*Barbers* cut, trim, shape, wash, style, tint, and bleach hair, and they trim and style mustaches and beards. They also fit hairpieces, give scalp treatments, facials, facial massages, and shaves, but they seldom care for skin or nails. They offer grooming advice and information about cosmetic products. Barbers work with razors, razor sharpeners, scissors, clippers, brushes, combs, tweezers, and hot towels. Some barber shops feature *shoe shine attendants, assistant barbers,* and *manicurists.*

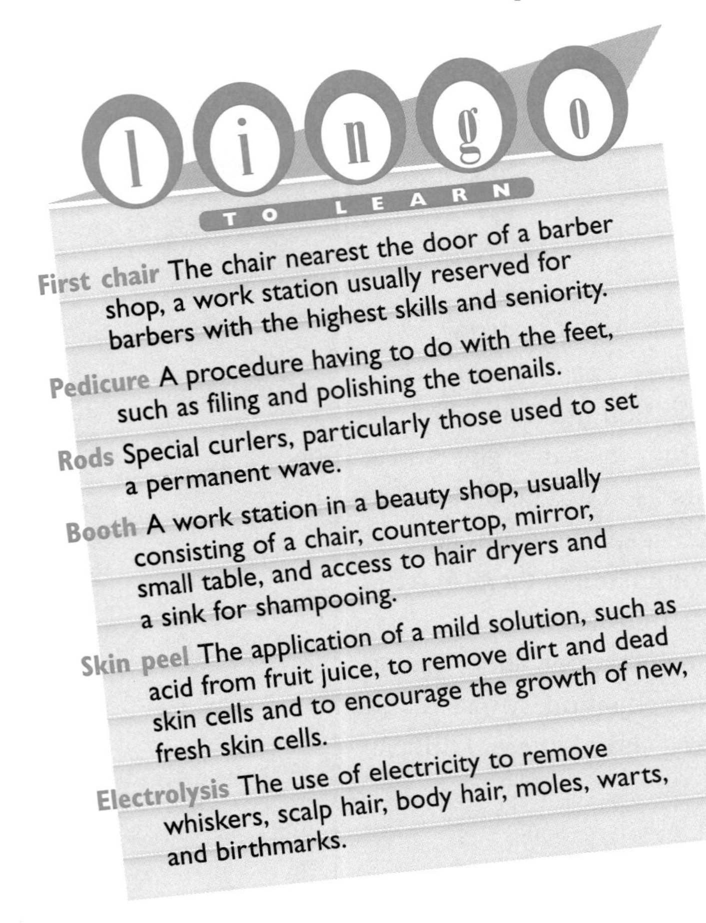


**First chair** The chair nearest the door of a barber shop, a work station usually reserved for barbers with the highest skills and seniority.

**Pedicure** A procedure having to do with the feet, such as filing and polishing the toenails.

**Rods** Special curlers, particularly those used to set a permanent wave.

**Booth** A work station in a beauty shop, usually consisting of a chair, countertop, mirror, small table, and access to hair dryers and a sink for shampooing.

**Skin peel** The application of a mild solution, such as acid from fruit juice, to remove dirt and dead skin cells and to encourage the growth of new, fresh skin cells.

**Electrolysis** The use of electricity to remove whiskers, scalp hair, body hair, moles, warts, and birthmarks.

*Cosmetologists* are usually licensed to perform all the services provided by barbers, with the exception of shaving men, and they perform some services that barbers do not. They help improve the appearance of their clients' hair, skin, and nails. They wash, condition, cut, curl, straighten, color, and style hair. They also give manicures and pedicures, apply scalp and facial treatments, shape eyebrows and eyelashes, apply makeup, clean and style wigs, and offer body care and skin care services. Some cosmetologists give massages, analyze the most effective type of makeup for each client, and offer advice about what products to use and how to apply them. Those who specialize as *hairstylists* may focus on an area such as permanent waving, cutting hair, or styling hair in the most difficult fashions. Cosmetologists work with scissors, razors, brushes, combs, curlers, clippers, manicure equipment, hair dryers, reclining chairs, and cosmetic aids.

Some cosmetologists cut only children's hair, but usually their clientele includes people of all ages, both women and men. There are many areas of specialization within the field of cosmetology. *Nail technicians,* also known as *manicurists,* are experts in the care of nails on hands and feet. They trim cuticles, shape nails, apply polish, decorate nails, and attach artificial nails, sometimes using linen or silk wraps. *Estheticians* focus on skin care, body care, and makeup. An esthetician might help clear up acne, for example, by applying a mild acid solution, such as an alpha hydroxy product, to remove dead skin cells. Estheticians sometimes work with *dermatologists* and *plastic surgeons,* who perform cosmetic surgery and other advanced procedures.

Both cosmetologists and barbers must ensure that their tools and surroundings are clean; some items must be sterilized. In small shops they schedule appointments and clean equipment and their work spaces. They also sell a variety of shampoos, conditioners, and other products designed to enhance a person's appearance. To improve their understanding of these products, they read trade magazines and other publications, and they attend trade shows, seminars, and other events where products are presented.

A large number of people in this profession are self-employed and must be proficient in business administration. That involves paying bills, keeping financial records, making sure equipment is in working order, purchasing supplies, and perhaps hiring, firing, and paying employees.

## what is it like to be a barber or cosmetologist?

Colleen Lyon has been in the cosmetology business for thirty-two years and now operates a beauty salon in her home in Missoula, Montana. Her work room is equipped with chairs that swivel and recline, hair dryers, curling irons, special sinks for washing hair, and an array of curlers, brushes, hair sprays, scissors, and other tools and products. Colleen performs a variety of services for men, women, and children, but hair care is her main focus. She applies conditioners and shampoos, and almost every day she colors someone's hair—a process that can involve anything from a total change of color to applying a "frost" or "streaking" to create the illusion of highlights. "I do a lot of hair cuts. I like it when I have a lot of permanent waves," she says.

Colleen usually begins work by about eight-thirty or nine in the morning and tries to finish by five or six in the afternoon, but sometimes she's still working at eight in the evening. "I try to work around the customer's schedule," she explains. She occasionally works on Saturdays but never on Sundays. Sometimes she performs her services for clients in their homes, most often for elderly women.

She has built up a loyal clientele over the years. Many of them come to the shop to have their hair done every week, and some have been coming for twenty years, following her each of the three times she has moved the business to a new location.

"In this business you never know from week to week what your income is going to be," she remarks. "You stagger them out so you're busy all the time. And you're trying to build your clientele. Word of mouth is the best way to get clients." It's important for a beautician to build up a clientele; if you're self-employed, you need to know you can count on certain people to patronize the business, and if you're employed at a salon, you'll probably be receiving commissions as part of your pay.

Colleen was an employee in several salons before she opened her own business, years ago. She enjoyed interacting with other beauticians but prefers working alone, because she is in charge here. She says, "The big reason I like running my own salon is you don't have the conflicts with other operators, the race to the telephone for appointments," and other disagreements. "I spent a lot of years in the three salons I worked in," she recalls, adding that she was not very happy with her last employer. "The hours she made me work, and her attitude, were what made me get my own salon. I took some business schooling, and my sister owned a salon. We sat down one night and did some [financial planning] books. Most of it is common sense and treating the customer with respect."

Now that she has her own shop, Colleen keeps in touch with developments in her field by reading how-to books and magazines such as *American Salon* and *Modern Salon*. She says she learns tips on colors, cuts, and the latest trends there, but many of the styles in magazines of that type are not very helpful for her, since they're too unusual for Montana. She isn't worried that she won't be able to style hair to suit her clients, though. "There's only so far they can go with hair," she comments, "so it seems like it makes a big revolution" from straight to curly, natural to wild, and back again every few years. A beautician who knows the basics, she says, can handle most trends as they become popular again.

## have I got what it takes to be a barber or cosmetologist?

A cosmetologist or barber needs to be pleasant and friendly but able to maintain a professional demeanor. Creativity is important, since customers will rely on you to help them choose styles best suited to their individual looks, and some will want new, trendy fashions. You should enjoy working with hair, skin, and nails; this profession revolves around personal grooming services that require close contact with clients. The job requires patience, because customers can be difficult.

"Sometimes I get tired of some of the people," Colleen says, "but I just try to overlook it. You have to be a person that likes people. If you're quiet and shy, you won't build up a clientele."

Since Colleen operates her own salon, she has been able to tailor much of the work to suit her personality. "I pretty much set my own hours," she notes. "When you get your own shop, you can do that." The down side is that she sometimes is tempted not to work diligently, since she has no supervisor. "No matter what job you have, you have to have discipline," she remarks.

Another beautician in Missoula,. Judy King, says that patience and communication skills are essential for a cosmetologist: "The biggest challenge is people who can't be satisfied. It's a challenge every day. You have to be real clear what they want. But it's fun," she's quick to add. Any cosmetologist has to communicate with customers to decide what tasks should be done. For example, it's important not to misunderstand how long a woman wants her hair to be. Once the hair is cut, an error cannot be remedied. The key, Judy says, is to discuss everything ahead of time and be sure you know what the customer is expecting.

Judy manages The Salon, a beauty shop within a large drugstore. The cosmetologists at The Salon stay busy throughout the day; they pierce ears, manicure nails, color and cut hair, sell retail products, and offer many other services. "You do all those things in the same day. We just give 'em whatever they want," Judy says. Her typical day, she adds, is: "Hectic! You get all types of people. Some of the clients have such a sense of humor. That's my favorite part."

Like Colleen, Judy King grew up in western Montana and earned her credentials in Missoula, but at Mr. Rich's Beauty College. She had raised two children and had been a seamstress, but when the work began to damage her hands, she changed careers. "I was already trying to cut my kids' hair, and I was always interested in it. It was something I knew I'd enjoy going into, even though I knew I wouldn't make much money," she says. Judy has been a cosmetologist for seven years and has worked at The Salon for about two years.

## how do I become a barber or cosmetologist?

Colleen recalls, "I fixed kids' hair all through high school and grade school. After I got out of school, I hated books, and I liked making people look good. I still do. Everybody said I have a natural knack for doing hair." Colleen wanted a career but did not want to go to college, so she attended Modern Beauty School in Missoula for about a year and a half to become a cosmetologist. "In Montana you have to go to school longer and learn more than in some other states. You get more practical experience and book learning," she notes.

To be a successful barber or cosmetologist, you should

- Have good finger dexterity
- Have patience
- Be pleasant and friendly
- Be professional
- Be creative
- Enjoy working with hair, skin, and nails

## education

### High School

An eighth-grade education is a prerequisite for entry into most cosmetology schools, but a high school education is recommended and is required in some states. A high school education and, in many cases, a few more years of advanced schooling are required for advancement to some positions in the field. Some states require only a high school or even an eighth-grade education for barbers, but most require completion of barber school or an apprenticeship.

To prepare for a career in cosmetology or barbering, take science classes, particularly biology, physiology, and chemistry. These courses will help you understand how hair and nails grow, for example. Chemistry is important, because you'll be working with permanent wave solutions and other chemical products. Business courses, including bookkeeping, accounting, and marketing, will prove helpful if you ever open your own beauty shop or manage a salon. To deal with the wide variety of customers, Colleen suggests, "A psychology class might be good."

### Postsecondary Training

To practice cosmetology in any state or barbering in most states, you must complete special training. In some states separate schooling is required for estheticians, manicurists, and electrologists; this training usually takes much less time than barber school or cosmetology school. Nationwide, there are about four hundred private barber colleges and nearly four thousand cosmetology schools. Training is also available through vocational training programs. Before enrolling in any school, be sure it meets the requirements of your state.

Programs at barber schools usually take nine months to a year and provide one thousand to eighteen hundred hours of instruction. Course work often includes the use of tools and equipment, sanitation, skin and scalp diseases, the craft of barbering, business management, sales, advertising, ethics, and advanced hair coloring and styling.

Training for cosmetologists varies from state to state but is usually one thousand to two thousand hours of study that takes six months to about a year; in some states the program can take up to a year and a half. Some public training school programs last two to three years, because they also feature academic studies. Most programs include textbook studies in cosmetology and hands-on experience with mannequins, live models (including other students), or actual clients who receive reduced prices in exchange for allowing students to practice on them under close supervision.

Training for cosmetologists usually includes studies in the bone structure of the head, the nervous system, and other aspects of human anatomy. Students learn to cut, wash, color, and style hair; give facials, manicures, and pedicures; and apply makeup. They study skin care, body care, nail technology, chemistry, physiology, bacteriology, hygiene, sanitation, business, and applied electricity. Some knowledge of scientific subjects is necessary to understand how the skin, hair, nails, and other systems function, but the training in these areas tends to cover only basic facts. "Most of the chemistry classes at beauty school were very simple," Colleen comments.

## certification and licensing

All states require that barbers and cosmetologists be licensed. Some require separate licenses for estheticians and manicurists. In some states applicants must pass an examination to become junior cosmetologists, then work in the field for one year and pass a second examination to become senior cosmetologists.

When Colleen graduated from beauty school, she received an operator's license that allowed her to work under the oversight of a licensed operator for one year. She now has a shop license, which is required for her to operate her own beauty salon, and a manager-operator license, which is required for her to practice cosmetology. When her shop was within the city limits, she also had a city business license.

If you entered this profession and ever moved to another state, you might need additional testing or a refresher course to obtain a new license, but some states recognize the licenses of cosmetologists and barbers from other states. In some states licenses must be renewed annually.

> *"You have to be a person that likes people. If you're quiet and shy, you won't build up a clientele."*

## internships and volunteerships

In most states barbers must complete apprenticeships for a year or two before they can be licensed as journeyman barbers. Some cosmetology apprenticeships are also available, but few people enter the profession this way. Most cosmetology schools include hands-on experience, often in a beauty shop operated by the school. In some barber shops and beauty salons you can work a summer job, cleaning the shop and running errands; this would allow you to observe barbers at work and decide whether the career interests you.

## labor unions

Cosmetologists and barbers sometimes belong to labor unions, most often the United Food and Commercial Workers International Union. Sometimes a union can help its members find employment.

## who will hire me?

Colleen found her first job as a cosmetologist immediately after her graduation from beauty school. Her teachers put her in touch with a prospective employer, she was interviewed for the position, and at that time she was asked to cut a customer's hair as a test of her skills. The customer was surprised, and Colleen admits, "I was a little nervous, but I did a good job." She worked in that salon for four years.

Many barber colleges and cosmetology schools help their graduates find work. Local union offices sometimes also refer union members to employers. Newspaper advertisements, personal referrals, and state, city, or private employment services can also provide information about job openings. If you decide to approach shops directly by dropping off resumes, which is probably the most common way to find a job as a cosmetologist or barber, Colleen advises, "Don't call on the phone! Go in person and talk to different salons. Look nice. Ask if they have time to talk to you later, and set up an appointment." Colleen says she finds it annoying when applicants call her salon and ask, "Are you hiring anybody right now?"

Barbers and cosmetologists may work in small shops with one or two operators or in large salons with perhaps a dozen employees. You might find a job in a combination barber-and-beauty shop, or you could work in a beauty shop in a hospital, retirement home, nursing home, department store, drugstore, resort, or hotel. Some cosmetologists demonstrate cosmetics and hair styles in retail stores, fashion centers, photographic centers, and television studios.

## where can I go from here?

Colleen has achieved her goal of operating her own salon, and she has no plans to pursue other options. Opening your own shop requires experience, an understanding of business practices, startup money, and a clientele that will follow you to your new location. Highly successful entrepreneurs sometimes operate a chain of salons.

Judy King also worked her way up from an entry-level job as an employee in a salon, but instead of opening her own shop, she has become the manager of a salon with a number of employees. Whether you're working as an employee or renting a work station in someone else's salon, she says, the key to success is to build a clientele of loyal customers, a process that usually takes two to three years. A promotion into management will follow naturally if the salon's owners like your work, she adds: "If they know you've got the attitude and potential, it's easy to slide right in. When you have a good crew, it's not all that hard."

Since there are few management positions within small shops, cosmetologists and barbers with experience often move on to larger, more attractive shops with better equipment and more chances for promotion. They sometimes continue their education through graduate school and by attending functions sponsored by trade associations, such as professional fashion reviews. With knowledge from events of that type or from training programs offered through large cosmetic houses, you could be a demonstrator or sales professional in the cosmetic department of a large store. Cosmetologists with advanced skills can become beauty editors for magazines and newspapers or style hair and apply makeup for television personalities. Some beauticians specialize in cosmetology work for mortuaries. Some estheticians become makeup artists for motion picture studios or television studios.

"The one thing I've thought it would be fun to be is a beautician on a cruise ship. That's a type of thing that younger girls just going into the business might want to consider," Colleen suggests. Those positions are available only to unmarried cosmetologists within a certain age range.

## what are some related jobs?

Judy King says the best way to earn a significant amount of money in this field is by branching out into related areas. "If you're working for someone else, you're pretty much at a standstill. The smart thing to do is to go into different avenues of the business. I encourage all of the people who do this for a living to take as many classes as they can," she says.

Judy attends seminars and other events to learn about the latest styles and products, because she works in her spare time as a representative for companies that make cosmetics and personal care products. She says there many opportunities to earn thousands of dollars on weekends and days off and still work full-time or part-time in a salon. You could teach people about a company's products at educator shows. You could be a platform artist at hair shows, demonstrating on stage how to style hair in the latest fashions. You could be a company representative who sells products to beauty salons within a given territory. "But you've got to keep your license current," she warns, "or you don't qualify for anything."

The U.S. Department of Labor classifies barbers and cosmetologists under the heading *Hospitality Services: Barber and Beauty Services* (GOE). Also under this heading are people who perform electrolysis and give therapeutic massages.

| **Related Jobs** |
|---|
| Beauty editors |
| Beauty consultants |
| Makeup specialists |
| Wig specialists |
| Beauty supply distributors |
| Fashion models |
| Massage therapists |
| Electrolysis technicians |
| Health club directors |
| Radio and television stylists |

## what are the salary ranges?

Earnings in this field vary widely, depending on the employer, the employee's skill and experience, economic status of the clientele, and whether the shop is in a rural or urban area. Workers in exclusive salons in cities often earn much more than the average, particularly if they are experts or specialists. Many workers in this profession receive no fringe benefits; those who work for nursing homes, department stores, beauty salon chains, and other large organizations are more apt to receive benefits such as health insurance and paid vacations. Some operators are paid only a percentage of the amount their clients pay the salon; others earn a base salary plus commission. Others are paid by the hour. Tips are an important and unpredictable source of income in this profession.

Judy King says she knows cosmetologists who are making almost $400 a day, but most start at minimum wage plus commission and gradually earn more as they gain experience. Being self-employed, she explains, is an important way to increase your earnings, but only if you have your own shop or branch out into related areas. Judy says cosmetologists who work in smaller salons in Montana are usually self-employed, paying weekly rent for a work station, and making somewhere near the minimum wage.

Colleen points out that if you're self-employed, you keep 100 percent of what your clients pay, but you have to buy supplies and meet other expenses, including self-employment tax. "There is no retirement fund and benefits like that, unless you do it yourself. That's one of the drawbacks," she says. Colleen is an accomplished painter who helps make ends meet by using her salon as a gallery to display art for sale.

According to the U.S. Bureau of Labor Statistics, the median annual income for barbers and cosmetologists in 1996 was $15,080, much lower than the $25,480 median for all workers. (It's important to remember that many professionals in this field work only part-time, however.) According to the *Encyclopedia of Careers and Vocational Guidance*, experienced barbers earn $20,000 to $30,000 annually; most hairstylists and barbers with their own shops earn at least $30,000; the average cosmetologist earns $20,000 to $25,000; and beginning cosmetologists earn $12,000 to $13,000.

## what is the job outlook?

Employment for barbers is expected to decline, but employment for cosmetologists is expected to grow about as fast as the average for all occupations through the year 2006. The demand for services will increase as the population grows, as more men patronize salons, and as working women seek out cosmetic services more frequently. A particularly strong demand is expected for manicurists, for cosmetologists trained in nail care, and for workers who can perform a variety of services. Currently, there are not enough cosmetologists to fill the available openings.

The market is expanding in new shopping centers, suburban areas, and full-service shops. Colleen notes a new trend toward salons that offer a range of services, everything from manicures, pedicures, and body wraps for cellulite to tanning booths and even attached fitness centers: "I think the career itself is getting more into a healthy, total-body look instead of just hair and nails. It's changed that way in probably the last twelve years. Facials and hair color seem to be the trends right now." She notes that the change is good for business, because it keeps customers interested and encourages them to visit salons more often.

## how do I learn more?

### sources of additional information

Following are organizations that provide information on barbering and cosmetology careers, accredited schools and scholarships, and possible employment.

**American Association of Cosmetology Schools**
901 North Washington Street, Suite 206
Alexandria, VA 22314
703-683-1700
http://www.beautyschools.org/
ronsmith@ix.netcom.com

### Advancement Possibilities

**Salon Managers** purchase and oversee the stock of supplies, make sure equipment is operating properly, set appointments, keep financial records, and ensure that the salon meets legal regulations.

**Salon Owners** may employ other cosmetologists or perform all the work in the salon. The owner oversees the business, assuming a great deal of risk and responsibility but possibly making a sizeable profit.

**Cosmetology or barber school instructors** teach at cosmetology and barber schools and vocational training schools. These positions typically require experience in the profession, along with a general education that includes some college.

**National Accreditation Commission of Cosmetology Arts and Sciences**
901 North Stuart Street, Suite 900
Alexandria, VA 22203
703-527-7600
http://www.naccas.org
naacas@erols.com

**National Cosmetology Association**
3510 Olive Street
St. Louis, MO 63103
314-534-7980
http://www.nca-now.com
nca-now@primary.net

## bibliography

Following is a sampling of materials relating to the professional concerns and development of barbers and cosmetologists.

### Books

Aucoin, Kevyn. *The Art of Makeup*. Beauty tips, step-by-step instructions, sample makeovers, basic makeup combinations. New York: Harper Collins, 1996.

Batson, Sallie. *Great Hair!: Your Complete Hair Care Guide and Styling Guide*. Fully illustrated, expert advice on total hair care and styling. New York: Berkley Publishing, 1995.

Begoun, Paula. *The Beauty Bible: From Acne to Wrinkles and Everything in Between*. Features specifics on caring for 16 different skin types. Seattle: Beginning Press, 1997.

Lamb, Catherine. *Milady's Life Management Skills for Cosmetology, Barber-Styling, and Nail Technology*. Vocational guidance for the aspiring beauty professional. Albany: Milady Publishing Corporation, 1996.

Rudiger, Margit and Renate Von Samson. *388 Great Hairstyles*. An up-to-date compendium of styles. Northampton: Sterling Publications, 1998.

### Periodicals

*American Salon*. Monthly. Covers the latest advances and trends in cosmetology. Advanstar Communication, Inc., 7500 Old Oak Blvd., Cleveland, OH, 44130-3369.

*Cosmetic/Personal Care CPC Packaging*. Bimonthly. Magazine for packaging decision-makers in the cosmetic and personal care industries. O & B Communications Inc., 3 Paoli Plaza, Suite A, Paoli, PA, 19301.

*Cosmetic World Online*. Monthly. Comprehensive coverage of the cosmetics trade. http://www.cosmeticworld.com.

*Cutting Edge*. Monthly. A rundown of noteworthy events in the haircutting world. http://www.hairnet.com.

*Elle*. Monthly. Covers the fashion industry. Hachette Filipacchi Magazines Inc., 6133 Broadway, New York, NY, 10019.

*Hot Locks*. Daily. Index of hair resources and information. http://www.hotlocks.com.

*Professional Beauty*. Monthly. Covers variety of beauty topics from hair removal to aromatherapy. Professional Beauty/LNE, Fairfax House, 461-465 North End Road, Fulham, London, SW6 1NZ.

# biomedical equipment technician

**Definition**

Biomedical equipment technicians install, maintain, repair, and calibrate biomedical equipment used in hospitals, clinics, and other medical or laboratory facilities.

**Alternative job titles**

Biomedical electronics technicians

Biomedical engineering technicians

Biomedical instrumentation technicians

Clinical engineering technicians

**Salary range**

$26,000 to $35,000 to $51,800

**Educational requirements**

Associate's degree highly recommended

**Certification or licensing**

Voluntary

**Outlook**

About as fast as the average

GOE 05.05.11

DOT 716

O*NET 85908

## High School Subjects

Anatomy and Physiology
Chemistry
Mathematics
Physics

## Personal Interests

Figuring out how things work
Fixing things

**Responding** to a request for assistance, Brent Doyen walks into the operating room to find a roomful of eyes staring in his direction. The surgeon states, "Mechanic, fix the machine. This patient is going to die. Correction: the patient is dying. Fix the machine." Brent is then steered toward the faulty equipment—an aortic balloon pump designed to assist the heart is not functioning correctly. Brent can't very well take the machine apart in the operating room, and this leaves only one other option. "I'll be right back," he says. He exits the operating room, runs down two floors, grabs a new set of cables, and rushes back up. When he runs into the operating room, everyone is still staring at him. He plugs one end of the new cables into the machine, and it starts pumping as it should. He breathes an internal sigh of relief and leaves the operating room as the surgery team hurriedly resumes the operation. Brent heads back down to the equipment shop. Saving lives is all in a day's work for a biomedical equipment technician.

## what does a biomedical equipment technician do?

*Biomedical equipment technicians* (BMETs) are responsible for the maintenance, installation, calibration, and repair of biomedical equipment, electronic equipment designed to diagnose and treat medical conditions. This equipment may include anesthesiology machines, cardiac monitors, infusion pumps, defibrillators, radiology equipment, and ventilators and may range in size from a handheld unit to a machine that takes up an entire hospital room. BMETs, with their highly specialized training in electronics, are an important link between technology and medicine.

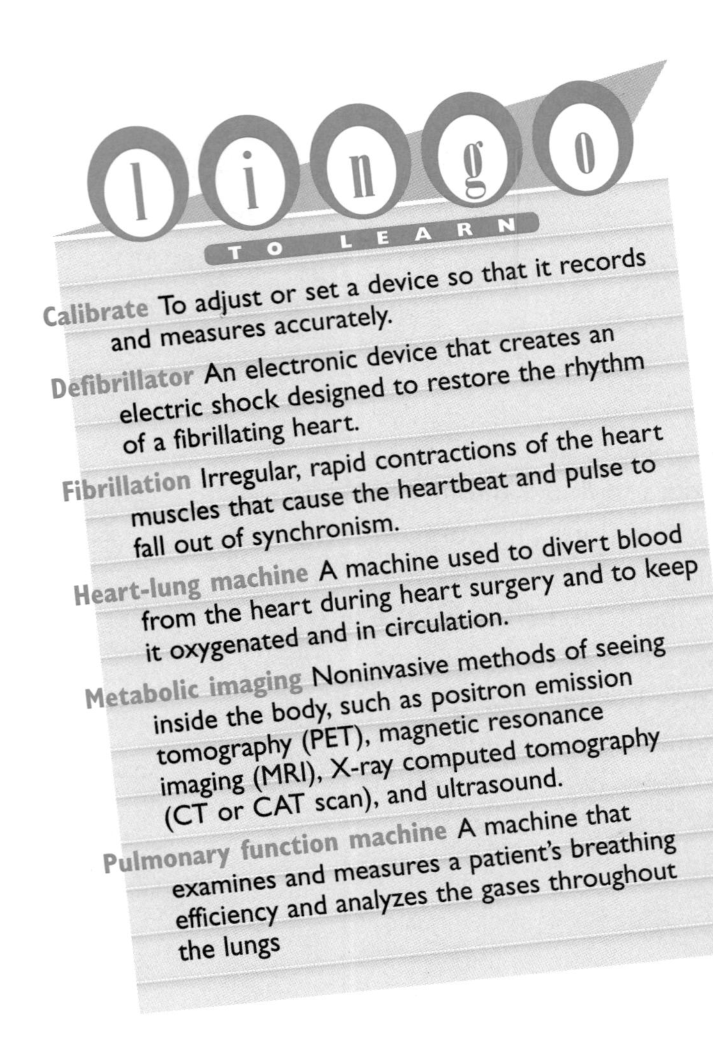


Biomedical equipment technicians spend a considerable amount of time on preventive maintenance. All biomedical equipment must undergo regularly scheduled preventive maintenance checks to ensure that everything is in top working condition. Maintenance includes cleaning and calibrating, or adjusting the machine so it works in a standardized manner, and conducting operational verification tests to make sure the machine is operating as designed. Technicians may test circuits, clean and oil components, and replace worn parts. BMETs use tools such as voltmeters, oscilloscopes, spectrum analyzers, and computers to make sure equipment is functioning properly. Detailed records of all preventive maintenance checks must be completed and kept by biomedical equipment technicians as well.

When equipment malfunctions, it's the job of the biomedical equipment technician to diagnose the problem and repair it. The BMET must determine whether the problem is due to operator error or whether the equipment is in need of actual repair. If the machine or instrument must be repaired, the BMET may refer to product manuals, test the equipment to try to pinpoint the problem, or speak with manufacturers about possible causes. The problem may be as elementary as a loose wire or as major as a defective motor. BMETs often take apart equipment and replace or repair parts, such as transistors, switches, or circuit boards.

Installing or upgrading equipment is also the responsibility of biomedical equipment technicians. BMETs follow manufacturer's guidelines to set up machinery then inspect and test it to make sure it complies with safety standards. The technician often trains those who will operate the equipment, such as nurses, doctors, and other health care personnel, and answers questions regarding equipment usage.

*Biomedical equipment technician I* is a junior-level or entry-level technician. These technicians generally work under heavy supervision, and the majority of their work is maintenance oriented. They are also capable of carrying out basic repairs on less-complicated equipment such as infusion pumps or defibrillators.

The *biomedical equipment technician II* is a senior-level technician and evenly splits time between preventive maintenance checks and repair work. These technicians work on equipment that is technically more demanding than the machinery entry-level technicians repair, including radiology equipment, laboratory analyzers, which involve robotics, pneumatics, and hydraulics, and anesthesiology equipment. Senior-level technicians may also oversee the installation of systems, such as nurse stations or heart monitor systems, while the junior-level technicians perform the physical tasks of installation. Technician IIs also evaluate new equipment and make purchasing recommendations.

The biomedical equipment technician who specializes in a particular area or type of equipment is known as a *biomedical equipment technician specialist.* Areas of specialization can include the catheter lab, pulmonary function machines, ultrasound, respiratory care, or X-ray equipment. Specialists must gain a solid foundation in biomedical equipment technology before focusing on a specialty. Many biomedical equipment technicians work in hospitals or clinics, taking care of all the equipment needs. Other BMETs work for third-party companies or manufacturers. Those working for manufacturers service and install equipment made by the manufacturing company. Frequent travel may be involved, and some technicians may be assigned a region covering several states.

"You shouldn't be in this job if you can't handle stress. You're quite often the front line for life support."

## what is it like to be a biomedical equipment technician?

The first thing Brent Doyen, clinical engineering supervisor, does when he arrives at work at St. Joseph's Medical Center in Tacoma, Washington, is to check the work orders that have been generated over the course of the night. The orders are prioritized according to type of equipment and urgency. For example, life support equipment will take precedence over a piece of equipment that is not in use or not being used to keep someone alive.

Once the work orders are handed out, the biomedical equipment technicians disperse to take care of repairs and routine preventive maintenance checks. Smaller equipment may be brought back to the shop area, while the large machinery will stay put. Brent and the other technicians wear pagers and frequently respond to calls throughout the day. "You stay pretty darn busy," Brent admits.

At St. Joseph's, preventive maintenance checks are conducted on a monthly basis and require thoroughness and precision. When Brent checks a heart monitor, he first walks into the room and observes the overall condition of the monitor—is the screen brightness at an acceptable level? Can he read everything clearly on the monitor? Is it in focus? If the monitor passes this initial check, Brent conducts an operational verification test. This entails hooking up a test device called a chicken heart to the machine. The test device simulates heart rate, blood pressure, temperature, and cardiac output and allows Brent to make sure the machine is functioning within specifications.

Brent then removes the heart monitor from the power source and prepares to examine the inside of the machine. He takes off the cover and cleans the inside, which attracts quite a bit of dust due to the heat of the components. "That's one of your biggest enemies when it comes to electronics," explains Brent. "The dust and the dirt generate heat and cause components to overheat and burn out. So we keep them clean." Brent also checks the power supplies for proper voltage levels, examines the wires to make sure none are loose, and looks for any signs of wear, such as discoloration of components. If everything looks in order, Brent reassembles the unit and performs an electrical safety check, which tests for leakage of current, resistance, and line voltage using an electrical safety analyzer. "Any time we open up a piece of equipment," Brent says, "the last thing we do after it's all completed, before we give it back, is electrical safety." The entire preventive maintenance check can take anywhere from thirty minutes to an hour to complete.

If the equipment does not pass all of the tests and is determined to be faulty, Brent removes it from service and transports it to the shop. He troubleshoots and evaluates the unit to determine the cause of the problem. Brent tries to narrow down the problem to a specific circuit board or component and then must decide how to remedy the problem. "You have to constantly be thinking about what you're doing," says Brent. "Is this cost effective to put my time into this to try and repair this, or is it more cost effective to just buy a board? That's the big question—which is the best way to do this so it's cost effective?" As long as Brent considers the most economical approach to fixing problems and keeps the biomedical equipment in top working condition, his employer is happy.

Although Brent generally works a forty-hour workweek, he is on call on a rotational basis and may have to report to the hospital on weekends or holidays. He may also have to stay late on occasion to repair equipment that is needed for the following morning, but this doesn't happen very often, and Brent feels the hours balance out.

## have I got what it takes to be a biomedical equipment technician?

If you are interested in becoming a biomedical equipment technician, you should have technical aptitude for working on a variety of electronic equipment. You must also be detail oriented, enjoy working with your hands, and have excellent troubleshooting skills. Stamina and patience are also important, and you must be able to see projects through to the finish—there are times you may be stumped by a problem, but you need to persevere and follow through.

Although biomedical equipment technicians are trained to fix and service electronic equipment, they must also communicate and work with others, so people skills are crucial. You have to be adept at listening to others as they explain problems with machinery, and you need to be able to communicate clearly and tactfully when you are training people or correcting operator error.

"You shouldn't be in this job if you can't handle stress," Brent adds. "You're quite often the front line for life support. When the equipment fails, you're it." A person's life may depend on whether or not biomedical equipment is functioning properly, and occasionally you may be called upon to repair life-sustaining machinery on the spot, so you must be able to work under pressure. And if you have a weak stomach, you might want to consider another job. "You do see a lot of blood," Brent says. "There's equipment that has, shall we say, high-protein substance on it?" You may also be exposed to hazardous substances, including chemicals and blood, so you must be careful and take precautions.

If you can handle the pressure and enjoy working with electronic equipment, however, biomedical technology can be very rewarding. One thing Brent particularly enjoys is the continuing education. "You're staying current with technology," explains Brent. "As technology advances, so does your education. You're always learning something new." He also enjoys the nationwide camaraderie with other biomedical equipment technicians and notes, "It's kind of like we're a big family."

**To be a successful biomedical equipment technician, you should**

- Enjoy working with electronic equipment
- Have problem-solving skills and be detail oriented
- Possess communication skills and enjoy working with others
- Have a professional attitude
- Be able to follow through on projects
- Have a calm temperament and a strong stomach

## how do I become a biomedical equipment technician?

Brent was an electronics technician before he became a biomedical equipment technician. He worked on repairing and servicing amusement games and equipment, such as video games and jukeboxes. Brent had been working on a pinball machine in a bar one evening. "I came across a person who was having a beer watching me work on stuff, and he said, 'Hey, I do the same thing you do, but I get paid more.' It turns out he was a biomed tech," Brent recalls. The prospect of earning more money and being able to work indoors instead of moving pool tables in the snow appealed to Brent, and he looked into biomedical equipment technology.

## education

### High School

If you're thinking about becoming a biomedical equipment technician, it's never too early to start preparing. In high school, you should take mathematics classes as well as science courses. Brent advises, "I would recommend taking any kind of electronics classes offered in high school, as well as math classes." He stresses that students shouldn't be scared off by the math—if math is not your forte, don't worry. Math can help your understanding of electronic processes and equipment, but it is not crucial to be a successful biomedical equipment technician. Shop classes can help you develop skills working with various tools, and if an electronics shop class is available, you should definitely enroll.

Computer science classes are helpful as well. As biomedical equipment becomes increasingly computerized, having an understanding of how computers function is important. Health science classes will acquaint you with medical terminology and basic anatomy, both very important in the realm of the BMET. Not only must you understand the electronic equipment, but you must also know how the equipment affects or works with the patient.

If there is an opportunity to join an electronics club in your school or community, you should. Many high schools also participate in statewide or nationwide technical or science fairs, which give students an opportunity to build various objects and compete against other schools. These fairs are an excellent opportunity for you to gain some experience seeing projects through to the end, working with hand tools, and troubleshooting.

### Postsecondary Training

Although a college degree is not absolutely mandatory to become a biomedical equipment technician, it is highly recommended, and many employers list a degree as a hiring requirement. Brent believes, "You would definitely have to have an associate's degree in biomed to pursue a good job." According to the Association for the Advancement of Medical Instrumentation (AAMI), there are currently sixty-five accredited two-year programs in biomedical technology offered in the United States. These two-year programs are available at both community colleges and technical schools. Training is also available through the Armed Forces.

**Advancement Possibilities**

**Biomedical engineers** design medical apparatus, including pacemakers, artificial organs, and ultrasonic imaging devices, by applying engineering principles.

**Clinical engineers** design and evaluate biomedical systems and are involved with technology management.

**Regional service managers** represent manufacturers or third-party companies. They supervise field offices and teams of technicians and may also develop customer relations and provide training to customers.

A two-year degree in electronics is sometimes acceptable, but because biomedical technology is rather specialized, it is preferable to find a biomedical technology program. Brent already had an electronics degree when he decided to pursue biomedical technology. He thought he could waive some of the classes in the biomedical technology program at Spokane Community College in Spokane, Washington, but decided not to after speaking with the advisor. Brent is glad he decided to start from scratch when he entered the program, explaining, "Those classes are something you definitely have to be dedicated to. We started out with twenty-four students in the first year. By the time we finished the second year, there were only ten of us left. The courses that he covers are very in-depth, and there's no time for monkeying around."

Courses in biomedical technology programs can include safety, including hospital and patient safety, medical terminology, medical instrumentation, physiology, circuits and devices, and digital electronics.

## certification

Certification is generally not required, but some institutions only hire certified biomedical equipment technicians. At Brent's workplace, you cannot become a senior technician without certification. Brent also believes that certified technicians command higher wages and that certification is important for the field. "It's a way for the biomedical community to police themselves."

Operating under the direction of the International Certification Commission for Clinical Engineering and Biomedical Technology (ICC), the Board of Examiners for Biomedical Equipment Technicians, which is affiliated with AAMI, maintains the certification programs. Certification as a certified biomedical equipment technician (CBET) can be attained after passing a rigorous examination and meeting the education and experience requirements. The candidate must have an associate's degree and/or proper work experience to meet the eligibility requirements. The examination tests the applicant's knowledge of anatomy and physiology, safety in the health care facility, electricity and electronics, medical equipment function and operation, and medical equipment problem solving. Two areas of specialization are also available: the certified radiology equipment specialist (CRES) and the certified clinical laboratory equipment specialist (CLES).

## scholarships and grants

Technical schools and community colleges with biomedical technology programs may have scholarship opportunities available. These schools may also have general scholarships open to the entire student population that you may wish to explore. Contact the financial aid office or your department advisor for further information.

Other avenues to investigate include professional associations involved in the biomedical or health care fields, manufacturing companies, or large health care organizations. Companies that manufacture medical instruments and equipment may offer scholarships to aspiring biomedical equipment technicians. Professional associations such as AAMI may either sponsor scholarships or provide lists of award opportunities to members. Large hospitals and health care organizations may also grant scholarships. Searching on the Internet and contacting organizations directly may lead you to some promising possibilities.

## internships and volunteerships

Internships are an excellent way to gain experience, skills, and connections in the biomedical field. Internships are often required for students in associate's degree programs and can lead to job opportunities after graduation. They are usually set up through the placement department and are without pay. "You're compensated slightly somehow," says Brent. "I know one facility that will give interns living quarters. Here you get a lunch every day." Compensation varies from facility to facility.

Volunteer opportunities in medical facilities are plentiful as well. Brent usually brings in an intern from one of the two biomedical techno-

## in-depth

Biomedical technology began in the 1970s, when consumer advocate Ralph Nader publicized a document that suggested that people were being killed by microshock, the leakage of an electrical current whose level is below the sensation of feel and is therefore almost impossible to detect. The leakage can be caused by improper grounding, a loose wire inside the instrument, or leaking components.

According to the document, microshock was occasionally causing patients' hearts to fibrillate, which is similar to a heart attack.

The awareness of microshock and its potential hazards created a need for technicians who could test the electronic equipment to ensure proper grounding and minimal leakage of current. And this is where the biomedical technology industry began.

logy programs in Washington, but, he recalls, "This last summer I had a high school student come to me and ask me if he could work with us, stay out of the way and just observe. And I said sure, and he turned out to be a real help." The student recently paid Brent a visit and told him that much of what he observed over the summer hadn't made sense to him, but now that he is in chemistry and physics, things are starting to click. Volunteering can give you some exposure to the industry and to the health care field in general. Brent suggests, "Call and ask if you can volunteer a few hours a week."

Employers are fond of internships because it provides them a chance to teach aspiring technicians about the field. It is also a means to seek job candidates. Brent is involved with hiring personnel at St. Joseph's, and many of the former interns are now employees. "What I use it for is to look at potential employees in the future," Brent explains. "It's kind of like a three-month interview."

## labor unions

Union membership depends on the employer. Brent believes that there are currently more nonunion biomedical equipment technicians. There is no union specifically for biomedical equipment technicians, which means that technicians must usually join the union that represents the majority of the other health care workers in the facility. Some of the unions BMETs can join include the International Brotherhood of Electrical Workers, the International Union of Operating Engineers, and the Service Employees International Union.

## who will hire me?

When Brent graduated from Spokane Community College's biomedical technology program, St. Joseph's Medical Center, where he had completed an internship, did not have a job opening. Brent sent out 100 resumes and found a job working for a third-party company that overhauled ventilators. He worked there for about a year when a position opened up at St. Joseph's. "Because of the internship," Brent feels, "they knew I would mix with the other employees there, so the internship did get me my job."

Many biomedical equipment technicians are employed by hospitals of all sizes. The federal government is another employer of BMETs, primarily through the Veterans' Administration Hospitals and medical centers on army bases. Technicians working for manufacturers often specialize in the repair of machinery. It is commonplace for manufacturing companies to provide maintenance agreements on new equipment, and biomedical equipment technicians are equipped to service the machinery. They may also install the equipment and train the operators or in-house technicians on its functions.

Independent service companies, or third-party companies, also service equipment. Hospitals who do not employ in-house biomedical equipment technicians may use the services of these third-party companies for repair, maintenance, and installation of equipment. Research and development departments within companies may also employ technicians to help test new equipment.

You may have to move a few times to find work as a biomedical equipment technician. "Basically, you probably won't find a job where you think you want to find a job," says Brent. Trade journal publications are an excellent source for job prospects. Magazines such as *Biomedical Instrumentation and Technology* and *Journal of Clinical Engineering* list job opportunities. Brent also suggests becoming involved with local biomedical associations and attending the meetings to find out about what is happening in the biomedical community and to develop some relationships and connections.

The Internet may provide some leads on job openings, and looking through the classified advertisements in the newspaper might be helpful as well. Many large hospitals, manufacturers, and health care organizations have job hotlines that announce new openings. These are often updated weekly. You might also send resumes and cover letters to all the facilities in the state you wish to live in to inquire about job possibilities.

## where can I go from here?

Brent is content with his current job as a supervisor, but he thinks a regional position at some point in the future might be interesting. His employer is now part of a nationwide network of medical facilities, and if Brent's boss moves into a national position, there might be an opportunity for Brent to assume a regional administrative position. If there's one thing Brent is sure of, it's that he would like to stay with his current employer. "I really, really like working for the company I work for. They're very aggressive in their technologies and the business side of it, too. They're not asleep at the wheel. They're aware of what's happening within health care, and it's a real honor to work for them."

As biomedical equipment technicians gain more experience, they begin working more independently and on more technically demanding equipment. They may move into supervisory positions, training entry-level technicians and overseeing the daily operation of facilities. Experienced technicians may also choose to specialize in one type of equipment.

With a four-year degree in biomedical engineering, technicians may become biomedical or clinical engineers and assist in the research and design of new equipment and processes. Clinical engineers are engineers who assess and repair biomedical systems and may be involved in technology management. They are more concerned with the big picture than with individual pieces of equipment. Biomedical engineers, on the other hand, design medical equipment and instruments by applying engineering principles.

Biomedical equipment technicians who enter the industry as field service technicians with manufacturers or third-party companies can move into regional service management positions. Regional service managers supervise biomedical equipment technicians and other staff and may oversee a number of field offices or service centers. Managers may also provide training to customers and solicit new clients.

## what are some related jobs?

Because of the electronics background necessary in biomedical technology, technicians may transfer their skills to related fields and work as, for example, electronics technicians.

The U.S. Department of Labor classifies biomedical equipment technicians under the headings *Occupations in Fabrication and Repair of Engineering and Scientific Instruments and Equipment, Not Elsewhere Classified* (DOT) and *Craft Technology: Scientific, Medical, and Technical Equipment Fabrication and Repair* (GOE). Also under this heading are people who repair technical equipment, such as optics technicians, dental laboratory technicians, orthotists and prosthetists, and musical instrument repairers.

| **Related Jobs** |
|---|
| Optics technicians |
| Dental laboratory technicians |
| Orthotists and prosthetists |
| Musical instrument repairers |
| Industrial radiological technicians |
| Watch and clock repairers |
| Photographic equipment technicians |

## what are the salary ranges?

Salaries for junior-level biomedical equipment technicians averaged $26,126 a year in 1996, according to a nationwide salary survey conducted by the *Journal of Clinical Engineering.* Those with more experience earned an annual salary of $34,687, with the median range spanning from $31,000 to $39,300. The highest 25 percent of biomedical equipment technicians earned between $39,400 and $51,800. Biomedical equipment technician specialists made an average salary of $46,131 a year.

Regionally, entry-level technicians fared the best on the West Coast, earning an average of $32,000. Biomedical equipment technicians beyond the junior level made a salary of $40,462 in the Northeast, while those in the Southeast earned $31,458. Specialists were compensated highest in the East with a salary of $50,526.

Biomedical equipment technicians generally receive generous benefits packages with medical benefits, pension plans, and more. Employers may also finance continuing education courses and seminars.

## what is the job outlook?

The *Occupational Outlook Handbook* indicates that jobs for biomedical equipment technicians will grow about as fast as the average. Technological advances will affect the health care industry, and qualified biomedical equipment technicians will be needed to install, maintain, and repair equipment, as well as train operators on proper usage and care. Equipment will rely more heavily on microprocessors and computers, which will also create a need for skilled technicians. New instruments and machines are developed and manufactured on a regular basis, and technicians are qualified to evaluate, test, and make recommendations from both the purchasing end and the design end.

The state of health care may influence the outlook for biomedical equipment technicians. As the trend toward health maintenance organizations (HMOs) increases, medical facilities will be persuaded to adopt cost-cutting measures. Biomedical equipment technicians will therefore be in demand to keep the existing equipment in top working condition.

Brent has noticed that many technicians who entered the field in the 1970s are nearing retirement, which means job openings will arise. And though he feels the biomedical community should have been more aggressive about presenting itself as a cost-saving option, he feels the future looks good. "Institutions like ours that realize there's a significant value in biomedical and clinical engineering are expanding the role there and developing it, so there are things on the horizon that will keep it going," Brent believes.

fyi

Depending on where you work, and what your assignment is for the day, you could be dressed in a neatly ironed shirt and pants with a lab coat, or comfortable cotton hospital "scrubs."

# how do I learn more?

## sources of additional information

Following are organizations that provide information on biomedical equipment technician careers, schools, and employers.

**Association for the Advancement of Medical Instrumentation (AAMI)**
3330 Washington Boulevard, Suite 400
Arlington, VA 22201-4598
703-525-4890
http://www.aami.org

**American Society for Healthcare Engineering (ASHE)**
One North Franklin
Chicago, IL 60606
312-422-3800
http://www.ashe.org

**American College of Clinical Engineering (ACCE)**
5200 Butler Pike
Plymouth Meeting, PA 19462
610-825-6067
http://info.lu.farmingdale.edu/~acce

**Engineering Medicine and Biology Society (EMBS)**
National Research Council of Canada
Building M-55, Room 382
Ottawa, Canada K1A 0R8
613-993-4005
http://www.bae.ncsu.edu/bae/research/blanchard/www/embs

**Biomedical Engineering Society (BMES)**
PO Box 2399
Culver City, CA 90231
310-618-9322
http://mecca.mecca.org/BME/BMES/society/bmeshm.html

## bibliography

Following is a sampling of materials relating to the professional concerns and development of biomedical equipment technicians.

### Books

Pacela, Allan F., and Linnea C. Brush, eds. *Biomedical Careers Book: Careers in Biomedical or Clinical Engineering, Technology, Supervision and Management.* Brea, CA: Quest Publishing Co., 1993.

## Periodicals

*AAMI News*. Monthly. Newsletter providing association members, including biomedical equipment technicians, with reports of association activities and news of legislative and regulatory developments. Association for the Advancement of Medical Instrumentation, 3330 Washington Boulevard, Suite 400, Arlington, VA 22201-4598, 703-525-4890.

*Biomedical Instrumentation & Technology*. Bimonthly. Presents technical and vocational information for biomedical engineers and technicians. Hanley and Belfus, Inc., 210 South 13th Street, Philadelphia, PA 19107, 215-546-7293.

*Biomedical Technology Information Service*. Semimonthly. Focuses on developments in biomedical engineering and medical device technology and covers recent federal regulations affecting the field. Lippincott-Raven Publishers, 227 East Washington Square, Philadelphia, PA 19106-3780, 215-238-4200.

*Journal of Clinical Engineering*. Bimonthly. Addresses both technical and professional subjects for those who work in the field of clinical or biomedical technology and features articles on new medical technology, education, and certification. Lippincott-Raven Publishers, 227 East Washington Square, Philadelphia, PA 19106-3780, 215-238-4200.

# broadcast engineer

**Definition**

Broadcast engineers set up, operate, and maintain the electronic equipment used to record and transmit radio and television programs.

**Alternative job titles**

Audio engineers
Broadcast operators
Broadcast technicians
Maintenance technicians
Sound editors
Video technicians

**Salary range**

$16,500 to $33,250 to $100,000

**Educational requirements**

Some postsecondary training

**Certification or licensing**

Voluntary

**Outlook**

About as fast as the average

GOE 01.02.03

DOT 194

O*NET 34028B

## High School Subjects

Computer science
Mathematics
Physics

## Personal Interests

Broadcasting
Figuring out how things work
Film and Television
Fixing things

**How do you** recreate a sound that no longer exists? That's what Chris Scarabosio is trying to figure out. He's working on a scene from a movie set in the past. A wooden lifeboat full of dozens of panic-stricken passengers is being lowered to the water. There are two crewmen on either side of the boat, clutching ropes and trying to ease the boat down as evenly as possible. The boat is heavy, and the rope is strained, putting pressure on the pulleys that are designed to help maintain the boat's balance on its descent. The pulleys squeak, and the rope makes stuttering noises as it jumps through the pulley, first on one side, then on the other. Chris must mimic the sound of the ropes and the pulleys. He edited together bits and pieces of sound samples, but the director didn't think the end result sounded realistic enough, so it was back to the drawing board.

One day, the sound designer, the person responsible for creating special sounds and effects, arrives looking rather dazed. He had been standing on an eighteen-foot-tall pole, holding a rope, trying to get the correct tension on the rope to create the sound, when he fell onto the ground. "I almost killed myself," he says, "but I think I got it." He plays back the sound, and sure enough, it works. Chris edits together the final sounds, and the director says it's perfect. "And that," concludes Chris, "is the lifeboat saga."

## what does a broadcast engineer do?

*Broadcast engineers*, also known as *broadcast technicians*, set up, operate, and repair the electronic equipment used to record and transmit radio and television programs. They work with equipment such as television cameras, microphones, tape recorders, antennas, transmitters, computers, and more. Broadcast engineers generally work for television or radio stations in specific departments, such as the news, or for individual programs. They may work in controlled environments such as studios, or they may work from remote locations. Some broadcast engineers work for motion picture production studios developing sound tracks.

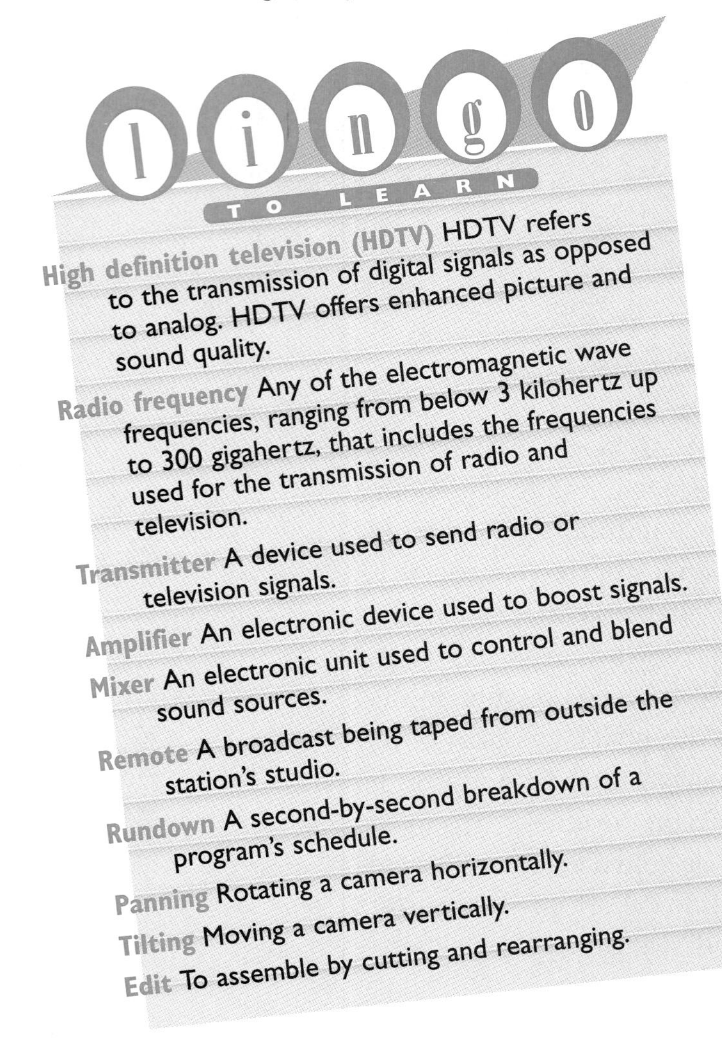


*Transmitter operators* or *transmitter technicians* operate and maintain the transmitters that broadcast television and radio programs. In other words, they beam the signals from the broadcasting station to the public. The operator must regulate and log the strength of outgoing signals and adjust the transmitters to the station's assigned frequency. Transmitter operators are also responsible for diagnosing and remedying transmitter problems.

The engineers who are primarily responsible for the installation, adjustment, and repair of the electronic equipment are *maintenance technicians*. They must make sure that every camera, microphone, transmitter, amplifier, and any cables used by the station are in proper working order.

*Video technicians* usually work in the control rooms of television stations and are responsible for adjusting the quality, brightness, and content of the visual images being recorded and broadcast. The video technicians who are primarily involved with broadcasting programs are known as *video control technicians* or *video control engineers*. They monitor on-air programs, regulate the picture quality, and control individual studio cameras through a camera control unit in the control room. *Video recording technicians* or *videotape engineers* are more involved with videotaping programs rather than broadcasting them. They record performances on videotape using video cameras and sound recording equipment, then edit together separate scenes into finished programs or segments for replay during a live news broadcast.

It is often necessary to broadcast live or tape a story at a site outside the studio, which is what the *field technician* does. Field technicians are responsible for setting up and operating television and radio transmitting equipment from remote locations. They obtain a link to the station through telephone wires or microwave transmitters, then hook up microphones and amplifiers for the audio.

Technology advancement and the introduction of robotic cameras in television studios have created a new type of broadcast engineer known as the *video-robo technician*. Video-robo technicians direct the movements of the robotic cameras from a control room computer, using joy- sticks and a video panel to tilt and focus each camera. With robotic cameras, one person can perform the work of two or three camera operators.

Broadcast engineers who focus on the sound aspect of television and radio transmission are *audio engineers*. Audio engineers set up, operate, and maintain equipment that regulates the quality and relative level of sound. They set up microphones of all shapes and sizes to capture specific sounds and operate mixing consoles that switch between audio from microphones, studios, prerecorded music and sound effects, and remote broadcasting locations. Audio engineers take care of anything that has to do with sound, such as dialogue, music, narration, and sound effects. *Audio control engineers* control the strength and quality of sound being recorded through the use of amplifiers, microphones, and other audio equipment. *Audio recording engineers* operate the controls of the recording equipment in a recording studio, often under the direction of a music producer. Audio engineers are sometimes responsible for manipulating and processing sound to produce special effects or to enhance existing sounds. These audio engineers are frequently called *sound editors* and usually work for motion picture production companies. *Sound mixers* also work in the motion picture industry, adjusting the volumes of the music, the dialogue, and the background sounds to produce the final sound track.

**To be a successful broadcast engineer, you should**

- Enjoy working with others and be a team player
- Be a problem solver
- Have a keen eye and/or ear
- Have a calm temperament with an ability to work under pressure
- Be attentive to detail

## what is it like to be a broadcast engineer?

Chris Scarabosio is a freelance sound editor in the motion picture industry. The majority of his work is for Skywalker Sound, a division of Lucas Digital Limited, in northern California. As a sound editor, Chris deals with creating and editing sound effects. "A lot of times we want to enhance the drama of a film," he explains, "and one way we do that is with sound effects."

Chris generally does not work a normal, forty-hour week. Film work is seasonal and sporadic, so he might work ten-hour days for three months, take a few weeks off, work on a two-week project, and so on. The days, however, are relatively standard—he edits. After collaborating with the sound supervisor and receiving direction, Chris sits down at his computer, ready to tackle and determine all the sounds, except for the dialogue and music, that should be included in the film.

At the computer, Chris watches a videotape of the film. The tape runs in synchronization with the computer, which allows him to match up the sound to the video image. When he needs a particular sound, like a distant tugboat, he searches through either the digital library database or sound network. Chris states that his work facility has "a pretty massive library," so most any sound Chris needs is at his fingertips. Chris lays down numerous tracks, such as all the vehicle sounds on one set of tracks, all animal sounds on another set, and ambient, or atmospheric, noises on another. This allows flexibility when sounds need to be changed or removed.

Sound editing is a precise process and major undertaking. "You try and make it as realistic as possible in very subtle ways," Chris says. He must consider and add all possible background noises, from birds chirping (and they must be seasonally appropriate birds) to water lapping to cars honking off screen. "One of the keys is to not take away from the story but to enhance it," explains Chris. "We're trying to find the perfect sounds to make a sound track come alive so everything is adding to the drama, to the whole fantasy of the film, so when you're in there, you're really part of the film."

Dealing with reality rather than fantasy, Walt Ward is the chief engineer of the maintenance department at KSBY Television, the NBC affiliate in San Luis Obispo, California. KSBY has a maintenance engineering department and a separate operations department that handles the broadcasting tasks. Walt's crew is responsible for repairing equipment and performing preventive maintenance checks on everything from tape machines to cameras to satellite dishes. Walt says, "We're also responsible for handling the day-to-day brush fires, as we like to call them, which are immediate concerns, like if a piece of equipment is malfunctioning and they need it right away." Walt makes sure there are plenty of spare parts and backup equipment, since he must be prepared for anything. "Generally," Walt notes, "when maintenance is required, it's usually an emergency."

At the beginning of the day, Walt reviews reports filled out by operators and others in the news department regarding malfunctioning or faulty equipment. He prioritizes the jobs and spends the rest of the day fixing equipment, carrying out routine maintenance tasks, and making sure the work orders are executed effectively. He also checks transmitter readings, visits the transmitters to make sure they are functioning properly, and reviews the station logs, which are required for compliance with the Federal Communications Commission (FCC). As the news department begins preparing for the evening broadcasts, the tempo picks up in engineering because more equipment is in use. Engineers may also assist with live broadcasts out in the field, setting up shots, running the cables, and checking each piece of equipment.

## have I got what it takes to be a broadcast engineer?

Anyone who wishes to become a broadcast engineer should be detail oriented, enjoy solving problems, and possess technical aptitude. Broadcasting, whether it's radio or television or audio or visual, does not leave much room for error, which is why you must be able to troubleshoot and diagnose problems and repair them quickly and efficiently. You must also have communication skills and work well with others. As Walt says, "People skills—I can't emphasize how important that is." Because the job of a broadcast engineer can often be stressful, with constant deadlines and harried co-workers, a calm temperament is essential.

If you want to focus on the visual aspect of broadcast engineering, it's important to have excellent vision and color perception, and if you're more interested in sound, it's crucial to have a keen ear and know what sounds will work. "It's not always about picking the sound that sounds the best," says Chris, "but it's picking the sound that you know will be heard." Chris stresses the importance of being computer literate as well. With technological advances and an increase in automation in broadcasting, more equipment will be controlled by computers.

Although broadcast engineers generally work a forty-hour week, those hours may not be standard. Newscasts may air early in the morning or late at night, and most stations are on the air around the clock, regardless of holidays. Engineers may also work overtime in order to meet deadlines, whether it's for a television program or a motion picture. Chris finds that the long hours and deadlines can be stressful. "There are times when you get so inundated with a project that you're basically almost nonexistent. I think that's definitely the hardest part," Chris says. Chris notes, however, that he doesn't work long hours all the time, and that it all balances out in the end.

Broadcast engineering can be a rewarding job, and engineers are able to see or hear the fruits of their labors. "You see that your work is making the movie better," Chris says, "and that's a good feeling. It's the type of work where people can hear what you do. That's rewarding."

## in-depth

### Going digital

High definition TV will offer approximately twice the vertical and horizontal resolution of your current television, which means the picture will be twice as clear. HDTV will also enhance the sound quality of programming. Also with HDTV, you'll have access to four times as many channels. Sounds too good to be true, doesn't it? It may be too early to tell, but going digital might have a few drawbacks. As television broadcasting switches from analog transmission to digital transmission, your analog television set is going to need an update. This will probably begin with analog-to-digital converter boxes, which may cost anywhere from fifty dollars to several hundred. Eventually, you will probably need to purchase a new television set that can accommodate HDTV. The first HDTV sets retailed in Japan for $28,000. The prices have since come down as low as $2,800, but that is still a considerable price for a television. Let's hope that by the time the digital revolution hits the United States, the prices will have dropped even lower.

# how do I become a broadcast engineer?

> *You see that your work is making the movie better and that's a good feeling.*

Chris has always been fascinated by sound, music, and how equipment works. As a child, he often took apart radios and tape recorders to find out what was inside and how they functioned. "I still think it's utterly incredible that you can have a blank tape, put it into a cassette player, hit record, and now you have that sound on that cassette."

## education

### High School

Classes to take if you're interested in becoming a broadcast engineer include mathematics, the sciences (physics in particular), and computer science. If possible, you should take electrical or electronics shop courses. For audio engineering, music classes would be helpful, and if you're particularly interested in working in the motion picture industry, try to enroll in a film appreciation class.

Many high schools now offer video production classes, and some may have television or radio stations on campus. It would definitely be to your benefit to enroll in any of these classes or to join radio, television, or audiovisual clubs. You can also gain some skills by volunteering to be a stagehand for the drama club, where you can work with lighting, sound, and videotaping equipment. Becoming a member of an amateur radio club will give you experience working with electronics equipment and help you gain an understanding of frequencies and broadcast technology.

Part-time or summer jobs at local broadcasting stations may be available. Some television and radio stations have programs produced and written by high school students, which is an excellent way to gain experience and make some connections in the industry. Summer classes in broadcasting or filmmaking may be offered through extended education or recreational programs as well. If, however, you are unable to locate any opportunities, you can gain some skills on your own. If you have a video camera, you can practice filming and editing your own segments.

### Postsecondary Training

Although a college degree is not mandatory for broadcast engineers, some postsecondary training is necessary in order to gain the requisite technical skills. Many technical schools and community colleges offer programs and classes in broadcast technology or electronics. The military is another source for training, and Walt, who handles the hiring for his department, finds military graduates to be highly qualified. Walt finds that vocational training or military schooling offers the best preparation for broadcast engineers. "I've interviewed a lot of people with college degrees who can't solder two wires together," Walt states matter-of-factly. "They can tell you why the two wires ought to be soldered together, but that's not what we need. We need people who can do it."

Some of the courses in a broadcast technology program include video technology, broadcast operations, electronic field production, physics, math, and audio technology and theory. Many programs offered by community colleges will lead to an associate's degree.

Although Chris attended San Francisco State University and enrolled in broadcasting and audio recording classes, a college degree is not required for those working in the motion picture industry, and it's difficult to find specialized training. "Our industry is more experience based," Chris explains, and on-the-job training is how he gained his skills.

## certification or licensing

Certification is not required for broadcast engineers. In fact, the FCC eliminated the licensing requirement for broadcast technicians in 1996. The Society of Broadcast Engineers (SBE), however, offers a certification program, and certification can sometimes lead to higher salaries and provide you with an advantage when seeking a job. Applicants must pass a written examination and provide proof of experience. There are four classes of certification: Broadcast Technologist, Broadcast Engineer, Audio or Video Engineer, and Senior Broadcast Engineer, each requiring more experience. Special endorsements in either television or AM/FM are available. Certification is valid for five years, after which renewal is necessary.

## scholarships and grants

Several organizations provide scholarships for students wishing to pursue careers in the broadcasting industry. Professional associations may also offer lists of scholarship or grant opportunities. The Broadcast Education Association, American Women in Radio and Television, National Association of Broadcasters, the Society of Broadcast Engineers, Inc., and Radio-Television News Directors Association are some associations to contact for information (*see* "Sources of Additional Information").

Technical schools and community colleges with programs in broadcast technology or broadcast engineering may award scholarships to students. Contact the financial aid office or your departmental advisor for details. These schools may also offer general scholarships for which you may qualify.

## internships and volunteerships

Internships are common in the broadcast technology field and are highly recommended. In fact, it is often difficult to find jobs without some internship experience. Chris participated in many internships and says, "If I hadn't done any internships, I don't know if I would have ever done this." Most internships are unpaid, but the experience you can gain is priceless.

If you are enrolled in a technical school or college, it is likely that participating in an internship is a graduation requirement. The school placement office or your advisor can assist you in locating appropriate internships. You may be able to earn college credit for your internship as well.

Many television and radio stations provide internship or volunteer programs of their own, so you should be able to find some opportunities even if you are not attending school. Don't discount the public-access cable television stations or the public radio stations, as they may welcome volunteers more readily than the network affiliates.

## labor unions

Union membership varies from station to station. KSBY, where Walt works, is in a small market, and union membership is not required. Stations in larger markets, however, are often unionized. The largest broadcasting union is the National Association of Broadcast Employees and Technicians. Union regulations often dictate what a broadcast engineer can or cannot do. For instance, because KSBY is not unionized, the maintenance engineers may occasionally fill in for operators, and job duties may overlap. This would generally not be allowed at a unionized station.

### Advancement Possibilities

**Chief engineers** supervise and assist broadcast engineers. They handle administrative duties such as the acquisition of broadcasting equipment, preparation of work schedules, and enforcement of station policies and procedures.

**Producers** are responsible for the overall planning, organization, and production of radio, television, and cable television programs.

**Program directors** coordinate the activities of all staff members involved in the actual production of television or radio programs.

**Technical directors** supervise personnel in radio and television studios to ensure the technical quality of picture and sound for programs.

**Sound supervisors** oversee a staff of sound editors. They collaborate with the motion picture's director and picture editor, receiving direction about the sounds they desire.

**Sound designers** design and create specialized sounds or sound effects. They may also contribute to creating a signature sound or sound environment for films.

Union membership is not mandatory for sound editors in the motion picture industry, but most established editors belong to one union or another. Chris belongs to the International Alliance of Theatrical Stage Employees, but notes that sound editors in the Los Angeles area belong to a different union altogether. Membership is often determined by the employer and the region in which you work.

## who will hire me?

Chris found his first job in the industry through an internship, which is why he considers internships so crucial. Even if you do not find a job through an internship, you can gain valuable contacts that may help you in your career. Chris was interning at a studio that had more work than it could handle, so Chris "volunteered" to take the job editing sound for an animated cartoon show. He believes he got the job because of his enthusiasm and his ability to get along with the other staff members. "Who you know is key, and that's why you have to be able to get along with others," he emphasizes.

The motion picture industry is an extremely difficult industry to enter, which is why Chris warns, "You've got to really want to do it." You should have a talent for it, as well as a passion for sound, because the field is highly competitive. Chris has won an Emmy award and worked on many Oscar-winning films, and yet, "There are absolutely zero guarantees that I will get hired," he states. "I have to make sure that I get hired. I'm on the phone, and I make my visits, and I try and maintain as many relationships as possible." Almost all sound editors are independent freelance editors, which means pounding the pavement is a normal part of the job.

Broadcast technology is also a difficult industry to break into. "To start out," Walt suggests, "the best thing to do would be to start sending resumes and contacting stations." Perseverance is key. Walt also advises reading the trade magazines, such as *Broadcast Engineering*. The back of the magazine lists job openings and also contains advertisements for employment agencies that specialize in broadcasting.

Walt also believes in maintaining good relationships with others in the industry. "Who you know is certainly the best," he states. "In fact, that's how I've done it my entire career. I've never, ever applied for a job. It's always been somebody calling me." But, he contends, it takes years of experience and time building a reputation before most will be in a similar position.

Professional organizations frequently provide lists of job opportunities as well, and most of them have web sites with the information. Some broadcasting jobs may be located in newspaper classified advertisements or on the Internet. Motion picture jobs, however, are usually not advertised.

The primary employers of broadcast engineers are television and radio broadcasting stations, cable and other pay television companies, and motion picture production companies. Entry-level broadcast engineers generally start in smaller markets, and the chances of finding this first job in your hometown are slim. This gives the engineer an opportunity to gain experience and decide on a field of specialization. After years of training, engineers can work their way up to larger stations in larger markets or cities.

## where can I go from here?

As broadcast engineers move up the ranks and gain more experience, they can move to larger markets and bigger stations or assume supervisory positions. *Chief engineers* supervise the work of other broadcast engineers. They also assume management duties and are responsible for budgeting and planning radio or television broadcast activities. They may also prepare work schedules, take care of hiring and personnel issues, and make decisions about equipment purchases. A college degree in engineering is generally needed in order to become a chief engineer at a large TV station.

Broadcast engineers can also advance to the rank of *program director*, handling all aspects of the broadcast and directing and coordinating the activities of personnel engaged in the production of television or radio programs. This includes writing and rehearsing scripts. Like the chief engineer, program directors may also be responsible for personnel issues and monitoring department budgets. *Technical directors* coordinate the activities of radio or television staff to ensure the technical quality of picture and sound for programs originating in the studio or from remote field locations. *Producers* are responsible for the overall organization and planning of a program. The producer researches ideas and develops scripts or working ideas. They may also coordinate the production schedule and make sure the program is within budgetary constraints.

For sound editors, there are a number of advancement possibilities. *Sound supervisors* oversee the work of sound editors. They meet with the director and picture editor for direction, then pass the information on to the sound editors. *Sound designers* create specialized sounds and conceptualize the overall sound environment for films. They may also work for, say, theme parks, developing sound effects.

## what are some related jobs?

Depending on the size of the station, broadcast engineers can work as field technicians, transmitter operators, or audio control operators. Chris believes that with his background and qualifications, it might be possible to work in music as a recording engineer, do some location recording in the field, go into picture editing, or switch industries and become an audio engineer for a news broadcasting station. Sound editing is, however, rather specialized, requiring years of dedication and perseverance, so it might be difficult sliding into a different job.

The U.S. Department of Labor classifies broadcast engineer under the heading *Sound, Film, and Videotape Recording, and Reproduction Occupations* (DOT) and *Visual Arts: Commercial Art and Crafts: Reproduction* (GOE). Also under this heading are photographic laboratory technicians and those working in commercial art, including graphic designers, art directors, camera operators, film and television directors, medical and technical illustrators.

| Related Jobs |
| --- |
| Photographic laboratory technicians |
| Graphic designers |
| Art directors |
| Camera operators |
| Film and television directors |
| Medical and technical illustrators |
| Merchandise displayers |
| Costume designers |

fyi

The Federal Communications Commission (FCC) was created in 1934 and is an independent agency of the U.S. government. It regulates radio, television, telephone, and telegraph systems, except those of the federal government. It licenses and regulates radio and television stations and assigns radio wavelengths and television channels.

## what are the salary ranges?

According to the 1998–99 *Occupational Outlook Handbook*, average earnings for broadcast technicians in the radio industry were $30,251 in 1996. In television, operator technicians earned an average of $24,260 a year, with salaries ranging from $16,422 to $45,158. Technical directors made an average of $25,962, and maintenance technicians earned $32,533 a year. Their salaries ranged from $24,210 to $50,235. Chief engineers earned an annual average income of $53,655, with salaries ranging from $38,178 in the smallest markets to $91,051 in the largest. The reported earnings in the *Occupational Outlook Handbook* were taken from a survey by the National Association of Broadcasters and the Broadcast Cable Financial Management Association.

A 1997 salary survey conducted by B*roadcast Engineering* magazine reports that staff engineers in broadcasting in the large markets earn an annual median salary of $49,090. Those working in the smallest markets make a median salary of $33,333 a year. In cable television, staff engineers earn an annual median salary of $47,000 while those working in production make $44,999 a year. The salary survey also indicates that engineers who are certified by the Society of Broadcast Engineers earn a higher wage.

Earnings in the motion picture industry vary widely and depend on the skill, reputation, and experience of the editor. Wages also vary by location and by union. The *Occupational Outlook Handbook* reports that earnings range from $20,000 to $100,000 a year, which Chris agrees is a wide but accurate range.

## what is the job outlook?

Although competition is expected to be strong in large metropolitan areas, the job outlook, according to the 1998–99 *Occupational Outlook Handbook*, is positive for broadcast engineers. The construction of new radio and television stations, along with an increase in programming hours, should ensure that the employment of broadcast engineers will grow about as fast as the average through the year 2006. The cable industry should also experience growth as technology advances. Cable modems will allow high-speed access to the Internet, and digital set-top boxes will provide better quality sound and pictures.

Growth may be inhibited somewhat by the increasing automation of radio and television stations. Cameras, playback, program recording, and other procedures may soon become completely automated and controlled by computers. As Walt explains, "Radio has been doing that for several years. You go into a radio station, and you'll be lucky if you find one person. Those things are running on autopilot. And TV is just starting to do that." Currently, stations such as KSBY record a large percentage of their programs from a satellite feed. Walt must manually place a tape into the machine, press the "record" button, rewind the tape when it is complete, catalogue it, place it on a shelf, then remove it from a shelf when it's time to playback the tape. Soon, all this will be performed by a computer.

Despite the *Occupational Outlook Handbook's* moderate outlook and the increase in automation, Walt believes this is a great time for broadcast engineers. "The job outlook for broadcast engineers in television is absolutely exceptional," he states emphatically. "Right now the industry is wide open, more than it has ever been. We can't get engineers fast enough." Many engineers are reaching retirement age, and the television industry is going through a revolution—by the year 2006, every station in the United States must have switched from analog to high definition television (HDTV). "This is, hands down, much bigger than color was. It's going to completely revolutionize the industry," says Walt. He feels this digital revolution will open up unlimited job openings in broadcasting and with companies who supply equipment or services to the broadcasting industry.

Employment in the motion picture industry, according to the *Occupational Outlook Handbook*, is expected to grow faster than the average. When the economy is strong and families have more disposable income, the motion picture industry generally does well.

## how do I learn more?

### sources of additional information

The following organizations provide information on broadcasting engineering careers, schools, scholarships, and employers.

**American Women in Radio and Television**
1650 Tysons Boulevard, Suite 200
McLean, VA 22102
703-506-3290
http://www.awrt.org

**Audio Engineering Society (AES)**
60 East 42nd Street
New York, NY 10018
212-661-8528
http://www.aes.org

**Broadcast Education Association**
1771 N Street, NW
Washington, DC 20036
202-429-5354
http://www.beaweb.org

**National Academy of Recording Arts and Sciences**
3402 Pico Boulevard
Santa Monica, CA 90405
310-392-3777
http://www.grammy.com

**National Association of Broadcast Employees and Technicians**
501 Third Street, NW, 8th Floor
Washington, DC 20001
202-434-1254
http://www.nabetcwa.org

**National Association of Broadcasters**
1771 N Street, NW
Washington, DC 20036
202-429-5300
http://www.nab.org

**National Cable Television Association**
1724 Massachusetts Avenue, NW
Washington, DC 20036
202-775-3669
http://www.ncta.com

**Radio-Television News Directors Association**
1000 Connecticut Avenue, NW, Suite 615
Washington, DC 20036
202-659-6510
http://www.rtnda.org/rtnda/

**Society of Broadcast Engineers, Inc.**
8445 Keystone Crossing, Suite 140
Indianapolis, IN 46240
317-253-1640
http://www.sbe.org

**Society of Motion Picture and Television Engineers**
595 West Hartsdale Avenue
White Plains, NY 10607
914-761-1100
http://www.smpte.org

# bibliography

Following is a sampling of materials relating to the professional concerns and development of broadcast engineers.


## Periodicals

*Broadcast Engineering.* Thirteen times per year. Directed at technicians who install, operate, and maintain broadcast equipment in the radio, television, and recording industries. Intertec Publishing, 9800 Metcalf, Box 12901, Overland Park, KS 66212-2215, 913-341-1300.

*Radio World.* Semimonthly. Newspaper directed at management and technical professionals in the radio industry. Industrial Marketing Advisory Services, Inc., 5827 Columbia Pike, Suite 310, Falls Church, VA 22041, 703-998-7600.

*SMPTE Journal.* Monthly. Focuses on various technical areas of television and motion picture industries. Society of Motion Picture and Television Engineers, 595 West Hartsdale Avenue, White Plains, NY 10607-1824, 914-761-1100.

*Television Broadcast.* Monthly. Discusses recent developments and the application of television technology affecting engineering, production, and management operations. Miller Freman, PSN Publications, Inc., 460 Park Avenue South, 9th Floor, New York, NY 100167315, 212-378-0400.

*TV Technology.* Monthly. Addresses the technical personnel working at television stations, production firms, and corporate and industrial television facilities. Industrial Marketing Advisory Services, Inc., 5827 Columbia Pike, Suite 310, Falls Church, VA 22041, 703-998-7600.

# cable TV technician

**Definition**

Cable television technicians install, maintain, and repair cable TV lines and equipment.

**Alternative job titles**

Bench technicians

Cable TV line technicians

Installers

Trunk technicians

**Salary range**

$15,000 to $22,000 to $30,000+

**Educational requirements**

High school plus on-the-job or technical school training

**Certification or licensing**

Voluntary; driver's license required to operate company vehicles

**Outlook**

Decline

GOE
05.10.03

DOT
821

O*NET
85702

## High School Subjects

Computer science

English (writing/literature)

Shop (Trade/Vo-tech education)

## Personal Interests

Figuring out how things work

Film and Television

Fixing things

thinks the cable sticking out of his TV ends somewhere behind the wall. He's wrong. Another customer thinks the cable TV company signed her up, plugged her in, and now sits back and collects her monthly subscription fee. She's wrong, too.

Somewhere out there, a cable TV technician is doing low-voltage electrician work while dangling from a pole thirty feet or more above the ground. Swaying gently in the breeze, and ignoring the empty space that yawns below him, he's checking on a potential line problem. Another tech is deep underground, running an electronics test on the part of the cable that snakes under the street. Still another is ending a ten-hour shift at the electronic "nerve center" of the cable TV system, where part of her time was spent carefully monitoring the quality of transmission signals.

# what does a cable TV technician do?

Running along poles in rural and suburban areas and in tunnels under the ground in cities are miles upon miles of cables that help bring TV channels like MTV, VH-1, CNN, ESPN, and others to American households. Cable TV is a "closed path" system that eliminates the need to receive transmissions over the air. Signals from broadcast airwaves, microwave transmitters, or satellites are picked up by antennas at the local cable company's electronic control center plant, or headend, and then sent through coaxial and/or fiber optic cables into peoples' homes. *Cable TV technicians* take care of these cable systems. Throughout the system, they make repairs, run tests to check for potential problems, replace worn-out components, add new technology to enhance reception, and perform other work to maintain and improve the system. They also bring the cable line into homes—the part of the technician's job that most people see—and may help to develop new cable systems.

**Amplifiers** Spaced throughout the cable system, these devices increase the strength of the electronic signal for clear reception to all customers.

**Channel capacity** The maximum number of channels that a cable system can carry at the same time.

**Coaxial cable** A transmission line for carrying television signals; the conducting material is copper or copper-coated wire that is insulated and encased with aluminum.

**Converter** A device that can increase the number of channels that can be received by the TV; sits on top of the set.

**Drop lines** Cable distribution lines that feed directly into customers' homes.

**Feeder lines** Cable distribution lines that connect the trunk line or cable to drop lines.

**Fiber optics** Transmission technology that uses a very fine, bendable tube of glass or plastic to carry frequencies.

**Trunk line** The main artery of a cable system; it is strung along main streets or highways of a city to the system's plant area.

Almost 60 percent of American homes with a TV now get cable, according to the National Cable Television Association (NCTA). The cable TV industry has grown from 25,000 employees in 1975 to more than 90,000 employees in 1995. The industry is made up of three broad groups: the local cable systems, which provide cable service to homes; the multiple system operators (MSOs), which are corporate groups of individual cable systems; and cable networks, which create the programming found on cable channels. Technicians work for the local cable systems.

*Installers* are the entry-level technicians. They get homes ready to receive cable by running a cable from a utility pole to the TV, installing a terminal device called a converter, or performing other work. After an installation, they may explain to customers how the system works and what options in programming and channels are available. Depending on their employer, installers also may upgrade or downgrade services, disconnect service when the customer has canceled his or her subscription, and take care of problems with feeder or drop lines. Feeder lines are the cable lines that run from the street to a group of homes. Drop lines are the direct lines to homes. Some installers also respond to customer calls about problems with service.

*Trunk technicians* fix electronic problems on the trunk line, which connects the feeder lines in the street to the headend. The work includes fixing electronic failures in the feeder amplifiers. Amplifiers increase the strength of the electronic signal for clear reception to everyone and are spaced throughout the cable system.

*Service technicians* respond to problems with customers' cable reception. Their work takes them into customer homes as well as all over the cable system. They also electronically scan the system from time to time to find potential problems.

*Bench technicians* are electronics specialists who work in the cable system's repair facility. They examine and repair broken or malfunctioning equipment, such as cable converters.

*Chief technicians* are the most highly skilled technicians and head up the technical staff. Their main responsibility is to ensure good, clear delivery of signals to the headend. This "nerve center" includes the antennas for receiving signals as well as the signal processing equipment. It is very sensitive to temperature and humidity and needs constant monitoring.

In addition, cable companies that have their own construction crew (for building a new system, or expanding or upgrading an old one) may employ a technical specialist known as a *strand mapper.* The strand mapper surveys the geographic area, figures out how the cable can be laid out, and then draws diagrams showing the path for the cable and where amplifiers or other necessary system components should go.

**To be a successful cable TV technician, you should**

- Like people and have strong interpersonal skills
- Have physical agility
- Be able to work at heights or in confined spaces
- Be able to work as part of a team
- Feel at ease with electrical equipment and electricians' tools

## what is it like to be a cable TV technician?

"When you first start, it scares you to death," says Jonathan Easton, technician with Chicago Cable TV, of the climbs up utility poles that technicians routinely make. "But we wear a safety belt for climbing, and if you use a ladder there's a safety strap to keep the ladder from sliding. We also wear climbing boots with a half-inch heel. Once you're up there and you realize you're secure, the fear goes."

Jonathan is an installer and, with four years' experience under his belt, knows the ropes. He works the second of two shifts at CCT, from 12:00 PM until 9:00 PM. (The Bureau of Labor Statistics said in 1993 that cable TV employees worked an average of about thirty-nine hours a week; some work five approximately eight-hour days; others work four ten-hour days.) As a second-shifter, Jonathan says, "I pick up the jobs involving customers who work until 5:00 PM or 6:00 PM." He is one of nineteen technicians at the company, only one of whom is a woman.

Like most installers, Jonathan spends the better part of his day out in the field, visiting customer sites. "When we come in at 12:00 PM, the first thing we do is grab a route," Jonathan says. "A route generally consists of three jobs. Then we check to see what supplies we might need. This might include channels, which are filters to bring in channels like 14 or 15; meters and boxes, special devices for people who don't have cable-ready TV or are getting pay-per-view channels; and splitters and wire for running additional lines from the house to the utility pole."

"As a technician, you're basically a low-voltage electrician," he continues. "You take safety precautions when working with the electrical equipment. You wear special clothes, gloves, and a hard hat."

Jonathan says that the job does not call for mechanical aptitude so much as it does for electrical/electronics interest. "Some young men like to work on cars and want to be mechanics," Jonathan says. "And some like audio equipment, stereos, electronics. That's different. And that's what's more helpful on this job."

For installers, the greatest challenge of the job may not be the heights, or the technology, or any other technical aspect. What else could it be? "The customers," says Jonathan with a laugh. "Patience is a virtue! If you don't have it, you can't do the job."

"This is a customer service position," he continues. "You're dealing with the public, going into peoples' homes. You encounter all kinds of personalities. In some scenarios, it can be difficult. Peoples' moods can change like the wind. They can be extremely hostile. Some people want more than you can give, and they can get irate when you can't do it. Some people have had a problem with a previous serviceperson and want to take it out on you. I just say to them, 'I don't know who was here before, but let's you and me talk and we'll straighten this out together.' The customer usually calms down then and agrees. "I always say this is not just a job," adds Jonathan, "it's an adventure."

## have I got what it takes to be a cable TV technician?

Strong interpersonal skills, physical agility, the ability to work at heights or in confined spaces, and the capacity to work as part of a team are all good personal traits for cable TV technicians. In addition, it is helpful to feel at ease with electrical equipment and electricians' tools.

Cable TV technicians who pay service calls on customer homes are acting as representatives of the cable TV system and company. They must be able to project a helpful, courteous, pleasant image. They may need to explain cable system operation and costs to customers, answer questions, and analyze customer descriptions of problems so repairs or other work can be done. The ability to communicate with others is essential.

Technicians may climb utility poles to check a line, work in tunnels or underground cable passageways to inspect cables for evidence of damage or corrosion, or perform other work that demands physical agility and comfort in confined spaces. Some heavy lifting (up to fifty pounds) may be required. Technicians must stay alert when working around the electrical conductors in order to avoid electrical shock. Those who work outdoors, such as trunk or service technicians, may have to do so in all kinds of weather. Bench technicians, who stay at the plant, must be able to concentrate for long periods of time at the bench.

It is a team effort to maintain and improve the cable lines and other parts of the system. other technicians, supervisors, and engineers and see their job as contributing to the overall team effort.

People who pay a monthly fee for cable don't want to settle for "almost good" reception; therefore, cable TV technicians must care about helping make sure that the system is operating as well as it can and that customers are getting the best reception possible.

FYI

In the early days of the cable television era, cable technology was used primarily to transmit television to places that couldn't be reached by regular television transmissions—such as places located in valleys or hilly regions, or in big cities where tall buildings blocked transmission. Just one coaxial cable can carry sixty or more television signals; it was only a matter of time before programming began to be developed to take advantage of all those channels. By the late 1970s, there were about 3,500 local cable television systems in the country; just ten years later, half the households in the United States that owned a TV—representing millions and millions of people—subscribed to cable.

## how do I become a cable TV technician?

Jonathan recalls the work he did with audio equipment at home as being more helpful than any specific high school classes. "But if your school has classes that can give you electrical/electronics experience, take those," he says. "And I'd encourage people interested in this field to get experience with computers. We do data processing, entering information about the jobs. You need to be comfortable with computers.

"I mostly trained on the job," he adds. "I had one month of classes, which taught me how to install the cable and troubleshoot the system. When I first started, I'd go out with experienced technicians."

## education

### High School

As Jonathan noted, classes that give you experience with electrical/electronics systems and computers are helpful. Those considering the most highly technical positions, such as chief engineer, should get a good background in math and science. For all technicians, English and related classes are helpful for building strong communication skills.

### Postsecondary Training

Some entry-level technicians just train on the job; others have studied cable television repair and maintenance in a one- or two-year program at a technical school or community college. Courses may include electrical wiring, electronics, broadcasting, blueprint and schematic diagram reading, and physics. Alternatively, some cable TV companies have their own special training schools for technicians.

Positions higher than entry level require more specialized training, says NCTA. For example, qualifications for trunk, service, and bench technicians include some electronics training. Chief technicians need an industrial background, electronics training, and lots of hands-on experience. A bachelor's or master's degree in electrical/electronic engineering may be needed for highly technical positions.

**Advancement Possibilities**

**Bench technicians** are electronics specialists who work in the cable system's repair facility, examining and repairing broken or malfunctioning equipment.

**Chief engineers** are responsible for all technical concepts of cable system design, equipment planning, layout for cable communications service, specification of standards for equipment and material, construction of facilities, equipment installation, and technical advice and counsel to staff and system operating managers.

**Chief technicians** are the highest-skilled technicians and head the technical staff.

**General managers** head the cable system office and are in charge of operations, company policies, and coordination of all functions of the cable system.

**Service technicians** respond to problems with customers' cable reception, may visit customer sites and may work on the cable system to find and fix the problems.

**Strand mappers** survey the geographic area to be served by cable system; determine necessary utility pole hookups; diagram cable system layout, showing where amplifiers or line splitters should go.

**Trunk technicians** fix electronic problems on the trunk or main cable line, including problems with the system amplifiers.

Technical training through self-study and correspondence courses is available from the National Cable Television Institute (NCTI). People in the industry, or thinking about joining it, can enroll. One program includes five courses on technical training from installer to advanced technician (*see* "Sources of Additional Information").

Cable TV technicians never stop learning, because technology is constantly changing. "You have to do ongoing training," notes Jonathan.

## certification or licensing

Technicians who drive a company vehicle must have a driver's license.

Certification is voluntary. One professional cable association, Women in Cable and Telecommunications, has a joint program with the University of Denver that awards a Certificate of Cable Management for completion of courses in just about every area of the cable TV business (*see* "Sources of Additional Information").

## scholarships and grants

Nine- to twelve-month fellowships are available from the Walter Kaitz Foundation for minorities with a college education who have at least three years of professional experience in an unrelated field. Submit a resume to the Walter Kaitz Foundation, 660 13th Street, Suite 200, Oakland, CA 93612, 415-451-9000.

## internships and volunteerships

You may be able to get an internship with your local cable TV system. Write or call to see if opportunities are available. For cable TV system companies, look in the phone book under "Television—Cable" or write to the NCTA for a free copy of *The Cable Development Book,* which lists all of the cable systems in the United States and Canada (*see* "Sources of Additional Information").

## who will hire me?

"When I came out of high school, I didn't really know where I was going," remembers Jonathan. "I had worked in the audio equipment department of a store through a work-study program in high school, and I really liked that; I knew I was interested in electronics. I had a friend here at the company who was a supervisor, and he told me about the job. It was a natural for me."

People who've completed a cable TV training program at a technical school or community college can get help from the school's placement division. Otherwise, a good bet is to check local newspaper want ads or contact your local cable TV system directly. To find the name of local systems, look in the Yellow Pages or the *The Cable Development Book* published by NCTA.

NCTA's "Careers in Cable Television" brochure lists the names of the country's fifty largest multiple system operators (MSOs), which are umbrella companies for individual cable systems. Those interested in finding branches of these MSOs in their area might start by contacting the MSO's corporate office. Contact NCTA for a copy of the brochure (*see* "Sources of Additional Information").

## where can I go from here?

Right now, Jonathan is the lead man in his department. "I'd like to move up to service technician next, and then trunk and line technician," he says. "I'll need to get electronics training for that."

Most installers have similar opportunities. With additional electronics training, they may rise to service, line, and trunk positions. Bench technician or supervisor of other technicians also are possible advancements. People with top technician and managerial skills plus good on-the-job experience may move into the role of chief technician.

Beyond that, the highest position in the technical department is that of the chief engineer. This person is in charge of all technical aspects of cable system design, equipment planning, cable layout, standards for equipment and material, construction of facilities, equipment installation, and technical advice to the other personnel. Chief engineers typically are very important in the development of new services. A degree in electrical engineering or equivalent experience usually is required, says the NCTA.

The construction crew's strand mapper should have knowledge of cable systems and utility construction.

Other opportunities might be found with companies that supply equipment to cable TV system companies.

## what are some related jobs?

"The main interests of this field, electronics and engineering, coincide with those of the phone company industry," notes Jonathan. There also are many related technician positions with utility companies.

The U.S. Department of Labor classifies cable TV technicians under the headings *Occupations in Assembly, Installation, and Repair of Transmission and Distribution Lines and Circuits* (DOT) and *Craft Technology: Electrical-Electronic Systems Installation and Repair* (GOE). These headings cover a wide range of jobs for erecting and repairing electrical power lines and circuits, and working with related equipment and structures.

Some of these occupations include that of utility company line supervisor, who oversees the construction and repair of overhead and underground power lines; emergency service restorer, who restores street railway service (like trolleys) after power failures, accidents, or equipment breakdowns; utility company troubleshooter, who finds the source of electric

Top Five Pay Cable Services (in thousands of subscribers), 1993
Source: NCTA, April 1994

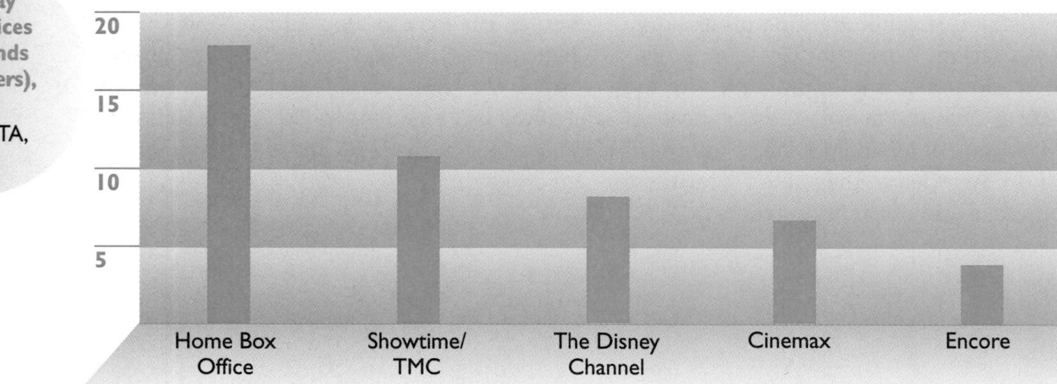

power line disturbance or failure and corrects the problem; utility company voltage tester, who tests electric power lines and auxiliary equipment to compile data about circuit conditions affecting efficiency of electric service and accuracy of meters; and tower erector helper, who assists in the erection of transmission towers, installation of tower hardware, and stringing of transmission lines or cables.

| Related Jobs |
|---|
| Emergency service restorers |
| Line maintainers |
| Tower erector helpers (utilities or construction) |
| Utility cable installer-repairers |
| Utility company line supervisors |
| Utility company troubleshooters |
| Utility company voltage testers |
| Utility company wireworker supervisors |

fyi

**Cable TV Competitor: "Phone" TV**

What's an alternative to cable? One is TV programming sent over telephone companies' fiber optic telephone lines. At this writing, a few regional phone companies are putting in place a fiber optic network, separate from the regular telephone lines, to carry TV programming. They have struck a deal with the Walt Disney Company to help develop new programming for the system. Other programming will probably be developed, too. Eventually, so-called "interactive television" transmitted over the phone companies' lines will allow customers to, for example, shop for groceries from the local market while they're at home.

## what are the salary ranges?

Salaries increase as you move up the ranks of technicians. A new installer may make about $7 an hour, or approximately $15,000 a year; a more experienced technician may make about $11 an hour, or $22,000 to $24,000 a year; a highly experienced technician or supervisor may make about $15 an hour, or $30,000 to $31,000 a year. Exact pay varies depending on the cable TV system company and geographic location. Also, depending on the company, the job may include paid vacation, medical insurance, and dental insurance.

## what is the job outlook?

According to the Bureau of Labor Statistics, the job outlook for cable TV line installers and repairers is expected to decline sharply through the year 2006. Improvements in the technologies used in the cable system, including underground fiber optic cables, are making systems more efficient and reducing maintenance needs. Better systems also mean fewer technicians may be needed. Also, the biggest need for installers came in the 1980s, when millions and millions of American homes were being wired for cable for the first time.

A different kind of TV transmission technology, involving transmission over phone companies' fiber optic networks, is coming. Its exact effect on the cable TV system business remains to be seen.

Still, trade groups like NCTA remain confident that cable TV technicians will continue to be needed. Existing cable must be extended to new areas, rebuilt, or enhanced. And ongoing service for existing customers will continue to be necessary.

## how do I learn more?

### sources of additional information

Following are organizations that provide information on cable TV technician careers, education, and employers.

NCTA is primarily a lobbying group for the cable TV industry, but it offers an informative thirty-six-page brochure, "Careers in Cable Television," on the structure of and career opportunities in the business.

**National Cable Television Association**
1724 Massachusetts Avenue, NW
Washington, DC 20036
202-775-3550

National Cable Television offers self-study and correspondence courses in technical areas.

**National Cable Television**
801 West Mineral Avenue
Littleton, CO 80120-4501
303-797-9393

With the University of Denver's Center for Management Development, Women in Cable and Telecommunications awards a Certificate of Cable Management.

**Women in Cable and Telecommunications**
230 West Monroe Street
Chicago, IL 60606-4703
312-634-2330

# bibliography

Following is a sampling of materials relating to the professional concerns and development of cable TV technicians.

## Books

Bone, Jan. *Opportunities in Cable Television Careers.* Revised. Lincolnwood, IL: VGM Career Horizons, 1992.

## Periodicals

*Cable TV Technology.* Monthly. Covers advances in technology. Paul Kagan Associates, Inc., 126 Clock Tower Place, Carmel, CA 93923, 408-624-1536.

*Cable World.* Weekly. Cowles Business Media, Inc., 1905 Sherman Street, Denver, CO 80203, 303-837-0900.

*Cable TV and News Media Law & Finance.* Monthly. Discusses the legal and regulatory aspects of the cable industry. New York Law Publishing Company, 345 Park Avenue South, New York NY 10010, 212-545-6170.

# calibration laboratory technician

**Definition**

Calibration laboratory technicians test, calibrate, and repair electrical, mechanical, electromechanical, electronic, and other instruments used to measure, record, and indicate voltage, heat, magnetic resonance, and numerous other factors.

**Alternative job titles**

Engineering laboratory technicians

Quality assurance calibrators

Standards laboratory technicians

Test equipment certification technicians

**Salary range**

$28,000 to $33,000 to $38,000

**Educational requirements**

High school diploma; two years of military or technical school training, or associate's degree

**Certification or licensing**

Voluntary

**Outlook**

About as fast as the average

GOE
02.04.01

DOT
019

O*NET
22505B

## High School Subjects

Computer science
Mathematics
Physics

## Personal Interests

Computers
Figuring out how things work
Fixing things

Colorado, site of the National Institute of Standards Technology (NIST), there's a room where scientists establish the time standard for the entire country, maintaining a kind of "master clock" against which other clocks are set. By taking continuous readings from twenty to thirty special clocks, using complex mathematical formulas, and comparing the results to the international time standard and other sources, they tell the rest of us exactly what time it is to within an infinitesimal fraction of a second.

Consumers feel confident that a pound of butter in New Jersey will weigh exactly the same as a pound of butter in California. Why? Because in the United States, laws require commerce and industry to use standard weights and measurements. Every scale, for example, is supposed to be calibrated to the national standard, that is, checked against the "master scales" of NIST in Boulder and Washington, DC. This way, the government makes sure all stores are measuring out the same weight, and customers know they're getting their money's worth.

Similarly, there are standards for many aspects of U.S. manufacturing. During the production process, for example, different instruments are used to test a part or equipment as it's being made. Then, from time to time, the instrument itself must be checked to be sure it's still producing an accurate reading. Just as there is a time standard by which to set clocks, and a weight standard by which to calibrate scales, there are standards by which instruments used in manufacturing can be checked.

## what does a calibration laboratory technician do?

This is where *calibration laboratory technicians* come in.

People who develop the standards and check the instruments are called *metrologists.* Metrology is the science of measuring things. Calibration laboratory technicians are metrologists who work in calibration laboratories, running tests on various instruments. At higher levels, calibration laboratory technicians may help to develop new standards, such as for new instruments, and design the test procedures required to see if the instruments meet the standards.

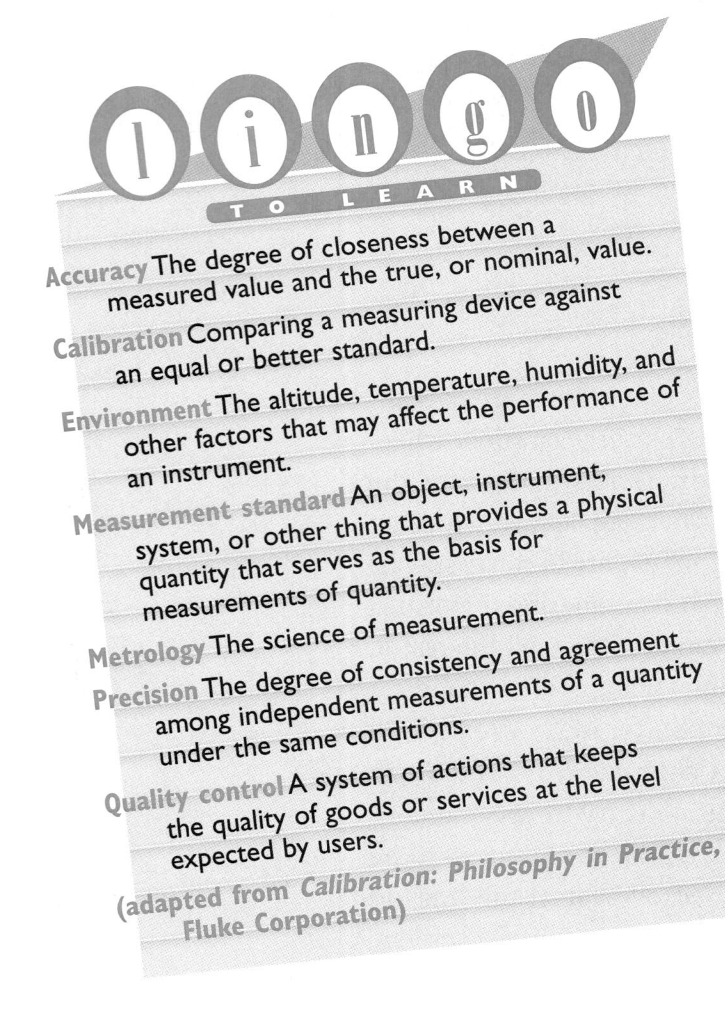


U.S. manufacturers, however, are not required by law to check their instruments against NIST or other standards. To do so is generally voluntary. One exception to this involves drug manufacturing. "The U.S. Food and Drug Administration (FDA) started to enforce new testing guidelines for pharmaceutical companies," explains Tom Kimbrell, director of the metrology program at Community College of Aurora, in Boulder, Colorado. "Previously, they were not required to integrate metrology into their research and development (R&D); now they have to."

But more and more U.S. manufacturers are working to standards—and using metrologists to help them achieve those standards—anyway. There are two important reasons for this.

First, meeting standards is a way of proving good quality control, and quality control is increasingly seen as a way to set yourself apart from the competition. If you can say, "We use quality control in our production process," you're more likely to stand out from other competitors for a job. If you can say, "Our quality control procedures include using trained metrologists to test our instruments," that's an added leg up on the competition.

In the global marketplace, ISO 9000 is further encouraging the use of metrology. This is a set of voluntary guidelines for helping to establish good quality control procedures and is recognized internationally. It's a mark of distinction to be able to say your company has received ISO certification and an important competitive edge in global business. One part of ISO 9000 specifies how to establish metrology labs and train metrology personnel.

Second, in some cases, a manufacturer is required by a customer to meet specific manufacturing standards. If it doesn't, it loses the contract. The U.S. Department of Defense (DOD), for example, and some private businesses require their vendors to manufacture according to strict standards as a condition of winning the job. If the end-product doesn't meet those standards, they can reject it and the manufacturer is out a lot of money.

A company may have its own internal standards. This is especially true when the manufacturing technology is so advanced that independent standards, like those from NIST, haven't been developed yet. More often, manufacturers work to standards designed by someone else, such as those developed by the military, NIST, or the International Standards Organization (ISO).

Metrologists help make sure manufacturers meet a given set of standards. For example, they might run a series of tests on oscilloscopes, which are used to measure electronic or electrical factors like voltage. The metrologist can put out a known quantity of voltage and see how close the oscilloscope comes to making the measurement. If it's close enough to conform to standards, it passes. If it's not, the metrologist tries to figure out if there are error sources present and if these are affecting the results. He or she also may have to decide how to remove, report, or reduce those errors. If the instrument's readings do not conform to the standard, the instrument might be pulled and sent to another lab for repair.

According to Kimbrell, currently about 50 percent of metrology work is in the area of electronics testing and calibration; about 40 percent is in the physical/dimensional area; and about 10 percent is in the mechanical. Specialized areas include laser, fiber optics, and microwave testing.

A lab for metrology or calibration may be part of a manufacturer's quality assurance department or the engineering, facilities, plant engineering, or manufacturing divisions. If the manufacturer doesn't have its own metrology lab, it may send its instruments to a specialized firm that does nothing but test other companies' instruments.

A calibration lab tends to be extremely clean and environmentally controlled. Factors like temperature, humidity, vibration, and static are kept at constant levels to help make sure variances don't interfere with the accuracy of the tests.

Besides running tests and developing standards and testing procedures, calibration laboratory technicians also document and report on their findings.

## what is it like to be a calibration laboratory technician?

Billy Burnam is a calibration laboratory technician Level I at the Johnson Space Center in Houston, Texas, a huge NASA facility employing fourteen thousand people. He's worked there for about a year and a half, since graduating from Texas State Technical College with a two-year associate's degree in metrology. "I've known about NASA since I was a little kid," says Billy, who grew up in Amarillo, Texas, six hundred miles from Houston. "The idea of working for them—just thinking I had an offer from them—was very exciting to me."

> "Standards are required for anything flight-related, like the power supplies used to charge the life support systems in space."

Like the DOD, NASA requires its vendors to meet specific manufacturing guidelines. This includes making sure that various instruments used in the production process are up to par. Periodically, the manufacturers send their instruments to NASA for testing. Billy runs the tests and determines such things as how many more weeks an instrument should be able to operate before it must be calibrated. "The goal of this testing is to verify that test instruments from NASA vendors are meeting NASA's standards," Billy says.

Accuracy and attention to detail are crucial. "Standards are required for anything flight-related," Billy says, "including things like the power supplies used to charge the life support systems required in space."

Typically, calibration laboratory technicians at Johnson work in one of three labs: an electronics lab, a physical/dimensional lab, or a reference lab. (A fourth lab, for fiber optics, is being developed.) Billy probably will have a chance to work in each of these labs during his career with Johnson.

But for now, he primarily works in the electronics lab, where oscilloscopes and other instruments used to measure electronic or electrical factors like voltage, current, and resistance are tested. The lab is kept very clean and at a temperature of about 68 degrees (+/- 2 degrees). "A variance in temperature could seriously affect your testing results," Billy notes.

Johnson technicians can choose to work either five eight-hour days (7:30 AM to 4:30 PM) or four ten-hour days (6:30 AM to 5:30 PM). Billy works a four-day shift, rotating his day off. Right now, there is no overtime.

Because he's still fairly new, Billy spends most of his time running automated tests in the lab. Computers help him select and run the appropriate test sequence, troubleshoot, and record results.

On a typical day, Billy might check a batch of five to ten oscilloscopes that have come in for testing from various manufacturers, using one of several Fluke 5700 multifunction calibration stations in the lab. These are automated stations capable of running a variety of different tests on the instruments. For example, they can provide dc voltage and current, resistance, low frequency and RF ac voltages to 30 MHz, and low frequency ac current. After warming up the system, Billy tells the computer what kind of instrument is to be tested, automatically starting the test procedure. "It controls all your standards automatically," he says. "For example, it puts out 10 volts, instead of you having to put out 10 volts."

Testing may take about thirty minutes to an hour for each instrument, "longer if you need to make adjustments or repairs," he says. "Some tests can take up to four hours." (In other industries, tests may take days or even weeks.) If Billy finds that a reading is out of tolerance, he refers to a standards manual describing how to adjust it. "I may 'tweak it' until it's back to where it should be," he says. He primarily traces to NIST standards, although NASA also has some specialized standards associated with its space flight technology.

After the tests, Billy enters the results in the computer and compiles a report. "Every instrument is on a variable schedule," he says. "If it performs well, we'll increase the number of weeks it can remain in operation. If it doesn't, we'll decrease the number of weeks it can stay in operation or pull it for repair." The computer calculates the date when the instrument is due back for another test, and the instrument is stamped with that due date. "We put the instruments in bins by category, and someone takes them back to the customer," Billy says.

From time to time, Billy does "in-place testing" at a manufacturer's plant, generally one located in the Houston area. This is required when an instrument is part of a large testing station and can't be brought into the calibration lab. In the future, Billy also will have the opportunity to develop the actual standards and test procedures followed by technicians.

**To be a successful calibration laboratory technician, you should**

- Be patient, persistent, attentive to detail, and have a strong sense of ethics
- Enjoy problem-solving using mathematical concepts
- Have manual dexterity and good hand-eye coordination
- Be able to communicate to others, both orally and in writing

## have I got what it takes to be a calibration laboratory technician?

Patience, persistence, curiosity, integrity, a healthy sense of ethics, and the desire to be truly accurate—not just close—are all good traits for a calibration laboratory technician. Accuracy is extremely important because, as with the aerospace-manufacturing testing done at Johnson, a test done poorly can endanger a person using the end product. "Depending on the business, there can be lives involved," notes Kimbrell. "It's a big responsibility."

"It's helpful to have a certain amount of perfectionism," Billy says. "And also a desire to make everything work better."

In addition to technical prowess and problem-solving ability, the technician should have good interpersonal and communication skills—like the ability to admit mistakes and ask questions when you don't understand something, and to work well with others. A strong aptitude in math and electronics, good decision-making skills, and the ability to communicate technical information in reports also help.

Says Don Dalton, vice president of education and training for Fluke Corp., Seattle, "Metrologists tend to be introverts. Very methodical, very unrushed. They need a high degree of patience. For example, they may have to take ten readings of something and average them together. There typically is a lot of mathematics, a lot of statistics, involved."

Kimbrell emphasizes the importance of ethics. "This is a point of pride in the metrology community," he says. "You have to be right—even when there's another job waiting in the wings for you, some other test or report; even when someone's breathing down your neck for results. You're under pressure and the temptation may be to let things slide, but you can't." Kimbrell tells students in his introductory course to metrology that sometimes "you have to be hard-headed" with supervisors or co-workers who may be pressuring you to hurry up.

Kimbrell has been in the field for thirty-five years and has an obvious love for metrology. "In my experience, people either love it or hate it," he says. "It's very exacting work. You have to be very inquisitive, to want to know what's going on around you."

## in-depth

### What Is ISO 9000?

ISO 9000 is part of a worldwide system of standards for measurements, products, and services. As more U.S. companies turn their attention to the "global marketplace," they are increasingly interested in understanding and conforming to these standards as a way of gaining a competitive edge.

International standards organizations fall into two basic types: those for physical standards ("treaty" organizations) and those for paper standards ("nontreaty" organizations).

Physical standards are specific references or artifacts for making measurements. Countries that agree to trace their calibrations to these standards (chiefly, technologically advanced countries) have actually signed a treaty (the Treaty of the Meter) saying they'll do so. Thus, organizations that establish and maintain these international physical standards are called "treaty" organizations.

Paper standards are more like guidelines for quality control, suggesting "what to do and how to do it." For example, they might recommend that companies have a metrology lab and employ trained personnel to run them. Paper standards are more informal, and they're mostly voluntary—i.e., no one signed a treaty agreeing to accept them. Hence, the organizations that establish and maintain these standards are known as "nontreaty" organizations.

The International Standards Organization (ISO), then, is one of several international, nontreaty organizations for paper standards. Some of its counterparts focus on specific disciplines: for example, there's an organization for electrical metrology, and an organization for law enforcement-related metrology. But ISO writes standards for many different areas. ISO 9000 is just one of the sets of mainly voluntary, nontreaty, paper standards from the ISO (there also is an ISO 9001, 9002, and so on).

ISO 9000 is not concerned solely with metrology; it also describes manufacturing standards, contract standards, and so forth. "It's mainly about total quality management (TQM), of which calibration is a part," says Wilbur Anson of NCSL.

If ISO 9000 is voluntary, how come so many U.S. companies care so much about it? "It's not a requirement, but it is a 'pressure,'" Anson says. "It's becoming more important in the 1990s as one aspect of surviving economically and being competitive, both here and abroad. More companies want to penetrate the European market or the world market; the world market involves ISO; those who want to sell abroad feel a 'pressure' to conform to these standards. Even at home, it's good to be able to put in a contract; it gives a company a competitive edge."

ISO and other international standards groups don't just make things up arbitrarily: they get input from all the national labs of different countries, and then write the standard. For the U.S., the American National Standards Institute (ANSI) is the liaison between the U.S. technical community and ISO: it relays U.S. ideas about standards to the ISO. Once the ISO standard is established, ANSI supports and helps spread it in the United States.

"There is a prestige that comes with the job," he adds. "Your work is vital. Engineers and others respect you because of your knowledge."

Good hand-eye coordination and manual dexterity are helpful physical traits in performing the work. "It all comes into effect," Billy says. "You also have to pay attention; it can be dangerous at times. For example, for a test, you could be putting out 14,000 volts. You have to be careful."

Other helpful skills include having a good understanding of how to read specifications. "Manufacturers can bend and twist specifications in many different ways," Billy says. "You have to be able to determine what they're saying." And, as with many other jobs, he adds, "the ability to learn on the job is very important."

Billy says that the hardest part of his job came at the beginning, when he was getting used to all the standards and instruments. "There is so much equipment and thousands of standards," he says. "Every instrument goes by a number and has different specifications; someone will just say something like, 'Bring me the 5100,' and you have to know what they're talking about."

Billy says he sometimes wishes he knew more specifically what the instruments he tests are used for; he's curious about what goes on outside the lab. He's also anxious to explore more areas of metrology at Johnson. "They're considering having me work in all of the different labs, moving me around, and I'd love that," he says. "I want to see what's going on as much as possible."

## how do I become a calibration laboratory technician?

"I'd never even heard of metrology," says Billy, until he visited Texas State Technical College and heard a talk given by Kimbrell, who taught there at the time. Billy had been studying electronics at another college, but "[Kimbrell's] enthusiasm for metrology sparked my enthusiasm. The field seemed exciting and to fit in well with my interests," which include computers and electronics. Billy transferred to TSTC and became one of the first people to earn a metrology degree from the school.

Billy recalls a session in statistics during a quality control class at TSTC. "We used calipers to test the width of a number of nickels and mapped out the results on a graph," he says. "For the most part, you could see there was little variance in the width of the nickels, which shows how well they're made. But one guy's readings started out really bad, though they gradually got better. As a class, we discovered the variance came from the fact that this guy wasn't experienced at using calipers. This was reflected in his test results: they started out badly and gradually improved as his skill with the calipers improved."

FYI

The recent focus on quality control means a wider range of calibration jobs in U.S. manufacturing industries, as companies learn that high standards improve profitability.

## education

### High School

"Automation is creating a new work force," Kimbrell says. "You can teach someone from high school to pull up a program on the machine and run the test. But when a problem arises, these kinds of operators usually don't know how to fix it. That's why we teach things like programming and troubleshooting the system" in two-year associate's degree programs at TSTC and the Community College of Aurora. "Industry is requiring us to do this."

"When I was in high school," Billy says, "I really liked math. I'm not one of those guys who can look at a problem and instantly know the answer. But I can work at it until I get it right, and I like the challenge." Algebra and trigonometry have been the most useful on the job for him so far, but "I'm also starting to use statistics more," he says. "I also took a very, very basic class in electronics in high school," he adds, "just enough to get me interested in it."

English classes helped, too. "We do a lot of technical writing on the job," Billy says. "If it's a new instrument that's never been tested before, we have to write a four- to five-page procedure for it. Or sometimes we have to update an old procedure."

Other subjects to study in high school include physics and computer science. "Computers are very important in this field," Billy says. He has one at home, and he enjoys taking it apart to make repairs and see how it works.

### Postsecondary Training

Traditionally, postsecondary training for metrologists has been from the military or technical schools. These options are still viable, although military training programs have been cut back in recent years. Billy's college degree in metrology is still somewhat unusual in the field, but this method of training may represent a growing trend. "While it's still fairly simple or straightforward to enter the field," says Wilbur Anson, business manager for NCSL, "things are starting to change. Today's technicians really need a higher level of education. They should have the equivalent of college sophomore physics and know calculus concepts, trigonometry, and a smattering of statistics."

College degree programs in metrology typically are two years in length. "Metrology by itself could never be a four-year degree," Anson says. However, he adds, more four-year degree programs in engineering, physics, computer science and other areas are starting to incorporate metrology classes in the curriculum.

No matter where the technician is trained, experts say, it's very important to receive as broad an education in metrology as possible. Technicians should study all of the primary areas of the field, including electronic, physical/ mechanical, and dimensional (including fiber optics). "You need the flexibility," Anson says. "That way, if your company downsizes and your particular lab is cut, or if your industry is hit by bad economic times, you'll still be employable. The day is gone when you can expect to start a career in one field and do the exact same thing for thirty years—or stay with the same company for ten years. In order to survive, a calibration laboratory technician needs as good a basic knowledge as he or she can get, with an eye toward being flexible in the future."

"The calibration technician needs to have knowledge that can be used throughout the plant," Anson adds. "With today's distributed measurement, he or she must be able to go out into the plant, troubleshoot the systems, and prescribe a solution."

Kimbrell helped to design the metrology degree curriculum for Texas State Technical College and the Community College of Aurora, and he agrees that students should receive as broad an education in the field as possible. "At TSTC, for example, they cover electronics, physical/ dimensional and mechanical skills, as well as repair," he says.

People in the field also make a point of getting ongoing training in order to keep up with new technology. Manufacturers like Fluke and Hewlett Packard might offer one- or two-week training seminars in metrology practice or in new equipment. Associations, consultants, and colleges also provide ongoing training programs. NCSL's *Training Information Directory,* published annually, describes classes, self-study courses, videos, textbooks, and reference materials for ongoing education in the areas of metrication and metrology management; general metrology and digital technology; electrical metrology; physical metrology; and dimensional and optical metrology.

## certification or licensing

Currently, there is no industry-wide mandatory certification or licensing system for technicians. NCSL and other groups are exploring the possibility of voluntary accreditation for laboratories.

## internships and volunteerships

Billy did not complete an internship while going to college. "Probably because the metrology program there is still relatively new," he says. "I imagine they'll be putting more of those kinds of programs in place in the future." Billy did, however, work in one of the school's labs while earning his degree.

Private companies may offer internships on their own. NASA's Sharp Program, for example, gives two or three young people a year the opportunity to work in the Johnson facility during the summer. At the end of the summer, they report to managers on what they did and what they learned. Students should contact individual companies for more information on possible internships.

## who will hire me?

As mentioned earlier, there is a drive in U.S. manufacturing to increase quality control, and that's boosting interest in metrology. Throughout a wide range of industries, manufacturers are finding that exacting standards can help them improve profitability: There are fewer rejected products, and less wasted time and materials. Pressure to achieve rigorous standards also comes from external factors like increased competition, customer request, FDA guidelines (for biomedical-pharmaceutical companies), the move toward ISO 9000 international certification, and so forth.

Therefore, calibration laboratory technicians are much more likely to be found in a wider range of manufacturing industries than their counterparts of a generation ago. Aerospace-defense and electronics manufacturers, the military, and the government are the traditional employers of metrologists and still use them today. However, technicians also are being used more and more by automotive, biomedical-pharmaceutical, chemical-process, nuclear energy, and other types of companies. They also are employed in university research and development facilities and by consultants.

A look at some of the participants in a recent NCSL conference illustrates the diversity of employers potential. They included companies like Abbott Labs, AT&T Capital Corporation, Martin Marietta Energy Systems, 3M Metrology Laboratory, Baxter Healthcare, Eastman Kodak, Fluke Corporation, and Hewlett Packard, in addition to Lockheed Missiles & Space, NASA, the U.S. Air Force, Army, and Navy, and the U.S. Department of Agriculture.

**Advancement Possibilities**

**Metrology engineers,** or **metrologists,** develop and evaluate calibration systems that measure characteristics of objects, substances, or phenomena, such as length, mass, time, temperature, electric current, luminous intensity, and derived units of physical or chemical measure.

**Quality control managers** plan, coordinate, and direct quality control programs designed to ensure continuous production of products consistent with established standards.

**Standards engineers** establish engineering and technical limitations and applications for items, materials, processes, methods, designs, and engineering practices for use by designers of machines and equipment.

Billy interviewed with Johnson and took the job on graduating from Texas State Technical College. "I had offers from two or three medical companies, too," Billy says, "but I really wanted to work for NASA." Other methods of entering the field include placement following training in the military or in technical colleges. "Most of the guys in the lab here at Johnson came out of the military or a technical school," Billy says.

Associations like NCSL (whose members are companies, rather than individuals) are good sources of information about employers and employment opportunities. Trade magazines and newspapers also may have classified ads describing available positions (*see* "How Do I Learn More?").

## where can I go from here?

At Johnson Space Center, Billy currently is classified as a Technician I. The levels go up to Technician IV, and then manager/supervisor. Advancement is based on skill and seniority. Johnson pays for additional schooling, and Billy says he'd consider getting a four-year degree. "I'd probably get a major in physics and a minor in math," he says.

This is a good strategy for calibration laboratory technicians who wish to advance. Professionals like engineers and physicists are more likely to design the standards and develop the testing procedures executed by the calibration laboratory technicians. This is not only more interesting to some people but also makes the technician more valuable to employers who need to develop their own internal standards.

"Today, companies like Lockheed, Hewlett Packard, and so forth will have engineers in their metrology department," Anson says. "There's pressure today for even more stringent standards than those provided by NIST. So there's a need for engineers who can extrapolate from NIST standards and help to develop and integrate these more stringent internal standards." Thus, technicians who go back to school and earn a four-year degree become more valuable to their employers.

## what are some related jobs?

As noted earlier, calibration laboratory technicians are employed in a wide range of manufacturing industries. An example of a related job is an automobile manufacturing laboratory technician, who tests the chemical and physical properties of materials used in the production of cars.

The U.S. Department of Labor classifies calibration laboratory technicians under the headings *Laboratory Technology* (GOE) and *Occupations in Architecture, Engineering, and Surveying, Not Elsewhere Classified* (DOT). Under these headings are people who apply equipment design and calibration/test technology in agricultural, automotive, biomedical, and other industries; technicians who build and test laser devices; and technicians who design and implement various in-line quality control testing systems.

**Related Jobs**

- Aeronautical and aerospace technicians
- Agricultural equipment test technicians
- Avionics technicians
- Biomedical electronics technicians
- Biomedical engineering technicians
- Biomedical equipment technicians
- Electronics test technicians
- Instrumentation technicians
- Laboratory technicians, auto manufacturing
- Laser technicians
- Quality control technicians

## what are the salary ranges?

According to a 1995 benchmark study by the NCSL, salaries are holding steady or rising across most of the metrology field. The range is about $28,000 for a beginner to between $34,000 and $38,000 for experienced technicians. Managers/supervisors can earn more than $50,000. Technicians who go on to earn degrees in engineering, physics, or other related disciplines also may earn $40,000 to $50,000 or more.

Salary varies by region and by type of industry in which the technician works. NCSL's survey turned up the highest salaries in the western United States. By specialty, aerospace-defense, government, and other traditional employers of metrologists reported salaries that were above the national averages for all industries. Technicians in the nuclear energy industry showed some of the highest rates versus all other industries.

## what is the job outlook?

As noted earlier, calibration laboratory technicians are becoming increasingly important as manufacturers work to achieve such quality standards as ISO 9000. However, somewhat offsetting this are cutbacks in areas of the defense industry and other traditional employers of calibration laboratory technicians as well as an increase in automation, which permits tests to be run faster and with fewer people. "Automation definitely has an impact," Billy says. "A test on an oscilloscope, for example, that used to take three or four hours on the bench," that is, a manual test, "now can be done in thirty to forty minutes."

While some tests continue to be done manually, such as for certain dimensional standards, automation will continue to grow. One new development is the so-called "smart instruments" that can run tests on themselves and signal when they need to be calibrated.

Still, there are opportunities for calibration laboratory technicians. Much depends on the technician's breadth of education in the field and choice of industries. According to Billy, Kimbrell, and other industry insiders, the biomedical-pharmaceutical industry promises some of the best prospects for growth. The industry as a whole is well paid, and FDA plans to enforce guidelines for these companies' research and development standards will mean increased employment of metrologists. "If any area is growing, it's that one," Kimbrell says. "If I had to recommend any one area for new people, it would be pharmaceutical-medical," agrees Billy. "Health care's going to be a good area."

Newer specialties like fiber optics testing also should provide opportunities. Fiber optics are key in advanced communication systems. Lighter and faster than wire, fiber optics are attractive to a wide range of manufacturers, including NASA. The technology for it is still new enough that NIST and private companies are still developing the standards for testing and calibrating fiber optics communication test instruments. These include things like optical power meters, attenuators, tunable lasers, and other highly sophisticated measuring devices. Testing may involve such things as sending signals to test the breakdown point, seeing how it acts at different temperatures, doing stress tests, and so on.

Generally, calibration laboratory technicians should continue to have opportunities in manufacturing industries where quality control and extremely precise, documented manufacturing standards are required. "There's going to be a need long-term for calibration skills," Anson says. "But as with all technical careers, the calibration laboratory technician is going to have to have a broad area of expertise."

## how do I learn more?

### sources of additional information

Following are organizations that provide information on calibration laboratory careers, accredited schools, and employers.

**Institute of Electrical and Electronic Engineers, Inc. (IEEE)**
1828 L Street, NW, Suite 1202
Washington, DC 20036-5104
202-785-0017
http:///www.ieee.org/usab

**Instrument Society of America (ISA)**
67 Alexander Drive
Research Triangle Park, NC 27709
919-549-8411

**National Conference of Standards Laboratories (NCSL)**
1800 30th Street, Suite 305B
Boulder, CO 80301-1026
303-440-3339
http://www.ncsl-hq.org

# bibliography

Following is a sampling of material relating to the professional concerns and development of calibration laboratory technicians.


## Books

Fluke Corporation. *Calibration: Philosophy in Practice.* Everett, WA: Fluke Corporation, 1993.

## Periodicals

*Control Engineering.* Fourteen times per year. Provides information on control and instrumentation systems and equipment. Cahners Publishing, 1350 East Touhy Avenue, Des Plaines, IL 60018-5080, 847-390-2780.

*IEEE Transactions on Instrumentation and Measurement.* Bimonthly. Discusses measurement and instrumentation involving electrical and electronic means. Institute of Electrical & Electronics Engineers, Inc., 345 East 47th Street, New York, NY 10017-2934, 732-981-0060.

*Instrumentation Science & Technology.* Quarterly. Technical information on numerous subjects. Marcel Dekker Journals, 270 Madison Avenue, New York, NY 10016, 212-696-9000.

*Measurements & Control.* Bimonthly. (Measurements) Measurements & Data Corp., 2994 West Liberty Avenue, Pittsburgh, PA 15216, 412-343-9666.

*National Conference of Standards Laboratories Newsletter.* Quarterly. National Conference of Standards Laboratories, 1800 30th Street, Suite 305B, Boulder, CO 80301, 303-440-3339.

*Test & Measurement World.* Thirteen times per year. Covers test, measurement, and inspection products and services in the electronics industry. Cahner's Publishing Co., 275 Washington Street, Newton, MA 02158-1630, 617-558-4367.

# cardiovascular technologist

**Definition**

Cardiovascular technologists support physicians in the diagnosis and treatment of heart and related blood vessel ailments.

**Alternative job titles**

Cardiac monitor technicians

Cardiology technologists

Echocardiography technologists

Electrocardiograph (EKG) technicians and technologists

Holter monitor and stress test technologists

Vascular technologists

**Salary range**

$15,000 to $22,000 to $32,000

**Educational requirements**

Some postsecondary training

**Certification or licensing**

Voluntary

**Outlook**

Little change or more slowly than the average

GOE 10.03.01

DOT 078

O*NET 32925

## High School Subjects

Anatomy and Physiology
Biology
English (writing/literature)
Health
Sociology

## Personal Interests

Helping people: emotionally
Helping people: physical health/medicine
Science

A major vessel in the patient's heart is blocked, potentially endangering the patient's life; he must undergo a special procedure that involves forcing a tube through the artery to "unblock" the obstruction. A team of health care professionals who specialize in diagnosis and treatment of heart ailments is assembled to carry out the procedure: On hand are a cardiologist (heart doctor), nurse, and various tech support people, including a highly trained cardiology technologist, which is a type of cardiovascular technologist.

After the patient is prepped, this special team begins its work. The physician begins by inserting a tube into the patient's leg. Slowly, carefully, the fine tube is woven up through the arteries and into the patient's heart. As she works, the physician sees an "internal view" of the tube making its way to the heart on the monitor of a special imaging device that's taking a video of the procedure. The cardiology technologist is standing by, all senses alert, checking the view on the monitor, making adjustments to the camera as needed, entering information about the procedure into a computer, and providing other support. Afterward, she will process the film obtained from the camera for use by the doctor.

This procedure is not without risk; sometimes, it doesn't work, and about 2 percent of patients—primarily older patients, weak patients, or those with very bad heart disease—may suffer from an infection, heart attack, or stroke while undergoing it. However, this patient is lucky: The obstruction is successfully cleared, and he is spared the need for open-heart surgery.

## what does a cardiovascular technologist do?

Congenital heart disease. Acquired heart disease. Coronary artery disease. Peripheral vascular disease. Heart disease of all kinds is still the leading killer of men and women in this country, despite increased awareness in recent years of the ill effects on the heart of stress, poor diet, lack of exercise, smoking, and other unhealthy behaviors. As the Baby Boomer generation ages, health professionals are expecting to see the number of coronary patients increase.

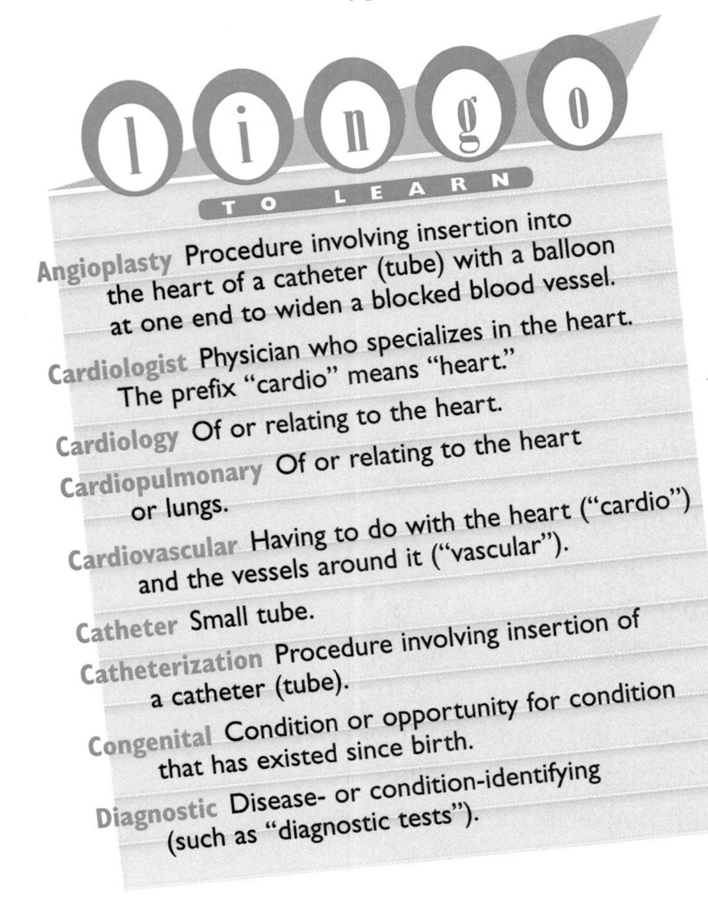


Technologists who assist physicians in the diagnosis and treatment of heart disease are known as *cardiovascular technologists.* ("Cardio" means heart; "vascular" refers to the blood vessel/circulatory system.) They include *electrocardiograph (EKG) technologists, Holter monitoring and stress test technologists, cardiology technologists, vascular technologists* and *echocardiographers* (both ultrasound technologists), *cardiac monitor technicians,* and others. The services of EKG technologists may be required throughout the hospital, such as in the cancer wards or emergency room; there may be a separate department for these EKG professionals. Increasingly, however, hospitals are centralizing cardiovascular services under one full cardiovascular "service line," all overseen by the same administrator. According to a spokesperson at the American Academy of Medical Administrators, "This is because cardiology services is the hottest area in health care today. At the present time, it is continuing to emerge, unfold, and expand."

In addition to cardiovascular technologists, the cardiovascular team at a hospital also may include radiology (X-ray) technologists, nuclear medicine technologists, nurses, physician assistants, respiratory technologists, and respiratory therapists. For their part, the cardiovascular technologists contribute by performing one or more of a wide range of procedures in cardiovascular medicine, including invasive (enters a body cavity or interrupts normal body functions), noninvasive, peripheral vascular or echocardiography (ultrasound) procedures. In most facilities they use equipment that's among the most advanced in the medical field; drug therapies also may be used as part of the diagnostic imaging procedures or in addition to them. Technologists' services may be required when the patient's condition is first being explored; before surgery; during surgery (cardiology technologists primarily); and/or during rehabilitation of the patient. Some of the work is performed on an outpatient basis.

Depending on their specific area of skill, some cardiovascular technologists are employed in nonhospital health care facilities. For example, EKG technologists may work for clinics, mobile medical services, or private doctor's offices. Their equipment can go just about anywhere. The same is true for the ultrasound technologists.

Some of the specific duties of cardiovascular technologists are described in the next sections. Exact titles of these technologists often vary between medical facilities because there is no standardized naming system.

### Electrocardiograph (EKG) Technologists

*Electrocardiograph (EKG) technologists* use an electrocardiograph (EKG) machine to detect the electronic impulses that come from a patient's heart during and between a heartbeat. The EKG machine then records these signals on a paper graph called an *electrocardiogram.* The electronic impulses recorded by the EKG machine can tell the physician about the action of the heart during and between the individual heartbeats. This in turn reveals important information about the condition of the heart, including irregular heartbeats or the presence of blocked arteries, which the physician can use to diagnose heart disease, monitor progress during treatment, or check the patient's condition after recovery.

To use an EKG machine, the technologist attaches electrodes (small, disk-like devices about the size of a silver dollar) to the patient's chest. There are wires attached to the electrodes which lead to the EKG machine. Up to twelve leads or more may be attached. To get a better reading from the electrodes, the technologist may first apply an adhesive gel to the patient's skin that helps to conduct the electrical impulses. The technologist then operates controls on the EKG machine or (more commonly) enters commands for the machine into a computer. The electrodes pick up the electronic signals from the heart and transmit them to the EKG machine. The machine registers and makes a printout of the signals, with a stylus (pen) recording their pattern on a long roll of graph paper.

During the test, the technologist may move the electrodes in order to get readings of electrical activity in different parts of the heart muscle. Since EKG equipment can be sensitive to electrical impulses from other sources, such as other parts of the patient's body or other equipment in the room, the technologist must watch for false readings.

After the test, the EKG technologist takes the electrocardiogram off the machine, edits it or makes notes on it, and sends it to the physician (usually a cardiologist, or heart specialist). Physicians may use computers to help them use and interpret the electrocardiogram; special software is available to assist them with their diagnosis.

EKG technologists don't have to repair the EKG machine, but they do have to keep an eye on it and know when it's malfunctioning so they can call someone to fix it. They also may keep the machine stocked with paper.

Of all the cardiovascular tech positions, EKG technicians/technologists are the most numerous. They made up about half of the approximately thirty-two thousand cardiovascular technologists employed in 1996.

FYI

**Women and Heart Disease: A Hidden Problem**

Were you surprised to read that heart disease is the leading killer of both American men and women? You're not alone. The perception is that heart disease primarily haunts men; everyone thinks of the middle-aged, overweight Type A business executive as being the most likely to have a heart attack. But the truth is that women have about half of the cases of heart disease in this country and suffer from half of the heart attacks (which kill six times more women than breast cancer every year).

Men tend to have heart attacks beginning around age forty, while women are more likely to have one after age sixty. Physicians think this may be due to the loss of estrogen in women after menopause. The last several years have seen the medical profession begin to make a stronger effort to include female subjects in their heart disease studies and otherwise recognize that heart disease is an equal-opportunity killer.

### Holter Monitor Technologists and Stress Test Technologists

Holter monitoring and stress testing may be performed by *Holter monitor technologists* or *stress test technologists,* respectively, or may be additional duties of some EKG technologists. In Holter monitoring, electrodes are fastened to the patient's chest and a small, portable monitor is strapped to the patient's body, at the waist, for example. The small monitor contains a magnetic tape or cassette that records the heart during activity—as the patient moves, sits, stands, sleeps, etc. The patient is required to wear the Holter monitor for up to twenty-four to forty-eight hours while he or she goes about normal daily activities. When the patient returns to the hospital, the technologist removes the magnetic tape or cassette from the monitor and puts it in a scanner to produce audio (sound) and visual representations of heart activity. (Hearing how the heart sounds during activity can help the physician diagnose a possible heart condition.) The technologist reviews and analyzes the information revealed in the tape. Finally, the technologist may print out the parts of the tape that show abnormal heart patterns or make a full tape for the physician.

Stress tests record the heart's activity during physical activity. In one type of stress test, the technologist hooks up the patient to the EKG machine, attaching electrodes to the patient's arms, legs, and chest, and first obtains a reading of the patient's resting heart activity and blood pressure. Then, the patient is asked to walk on a treadmill for a certain period of time while the technologist and the physician monitor the heart. The speed of the treadmill is increased so that the technologist and physician can see what happens when the heart is put under higher levels of exertion.

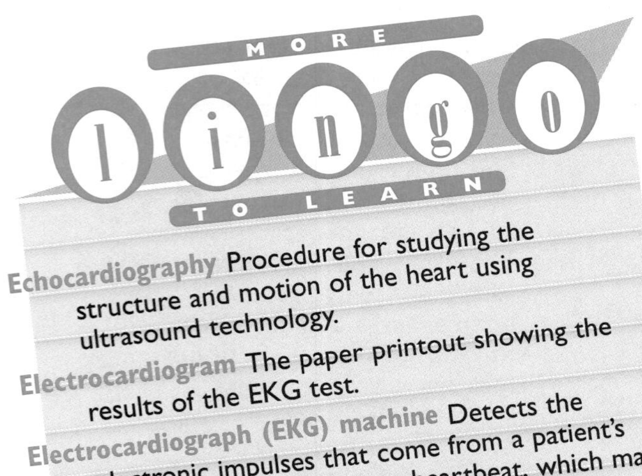


**Echocardiography** Procedure for studying the structure and motion of the heart using ultrasound technology.

**Electrocardiogram** The paper printout showing the results of the EKG test.

**Electrocardiograph (EKG) machine** Detects the electronic impulses that come from a patient's heart during or between a heartbeat, which may reveal heart abnormalities, and records that information in the form of a paper graph called an electrocardiogram.

**Electrode** Device that conducts electricity.

**Holter monitor** Cardiac-function monitoring device.

**Invasive** A medical procedure that penetrates into a body cavity or interrupts normal body functions; examples in cardiology include cardiac catheterization procedures.

**Noninvasive** A medical procedure that does not penetrate into a body cavity or interrupt body functions; examples in cardiology include ultrasound tests.

**Phonocardiograph** Sound recordings of the heart's valves and of the blood passing through them

**Radiographs** X rays.

**Vascular** Relating to the blood vessels. Vascular technologists are concerned the blood vessels around the heart.

## Cardiology Technologists

*Cardiology technologists* specialize in providing support for cardiac catheterization (tubing) procedures. These procedures are classified as "invasive" because they require the physician and attending technologists to enter a body cavity or interrupt normal body functions. In one cardiac catheterization procedure—an angiogram—a catheter (tube) is inserted into the heart (usually by way of a blood vessel in the leg) in order to diagnose the condition of the heart blood vessels, such as whether there is a blockage. In another procedure, known as angioplasty, a catheter with a balloon at the end is inserted into an artery to "widen" it. Angioplasties are being performed with increasing frequency: about three hundred thousand in 1997, compared with about six thousand in 1983. Cardiology technologists also perform a variety of other procedures.

Unlike some of the other cardiovascular technologists, cardiology technologists are actually in on surgical procedures. They may assist in surgery by helping to secure the patient to the table, setting up a 35mm video camera or other imaging device under the instructions of the physician (to produce images that assist the physician in guiding the catheter through the cardiovascular system), entering information about the surgical procedure (as it is taking place) into a computer, and providing other support. After the procedure, the tech may process the angiographic film for use by the physician. Cardiology technologists may also assist during open-heart surgery by preparing and monitoring the patient, and may participate in placement or monitoring of pacemakers.

## Vascular Technologists and Echocardiographers

These technologists are specialists in noninvasive cardiovascular procedures using ultrasound equipment to obtain and record information about the condition of the heart. Ultrasound equipment is used to send out sound waves to the part of the body being studied; when the sound waves hit the part being studied, they send back an echo to the ultrasound machine. The echoes are "read" by the machine, which creates an image on a monitor, permitting the technologist to get an instant "picture" of the part's condition.

*Vascular technologists* are specialists in the use of ultrasound equipment to study blood flow and circulation problems. *Echocardiographers* are specialists in the use of ultrasound equipment to evaluate the heart and its structures such as the valves. (Ultrasound also is used in other medical procedures, perhaps most familiarly in capturing images of a fetus to check its condition and learn what sex it is; see the chapter "Sonographer.")

### Cardiac Monitor Technicians

*Cardiac monitor technicians* are similar to, and sometimes perform some of the same duties as, EKG technologists. Usually working in the intensive care unit (ICU) or cardio-care unit of the hospital, cardiac monitor technicians keep watch over all the screens that are monitoring the patients to detect any sign that a patient's heart is not beating as it should.

Cardiac monitor technicians begin their shift by reviewing patients' records to familiarize themselves with the patient's normal heart rhythms, the current pattern, and what types of problems have been observed. Throughout the shift, the cardiac monitor technician watches for heart rhythm irregularities that need prompt medical attention. Should there be any, he or she notifies a nurse or doctor immediately so that appropriate care can be given.

In addition to these positions, there may be other cardiovascular technologists, depending on the specific health care facility. For example, a *cardiopulmonary technologist* specializes in procedures for diagnosing problems with the heart and lungs. He or she may conduct electrocardiograph, phonocardiograph (sound recordings of the heart's valves and of the blood passing through them), echocardiograph, stress testing, and respiratory test procedures.

Cardiopulmonary technologists also may be in on cardiac catheterization procedures, measuring and recording information about the patient's cardiovascular and pulmonary system during the procedure and alerting the cardiac catheterization team of any problems.

*Nuclear medicine technologists*, who use radioactive isotopes in diagnosis, treatment, or studies, may be in on diagnosis or treatment of cardiology problems. *Radiology, respiratory,* and *exercise technicians and therapists* also may assist in patient diagnosis, treatment, and/or rehabilitation.

## what is it like to be a cardiovascular technologist?

"Every day is different" in the life of a cardiovascular technologist, according to EKG technologist Larry Taylor at Illinois Masonic Hospital, Chicago. "We do EKGs of cardiac patients as needed, whether presurgery, postsurgery or, on an out-patient basis, in cases where there's a suspected heart problem," Larry says.

The hospital has a separate cardiology department, and Larry spends part of his time there, but his work actually takes him all over the hospital. "I also work in the cancer ward and AIDS ward, because these patients often need EKGs, too. We're also designated as a trauma unit and I'm in on emergency work. Some of that can be quite messy—things like gunshot victims or victims of multiple motor-vehicle [accidents]." At Illinois Masonic, separate technicians perform the different procedures for cardiovascular care such as stress tests, Holter monitoring, and 2D echo (two-dimensional echocardiograph).

FYI

Average time elapsed between a heart attack and the arrival of the patient at a hospital: four hours.

"The EKG we use is actually a microprocessor, from Hewlett-Packard," explains Larry. "It senses the electrical impulse that comes from the heart and produces a tracing that looks like sound waves. When there's a problem, the tracing looks different; we're trained to watch out for these irregularities.

"Sometimes we'll talk directly with the physician about the EKG, but that's not very common," he adds. "More likely, we'll make notes on the EKG printout. For example, when a person comes out of surgery all bandaged up, or when the patient has an extreme amount of pain, we may have to put the electrodes someplace on the chest other than where they would normally go. We'll put a note on the electrocardiogram to explain to the physician why we had to do this."

Though working with severely ill or injured patients can be emotionally difficult, says Larry, "in general, I like the working environment. I'm more like an independent contractor; there's no one constantly looking over me. You know what you have to do, and you do it. You can take the initiative and make the job your own."

Larry is on the first shift of two at the hospital, getting off work at 3:30 PM each day. He's been doing the job for three years and plans to continue in the profession.

Do patients get nervous about the procedures? "Not usually," says Amina Moten, who does EKG, stress testing, and Holter monitoring procedures in the cardiology unit at Rush North Shore Hospital, Skokie, Illinois. "But if patients are nervous, you can just talk to them and calm them down." Like Larry, Amina performs a varying number of the procedures each day, depending on the patient work load and number of technicians on hand. Typically, there are five working each shift.

Amina works the first shift plus every other Sunday. She's been doing the job for about four years, since coming to the United States from Pakistan and going through a training program. "I like the work, and I like the money," she says. "There's really nothing about the job that I don't like."

Evangelista (Eva) Torres performs Holter monitoring and stress testing at St. Joseph Hospital and Health Care Center in Chicago. She had training in Holter monitoring first, and then "stress testing came about a year later," she says.

What can Holter monitoring do that can't be done in a hospital? "Mainly we're looking for rhythm problems," says Eva. "It's not for a person who has had a heart attack; it's mainly for bad rhythm." Eva's patients wear the monitor for twenty-four hours.

"When the patient comes back and we get the monitor tape, we put it in a computer to analyze it. The computer is looking for abnormal beat, plus other related problems such as lack of oxygen. We go over the analysis and make changes as needed. For example, we may 're-label the beats' if the computer's done it wrong."

The computer has done a morphology (that is, a study of the structure and form) of the data and "tags the beats" on-screen to show what it calculates as incorrect; the technologist then looks at these and, through his or her knowledge of morphology, decides whether the computer's analyses are correct or not. "As trained technicians, we can tell," Eva says. "The doctor makes the actual diagnosis, but we write in our comments about what we see. If we're wrong, he'll tell us."

"We also look through the patient diary," says Eva. The Holter monitor patients are given a diary in which they record any problems, such as "unusually strong chest pains, shortness of breath," she says, "as well as the time the problem occurs."

Eva says she performs an average of two to three Holter tests in each eight-hour shift. Various other technicians in the department may perform the test, too.

As for the stress testing, Eva notes that it goes beyond the familiar treadmill test that most people know of. "There's a persantine thallium test, for example," she says. "The patient is injected with a dilator drug at the beginning of the test. They're asked to do some exercises with their hands while lying down on a bed—such as hand-grip exercises—while being monitored by an EKG machine." A camera is trained on their heart. Then, the patient is injected with thallium, a soft metal-based drug.

"When a person is exercising, the heart works faster, harder. The camera takes pictures of the heart during activity," explains Eva. "With the drug, we can see on-screen any changes in the heart during activity—in rhythm, rate, in certain critical parts of the heart's structure. Throughout the test, we ask the patients if they have any chest pain or other discomfort. The test itself takes about an hour, and we also have to prepare the patient beforehand; so all together, the whole procedure takes about an hour and a half."

Again, do people get nervous? "No," says Eva. "There's a doctor, nurse, EKG technician, and nuclear technician there. They know they're getting good care."

**To be a successful cardiovascular technologist, you should**

- Have mechanical/technological aptitude and feel comfortable using computers
- Be able to empathize with others and project a calm and reassuring manner
- Have analytical and problem-solving skills; be able to measure, calculate, reason, evaluate, and synthesize information; and demonstrate good judgment
- Be detail-oriented
- Be flexible and adapt well to change; be unfazed by unusual personalities or problems and improvise when the usual procedures don't work
- Be able to work under pressure, if necessary

## have I got what it takes to be a cardiovascular technologist?

Cardiovascular technologists need a combination of mechanical/technological, analytical, and people skills. They should be able to follow instructions and communicate what they know to others. Cardiovascular technologists should be detail-oriented; test results are very important to accurate diagnoses. The ability to work under pressure, including medical emergencies, also is helpful for some of these positions.

The mechanical/technological aptitude is helpful because the technologists in many of these positions use sophisticated diagnostic imaging equipment. Almost all of this equipment is computerized, so it helps to feel comfortable with the computer. Technologists also may have to calibrate or run tests on their equipment to make sure it is operating correctly.

All of the technologists need analytical skills. They have to be able to measure, calculate, reason, evaluate, and synthesize information. Problem-solving ability, flexibility, and good judgment are crucial. Technologists must be able to handle "nontraditional" patients or problems (those who don't fit the "textbook case" mold) and use alternative methods when the regular ones won't work.

The ability to think "spatially"—that is, comprehend two-dimensional (2-D) or three dimensional (3-D) relationships, for example, or visualize the relationships between the imaging equipment and the part of the body being studied—also is very useful for the imaging technologists.

At the same time, human skills—the ability to empathize with patients, calming their fears, explaining the procedure, and answering any questions if necessary—also are important for the technologist. Patience, and a calm, reassuring, confident manner are helpful.

Technologists don't necessarily have someone looking over their shoulder all day long, as Larry noted. However, they do need to be able to follow written or verbal instructions from the physician or a supervisor as needed.

Finally, cardiovascular technologists should be able to adapt well to change. Every day brings different patients with different specific problems. Employers may tinker with positions or departments as they strive to find a good balance between services and cost-efficiency. Constant medical advances mean that the technologist has to be flexible and keep up with changes in equipment and procedures.

EKG technologists usually work five-day, forty-hour weeks, but some may be on twenty-four-hour emergency call, and almost all work occasional evenings or weekends. Cardiology departments may be closed or run only a skeleton crew on the weekends. Technologists whose tests usually are scheduled in advance—such as stress tests or Holter monitoring—usually don't face emergency work loads and the need for overtime. Some of the other types of cardiovascular technologists, such as the catheterization professionals, may work longer hours and evenings, and also be on call for emergencies.

Cardiovascular technologist positions are typically salaried, but overtime pay (or extra time off in lieu of overtime) may be available. Technologists in large hospitals may work one of three shifts and may receive a higher rate of pay for taking second- or third-shift work.

"My father and brother both developed cardiac problems, and I wanted to know why."

## how do I become a cardiovascular technologist?

Each of the technologists described earlier got into the field through the influence of a family member or advice of a friend. "My father and brother both developed cardiac problems, and I wanted to know why," says Larry. "I went to a school for EKG in downtown Chicago for a year.

"I heard from a friend about the work, and I went to school for six months" to learn how to do EKG, stress testing, and Holter monitoring, says Amina. "I had gone to secondary school in Pakistan, and didn't know English before I came to this country. It was hard at first, but it worked out."

"I have a sister who does echoes (sonograms); she always talked about what she did, and about the heart, and I became interested," says Eva. "I had never really known anything about the health care profession before that. But I was curious about what she did. To me, it's very interesting to know what's going on inside a part of the body. My mother had high blood pressure, and I wanted to understand that. I took a class in Holter monitoring two times a week for six months several years ago. The stress testing came about a year later."

## education

In the past, EKG operators may have simply been trained on the job by an EKG supervisor. This still may be true for some EKG technician positions. However, increasingly, EKG technologists get postsecondary education before they are hired. Holter monitoring and stress testing may be part of the student's EKG training, or they may be learned through additional training. Ultrasound and cardiology technologists tend to have the most postsecondary schooling (up to even a four-year bachelor's degree), and to have the most extensive education and experience requirements for credentialing purposes.

As was true for Larry, Amina, and Eva, people can get into these positions without having had previous health care experience. However, it certainly doesn't hurt to have had some previous exposure to the business or even training in related areas. People with academic training or professional experience in nursing, radiology science, or respiratory science, for example, may be able to make the move into cardiology technology, if they wish.

### High School

At a minimum, cardiovascular technologists need a high school diploma or equivalent to enter the field. Although no specific high school classes will directly prepare you to be a technologist, learning problem-solving skills and getting a good grounding in basic high school subjects are important to all technologist positions.

During high school, students should take English, health, biology, and typing (data entry). They also might consider courses in the social sciences to help them understand their patients' social and psychological needs.

### Postsecondary Training

As a rule of thumb, the medical profession powers-that-be value postsecondary schooling that gives the student real hands-on experience with patients, in addition to classroom training. At many schools that train cardiovascular technologists, you will be able to work with patients in a variety of health care settings and train on more than one brand of equipment.

**EKG.** With some employers, EKG technicians are still simply trained on the job by a physician or EKG department manager. Length of time for this training varies, depending on the employer and the trainee's previous experience, if any; it is usually at least one month long and may be up to six months. The trainee learns how to operate the EKG machine and produce and edit the electrocardiogram, along with related tasks.

**EKG/Holter/stress.** Some vocational, technical, and junior colleges have one- or two-year training programs in EKG, Holter monitoring, or stress testing; otherwise, EKG technologists may obtain training in Holter and stress procedures after they've already started working, either on the job or through an additional six months or more of education. The formal academic programs give technologists more preparation in the subject than available with most on-the-job training, and allow them to earn a certificate (one-year programs) or associate's degree (two-year programs). The American Medical Association's Allied Health Directory has listings of accredited EKG programs (*see* "Bibliography").

**Cardiology technologists.** These technologists tend to have the most stringent education requirements of all: for example, a four-year bachelor of science degree or two-year associate's degree; or certificate of completion from a hospital, trade, or technical cardiovascular program for training of varying length. A two-year program at a junior or community college might include one year of core classes (math, science, etc.) and one year of specialized classes in cardiology procedures.

**Ultrasound (vascular and echocardiography).** These technologists usually need a high school diploma or equivalent plus one, two, or four years of postsecondary schooling in a trade school, technical school, or community college. Vascular technologists also may be trained on the job. Again, a list of accredited programs can be found in the AMA's Allied Health Directory; also, a directory of training opportunities in sonography is available from the Society of Diagnostic Medical Sonographers (SDMS) (*see* "Bibliography").

**Cardiac monitor.** These technicians need a high school diploma or equivalent, plus education similar to that of the EKG technician.

Cardiology is a cutting-edge area of medicine, with constant advancements, and medical equipment relating to the heart is always being updated. Therefore, keeping up with new developments is vital. Technologists who add to their qualifications through continuing education also tend to earn more money and have more opportunities. The major professional societies encourage and provide the opportunities for professionals to continue their education (*see* "Sources of Additional Information").

## certification or licensing

Right now, certification or licensing for cardiovascular technologists is voluntary, but the move to state licensing is in the air. Many credentialing bodies for cardiovascular and pulmonary positions exist, including ARDMS, CCI, and others, and there are more than a dozen possible credentials for cardiovascular technologists. For example, sonographers can take an exam from the American Registry of Diagnostic Medical Sonographers (ARDMS) to receive credentialing in sonography. Their credentials may be registered diagnostic medical sonographer, registered diagnostic cardiac sonographer, or registered vascular technologist. Especially at the level of cardiology technologist or ultrasound technologist, the credentialing requirements may include test-taking plus formal academic and on-the-job experience requirements. Professional experi-

## Advancement Possibilities

**Chief cardiopulmonary technologists** coordinate the activities of technologists who perform diagnostic testing and treatment of patients with heart, lung, and blood vessel disorders.

**Chief radiologic technologists** coordinate activities of the radiology or diagnostic imaging department in the hospital or other medical facility.

**Medical technologist teaching supervisors** teach one or more phases of medical technology to students of medicine, medical technology, or nursing arts, or to interns; organize and direct medical technology training program.

**Nuclear medicine chief technologists** supervise and coordinate activities of nuclear medical technologists engaged in preparing, administering, and measuring radioactive isotopes in diagnosis, treatment, or studies.

ence or academic training in a related field—such as nursing, radiology science, respiratory science—may be acceptable as part of these formal academic/professional requirements. As with continuing education, certification is a sign of interest and dedication to the field, and is generally looked upon more favorably by potential employers.

# internships and volunteerships

An internship as an EKG assistant is possible. Check local hospitals to learn of opportunities. Some employers prefer to hire a person who has already worked in the health care field, such as a nurse's aid. People interested in health care positions may consider being a volunteer at a hospital, which will give them exposure to patient-care activities. Or they may visit a hospital, clinic, or doctor's office where EKG or other cardiovascular procedures are used and ask to talk to a cardiovascular technologist.

# who will hire me?

Amina started out at a different hospital and then saw an ad in the paper for her present position; Larry and Eva followed similar paths. EKG technologists work in large and small hospitals, clinics, health maintenance organizations (HMOs), cardiac rehabilitation centers, cardiologists' or other physicians' offices, long-term care facilities, and nursing homes. Other types of cardiovascular technologists may work primarily in either community hospitals (less than 200 beds, 200-350 beds, or more than 350 beds) or teaching hospitals, in industry, for mobile medical service clinics, in academia, or in government hospitals (VA, armed forces, public health services, etc.), according to ASCP. Another possibility is finding employment with a manufacturer of EKG or other equipment.

EKG technicians, by far, outnumber the other specialists; there are fewer ultrasound technologists at this time, for example, because of the expertise required and the lower frequency of these tests. "Even in a large-size hospital where there may be a large radiology department with fifteen rooms and a twenty- to thirty-person staff, the ultrasound department may have a staff of six and maybe two rooms," notes Dennis King, chairman of the diagnostic medical imaging program at Wilbur Wright College, Chicago. "However, sonography is a growing profession nationally, even if some areas are 'soft' for new hires, or oversaturated, from time to time."

Check the want ads ("health care professionals" and related sections), contact national associations, and check with hospitals or other health care facilities in your area to learn of opportunities.

One of the services of ASCP for its members is a job hotline, which plays a recording of recent job postings. A recent call to this line revealed one job posting, for a fairly experienced individual; still, it's a resource you might find useful for your job hunts (*see* "Sources of Additional Information").

## where can I go from here?

"I want to do echoes," says Eva. "That's the way to go. I'll need to go back to school to do it." Larry has plans to move up, too. "I've applied to Triton [Community College]" to obtain training in some of the other procedures, such as stress testing and Holter monitoring, he says. Amina, too, has considered getting more education. "I'm not sure about moving right now, though," she says. "I have a child and may wait a few years."

Opportunities for advancement are best for EKG technologists who learn to do or assist with more complex procedures, such as Holter monitoring, stress testing, echocardiography, and cardiac catheterization. With proper training and experience, they may become cardiology technologists or cardiopulmonary technologists, another type of heart-related technology specialist. Besides specialist positions, other opportunities may be found in supervisory or teaching positions. At some hospitals there may be a *chief cardiopulmonary technologist,* for example, who coordinates the activities of technologists who perform diagnostic testing and treatment of patients with heart, lung, and blood vessel disorders. The *chief radiologic technologist* coordinates activities of the radiology or diagnostic imaging department in the hospital or other medical facility.

Besides those already mentioned, areas of special experience for cardiovascular technologists may include infant pulmonary function testing, blood gas studies, sleep disorder studies, pacemaker procedures, and other related procedures. Nuclear medicine, which uses radioactive isotopes in diagnosis, treatment, or studies, also is applied to cardiology problems.

## what are some related jobs?

The U.S. Department of Labor classifies EKG technologists and other cardiovascular technologists under the headings *Occupations in Medical and Dental Technology* (DOT) and *Child and Adult Care: Data Collection* (GOE).

**Related Jobs**
- Biochemistry technologists
- Cardiopulmonary technologists
- CT scan technologists
- Cytogenetic technologists
- Dental hygienists
- Dialysis technologists
- Electroencephalographic (EEG) technologists
- Electromyographic (EMG) technologists
- Magnetic resonance imaging (MRI) technologists
- Nuclear medicine technologists
- Ophthalmic technologists
- Orthotists
- Perfusionists
- Prosthetists
- Pulmonary function technicians
- Radiation-therapy technologists
- Radiologic (X-ray) technologists

## what are the salary ranges?

A survey done by the University of Texas Medical Branch of national hospitals and medical centers found an average salary of about $17,000 for EKG technologists, with a low of around $15,000 and a high of about $22,000. The average salary for all , cardiovascular technologists is around $32,000. This excludes extra pay for shift work (that is, extra earnings for taking on second- or third-shift work) or overtime. Specialists, whether in Holter monitoring, stress tests, ultrasound technology, or cardiology technology, make more than basic EKG technologists. EKG technologists, in turn, make more than EKG technicians. Also, the type of employer and demand in the employer's geographic area affect what salary you'll be able to command. A large urban hospital, for example, probably will pay you more than a small-town clinic.

Hospital benefits generally are good, including health and hospitalization insurance, paid vacations, sick leave, and possibly educational assistance and pension benefits.

## what is the job outlook?

The job outlook for cardiovascular technologists is a mixed bag. Sources say the employment of EKG technologists is on the decline, because the equipment and procedures have grown increasingly efficient and fewer technologists are required to do the work. Also, the equipment is increasingly easy to use, so hospitals can train other personnel, such as registered nurses or respiratory therapists, to do the job. EKG technologists who have experience with Holter monitoring or stress testing should fare better, though. Their multiple skills make them more attractive to employers trying to improve efficiency.

Ultrasound technologists should see slow but steady growth into the next century. Some areas of the country are currently "soft" for new hires, partly because of low turnover in the job, but in general this is a growing specialty.

## in-depth

### History

EKG machines are very "high tech," and equipment manufacturers are constantly coming up with new improvements to these vital parts of cardiology care. However, EKG technology is actually nearly one hundred years old.

The first EKG was invented in 1902 by a Dutch physiologist, Willem Einthoven (1860-1927). Before that, physicians had to rely mainly on their stethoscopes and their own perceptions; with the EKG, they now had a whole new insight into heart problems—especially irregular heartbeats and severe blockages.

The invention of the EKG came on the heels of the discovery of X-rays just seven years earlier, in 1895. Together, they made detecting and diagnosing heart ailments much more of a science—with objective data to help in the process—and much less of a hit-and-miss proposition.

Even today, EKGs can't spot everything; with women, for example, the breast tissue may make a reading more difficult, and a positron-emission tomography (PET) scan may be more useful. Also, EKGs still can't pick up subtle heart problems, although researchers are working on improving the EKG's ability to make more accurate and detailed readings of the heart. However, EKGs will continue to be in wide use in routine physicals, presurgical physicals, in diagnosing disease, and in monitoring the effects of surgery or drug therapy.

Cardiology technologists should fare the best of all, as more hospitals create cardiac catheterization units and as new procedures and drug treatments are developed. Some of the procedures in which cardiology technologists assist—notably, angioplasties—are still the subject of some controversy and ongoing research. This shouldn't hurt cardiology technologists' opportunities for employment, but you should be aware that procedures are constantly changing and that what you study in school now may need updating in the near future.

## how do I learn more?

### sources of additional information

Following are organizations that provide information on cardiovascular technician careers, accredited schools, and employers.

**American Registry of Diagnostic Medical Sonographers (ARDMS)**
600 Jefferson Plaza, Suite 360
Rockville, MD 20852-1150
301-738-8401

**American Society of Cardiovascular Professionals (ASCP)**
120 Falcon Drive, Unit #3
Fredericksburg, VA 22408
703-891-0079
Job hotline: 1-800-683-6728

**Cardiovascular Credentialing International (CCI)**
4456 Corporation Lane, Suite 350
Virginia Beach, VA 23462
800-326-0268

**Society of Diagnostic Medical Sonographers (SDMS)**
12770 Coit Road, Suite 508
Dallas, TX 75251
214-235-7367

### bibliography

Following is a sampling of materials relating to the professional concerns and development of cardiovascular technicians.

#### Periodicals

*American Society of Echocardiography Journal.* Bimonthly. Devoted to scientific articles of interest to cardiologists and technicians in the field of echocardiography. Mosby, 11830 Westline Industrial Drive, St. Louis, MO 63146-3318, 314-872-8370.

*Cardiology.* Monthly. Provides international articles concerned with the latest strategies in the prevention and treatment of heart disease. S. Karger AG Allschwilerstr. 10, PO Box CH-4009 Basel, Switzerland, 41-61-3061111.

*Cardiovascular Reviews and Reports.* Monthly. Concentrates on the practical application of cardiovascular therapy and on recent or controversial clinical issues. Le Jacq Communications, Inc., 777 West Putnam Avenue, Greenwich, CT 06830-5014, 203-531-0450.

*Circulation.* Monthly. Features clinical and laboratory research pertaining to cardiovascular disease. American Heart Association, 7272 Greenville Avenue, Dallas, TX 75231-4596, 214-706-1310.

*Journal of Cardiovascular Diagnosis and Procedures.* Quarterly. Serves individuals working in the field of cardiovascular medicine. Mary Ann Liebert, Inc., 2 Madison Avenue, Larchmont, NY 10538, 914-834-3100.

*Newspaper of Cardiology.* Monthly. Covers recent developments in the field of cardiology. McMahon Group, 148 West 24th Street, New York, NY 10011, 212-620-4600.

*Progress in Cardiovascular Diseases.* Bimonthly. Provides articles focusing on different aspects of cardiovascular diseases. W. B. Saunders Co., Curtis Center, Independence Square West, Philadelphia, PA 19106-3399, 215-238-7800.

# career information specialist

**Definition**

Career information specialists help students and clients with career planning and job hunting.

**Alternative job titles**

Career services coordinators
Employment counselors
Outplacement consultants
Personnel specialists
Placement coordinators
Recruiting coordinators

**Salary range**

$18,600 to $35,800 to $60,100

**Educational requirements**

High school diploma; bachelor's and master's degrees for advanced positions

**Certification or licensing**

Required by certain states for advanced counseling positions

**Outlook**

About as fast as the average

GOE
11.03.04

DOT
166

O*NET
21511C

## High School Subjects

Business
English
Psychology

## Personal Interests

Business management
Helping people: personal service

nice to have someone devoted to making your dreams come true? That's partly what Barbara McCaslin, a career information specialist, does for a living. Her job is to help people find jobs, and this can entail setting up interviews, tracking down potential employers, and offering interview and resume advice.

At 7:00 am on a typical morning, Barbara is already hard at work in her office. As she looks over the day's schedule on the computer log, she adds up the number of students interviewing that day.

"*Fifteen* interviews," she says out loud, amazed.

Fifteen interviews means fifteen student profiles she must prepare from the computer database. When she finishes, she checks to make sure the interview rooms are in order.

"Everything's ready, and just in time!" she says to herself as the phone rings.

"It looks like I'll be a little late," says the recruiter on the other end of the line, "so we'll have to reschedule a couple interviews for another day. And we decided that we want to interview Adam Sterling for sure. Can you arrange it?"

"I'll take care of everything," Barbara assures her pleasantly.

As she hangs up, Barbara is already making a mental list of everything she now has to redo. "Good thing I came in early," she says to herself; "it's going to be another hectic day."

## what does a career information specialist do?

For many people just graduating from high school or college, choosing the right career can be a confusing and difficult task. Even if you already know what field interests you, you may be intimidated by the entire job hunting process. Where do you apply? What is the best company for you? Is your resume well written? How should you act during an interview? Since finding a job is a full-time job in itself, people often seek the expertise of career information specialists, trained and experienced professionals who help answer these types of questions.

### lingo TO LEARN

**Career development** The process of evaluating career goals and clarifying the means to achieve them.

**Career Planning and Placement Office** An office whose function it is to help students find jobs.

**Computer database** A computer system that allows career planning and placement offices to store information about both students or clients and recruiting companies. It can also find matches between student profiles and company requirements.

**Headhunter** An employment specialist focused on recruiting personnel.

**Human resources** Refers to the division of a company that handles personnel issues.

**Outplacement** The process of assisting terminated employees in finding new jobs

**Recruiting** An active search made by companies to hire highly qualified individuals for specific jobs.

**Resume** A summary of a job hunter's education, work experience, community activities, and interests.

Career information specialists often work in the placement offices of universities and vocational schools. They may also work in commercial employment agencies or career counseling centers. Job responsibilities vary according to the size and focus of the office in which they work.

*Career information specialists* work very closely with students, alumni, and clients. To help each person make a wise decision about his or her future career, they may conduct interviews or administer tests designed to evaluate the job hunter's education, work history, interests, skills, and personal traits. The results of the tests can suggest careers in which the clients may be happy and successful.

Part of their time may be spent researching various companies and gathering resources in an informational library. The library offers a collection of printed and electronic resources and files to help clients and students research prospective employers and career paths. Career information specialists may also evaluate and make purchasing decisions regarding career services materials for the library, and they may spend time indexing and cataloging materials. They might also make a habit of scanning newspapers and trade magazines to be aware of the kinds of positions companies are trying to fill.

Clients often turn to career information specialists for guidance in developing job finding skills such as resume writing and interviewing techniques. The specialist may also be responsible for maintaining computer files on each client.

Career information specialists who work in commercial employment or personnel agencies have duties similar to those working in academic environments. Their clients, however, are generally the hiring companies, though job seekers are also clients. The career information specialist's primary goal is to find job candidates for the client companies.

Career information specialists in employment agencies may administer skills tests to job seekers to assess computer aptitude, typing skills, or specific subject knowledge. They act as the liaison between hiring companies and job seekers, setting up interviews and making sure the hiring company receives the job seeker's resume and background information. In some cases, the career information specialist interviews the job candidate on behalf of the client company. Matching the job seeker's goals and skills to the appropriate job opening is an important duty of the specialist.

Career information specialists can also specialize in recruiting. Recruiting coordinators work very closely with companies to recruit interns and full-time employees in a particular field or specialty. Recruiting coordinators, or recruiters, coordinate the hiring process for companies. Recruiters seek ideal candidates to fill employment openings for their client companies. Some large companies employ in-house recruiting coordinators to handle staffing needs.

Recruiters seeking candidates at schools post job profiles of the company's openings, request that qualified candidates send resumes, forward the information to the company, and help the company find potential matches. Since the number of interviews granted by one company is often very limited, recruiting coordinators establish priority lists and have some authority in final decisions regarding who receives an interview slot.

Recruiters may also actively search for qualified candidates. In some instances, this can entail luring a candidate away from a current employer. This is sometimes referred to as headhunting, although career information specialists prefer not to use the term because of negative connotations.

The work of recruiting coordinators is very detail oriented. They are in charge of setting up deadlines for both job seekers and companies. They are also responsible for keeping interview schedules. This task involves routinely sending confirmation letters or email to all concerned parties.

Although much of their work is performed over the telephone or on the computer, recruiting coordinators are often present for interviews. They may be in charge of making interviewers and interviewees feel welcome and comfortable.

A relatively new specialization for career information specialists is that of the outplacement consultant. Outplacement consultants help downsized or terminated employees find new jobs or career paths. They work on behalf of the companies that terminate employees. Outplacement is becoming a standard part of severance packages in the corporate world.

Outplacement consultants help people re-enter the work force or find jobs through guidance sessions, interview training, resume assistance, and career assessment counseling. Outplacement consultants also contact potential employers. They sometimes provide adjustment counseling and management training for remaining employees.

> You've got to be a people person. I always have to be friendly and accommodating, even when they complicate my job.

## what is it like to be a career information specialist?

Barbara McCaslin has worked in career services for Ohio State University for ten years. She currently works in the Career Planning and Placement Office of the College of Engineering. "Our office is always buzzing with activity," she says. "At any given moment, you're likely to find students going in every direction; a few are flipping through company literature, others are interviewing for full-time jobs, and still others are meeting with counselors. I like the hustle and bustle of it all."

Like most office employees, Barbara works a five-day, forty-hour week, arriving in the office between 7 and 7:30 AM every day. During the heavy recruiting seasons of fall and winter quarters, however, she may have to arrive even earlier and stay well after 5 PM to accommodate everyone's schedule.

Barbara spends most of her day in her office within the placement office. She is constantly on the telephone and computer terminal. The placement office is organized and operated very much like any office, with a waiting room and private offices. In addition, there is a library area filled with career materials, computer terminals accessible to both students and counselors, and several interview rooms.

When asked to describe her typical day, Barbara smiles and says, "I do everything!" She looks over schedules, sends letters, and updates computer files. She also greets company recruiters, up to fifteen per day, and sets up interview rooms.

Although she is not officially responsible for advising students directly about big career decisions (guidance counselors are largely responsible for that), a few seem to wander into her office now and then looking for help. She spends as much time as she can spare with them and at least tries to point them in the right direction.

As the liaison between recruiting companies and prospective employees, Barbara makes dozens of calls every day to company representatives and students. During these conversations, she is always bright, cheery, and enthusiastic. She calls company representatives to remind them about deadlines for job offering bulletins and confirm already scheduled interviews. She might also consult with them about the type of students they wish to interview for a certain job and recommend students who would be strong candidates.

Even though many deadline notices are done automatically through the computer, she occasionally calls students to remind them about getting their applications in on time and to inform them about job openings that may interest them. She also receives many phone calls, particularly from companies who want to schedule interviews. Because the interview calendar is always done one year in advance, this task can be quite confusing.

## have I got what it takes to be a career information specialist?

Career information specialists interact with people nonstop. "You've got to be a people person," explains Barbara. "I deal with dozens of people every day on a professional basis, and I always have to be friendly and accommodating, even when they complicate my job." The responsibility to satisfy everyone all the time often becomes a source of stress for Barbara, but she has learned how to balance the many factors to minimize the pressure. "We establish deadlines for very good reasons, and individuals come to me with excuses, just like students with late homework. I have had to perfect my skills at being firm yet not offensive." In fact, keeping people on schedule is the hardest and most stressful part of her job.

Barbara loves to work with young people and for this reason finds her work very rewarding. "I wouldn't do it if I didn't care about their futures," she says. "It's a great feeling to help students get started on the right foot in their careers."

Students can be very shy about discussing their resumes and interviewing skills since the whole process of job hunting can be intimidating. Barbara explains, "You've got to be the kind of person who can get people to trust you right away. I always explain that I have been through the whole job hunting thing myself and understand the stress." When students have confidence in her, they take constructive advice to heart and tend to interview better.

Organizational skills are extremely important for her job. At any one time, there are between six and seven hundred students registered in Barbara's placement office. One fall quarter, 1,046 interviews were conducted, and the year's total reached 3,508. "That's a lot of scheduling and scheduling changes, and I do it all," Barbara says. "If I weren't on top of every detail, the whole system would fall apart."

**To be a successful career information specialist, you should:**

- Be a people person, with genuine enthusiasm for others' success
- Have friendly and professional telephone skills
- Be able to inspire trust and confidence in people with whom you interact
- Have the ability to organize large amounts of very detailed material
- Be able to work on deadlines and make sure others do the same

## how do I become a career information specialist?

When Barbara was looking for a job, she knew she wanted to work closely with college students. She applied at nearby Ohio State University and was offered a job in a placement office, even though she had no formal training in the field. "You do have to know how to work with computers, at least at a basic level," Barbara says.

## education

### High School

High school courses in English, business, typing, and computer science will give you a head start in developing the skills a career information specialist should have. Any course or project that requires you to be very organized and work with many details also builds a strong foundation for success in this career.

Because the job of a career information specialist is so people-intensive, joining clubs and other organizations will give you the opportunity to work closely with others. Running for student body positions or assuming officer roles in clubs will teach you leadership skills and the art of negotiation and compromise.

Part-time or summer jobs in employment agencies, placement offices, or counseling centers can provide you with an in-depth look at the daily tasks of a career information specialist or counselor. Another method for gaining experience in guiding others and assessing their strengths and weaknesses is to volunteer in youth programs or community programs dedicated to helping others.

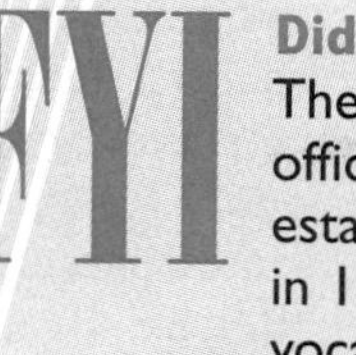

**Did you know?**
The first funded employment office in the United States was established in San Francisco in 1886. The first program of vocational guidance, designed to help young people select and enter appropriate careers, was founded in Boston in 1908.

### Postsecondary Training

Although a college degree is not required to be an entry- level career information specialist in a university or vocational school placement office, most schools will not hire new employees who lack college degrees. While there is no specific degree in career services, any postsecondary course in business management, human resources, psychology, or guidance counseling is helpful preparation. These types of courses are often offered at two-year or community colleges even if a degree is not granted in the discipline.

A bachelor's degree in a specific subject area can be helpful if you choose to specialize. For example, if you plan to work as a career placement specialist for the engineering department of a university, an engineering background or degree will provide you with the necessary tools for helping students and matching candidates with appropriate jobs. If you plan to work as a recruiter specializing in finance, an accounting degree will help you understand the job requirements and the particular skills to look for in a candidate.

Barbara graduated from college with a degree in education. She says many courses, like psychology and statistics, have made her job much easier. Education classes that dealt with relating well and communicating effectively with all different kinds of people also helped prepare her for her position.

Some career information specialists go to work in private industry. Private companies generally require new employees to have at least a bachelor's degree, but they tend to be flexible about the college major. For example, some career information specialists hold degrees in such diverse disciplines as English, political science, education, psychology, human resource management, and business administration.

## certification or licensing

There is no mandatory certification program for career information specialists. To be a career or guidance counselor, however, is a different story. About forty states require counselors to be licensed, certified, or registered in addition to holding a master's degree. The licensing process differs from state to state.

The certification process varies based on the specific kind of counseling an individual does. For example, if someone wants to be a school counselor, many states require that he or she have between one and three years' teaching experience. This means that the candidate also has to be a state-certified teacher.

Many counselors choose to be certified as National Certified Counselors by the National Board for Certified Counselors, Inc. (NBCC). This certification requires a master's degree in counseling and at least two years of experience. This is a voluntary program, but some states accept NCC certification in place of their own licensing program. Counselors in most states must participate in special workshops, continuing education courses, and institutes to maintain their certification.

Outplacement consultants can also seek certification, although it is not required by employers. Outplacement consultants can achieve the designation of a Certified Practitioner (CMP) by the International Board for Career Management Certification, a division of the International Association of Career Management Professionals (IACMP). This certification requires a bachelor's degree and work experience.

## scholarships and grants

Since there is no formal training requirement for career information specialists, there are no scholarships or grants specifically for those interested in these positions. If you plan to pursue a college degree, you should contact your school's financial aid office for a generalized list of scholarships and grants for which you may qualify.

## internships and volunteerships

If you are a student, you might try to get your foot in the door by getting a job in a placement office at your school. If not, you might try to get a clerical job at a career counseling center or employment agency in order to gain experience and demonstrate your availability when job openings arise.

Internships in the human resources department of large companies may be available, as well as internships or work-study opportunities at your school's placement office.

Volunteering at agencies or organizations that provide guidance or counseling services to people will give you an idea of whether or not you enjoy working with others. You might also try to work as a temporary employee through an employment agency. Employment agencies often handle both direct placement and temporary positions. If you demonstrate your reliability and skills as a temporary employee, and voice your interest in working as a career information specialist, chances are the agency will want to bring you on staff when a position opens up.

## who will hire me?

Many career information specialists work for career counseling and placement offices at universities and vocational schools. The number of such offices operating at one university depends on student enrollment. In large universities such as Ohio State and other Big Ten schools, every college or department has its own placement office, meaning there are between fifteen and twenty-five offices at one university. Placement offices serving students with very specialized or technical degrees, such as law, pharmacy, and engineering, tend to need fewer employees with counseling certification. This is because those students often have a clearer vision of their career goals.

Employers Extending Job Offers

54% Services

14% Manufacturing

32% Government/Nonprofit

Knowing she wanted to work with college students, Barbara went to the human resources department at Ohio State and found a position in the college of education's placement office. Her first job required her to take the civil service test, which is a test that the state of Ohio requires for all state employees. Large universities and businesses often publish a monthly list of job openings that is circulated internally among employees. While gaining experience, Barbara kept her eye on the job list and later applied for the position of recruiting coordinator in the college of engineering.

Other career information specialists work in commercial career counseling centers. Usually, however, these positions are reserved for individuals certified or educated as counselors. Consult your local telephone directory for a list of such businesses in your area. Employment or temporary agencies also provide a possible entry into the field; requirements about education and work experience will vary depending on the company. Federal, state, and local government agencies may have some jobs in this field, but may require training in social work or human services. Again, each agency will have different regulations, so you should contact the ones in your area for detailed information.

Large businesses typically have a human resources department that may have related career services or recruiting positions available. Such businesses will most likely require a college degree, but it may be worthwhile to check with the major employers in your area.

The National Association of Colleges and Employers (formerly the College Placement Council) provides information on job opportunities and development in career services positions in their journal, *Spotlight*. The American Counseling Association also publishes several journals devoted to the concerns of individuals working in all areas of career services, including details about job opportunities. (*See* "Sources of Additional Information"). Many companies and professional organizations list job opportunities and helpful information on the Internet. You may also find job openings at one of the many web sites devoted to the job search.

## where can I go from here?

Since she is already considered on equal footing with assistant directors in her office, Barbara would like to work toward promotion to director.

Her college degree makes her technically qualified for the position, but she believes that she would be a stronger candidate if she went back to school at the graduate level and took courses in counseling and psychology.

"Directors do a little bit of everything, so they like to hire people who have experience in everything, including career and psychological counseling and business management," she says.

Advancement from the position of career information specialist normally requires a college degree and depends on the size and specialization of the office. Some placement offices employ several associate or assistant directors, so chances of becoming one are higher. In addition, those people who work for an organization that operates numerous placement offices have a greater chance of promotion because they can transfer to other offices when openings occur.

Career information specialists working for educational institutions may choose to go to work for private industry, where they could make more money as recruiters or human resource managers. Over the past several years, Barbara has established hundreds of contacts in businesses all over the country and would probably be able to make this switch rather easily if she wanted to.

Other career information specialists go on to become career or guidance counselors. These positions require at least a bachelor's degree and often a master's degree in guidance counseling or psychology, since counselors spend most of their time performing psychological evaluations of students and clients who come to see them.

## what are some related jobs?

According to Barbara, career information specialists are qualified to work in a number of full-time positions in career planning and placement offices. In addition, they can work in the human resources departments of government agencies and private businesses. With the proper education, they can work in career and psychological counseling centers, school guidance offices, and social service organizations.

FYI

**Hot Job**

Who could imagine that finding people jobs could be so profitable? Technical recruiters find job candidates for executive positions in the ever-changing high-tech industry. Technical recruiters must possess strong sales skills and a background in the applicable subject area, whether it is networking, information systems, or relational databases. Technical recruiters work on a commission basis, and earnings can range from $100,000 to $450,000 a year.

The U.S. Department of Labor classifies a career information specialist under the headings *Personnel, Administration Occupations* (DOT) and *Educational and Library Services: Occupational* (GOE). Also under these headings are people who work as ergonomists, psychologists, demographers, sociologists, intelligence officers, anthropologists, and research assistants.

| Related Jobs |
|---|
| College and student personnel workers |
| Employee recruiters |
| Employment firm workers |
| Human resources workers |
| Occupational therapists |
| Rehabilitation counselors |
| School administrators |
| Training and employee development specialists |

## what are the salary ranges?

Because career information specialists work in a number of industries, salary standards are difficult to pinpoint. According to the 1997 career services survey conducted by the National Association of Colleges and Employers, counselors who guide students in employment and college major decisions earned an average income of $29,312 a year. All of these counselors had at least a bachelor's degree, and 83 percent held master's degrees.

According to the 1998–99 *Occupational Outlook Handbook,* educational and vocational counselors earned a median salary of $35,800 a year in 1996. The bottom 10 percent earned less than $18,600 a year, and the top 10 percent made more than $60,100 a year. The middle 50 percent made in the range of $25,600 to $48,500.

Human resource management salary information is more readily available than that of career information specialists. The average base salary for an entry-level generalist in the human resources department, according to a 1997 salary survey by the Society for Human Resource Management, is $34,900. A human resources manager averaged $74,300, while a recruiter earned $34,500. A similar survey by the management consultant group of Abbott, Langer & Associates indicates that recruitment and interviewing managers earned an average of $55,590 in 1996. Employee counseling specialists earned $41,974, and employment interviewing clerks earned a median income of $18,815 a year.

Most career information specialists and career services directors work for educational institutions or corporations, and their income is usually supplemented by a full benefits package including vacation, sick days, and insurance.

## what is the job outlook?

Overall employment of career services employees and counselors is expected to grow about as fast as the average for all occupations through the year 2006. The number of career information specialists needed in educational institutions depends in part on student enrollment, which has steadily increased in recent years.

More students, however, do not necessarily mean more job opportunities. When companies decide to put a hiring freeze in place, the need for career services employees is lower. During periods of economic hardship, the job duties of career information specialists may shift to include more research and marketing aimed at persuading more companies to interview students or clients.

When budget constraints become a problem in educational institutions, there may be little effort made to fill career services job openings or create new positions. Instead of hiring new people, employers tend to ask those already employed in career services to take on extra duties. However, career planning and placement offices are always considered an important part of the college because their reputation often

### Advancement Possibilities

**Career counselors** usually hold at least a master's degree in counseling or psychology and are licensed to work in private practice. They help clients choose careers well suited to their abilities, interests, and needs.

**Career services assistant directors** supervise a large section of a career placement office. They may organize and manage co-op or internship programs, recruit new companies to hire through their offices, or oversee computer automation of important office functions like interview scheduling and resume writing.

**Career services directors** manage all operations within the career placement office and have final say in decisions made by assistants and other employees. Typically, they hold at least a bachelor's degree and are qualified to meet with clients in one-on-one career counseling sessions.

depends on graduates' success at finding jobs. The reputation, in turn, affects student enrollment and tuition dollars, so universities like to keep placement offices well staffed.

Career information specialists working in the field of human resources in the private industry can anticipate a positive job outlook. When the economy is strong, more jobs are created, and career information specialists and recruiters are needed to fill these positions. Along the same lines, some industries tend to be quite volatile, and there can be major layoff cycles, which increase the need for qualified outplacement consultants.

# how do I learn more?

## sources of additional information

Following are organizations that provide information on career information specialists and other career services jobs, accredited schools, and possible sources of employment.

**American Counseling Association**
5999 Stevenson Avenue
Alexandria, VA 22304
703-823-9800
http://www.counseling.org

**Association of Career Management Consulting Firms International**
1200 19th Street, NW, Suite 300
Washington, DC 20036-2422
202-857-1185
http://www.aocfi.org

**Career Planning and Adult Development Network**
4965 Sierra Road
San Jose, CA 95132
408-441-9100
http://www.careertrainer.com/network.html

**International Association of Career Management Professionals**
PO Box 1484
Pacifica, CA 94044
http://www.iacmp.org

**International Board for Career Management Certification**
PO Box 150759
San Rafael, CA 94915
415-459-2659

**National Association of Colleges and Employers**
62 Highland Avenue
Bethlehem, PA 18017
610-868-1421
http://www.jobweb.org/NACE

**National Board for Certified Counselors**
Three Terrace Way, Suite D
Greensboro, NC 27403
336-547-0607
http://www.nbcc.org

**Society for Human Resource Management**
1800 Duke Street
Alexandria, VA 22314
703-548-3440
http://www.shrm.org

## bibliography

Following is a sampling of materials relating to the professional concerns and development of career information specialists.

### Periodicals

*Career Development Quarterly.* Quarterly. Journal for career counseling and career education professionals in schools, colleges, private practice, government agencies, personnel departments in business and industry, and employment counseling centers. American Counseling Association, 5999 Stevenson Avenue, Alexandria, VA 22304-3300, 703-823-9800.

*Journal of Career Planning & Employment.* Twenty-one issues per year. Devoted to career planning and employment of college students. National Association of Colleges and Employers, 62 Highland Avenue, Bethlehem, PA 18017, 610-868-1421.

*Journal of College Student Development.* Six issues per year. Journal covering ideas for improving student services, research and development theory, professional issues, and ethics. American College Personnel Association, One Dupont Circle, Suite 300, Washington, DC 20036-1110, 202-835-2272.

*Journal of Employment Counseling.* Quarterly. Journal devoted to state employee, vocational, college placement, education, business, and industry counseling. American Counseling Association, 5999 Stevenson Avenue, Alexandria, VA 22304-3300, 703-823-9800.

# chef

**Definition**

Chefs plan, prepare, and cook food in hotels, restaurants, institutions, and other food establishments.

**Alternative job titles**

Bakers
Chefs de cuisine
Cooks
Executive chefs
Pastry chefs
Sous chefs

**Salary range**

$25,000 to $35,000 to $100,000+

**Educational requirements**

Some postsecondary training

**Certification or licensing**

Voluntary

**Outlook**

About as fast as the average

GOE 05.05.17

DOT 313

O*NET 61099A

## High School Subjects

Business
Mathematics

## Personal Interests

Business management
Cooking

**It's Valentine's Day,** the busiest day of the year at Polaris, the revolving restaurant on the top floor of the Hyatt Regency in Milwaukee, Wisconsin. It also happens to be the grand reopening of the newly renovated restaurant. Rob McCrea, sous chef of Polaris, worked fourteen days in a row, twelve to fifteen hours a day, to ready the restaurant for the opening.

An order comes into the kitchen, and Rob shouts, "Ordering! Two chicken pasta dishes, one filet, medium!" They aim for a twenty-minute window on the food, and each of the five cooks knows exactly what he or she must do to accomplish this. The kitchen is bustling with activity, and it is hot and noisy. The six-burner stove has six saute pans on it. The grill has twenty steaks going. The broiler has five salmon steaks cooking away.

Racks of lamb are being placed into the oven. Sauces that must be made at the last minute are bubbling. Meals are readied, and garnishes finish off the plate. Lids cover the dishes, and the server whisks them away to the hungry diners. Everything happens rhythmically and systematically.

All the work Rob and his staff put into preparing for the evening paid off. They served 378 dinners that night, and it went by in a flash. As Rob recalls, "When the night came, it was wham, bam, and it was over, and everything went very smoothly."

## what does a chef do?

A *chef,* first and foremost, cooks. Chefs are primarily responsible for the preparation and cooking of foods, and other duties may be involved, depending on the size and type of establishment. Chefs who work in large hotels or restaurants often supervise cooks, plan menus, determine the prices of the meals, and order and purchase food supplies. Planning menus and ordering food requires the chef to be knowledgeable about the current costs of items such as fruits and vegetables, since the prices fluctuate. Chefs must know which food items are in season and readily available for reasonable prices. The chef must also have a good understanding about quantities of food to order—ordering too much may cause the items to spoil, which means the restaurant or establishment loses money. Chefs may also be responsible for determining portion sizes and creating new recipes.

**Saute** To fry in a small amount of fat.

**Julienne** To slice into thin, matchstick-size strips.

**Braise** To cook slowly in fat in a closed pot with little moisture.

**Buffet** A meal set out on a table or counter for self-service and ready access.

**Broil** To cook by direct exposure to heat.

**Dice** To cut into small cube-sized pieces.

**Garde manger** The pantry area where cold buffet dishes, including salads, meats, and fish, are prepared.

**Patisserie** The category of sweet baked goods that includes cakes, cookies, and pies. Also refers to the art of pastrymaking.

Chefs often have administrative duties in addition to cooking responsibilities. They may be in charge of personnel matters, including hiring and dismissing cooks and kitchen staff, organizing work schedules, and completing employee reviews. Paperwork is a regular part of the chef's workday.

Chefs use a variety of tools and perform a number of cooking tasks. They measure and mix ingredients for sauces, desserts, soups, salads, and other dishes. Chefs also prepare meats, poultry, fish, vegetables, and other foods for baking, roasting, broiling, sautéing, grilling, and steaming. They may use blenders, mixers, grinders, slicers, or tenderizers to prepare the food. Many chefs butcher their own meats. The chef may also place garnishes and final touches on orders and arrange food on serving plates. In essence, chefs must do whatever is necessary to serve good-quality, appetizing meals.

*Cooks* generally have less experience than chefs, although the terms are often used interchangeably. Cooks do not necessarily have any prior cooking experience unless they are *certified cooks.* Certified cooks must have a culinary education or ten years of work experience and must pass the American Culinary Federation's certification program. Certified cooks work in any one station in a food service operation, such as frying, broiling, or sauce cookery.

*Pastry chefs* are trained in preparing baked goods and desserts. They may prepare and cook pies, cookies, cakes, breads, rolls, ice creams, mousses, and other dessert items. They may also bake wedding cakes or special items requiring elaborate decorations and garnishes.

Chefs who work in institutions such as schools, hospitals, prisons, and industrial cafeterias are known as *institutional chefs.* Although they may not serve a large selection of foods, institutional chefs are skilled in cooking large quantities of items. They must also be knowledgeable about the dietary and nutritional values of foods, particularly if they are employed in hospitals or schools.

*Sous chefs* are chefs who are in charge of food production at a food establishment. Sous chefs oversee a shift, such as lunch or dinner, or stations—they may not be responsible for all food production. They supervise cooks and often are responsible for ordering supplies and planning menus. Some job titles that fit under the sous chef category include banquet chef and chef garde manger (chefs in charge of designing and preparing buffets).

The chef who is responsible for overall food production at a food establishment is known as the chef de cuisine. The *chef de cuisine* makes the final decisions regarding culinary operations and supervises kitchen staff.

*Executive chefs* are department heads and are responsible for management and supervision. They spend little time cooking or preparing food because they are busy with the overall operation of food establishments. Executive chefs must determine ways to increase business and profit and cut costs. They may spend time inspecting equipment or interviewing and hiring kitchen personnel or dining room employees.

"It's a high stress job, but you're feeding people, not saving lives."

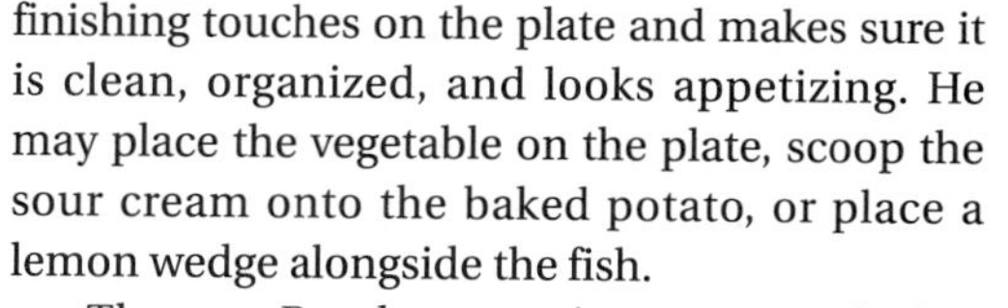

## what is it like to be a chef?

Rob has a long and busy day at the Hyatt Regency that begins around 8 AM. He spends the first part of the morning returning messages, getting some paperwork out of the way, taking an inventory of his cooler and freezer, placing orders, checking reservations, and walking through the main kitchen in the basement of Polaris, the restaurant he runs, which is twenty-two floors up.

Rob's restaurant runs two specials each evening, so around midmorning he starts thinking about some options while he readies fish, butchers meat, and takes care of other food preparations. Rob heads upstairs for the lunch shift around 11:30 AM. Polaris has its own full-service kitchen for cooking orders as they come in, but much of the preparation is done downstairs in the main kitchen. "Upstairs we have stoves, grills, broilers, ovens, and everything for cooking each entree that is ordered," Rob explains. "In the main kitchen in the basement is the banquet kitchen. Down there we have big kettles that we can cook 100 gallons of soup in."

As soon as lunch dwindles down, Rob is back in the basement getting ready for the dinner crowd. As the lunch crew cleans up and puts things away, Rob's dinner crew begins setting up for dinner. They start making sauces, putting the meat and fish in the proper trays and drawers, chopping lettuce for salads, cutting vegetables, and setting up the stations. Rob may help upstairs or work on paperwork downstairs. He may also help out the banquet department if they are especially busy.

Polaris opens for dinner at 5 PM, and at 5 PM, Rob says, "I'm back in the line again. And I basically stay until 9:30 or 10:00 PM." Polaris averages 75 to 125 dinners a night during the week and 200 to 250 on the weekends, and Rob often works as the expediter, which means he puts the finishing touches on the plate and makes sure it is clean, organized, and looks appetizing. He may place the vegetable on the plate, scoop the sour cream onto the baked potato, or place a lemon wedge alongside the fish.

Thomas Brockenauer is a pastry chef at Erna's Elderberry House in Oakhurst, California, a small inn and fine dining establishment. His hours are just as long as Rob's, and many of their duties are similar. Thomas checks on reservations to determine how many desserts he will need to prepare, takes an inventory of his products and supplies, and places orders. Erna's serves a seven-course fixed menu, which means every diner receives a dessert. This helps Thomas in planning ahead, but it also means he must make up to 150 servings a day, depending on how busy the restaurant is.

Making pastries is complex and time consuming. "It takes time to bake all the stuff," Thomas explains, "and to make the batters. Then maybe for a cake you must bake a biscuit or a sponge cake. You must make colorings, then you need a garnishing for the cake, and then you need another dessert, and then you need the garnishing for the plate, so it's very busy for the whole day." Baking requires precision, and each step takes time and patience.

The long hours can be a drawback to being a chef. As Thomas says, "You have to work a lot of hours, and your hours are not common." Chefs work in the morning, at night, and on the weekends and holidays. Being a chef can also be stressful—chefs must constantly think ahead and be prepared for any number of diners, order appropriate amounts of supplies, and make sure the food is perfect. Rob keeps things in perspective, however. "It's a high stress job," Rob says, "but you're feeding people, not saving lives." Thomas agrees and feels being prepared and experienced can help. "If something happens, you must find a way to fix it," he asserts, "because you cannot say, 'oh, my cake is bad today. So sorry.'"

## have I got what it takes to be a chef?

To be a successful chef, you must be creative, enjoy cooking and food, and have a keen sense of taste and smell. Being able to work with others is mandatory, since chefs work closely with others, sometimes in small quarters. You should also be able to work under pressure and have a calm temperament—kitchens can be loud and chaotic during busy hours, and you must be able to focus on your tasks without becoming flustered. Hand-eye coordination and a good memory will come in handy: you don't want to cut off your fingers, and you don't want to forget the pasta sauce on the pasta.

Rob stresses the importance of organization skills to make your kitchen function efficiently and to be prepared for anything. "You don't want to be caught with your pants down," says Rob, "but you don't want to have too much product ahead of time because it could spoil on you. You have to have a bit of vision." You should also be detail oriented, "but there's room for error," Rob concedes. "It's not like a house you're building or a surgeon opening somebody up." Rob also emphasizes basic math and communication skills—you need to be able to calculate quantities and measurements and to give instructions clearly and concisely.

If you want to be a pastry chef, however, there is little room for error, and you must be precise and pay attention to detail. It also helps to be artistic and have the ability to work with your hands, since much of pastry is creating delicate and intricate pieces. Thomas loves being a pastry chef and feels it's important to be dedicated and motivated and to think of pastry as a profession and a calling instead of "just a job."

**To be a successful chef, you should**

- Enjoy working closely with others
- Be well organized and detail oriented
- Be creative
- Have a calm temperament
- Love to cook

## how do I become a chef?

Rob always enjoyed cooking but didn't pursue it as a career until he had worked as a paramedic for ten years and decided it was time for a change. He made a list of his likes and dislikes, and everything pointed toward becoming a chef. He entered an apprenticeship program through the Culinarians of Colorado and gained hands-on experience in the hotel industry.

Thomas grew up in Germany, where most students enter vocational programs at sixteen. He knew he wanted to do something artistic, so he applied to both a photography program and a cooking school. Thomas wasn't accepted to the photography program, so he became a chef instead, with no regrets.

## education

### High School

Some high schools offer cooking programs complete with on-campus restaurants. If you're lucky enough to attend one of these schools, this is an excellent way to gain some valuable experience and to determine if the path of a chef is for you. If you aren't fortunate enough to go to one of these high schools, however, all is not lost. Home economics classes can teach you the fundamentals of food handling, cooking, and baking.

Other classes that will be helpful if you want to become a chef are mathematics, so you can calculate measurements of ingredients and supply quantities; chemistry, to understand the reactions of flours, eggs, proteins, and sugars; and business administration, so you can learn how to run a business and make it profitable. For pastry, Thomas recommends art classes, such as drawing.

Part-time or summer jobs in any type of food establishment are readily available to high school students and may be a way to gain experience and observe the work of chefs. Whether it's a fast-food restaurant, a fine dining hall, a bakery, or a school cafeteria, you will learn about the food industry and whether or not you're cut out for it. Volunteering at organizations that serve food or cater events is another option.

### Postsecondary Training

Although a college degree is not mandatory for chefs, if you wish to work for large corporations or advance into higher positions, a culinary education is highly recommended. There are several options for gaining training to become a chef. Two-year colleges and vocational schools often offer programs in the culinary arts. There are also private culinary schools that offer degree programs. For example, the Culinary Institute of America offers two associate's degrees in occupational studies: one in culinary arts and one in baking and pastry arts. The school also offers a bachelor of professional studies degree in either culinary arts management or baking and pastry arts management. The bachelor's program takes thirty-eight months to complete, while the associate's degree program is a twenty-one-month program. Most culinary programs focus on hands-on training in kitchens, but they also include courses in purchasing food, menu planning, sanitation, and food cost control.

Professional associations and trade unions sometimes offer apprenticeship programs. One example is the three-year apprenticeship program administered by the American Culinary Federation (ACF), which Rob completed. Rob worked at a sponsoring house on a full-time basis and also attended classes. When he graduated, he was a certified cook.

Other options for postsecondary training include the Armed Forces and training programs offered by some large hotels and restaurants. Many school districts offer workshops for cafeteria kitchen workers who wish to become cooks.

## certification

Certification is not required for chefs, but it can provide you with an advantage over others when searching for a job with a large corporation or hotel chain. The ACF offers a number of certification programs for chefs. They include the certified cook, certified pastry cook, certified working pastry chef, certified sous chef, certified chef de cuisine, certified executive pastry chef, certified executive chef, and certified culinary educator. Certified culinary educators can teach at accredited culinary schools or foodservice management programs. Education courses as well as experience and knowledge equal to that of a certified sous chef are required. Climbing up the certification ladder indicates more experience and supervisory duties. It also usually means the chef spends less time in the kitchen preparing food.

Eligibility requirements for the certification programs include experience and involvement in the organization and the industry. Mandatory courses include sanitation, nutrition, and supervisory management. The certification process includes a written examination.

## scholarships and grants

There are many scholarship opportunities for those interested in pursuing an education in the culinary arts. Professional organizations with members in the food or cooking industries may offer scholarships and/or lists of available scholarships. The International Association of Women Chefs and Restaurateurs offers scholarships to women in restaurant careers, and the ACF compiles a list of educational scholarships for those in the hospitality industry. The ACF also awards scholarships to its junior members (see "Sources of Additional Information").

Schools with culinary arts programs may also offer scholarships to students. Contact your advisor or financial aid office for further information.

## internships

Internships are common for those pursuing careers as chefs. In fact, internships are often a graduation requirement for students in culinary arts programs. Most internships are set up through a school placement office.

Large hotel chains or restaurants may also offer internship programs. You may wish to contact the personnel offices of corporations you are interested in working for if you are not enrolled in a culinary program and therefore cannot take advantage of a placement office.

Apprenticeship programs provide an opportunity to work while attending school. The most well known of these are those administered by the ACF's local chapters in conjunction with two-year colleges, vocational schools, and local employers of restaurant workers. This is the route Rob took, and he recommends the three-year program highly, saying, "It's total hands-on hours you've spent in the kitchen, and that's why I'm so fond of the apprenticeship program. The culinary schools have a great base, too, and there's room for both, but in the apprentice program you spend 6,000 hours in there just doing—learning the right way to do it, learning the profitable way to do it."

## labor unions

Participation in unions is generally not mandatory for chefs and depends on location and employer. Some chefs who work for large hotels and restaurants belong to unions, the most common being the Hotel Employees and Restaurant Employees International Union.

## who will hire me?

Rob's first job in the food industry was in a small diner when he was fifteen years old. "I peeled potatoes. I cooked hamburgers. I stocked shelves. I mopped floors. I peeled vegetables," Rob recalls. Rob worked in the restaurant business off and on over the years, "then one day I decided it was time to do this for real," he states. His first job as a trained chef was with the Hyatt Regency in Denver, Colorado, during his apprenticeship training, and he's been with the Hyatt ever since.

After a rigorous three-year cooking apprenticeship program in his native Germany, Thomas spent an additional two years concentrating on pastry. Thomas worked as a chef in Germany for three years to gain additional pastry experience, then spent four years in Switzerland as a pastry chef. Inspired by a colleague who had traveled abroad, Thomas looked through a catalog of hotels and inns, then wrote letters and sent resumes to twenty-five establishments in the United States and Canada. He selected Erna's Elderberry House because of its size and location.

Chefs can work in any establishment that serves food, such as restaurants, hotels, and institutions, including schools, airports, hospitals, and nursing homes. Chefs can also work for government agencies, factories, caterers, or private clubs.

Rob believes word of mouth is a common way to find out about chef opportunities and suggests being involved with professional organizations or clubs. "When I go to a chefs' meeting," Rob notes, "I'm always hearing about jobs that are open." And though Rob feels trained chefs are currently in demand, he emphasizes that persistence is necessary to find a good job. "A lot of the time you just have to go out and pound the pavement." Rob also stresses the importance of knowing the path you wish to take—do you want to work in a hotel, a resort, a small dining establishment, or a country club. He also advises trying different cuisines to gain skills and determine whether you wish to specialize.

Job opportunities can also be located through newspaper classified advertisements, the Internet, employment agencies, or local offices of the state employment service. Another method is to apply directly to restaurants or hotels like Thomas did. Many large hotels and eating establishments offer job hotlines to announce available job vacancies.

## where can I go from here?

Someday Rob hopes to open a small inn with a really great dining room, but for now Rob plans to continue working at the Hyatt Regency, building the business at Polaris and hopefully making it to executive chef. Rob will also pursue the various levels of certification through the ACF—he feels that certification lends credibility and support to his profession.

Thomas plans to return to Germany and attend hotel management school. After graduating, he wants to become a master chef and concentrate more heavily on the business administration aspect of being a chef. He plans to open his own pastry and coffee shop, which is what most trained pastry chefs in Europe eventually do.

Chefs can advance to supervisory or management positions or to executive chef positions in hotels, country clubs, or large restaurants. Chefs sometimes move from one job to another,

### Advancement Possibilities

**Executive chefs** are department heads who are responsible for all culinary units in a food establishment.

**Culinary educators** are chefs working as educators in colleges and other educational institutions.

**Master chefs** are chefs with the most experience and highest degree of culinary knowledge. They often work for large corporations or as educators. There are few certified master chefs.

**Caterers** prepare, deliver, and serve food for clients, often for special occasions and banquets.

acquiring higher-paying jobs with each move. Some chefs may go on to open their own restaurants, bakeries, or catering businesses. Becoming an educator is yet another option for chefs with considerable experience. Qualified teachers are needed in colleges, culinary arts institutes, high schools, and other educational institutions.

FYI

**Soup is hot**

According to the National Restaurant Association, soup is growing in popularity and turning up on more menus across the nation. In 1992, 87 percent of menus offered a soup selection, but in 1997 the number grew to 92 percent. In the Northeast, many soup-only establishments have been setting up shop. These small restaurants focus on high-volume, takeout sales.

The variety of soups served today ensures that all patrons will be able to find a version to their liking. In addition to the classics such as clam chowder and chicken noodle soup, you'll find borscht, Vietnamese pho soup, Mexican albondigas soup, Mediterranean roasted eggplant and curry soups, cold fruit soups for the summer, and many more. The classic comfort food has adapted to the nineties. Soup's on!

## what are some related jobs?

According to Rob, other jobs open to trained chefs include sales jobs for food purveyors or food stylist positions. Food stylists prepare food for photography shoots or demonstrations.

The U.S. Department of Labor classifies chefs under the headings *Chefs and Cooks, Hotels and Restaurants* (DOT) and *Craft Technology: Food Preparation* (GOE). Also listed under this heading are people who work as dietitians; dietetic technicians; and cooks, chefs, and bakers.

| **Related jobs** |
|---|
| Dietitians |
| Dietetic technicians |
| Cooks, chefs, and bakers |
| Food stylists |
| Food sales representatives |

## what are the salary ranges?

According to the 1998–99 *Occupational Outlook Handbook*, chefs' salaries vary tremendously and depend partly on the type of establishment and its location. Wages are generally highest in hotels and elegant restaurants, and executive chefs can earn more than $38,000 a year.

A salary survey by the *Washington Weekly* indicates that the median annual salary in 1996 for a sous chef was $25,000 and $30,000 for a chef. *Money* magazine states that being a chef is the hottest job of the 1990s, and that chefs earn an average yearly salary of $35,000. According to *Money* and Rob, it is not unusual for executive chefs in large establishments to earn more than $100,000 a year.

## what is the job outlook?

Rob feels the job outlook is extremely positive for chefs right now. "If you're trained and schooled," says Rob, "there are lots of headhunters out there now." Rob is frequently contacted by recruiters looking for qualified chefs. The 1998–99 *Occupational Outlook Handbook* confirms Rob's opinions. It indicates that job openings for chefs and other kitchen workers should be plentiful through the year 2006. Part of this growth can be attributed to increases in household income and dual income homes—more families will be able to afford dining out, and they may also find it more convenient.

Also according to the *Occupational Outlook Handbook*, the job outlook for pastry chefs and bakers is positive due to the popularity of freshly baked breads and pastries. The outlook for institutional chefs, however, is not as rosy. As more institutions opt to contract out their food services, demand for trained chefs may decrease, since many of the contracted companies focus on fast-food and thus employ more fast-food and short-order cooks.

There are plenty of areas in the cooking field where demand is high, however, so institutional chefs should be able to find employment elsewhere. As Rob indicates, "The hotel industry right now is just raging. The restaurant industry is going very hard. The chains are hiring." Restaurant chains are also expanding, which means more chefs will be needed.

Trained chefs are needed everywhere, from small resorts in remote spots to large dining rooms in urban locations, but job opportunities are generally more plentiful in areas with a large population. "Both the coasts have always been very good," says Rob, "because of the variety and the populations of the areas." Cities or regions experiencing rapid growth are also excellent places for trained chefs. For example, the Denver region is currently expanding greatly and is gaining a reputation for high-quality, innovative dining establishments.

# how do I learn more?

## sources of additional information

The following organizations can provide information on cooking careers, schools, and scholarship opportunities.

**American Culinary Federation**
PO Box 3466
10 Bartola Drive
St. Augustine, FL 32086
904-824-4468
http://www.acfchefs.org

**American Institute of Baking**
1213 Bakers Way
Manhattan, KS 66502
913-537-4750
http://www.aibonline.org

**Culinary Institute of America**
433 Albany Post Road
Hyde Park, NY 12538
914-452-9600
http://www.ciachef.edu

**International Association of Women Chefs and Restaurateurs**
110 Sutter Street
San Francisco, CA 94104
415-362-7336
http://www.ontherail.com/iawcr.html

**National Restaurant Association**
1200 Seventeenth Street, NW
Washington, DC 20036
800-424-5156
http://www.restaurant.org

**U.S. Pastry Alliance**
3349 Somerset Trace
Marietta, GA 30067
888-APASTRY
http://www.uspastry.org/index.html

## bibliography

Following is a sampling of materials relating to the professional concerns and development of chefs.

### Books

Child, Julia and Paul Child. *The French Chef Cookbook.* New York: Ballantine Books, 1998.

Dale, Charles, Aimee Dale, and Jock McDonald. *The Chef's Guide to America's Best Restaurants.* Selections and comments of 170 of America's top chefs.Aspen: The Chef's Guide, Inc., 1997.

Donovan, Mary Deidre, ed. *The New Professional Chef.* 6th ed. Textbook written by culinary experts. New York: John Wiley & Sons, 1995.

Dornenburg, Andrew and Karen Page. *The Becoming a Chef Journal. Insight into pursuing various cooking careers.* New York: John Wiley & Sons, 1997.

Rombauer, Irma S., Marion Rombauer, and Ethan Becker. *Joy of Cooking.* An authoritative cookbook. New York: Scribner, 1997.

### Periodicals

*American Connoisseur: The Internet Hub of Good Taste.* Daily. Features gourmet recipes, dining, and travel recommendations.
http://www.americanconnoisseur.com.

*Art Culinaire.* Monthly. Professional magazine for chefs. Culinaire Inc., 40 Mills Street, Morristown, NJ, 07960.

*Bon Appétit.* Monthly. Cooking magazine. Condé Nast, 350 Madison Avenue, New York, NY, 10017.

*Gourmet.* Monthly. Magazine covering cuisine and lifestyle. Condé Nast, 350 Madison Avenue, New York, NY, 10017.

*Cooking Light.* Monthly. Cooking for health-conscious clientele. Cooking Light Magazine, P.O. Box 830656, Birmingham, AL, 35283-0656.

*Star Chefs.* Daily. Cooking resource, including promotions, articles, cooking links, columns, and a dictionary. http://www.starchefs.com.

# chemical technician

**Definition**

Chemical technicians conduct physical tests, chemical analyses, and instrumental analyses for research, product development, quality control, and establishing standards. They plan, design, and set experiments, make samples and test them, document results of tests and analyses, and maintain the laboratory and equipment.

**Alternative job titles**

Analytical technicians
Associate chemists
Chemical engineering technicians
Chemical laboratory technicians
Research associates
Synthetic chemistry technicians

**Salary range**

$20,000 to $32,000 to $45,000

**Educational requirements**

High school diploma; associate's degree; bachelor's degree

**Certification or licensing**

None

**Outlook**

Faster than the average

GOE
02.04.01*

DOT
022

O*NET
24505A

## High School Subjects

Biology
Chemistry
Mathematics
Physics

## Personal Interests

Computers
Cooking
The Environment
Figuring out how things work
Science

a chemical laboratory and most likely you conjure up visions of smoke, bubbling liquids, test tubes, beakers—what seems to be magic. Perhaps that's more of a movie maker's idea of a science lab.

Today's real-life laboratories do use test tubes and beakers, but they can also contain pumps, pipelines, tanks, valves, and computers or other electronic equipment, such as dilatometers, spectrometers, and X-ray diffraction devices. Modern laboratories are also clean, tightly controlled, and in compliance with very high standards of safety and health.

But the magic is still there.

The magic of chemistry is what appeals to Connie Murphy. "In high school, I didn't choose a career as a *chemical technician* and take chemistry to achieve that goal," she says. "I chose chemistry. I enjoyed the lab experience—watching things change and react right before my eyes."

It wasn't until many years later that Connie became a technician, but now she spends most of her work day in the lab making samples of high-performance polymer film to be sent to another lab for testing. After fifteen years working on numerous projects for the same company, the laboratory still holds magic for Connie.

## what does a chemical technician do?

Products you use every day—and their solid, liquid, and gas components—have been invented, tested, and manufactured in laboratories. Your food, drugs, fertilizers, plastics, paints, detergents, paper, petroleum, and cement, for example, are first studied in the laboratory to determine strength, stability, purity, chemical content, and other characteristics.

There are four general areas of chemistry a technician may work in: analytical, organic, inorganic, and physical. Analytical chemistry is involved with determining the composition of substances. Inorganic chemistry is concerned with the properties and reactions of compounds that do not contain carbon. Organic chemistry studies the compounds that do contain carbon. Physical chemistry is concerned with the role of energy transformations in reactions.

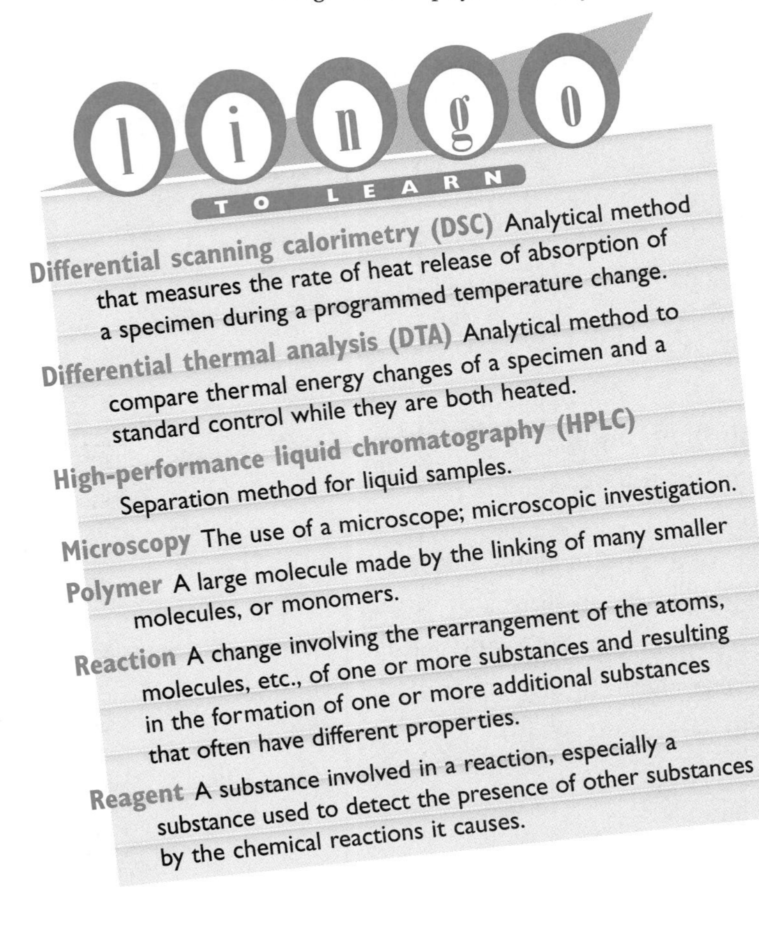


Chemical technicians work in research, new product development, quality control, or criminal investigation. Those in research and development often work in pairs or small groups with Ph.D. chemists and chemical engineers to develop chemicals, synthesize compounds, or develop new processes. Technicians who perform quality control tests often work together in groups under supervision. They make samples of new products or collect soil, water, or air samples. They study their composition or test certain properties. Other technicians conduct physical tests on samples to determine such things as strength and flexibility, or they may characterize the physical properties of gases, liquids, and solids and describe their reactions to changes of temperature and pressure.

Usually, a chemist or chemical engineer will plan and design an experiment. Technicians help them conduct research, locate resources, create a statistical design, describe procedures in writing, or design and run computer simulations. Technicians obtain or make samples and make observations. They gather, clean, and calibrate all necessary glassware, reagents, chemicals, electrodes, and other equipment; choose, check, and calibrate test equipment; and perhaps fabricate or modify the test equipment for a particular use. Others analyze the samples and report the results.

Chemical technicians work in a wide variety of settings. They might test packaging for design, materials, and environmental acceptability. Others might test and develop new plastic compounds for use in the manufacture of small appliances, or develop a colorfast dye for fibers. Some collect and analyze samples of ores, minerals, gases, soil, air, water, and pollutants. Chemical technicians might manage the laboratories at a school, ordering equipment and maintaining supplies. Some work in the petroleum industry to ensure the quality of gasoline, furnace oil, and related products. And some chemical technicians help develop and test new drugs and medicines.

Exposure to hazardous conditions is common in laboratory work. There may be toxic or flammable chemicals, or dangerous equipment like compression cylinders. Because of these hazards, technicians may be required to wear protective clothing, eye protection, or respirators. Following safety procedures is extremely important.

Chemical technicians usually work regular hours and spend most of their time in the laboratory, but they also are required to document the results of their tests and analyses. They compile data and keep accurate records, preparing charts, sketches, and diagrams, usually using a computer. Finally, technicians maintain the laboratory, its inventory, and equipment according to federal, state, and local safety and health regulations. They may be responsible for the transportation and disposal of materials and hazardous wastes in compliance with regulations.

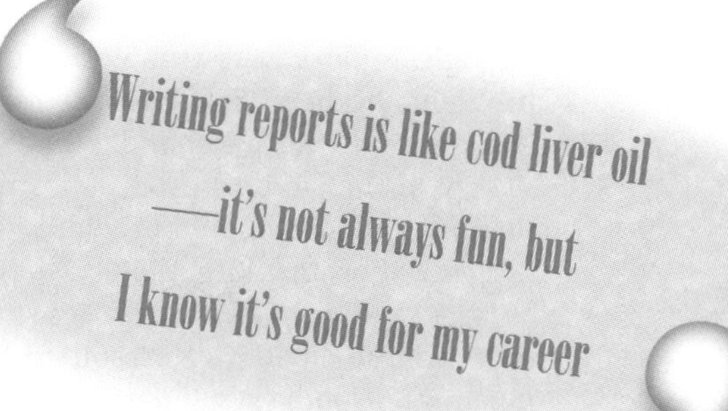


## what is it like to be a chemical technician?

"My job responsibilities are pretty varied," says Connie Murphy, a senior research technologist for a major U.S. chemical manufacturer. "I do experimental work. Sometimes I perform analytical work on my own samples, but I also send samples out to our analytical facility or another testing facility for evaluation. I keep records in my data book, and have to give oral reports, written progress reports, and research reports."

Connie's primary function is as a research lab technician, and she does process research for high-performance polymer film for electronic applications. She works directly with a Ph.D. scientist, but has a lot of independence. She has been working fifteen years in the same organization in basic and applied research, and though she works in a number of different project areas, most of her work is related to polymer synthesis, processing, and evaluation.

"My current project is making samples to be sent for outside evaluation. Most of my time is spent in the lab, but I also work at a desk. I describe in my data book how the samples are made and record all the information I need to track where the samples are sent. I sometimes handcast films, and sometimes use machines that mechanically cast the film. I analyze the samples with equipment, such as a differential scanning calorimeter."

Connie also helps other technicians in her group with synthesis work, which involves setting up glassware, running the reactions, working the product up, and writing results in the data book.

"The data book is very important," Connie says. "We have regular staff meetings once a week for technicians and Ph.D.s, and I am required to give an oral progress report once every six months." Connie is also taking her turn as waste coordinator for her building. About 10 percent of her time is spent taking care of paperwork and seeing that regulatory requirements are met.

Mike Dineen is a research technologist at the same company as Connie, but in a different department. He works in research development and production of plastics using microscopy. "I use electron microscopes to study morphology (particle size, particle shape, and structure), and I correlate that with product performance—why it has a certain color, why it fractures or fails, why it does or doesn't take paint well. We work on everything from small appliances to automotive dashboards," he says.

Mike spends about 80 percent of his time in the laboratory, mostly running samples. Like Connie, he attends occasional group meetings and gives oral and written presentations. Unlike Connie, he does not work closely with his manager or the people he supports. He sets his own schedules and priorities. "I have both scheduled and crisis work. I generally have a backlog of two to six weeks worth of work and I do things chronologically or based on priority of the project. Then quite often there's a client who has a problem that must be solved immediately."

Another technician in the plastics division at this company is in charge of the research lab where quality assurance, product characterization, and testing take place. She maintains and calibrates all the equipment, which includes such thermal analysis equipment as differential scanning calorimeters to measure melting point, crystallization, and glass transition and a thermal mechanical analyzer for measuring melting point and heat distortion under loads. There is machinery for penetration, tensile, flex, and fatigue testing, and rayometers for measuring flow rate of plastics. She researches new equipment, finds suppliers, handles the order, installs new equipment, and keeps it running.

Debra McCombs is a technologist at Zeneca Pharmaceuticals. She sets up and conducts studies, and works on research projects with her supervisor. "My job varies quite a bit, which is one of the reasons I really like it. I mostly work with high-performance liquid chromatography. We evaluate HPLC columns and look at different types of drugs on a particular column. We use biological matrices to troubleshoot separation problems in drugs, and then we research new and improved ways to handle those problems. We are also a support group for other departments in the company, providing them with drug metabolism information."

"It's exciting to be a part of research projects that are published and presented at national meetings," Debra says, but science is more than just a job to her. She is taking evening classes toward a bachelor's degree in plant sciences. "It's not chemistry related, and will not affect my job. It's just one of my own personal goals," she says. Debra also belongs to a local group in Delaware called the Science Alliance. It is made up of volunteers from science industries who talk to elementary school children about a variety of science subjects. "The kids are really fascinated with science and it's worth the time you spend putting a presentation together."

## have I got what it takes to be a chemical technician?

Chemical technicians spend anywhere from 50 to 80 percent of their time in the laboratory, but it's rare for them to work alone in a room full of test tubes and samples. "A chemical technician has to have excellent communication skills. That's probably number one," Connie says. "You have to be able to communicate the results of your work both orally and in writing."

Mike agrees. "Communication skills and people skills are important. I work with forty to sixty customers a year whom I see routinely. You must be highly flexible to work with the company, customers, and co-workers from a variety of educational and cultural backgrounds." Technicians have to carefully follow directions from chemists or engineers and often work in teams with other technicians and scientists.

To be a successful chemical technician, you should

- Be detail-oriented, precise, and meticulous
- Be able to communicate with a variety of people, both orally and in writing
- Be flexible to handle a number of tasks simultaneously
- Have analytical and problem-solving skills
- Be inventive and resourceful

"Technicians are taking on more and more responsibilities, doing the things Ph.D.s were doing fifteen years ago. There's a lot more responsibility and accountability than ever before," Connie says, "so technicians who work in research in particular have to pay a lot of attention to detail. It's the small details that very often are the most crucial."

Debra's job requires her to be very organized, pay attention to detail, be able to think analytically, and be able to troubleshoot. "The only difficult thing about my job," she says, "is being faced with a problem I can't figure out. But there's something new every day. It's not routine in any way."

Connie says she has experienced some monotony from time to time. "Sometimes you have to crank out samples or do repeated testing. I worked in adhesives for a while and a lot of that is making up samples and pulling them apart—but sometimes you just have to do the grunt work."

Mike finds it difficult sometimes to set priorities. "Writing reports is not a favorite part of my job, but at the same time it's good for you, sort of like cod liver oil—it's not always fun, but I know it's good for my career."

## how do I become a chemical technician?

Chemical technicians usually have an associate's degree in chemical technology, laboratory science, or another science specialty. Some employers hire high school graduates and place them in their own training programs, but that practice is becoming rare.

Many industrial employers would rather hire a graduate from a two-year technical program than one with a bachelor's degree in chemistry. They believe four-year programs don't prepare students to work as technicians because they don't include enough laboratory practice. The bachelor's degree traditionally has more theoretical studies and is designed as preparatory work for graduate school.

"For many people, working as a chemical technician is a second career choice," Connie says. "Many people entering the work force as chemical technicians are older. I think the average age for the first level technologist in my company is twenty-nine. Personally, I went to college right out of high school, got married, divorced, worked in a factory for a while, and did all sorts of other things. But my interest had always been in chemistry. Then I found out about the two-year program." She says it's a good road for someone who needs to get a job quickly and start making money. Debra also waited until ten or fifteen years after she graduated from high school before she earned her associate's degree.

**Coming Attractions...**
Although very few schools offer two-year associate's degree programs for *chemical plant operators* and *chemical engineering technicians*, the American Chemical Society indicates that industry recognition of and demand for these jobs will affect the number of schools offering programs. In the next several years, look for these jobs to be big.

## education

### High School

To prepare for a program in chemical technology, high school students should be interested in science and take courses in math, chemistry, physics, and computer science. Computers and computer-interfaced equipment are often used in research and development laboratories.

Students might find it valuable to participate in the extracurricular science clubs that many schools offer. Science contests are another fun way to apply principles learned in classes to a special project. You may work alone or in a team in competitions that are held within your own school, across the state, or even nationally.

### Postsecondary Training

Approximately eighty two-year colleges offer courses in chemistry or chemistry-related subjects. There are about forty two-year colleges in the United States that have chemical technology programs.

"The two-year program can be very difficult," Connie says. "Many people drop out because it's very intensive in math, and there are difficult courses like organic chemistry. You spend a lot of time in the laboratory and studying many of the same chemistry courses you would in a four-year program. It's not easy."

Debra also found her two-year program very challenging. She has an associate's degree in biotechnology. Although chemistry is what she uses now, she also studied such subjects as microbiology and hematology. "The technology degree is the best way to go. Earn an associate's degree from a school that has a chemical technology program," she recommends.

## scholarships and grants

Many technical schools offer scholarship programs and awards to their students. Connie received a couple of awards from Milwaukee Area Technical College. After her first year, she received the chemistry department award for being the outstanding chemistry student. After her second year, at graduation, she received an award from the chemical technology program for being the outstanding student. She thinks it was significant in getting her first job as a chemical technician.

## internships and volunteerships

Mike took a different track. He studied electron microscopy at a community college and then participated in a government exchange program. He spent a year in Germany studying microscopy and working in a German company doing microscopy. "The real ace in the hole for me was my overseas experience—six months in a foreign university and six months in a company in my field was really what opened all the doors for me." He saw a notice on a bulletin board at the community college he was attending about the exchange program, Congress Bundestag, which accepts fifty U.S. students and one hundred Germans. It is funded by the U.S. and German congresses. "The German company where I apprenticed was a world-renowned company in the field of microscopy. I think just the fact that an individual would take some risks and go to a foreign country to gain something helped me in my job search later," Mike says. "It wasn't an award or scholarship exactly, but considering that only fifty U.S. students were accepted, and adding the value of the foreign experience, I think of it as an award."

While working toward an associate's degree, some students work as co-ops or do summer work in their field. Co-ops are full-time students who work about twenty hours a week for a local company. Some co-op positions may be available to high school seniors.

Even after you earn your associate's degree and get a job as a chemical technician, there will always be more training on the job. Connie explains, "Most community college chemical technology programs give you a good background in basic laboratory skills, instrument skills, chemistry theory, inorganic chemistry, quantitative analysis, math, and physics. You have a good basic knowledge, but you're not ready to go into the lab and do things on your own. Much of what you do, you learn on the job. Most chemical technicians' jobs are so diverse, there has to be some on-the-job training."

## labor unions

Some chemical technicians belong to a union, but unionization varies drastically across the country. In some cases, the plant is unionized, but the laboratory is not. Many technicians have nonexempt status, meaning they are not exempt from federal wage and hour laws. That status determines the requirement for union membership. Overall, the percentage of technicians who belong to a union is very small.

## who will hire me?

Chemical technicians work for chemical companies, pharmaceutical companies, manufacturers, food processing companies, research and testing laboratories, schools and universities, petroleum refineries, biotechnology companies, agricultural organizations, and crime laboratories. Federal and state government agencies, particularly the federal Departments of Defense, Agriculture, Interior, and Commerce, also hire chemical technicians.

Some companies that have ongoing needs for chemical technicians work with local community colleges and technical schools to develop two-year programs. This ensures them a supply of trained chemical technicians. Other companies have joined forces to establish local, independent, technician training programs.

Most employers recruit locally or regionally for technicians. They have found that trained technicians aren't willing to relocate for a job. This is especially true if it's a second career choice. By the time they earn their two-year degree, technicians may be settled with a family in a community they like. Because companies hire locally and work closely with technical schools, placement offices are usually successful in finding jobs for their graduates.

Connie has attended several recruiter panel discussions. "Employers say they prefer to hire people with associate's degrees, but there aren't enough people available. There's been basically a starvation mode for appropriately trained technicians for a number of years."

Some companies hire high school graduates and train them on the job, but it is not a common practice. Companies generally prefer a graduate of a two-year program in chemical technology or another science. When hiring, they consider your grades, but companies are also attracted by special awards or efforts. Connie's two awards, for example, helped her land a job, and Mike's overseas experience was impressive on his resume.

When he returned from Germany, Mike went through technical journals and geographically selected places of employment from companies mentioned in articles. "It may not be the best way to go," he says, "but I sent resumes to about fifty authors. Since authors don't usually have anything to do with hiring, I imagine many of those resumes got tossed." He did, however, receive ten responses, interviewed with three firms, and was immediately hired by one of them. He has been with the same company for ten years.

Mike's company also hires co-ops to assist technicians and scientists. "Co-op work is highly regarded by employers," he says. "They can judge technical competence from grades and scholastic achievements, but it's difficult for them to judge people skills at an interview." Co-ops have the opportunity to prove their social skills to employers, which gives them an edge.

Contracting is another option. There are local agencies that place technicians with companies for special projects or temporary assignments. Many technicians work as contractors, from one month to as much as a year, and then are hired as full-time employees.

"Be flexible about where you want to work and don't limit yourself to a specific area of chemistry," advises Debra. "Flexibility is a key. My degree was in biotechnology. I worked in a toxicology lab for a month and then got a job with a pharmaceutical company. I've been with the same company for over thirteen years."

## where can I go from here?

Connie, Mike, and Debra have all been working for their companies for over ten years. And all three are happy. "I like what I do. The supervisors I work with have been very supportive, encouraging me to take on as much responsibility as I can handle," Connie says.

There are opportunities to work in the areas of technology development or technology management, with equal or comparable pay. Many companies offer the option to move to other departments, from research and development to manufacturing, for example. Technicians,

**Advancement Possibilities**

**Assayers** test ores and minerals and analyze results to determine value and properties of components.

**Chemical engineers** design equipment and develop processes for manufacturing chemicals and related products.

**Chemical research engineers** conduct research on chemical processes and equipment.

**Chemists** conduct research, analysis, synthesis, and experimentation on substances.

**Food chemists** conduct research and analysis concerning chemistry of foods to develop and improve foods and beverages.

however, can become specialists. Connie explains, "They can become very good at a specific area of technology, so the higher up a technician is, the more difficult it is to move to another area, because they basically create their own job."

That's true in Mike's case. "I'm highly specialized, but I lack a higher degree for moving up the ladder. Fortunately I like what I'm doing and management allows me to do different things. My company would support continuing education if I were interested."

Some companies have the same career track for all lab technicians, whether they have an associate's degree, a bachelor's degree, or are highly experienced with no degree. Other companies have different promotion systems for technicians and chemists, but will promote qualified technicians to the chemist level and allow them to progress along that track.

The associate's degree can be a stepping stone toward a bachelor's in chemistry. Some technicians earn their two-year degree, get into the work force, and then return to school to work on their bachelor's degree. Many companies encourage employees to continue their education, and some help with tuition.

## what are some related jobs?

Chemical technicians can shift to other science careers, such as in physics, geology, or biology. There are also opportunities in the health care industry, including pharmacy and dentistry. The U.S. Department of Labor classifies chemical technicians under the headings *Occupations in Chemistry* (DOT) and *Laboratory Technology: Physical Sciences* (GOE). Also under these headings are people who work as laboratory technicians in the aircraft-aerospace industry, metallurgy, utilities, photo-optics, and spectroscopy. Chemical engineering technicians are classified under the headings *Chemical Engineering Occupations* (DOT) and *Engineering: General Engineering* (GOE), which also includes people who work in automotive or structural engineering, the petroleum industry, agricultural engineering, electrical engineering, and mechanical engineering.

| Related Jobs |
|---|
| Agricultural engineers |
| Criminalists |
| Electronics engineers |
| Fiber technologists |
| Metallurgical technicians |
| Petroleum engineers |
| Pharmacists |
| Photo-optics technicians |
| Ultrasound technologists |

## what are the salary ranges?

In 1995 the median starting salary of bachelor's level industrial chemists employed full time as technicians was $25,000; with ten to fourteen years of experience, technicians earned $35,000, according to the 1995 American Chemical Society Salary Survey. Salaries are highest in private industry and lowest in colleges and universities.

Salaries vary according to the education and experience of the employee and the size and type of employer. The greatest variation in salary occurs between regions of the country. Starting salaries are highest in the Middle Atlantic and lowest in the East South Central region.

Most large companies that hire chemical technicians offer paid vacations, health insurance, and pension plans.

## what is the job outlook?

Many companies are cutting back on in-house training, preferring to hire those with degrees and proven skills, especially in the laboratory. Technicians are gaining more respect from employers. In the past they have been considered assistants to the "real" scientists, but their status is changing to one of valued, highly trained specialists who are indispensable in the work force.

More and more companies are opting to hire contract employees as technicians for temporary assignments, but there is still a need for full-time, competent laboratory technicians. Connie, Mike, and Debra have proven they can have long-term careers in the same company. Technicians who have training in a variety of skills will find it easier to make lateral moves within a company as projects, programs, and reporting structures change.

Douglas J. Braddock, an economist at the Department of Labor, says the number of science and math technicians in the United States is projected to grow 25 percent (from 244,000 to 304,000) between 1992 and 2005, compared with 22 percent growth for all occupations.

Connie says, "The job outlook is very good to excellent. As the country becomes more technologically advanced, there will be a need for more training beyond high school." Most recent graduates of two-year programs in science are not likely to have difficulty finding jobs in the next decade.

## how do I learn more?

### sources of additional information

Following are organizations that provide information on chemical technician careers, accredited schools and scholarships, and possible employers.

**American Chemical Society (ACS)**
1155 16th Street, NW
Washington, DC 20036-4800
202-872-4600

**American Institute of Chemical Engineers (AICHE)**
345 East 47th Street
New York, NY 10017-2395
212-705-7660
http://www.aiche.org

**Junior Engineering Technical Society (JETS)**
1420 King Street, Suite 405
Alexandria, VA 22314-2715
703-548-5387
jets@nas.edu

## bibliography

Following is a sampling of materials relating to the professional concerns and development of chemical technicians.

### Periodicals

*Accounts of Chemical Research.* Monthly. Covers basic research and applications involved in all aspects of chemistry. American Chemical Society, 1155 16th Street, NW, Washington, DC 20036, 202-872-4363.

*American Laboratory.* Monthly. Addresses chemists and biologists by providing articles on all areas of basic research and current laboratory practice. International Scientific Communications, Inc., 30 Controls Drive, Box 870, Shelton, CT 06484-0870, 203-926-9300.

*Chemical Engineering Progress—Student Edition.* Quarterly. Student magazine featuring technical articles and providing advice on study and careers. American Institute of Chemical Engineers, 345 East 47th Street, New York, NY 10017-2395, 212-705-8100.

*Chem Matters.* Quarterly. Popular chemistry magazine providing articles that relate chemistry to everyday life. American Chemical Society, Office of High School Chemistry, 1155 16th Street, NW, Washington, DC 20036, 202-872-4600.

*Chemical and Engineering News.* Weekly. Features coverage of the scientific, technical, educational, and professional aspects of the field. American Chemical Society, 1155 16th Street, NW, Washington, DC 20036, 202-872-4600.

*Chemist.* Eleven times per year. Covers professional, economic, social, and legislative topics of interest to chemists and chemical engineers. American Institute of Chemists, Inc., 7501 Wythe Street, Alexandria, VA 22314, 703-836-2090.

*Journal of the American Chemical Society.* Biweekly. Focuses on the many areas of chemical science. American Chemical Society, 1155 16th Street, NW, Washington, DC 20036, 800-333-9511.

# civil engineering technician

**Definition**

Civil engineering technicians work in the design, planning, and building of railroads, airports, highways, drainage systems, and many other structures and facilities.

**Alternative job titles**

Construction engineering technicians

Highway technicians

Water resources engineering technicians

**Salary range**

$15,500 to $32,700 to $54,800+

**Educational requirements**

Associate's degree

**Certification or licensing**

Recommended

**Outlook**

Faster than the average

GOE 05.03.02

DOT 005

O*NET 22121

## High School Subjects

Biology
Chemistry
English (writing/literature)
Mathematics
Physics
Shop (Trade/Vo-tech education)

## Personal Interests

Building things
Computers
The Environment
Figuring out how things work
Trains

**Before** the new school building can go into production, Mark Funkey has the responsibility of drawing the plans for the pieces of steel. The deadline is rapidly approaching and, if it's missed, the whole project will be thrown off schedule. Without Mark's plans, the steel fabricators can't make the steel, and can't deliver it to the school site.

Hovering over the drafting board, long past quitting time, Mark checks the placement of bolts and welds. The AISC manual in hand, he determines how much weight the bolts will support given the beam's size and span. He determines the maximum bending stress and where it will occur. Classes will be taught in this building, he thinks, and games will be played, many teachers and students going about their lives. Mark is dedicated to making sure the building is a safe one.

# what does a civil engineering technician do?

If you've ever ridden on a train, driven along a highway, or taken an elevator to the very top of a skyscraper, you've relied on the expertise of a civil engineering technician. Or if you've ridden a roller coaster, or crossed a bridge, or gone for a stroll through a city park. *Civil engineers* and *technicians* work together for the community, providing better, faster transportation; designing and developing highways, airports, and railroads; improving the environment; constructing buildings and bridges and space platforms. Civil engineering technicians get you to work or school on time, as well as take you around the world safely, quickly, and efficiently.

There are seven main civil engineering areas: structural, geotechnical, environmental, water resources, transportation, construction, and urban and community planning. The work is closely related, so a technician might work in one, or many, of these areas.

*Structural engineering technicians* assist in the design of the structures we use in our daily transportation, travel, work, and recreation. Technicians are involved in the planning of all kinds of structures: buildings, bridges, platforms, even amusement park rides. With their knowledge of building materials, and the effects of weather and climate on these materials, they help design the best structure for the purpose. Structural engineering technicians calculate the size, number, and composition of beams and columns. Sometimes their work involves geotechnical engineering.

*Geotechnical engineering technicians* analyze the soil and rock that will be needed to support both underground and above-ground structures. The construction of tunnels, dams, embankments, and offshore platforms requires understanding of the soil and rock, so the structures will remain stable. If soil pressures from the weight of the structures will cause excessive settling or some other failure, technicians design special piers, rafts, pilings, or footings to prevent structural problems.

Environmental engineering involves removing pollutants and contaminates from the water and the air. *Environmental engineering technicians* also work with wastewater and solid waste management. With their understanding of biological processes, these technicians help to preserve our natural resources.

Water resources engineering technicians gather data and make computations and drawings for water projects such as pipelines, canals, and hydroelectric power facilities. A water resources technician helps with the control of water to prevent floods and to manage rivers.

Highways, airports, and railroads are designed and constructed by transportation engineers and technicians. *Transportation engineering technicians* also work to improve traffic control and other systems that will allow for the faster, more efficient transportation of people and products. *Highway technicians* perform surveys and cost estimates as well as plan and supervise highway construction and maintenance. *Rail and waterway technicians* survey, make specifications and cost estimates, and help plan and construct railway and waterway facilities.

*Construction engineering technicians* help to bring a design project to the construction stages. They prepare specifications for materials and help schedule construction activities. They inspect work to assure that it conforms to blueprints and specifications. *Materials technicians* sample and run tests on rock, soil, cement, asphalt, wood, steel, concrete, and other materials. *Party chiefs* work for licensed land surveyors, survey land for boundary-line locations, and plan subdivisions and other large-area land developments.

*Urban and community planning technicians* work in the full development of a community. Technicians in this area coordinate the planning and construction of city streets, sewers, drainage systems, and refuse facilities, as well as identify recreation areas and areas for industry and residence.

**Cantilever beam** A beam that projects beyond its support.

**Catch basin** A structure with a sump below the pipes for the purpose of causing sand and gravel to settle by slowing the velocity of the water.

**Curvilinear** A pattern of curved streets and tee intersections.

**Development** Improvement to the land for the benefit of the public.

**Engineer's level** The instrument most commonly used for determining elevations in the field.

**Galvanizing** Coating a pipe with zinc for protection against corrosion.

**Grade** The steepness of a slope.

**Subdivision** A parcel of land divided into more than one section.

Civil engineering technicians also may specialize in areas of sales and research. A *research engineering technician* tests and develops new products and equipment, while *sales engineering technicians* sell building materials, construction equipment, and engineering services.

**To be a successful civil engineering technician, you should**

- Be able to communicate and work with others to solve problems
- Have an interest in building and planning
- Be curious about how things work
- Have an interest in the environment
- Have an understanding of math and the sciences

## what is it like to be a civil engineering technician?

Mark Funkey assists engineers by detailing the steel beams and columns used in buildings and other structures. He draws plans for the pieces of steel, and fabricators work from those plans to actually make the pieces. "We draw the pieces of the puzzle," Mark says, "and others make the pieces and put the puzzle together." He primarily works from an engineer's blueprints, transferring the information into a plan detailing what the steel piece will look like. He works in close cooperation with the engineer by phone and fax.

In his first detailing job five years ago, Mark worked for a manufacturing company, detailing steel plates for prefabricated buildings. Now, Mark works for a business that details steel for all sorts of structures all across the country. He has done work on university buildings, the new Denver airport, the Alaska Native Medical Center in Anchorage, churches, and coal and ethanol plants.

Mark works with eleven other detailers for a company called National Detail. The company subcontracts to steel fabricators who are contracted to builders. The detailers usually have two different projects going at at time, and it takes an average of four to five months to detail all the steel for one building. The amount of time taken to complete a detailing does vary from project to project, though; one building may take only a few weeks to finish, while another may take a full year.

In a typical week, Mark works forty hours, though it is sometimes necessary to work overtime, such as when a project is close to deadline. He does most of his work at the computer, and he is sometimes required to work at a drafting table, relying on a parallel bar or drafting machine for the quick drawing of angles and straight lines. He also uses mechanical pencils and triangles. Manuals assist him in his work; they show him how to connect pieces of steel, and they tell how much weight a bolt or weld will hold.

Every day, Mark tries to detail as many pieces of steel as he can. At National Detail, there are those who draw the steel pieces, and others who check to make sure the plans are drawn up correctly. Changes are suggested by the checker, and if agreed upon, the detailer alters the original drawing.

Once the steel pieces are detailed and planned, the work is sent on to the engineer. The engineer makes sure everything is drawn correctly and effectively. He or she also checks to see that the pieces will support the necessary weight, that there are enough bolts, and so on. If there are changes to be made, the plans go back to the detailer. The detailer then makes those changes, and the plans are sent on to the fabricator.

## have I got what it takes to be a civil engineering technician?

Civil engineering technicians are relied upon for solutions and must express their ideas clearly in speech and in writing. Good communication skills are important for a technician in the writing and presenting of reports and plans.

Mark stresses the importance of math skills. "I use math every minute of every day," he says. He also relies on his common sense—his ability to approach and clearly understand the demands of a project.

Mark's job also requires a lot of analytical thought. "There's more thinking to it than what you'd imagine," he says, referring to the amount of analysis that goes into detailing a piece of steel. Mark likes these analytical aspects of the job, but he doesn't like the stress that can result from deadlines. "There's a lot of pressure when you're in at the beginning of a project," he says. "You're under the gun to make the deadline, to get the plans to the fabricator so the fabricator can get the steel to the job site."

# how do I become a civil engineering technician?

Although Mark took math and science courses in high school, he didn't take mechanical drawing classes; he didn't become interested in drafting until after high school. He then enrolled at the Western Iowa Technical Community College in Sioux City, Iowa, and pursued an associate's degree in architectural construction engineering technology. This two-year program offered courses in every aspect of detailing, and prepared him for engineering and design work. His favorite courses were the math courses, such as trigonometry and precalculus. "I didn't have to study that much," he says, "but there was a lot more to it than I first thought. There are some engineering concepts that can be hard."

## education

### High School

Students interested in civil engineering technician work should take math and science courses, including at least two years of algebra, plane and solid geometry, and trigonometry. Courses in physics, chemistry, and biology can provide you with necessary lab experience. Civil engineering technicians often make use of mechanical drawings to convey their ideas to others, and neat, well-executed drawings are important for accuracy. A shop class, or a class in mechanical drawing, will prepare you for the drafting and designing requirements of the job.

Because computers have become essential for engineering technicians, a computer programming course can introduce you to the skills you'll need. Also, develop your English and language skills; you'll be working closely with engineers and their reports, as well as writing reports of your own.

### Postsecondary Training

Technical schools, or junior and community colleges, offer basic math and science courses that prepare students for courses in surveying, materials, hydraulics, highway and bridge construction and design, railway and water systems, soils, heavy construction, steel and concrete construction, cost and estimates, and management and construction technology. In addition to the course work, laboratory and field experience is required for students in technical school programs.

Civil engineering technicians may eventually want to enter an engineering program at a university. An engineering program would require courses in engineering science and analysis, engineering theory and design, as well as courses in the social sciences, the humanities, ethics, and communications. The Accreditation Board for Engineering and Technology (ABET) develops accreditation policies and criteria (*see* "Sources of Additional Information").

## certification or licensing

Becoming a Certified Engineering Technician can help a technician advance in professional standing. An experienced technician may also take an examination for licensing as a Licensed Land Surveyor. With this licensing, technicians can operate their own surveying businesses.

## scholarships and grants

Scholarships and other financial aid should be available through the technical school or community college you choose to attend. For students pursuing an engineering degree, the American Society of Civil Engineers (ASCE) offers scholarships, fellowships and grants (*see* "Sources of Additional Information").

### Advancement Possibilities

**Civil drafters** draft detailed construction drawings, topographical profiles, and related maps and specifications used in planning and construction of civil engineering projects.

**Highway-administrative engineers** administer statewide highway planning, design, construction, and maintenance programs.

**Structural drafters** draw plans and details for structures employing structural reinforcing steel, concrete, masonry, wood, and other materials.

## internships and volunteerships

A local construction company may offer you the best opportunity for hands-on experience, either through part-time summer work, or an internship. With a construction company, you can observe surveying teams, site supervisors, building inspectors, skilled craft workers, and civil engineering technicians.

## who will hire me?

The month before Mark even received his degree, he had a job as a steel detailer. His school had a good job placement program with good connections and arranged his job interview.

State highway departments employ many civil engineering technicians, as do railroads and airports. Technicians can also find work for city and county transportation services. The natural resources division of the United States Department of Agriculture also employs civil engineering technicians for conservation and water management projects. By contacting the USDA, transportation services, and other government agencies, you can receive information about job positions and opportunities. State and private employment services can also provide you with job listings and will sometimes arrange interviews.

Many schools have cooperative work-study programs with companies and government agencies. Students in these programs often move into permanent positions upon graduation. Also, by getting to know people who work in your chosen area, you'll learn about job and advancement opportunities.

A local college or university might sponsor an American Society of Civil Engineers (ASCE) Student Chapter, an organization that involves civil engineering students with engineering projects. Through the chapter, you can get to know the professionals in your area.

## where can I go from here?

Mark enjoys what he's doing, but wants to move on to better-paying work. "In a small company," Mark says, "it can be hard to move up. In a larger company, things are more structured, and there is more employee turnover." Mark was recently promoted, and is now in charge of distributing work among the other detailers, putting the right people on the right jobs. He plans to return to school to pursue a business degree.

Civil engineering technicians are constantly learning new things about the job as they gain more experience. They are required to learn new techniques, and how to use the latest equipment. Some technicians move on to supervisory positions, while others go on to earn degrees in civil engineering.

Some civil engineering technicians go on to become *associate municipal designers,* a position in which they direct workers to prepare design drawings and feasibility studies for dams and municipal water and sewage plants. *City* or *county building inspectors* review and then approve or reject plans for construction of large buildings. *Project engineers* supervise numbers of projects and field parties for city, county, or state highway departments.

## what are some related jobs?

The U.S. Department of Labor classifies civil engineering technician under the headings *Civil Engineering Occupations* (DOT) and *Engineering Technology: Drafting* (GOE).

| Related Jobs |
| --- |
| Aeronautical drafters |
| Architectural drafters |
| Field-map editors |
| Geological drafters |
| Mechanical equipment engineering assistants |
| Photogrammetric engineers |
| Specification writers |

## what are the salary ranges?

According to the U.S. Department of Labor, beginning civil engineering technicians in 1995 earned between $15,500 and $32,700 a year, while more experienced technicians earn between $28,000 and $33,000 a year. Those in supervisory positions, or with advanced education, earn $54,800, or more, a year.

Civil engineering technicians who operate their own construction, surveying, or equipment businesses can make thousands of dollars more a year than technicians in salaried positions.

# what is the job outlook?

Civil engineering jobs are greatly affected by the economy; new building projects are usually only initiated in times of financial stability. But civil engineering technicians are always needed for improvements on highways, railways, and structures. Technological advances allowing for faster, safer modes of transportation also result in jobs.

As more people and businesses become concerned with the preservation of natural resources, more civil engineering technicians will be employed in the treatment of water and air. Environmentally sound production systems, and other advances and improvements, will also lead to more opportunities for technicians.

# how do I learn more?

## sources of additional information

Following are organizations that provide information on civil engineering technician careers, accredited schools, and employers.

**American Society for Engineering Education**
11 Dupont Circle, Suite 200
Washington, DC 20036
202-331-3500
http://www.asee.org

**American Society of Certified Engineering Technicians**
PO Box 1348
Flowery Branch, GA 30542
404-967-9173

**American Society of Civil Engineers**
345 East 47th Street
New York, NY 10017
800-548-ASCE

## bibliography

Following is a sampling of materials relating to the professional concerns and development of civil engineering technicians.

### Books

Chrimes, Mike. *Civil Engineering, 1839-1889: A Photographic HIstory.* Wolfeboro Falls: Sutton, Pub., 1991.

Randolph, Dennis. *Civil Engineering for the Community.* New York: American Society of Civil Engineers, 1993.

Scott, John S. *Dictionary of Civil Engineering.* Revised 4th ed. New York: Chapman & Hall, 1993.

### Periodicals

*ASCE News.* Monthly. Gives civil engineering employment information, along with educational and practical topics. American Society of Civil Engineers, 345 East 47th Street, New York, NY 10017-2398, 212-705-7512.

*Civil Engineering.* Monthly. Discusses all forms of civil engineering, including bridges, irrigation, highways, and dams. American Society of Civil Engineers, 345 East 47th Street, New York, NY 10017-2398, 212-705-7288.

*Civil Engineering News.* Monthly. Directed toward civil engineer personnel. Civil Engineering News, 1255 Robert Boulevard, Suite 230, Kennsaw, GA 30244-3694, 770-499-1857.

*Engineering Structures.* Monthly. Elsevier Science, Box 945, New York, NY 10159-0945, 212-633-3730.

*Journal of Materials in Civil Engineering: Properties, Applications, Durability.* Quarterly. Relates current news regarding civil engineering materials. American Society of Civil Engineers, 345 East 47th Street, New York, NY 10017-2398, 212-705-7288.

*Journal of Professional Issues in Engineering Education and Practice.* Quarterly. (Issues in Engineering) American Society of Civil Engineers, 345 East 47th Street, New York, NY 10017-2398, 212-705-7288.

*Journal of Structural Engineering.* Monthly. Includes issues related to the analysis, safety, and components of different structures. (Structural Division Journal) American Society of Civil Engineers, 345 East 47th Street, New York, NY 10017-2398, 212-705-7288.

*NTIS Alerts.* Weekly. Includes equipment and materials topics. United States National Technical Information Service, 5285 Port Royal Road, Springfield, VA 22161, 703-487-4630.

*Professional Surveyor.* Bimonthly. Nontechnical approach to surveying and mapping issues related to the engineering field. American Surveyors Publishing Company, 1713 Rosemont Avenue, Suite J, Frederick, MD 21702-4199.

# computer and office machine technician

**Definition**

Computer and office machine technicians install, test, calibrate, clean, adjust, and repair office machines and computer terminals and other computer-related machines.

**Alternative job titles**

Bench technicians/engineers

Computer and office machine servicers

Computer service technicians

Field service technicians/engineers

**Salary range**

$20,000 to $30,100 to $50,000

**Educational requirements**

High school diploma; certificate or associate's degree preferred

**Certification or licensing**

Recommended

**Outlook**

Faster than average for computer technicians; about the same or slightly declined for office machine technicians

GOE
05.05.09*

DOT
633*

O*NET
25104

## High School Subjects

Computer science
English (writing literature)
Mathematics
Physics
Shop (Trade/Vo-tech education)
Speech

## Personal Interests

Computers
Figuring out how things work
Fixing things
Photography

has completed his paperwork from yesterday and is examining the service requests he must handle today—a computer needs wiring repairs, a fax machine keeps jamming, a copier needs cleaning, and a long list of other repair and maintenance jobs. He needs to finish as many of these tasks as he can this morning so he can attend an afternoon training workshop offered by the manufacturers of a new color laser printer his company recently purchased. Everett hopes to learn how the machine operates so he can help with repairs in future emergencies, instead of having to always rely on field service representatives.

"Things sure were much simpler back when I was hired," he says. "I started here fixing calculators and adding machines. Now I'm learning about color photography, laser printing technology and computers. It's really a whole different language!" Businesses have indeed become more reliant on sophisticated, computerized machines. When they break down and the normal flow of work is slowed or stopped, technicians like Everett must quickly step in and solve the problem.

Everett arranges his stack of service requests and pulls from his files the schematic drawings and service manuals he needs. Then, armed with his tool kit, he heads out to face the first downed machine of the day.

## what does a computer and office machine technician do?

Businesses make use of many different kinds of computers and machines in order to perform normal office tasks more efficiently and accurately. Like household appliances and cars, many of these machines require regular preventive maintenance. And, of course, like all machines, they develop problems or break down completely, often creating emergency situations in offices that depend entirely on their running well. *Computer and office machine technicians* are responsible for ensuring that machines are functioning properly at all times.

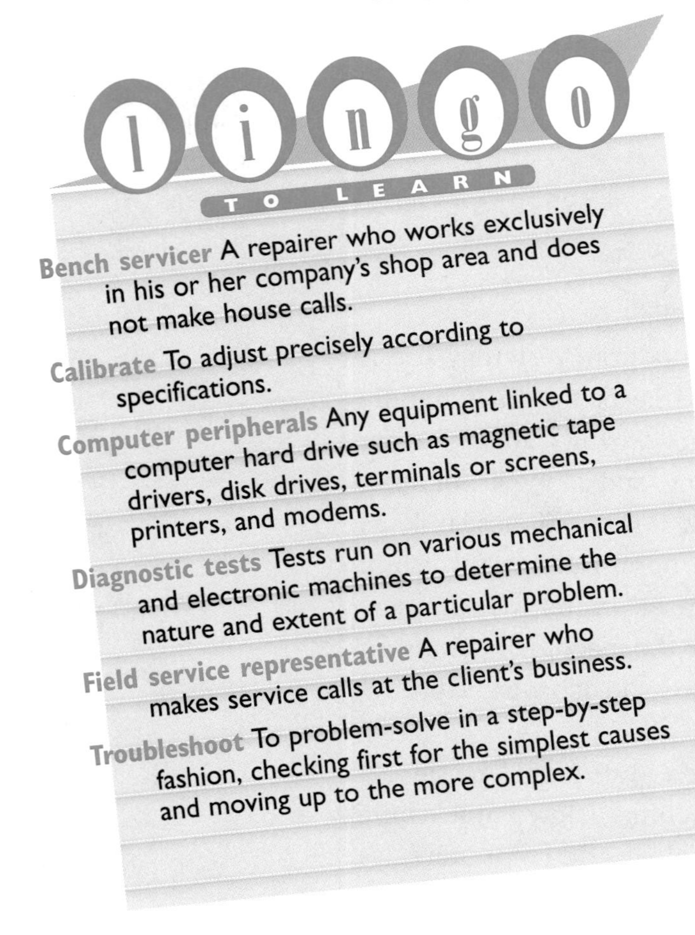


Most technicians, or *servicers* as they are sometimes called, basically perform the same duties: they install, calibrate, clean, maintain, troubleshoot, and repair machines. The main differences among them lie in the particular machines they are qualified to work on and the nature and scope of the business they work for.

If they work in a specialized repair shop, technicians might be experts in fixing one or two types of machines, like manual and electric typewriters, for example. Or, they might know how to repair all types of basic office machines like calculators, adding machines, printers, and fax machines. Other repair specialties include *mail-processing equipment servicers* and *cash register servicers*.

As the use of computers in business continues to grow, an increasing number of technicians are discovering that they can make more money and have more job security if they are well trained in the maintenance and repair of computers and computer peripherals. Many large computer companies provide a service contract with the purchase of their product. *Computer service technicians* are employed by computer companies to fulfill the obligations of the service contract. They install and set up computer equipment when it first arrives at the client's business. They follow up installation with regular preventive maintenance visits, adjusting and cleaning mechanical pieces. In addition, one technician is always on call in case equipment fails. They know how to run diagnostic tests with specialized equipment in order to determine the nature and extent of the problem. They are usually trained in replacing semiconductor chips, circuit boards, and other hardware components. If they cannot fix the computer on site, they bring it back to their employer's service area.

Some technicians are employed in the maintenance department of large companies. They generally have a good working knowledge of mechanics and electronics and then are trained further by the company to work with the specific equipment the business uses most. In-house training is often provided by industrial and manufacturing companies as well as those corporations that depend on big machinery like mail sorters and check processors.

Employed either by a specialty repair shop, machine manufacturer, or product-specific service company, *field service representatives* are technicians who travel to the client's work place to do maintenance and repairs. Their duties include following a predetermined schedule of routine maintenance. For example, they might change the toner, clean the optic parts, and make mechanical adjustments in photocopiers and printers. They also must make time each day to respond to incoming requests for emergency service. Even though supervisors are usually responsible for prioritizing maintenance and repair requests, juggling the variety and number of responsibilities can be hectic. In addition, technicians are required to keep detailed written explanations of all service provided so that future problems can be dealt with more effectively.

Sometimes machines need major repairs that are too complicated and messy to be handled in the office or workplace. These machines are taken to a repair shop or company service area to be worked on by *bench servicers*, that is, technicians who work at their employer's location.

Some very experienced servicers open their own repair shops. Often, these *entrepreneurs* find it necessary to offer service for a wide range of equipment in order to be successful, particularly in areas where competition is tight. To further supplement their income, they might start selling certain products and offering service contracts on them. Business owners have the added responsibility of normal business duties, such as bookkeeping and advertising.

## what is it like to be a computer and office machine technician?

Everett Blevins has worked in the equipment maintenance department of Nationwide Insurance for the past ten years. "I have always been the type of guy that goes around tinkering with everything. If anyone in my family needs something fixed, they know to bring it to me."

Employed by a large corporation, Everett enjoys regular office hours, 8 AM to 4:30 PM, five days a week. He receives full benefits and is paid time and a half for any overtime he puts in during emergency situations. Such emergencies don't occur that often, but usually accompany the installation of a new printing press, mail sorter, or other important (and big) piece of equipment. He might also stay after hours to work on machines that still function but are in need of minor repair. "If I work on that stuff after hours, it lets people keep up with their work during the business day," Everett says.

His work area is located in a large, modern office building and is well lit and generally quite clean considering the sometimes messy work he does. "A lot of these machines need to be oiled and lubed a lot, so my clothes can get pretty greasy. That's why I like our maintenance uniforms." Everett has to do some heavy lifting, but, since he works in a department with many technicians, he can always find someone to help him with the biggest loads.

> "If anyone in my family needs something fixed, they know to bring it to me."

His typical day starts off with finishing up the paperwork from the day before. Like all technicians, he is required to keep very detailed records of all adjustments and repairs done on each and every machine. "Sometimes writing the reports up can be a pain, but later if a machine breaks down completely, the reports really help us figure out what's wrong," Everett explains. The records also help maintenance supervisors and managers find breakdown trends in different name brands and thus help them decide which machines to buy and which to avoid in the future.

Next, Everett examines the requests for service made late the previous day, after he left for the evening, and early in the morning. "Believe it or not, you can be gone for what seems like a very short time and come back to a huge stack of service calls to make." Since Nationwide employs about eight thousand people in his building complex alone, Everett usually has a wide assortment of machines that need to be fixed.

He arranges the requests in order of priority and starts on his way. Some repairs, like those to electrical, telephone, or computer wiring as well as minor repairs to small desk machines, are done on the spot. More complicated problems that require more sophisticated diagnostic equipment and tools are dealt with in his shop.

Everett often has to explain to co-workers what the particular problem is, how it happened, and how to prevent it from happening in the future. For more complex problems that he might not be able to handle on his own, he might consult with fellow technicians or a manufacturer's customer service representative in order to determine the exact cause of the problem and the repair action that should be taken to resolve it. These cases are especially true of the newer, more advanced computer systems.

Occasionally, Everett might attend a manufacturer's workshop on a new system or machine. These workshops or training seminars can take place on site, at a corporate training center just outside of town, or at a location chosen by the manufacturer. Everett likes to participate because they offer him a jump start on getting acquainted with new equipment as well as an opportunity to keep up with the latest technology.

## have I got what it takes to be a computer and office machine technician?

Computer and office machine technicians should have a solid grasp of the basics of mechanics, electronics, photography, and computers. This foundational knowledge is important for several reasons. First of all, it allows the technician to handle simple repairs easily. Second, it provides the background necessary for further training on more complicated machinery. A technician cannot learn the intricacies of a mail sorter, for example, if he or she does not already understand basic mechanics.

Third, such knowledge adds to an individual's willingness to learn about new equipment and allows the individual to be flexible about the kinds of machines he or she can work on. In fact, one of the hardest (and most enjoyable) parts of his job, Everett says, is the sheer variety of his duties. "One day I'm working on a calculator and the next, it's a big printing press."

Technicians are often those who, like Everett, have long been interested in figuring out how things work and who have the manual dexterity to tinker around with gadgets, tools, and household machines. Coomputers and office machines often need repairs in spots that are hard to get to, where the parts and tools used are small, even tiny.

To be a successful computer and office machine technician, you should

- Have superior manual dexterity
- Be able to follow complex written instructions and diagrams
- Like to learn about new machines and technology
- Work well alone, without direct supervision
- Be able to communicate difficult technical ideas effectively to peope with different levels of experience

While manual dexterity is a must, it is useless without aptitude in following technical diagrams and explanations. "We have shelves and shelves of equipment manuals," Everett says. And as anyone who has ever installed a household appliance or assembled a piece of furniture by following a technical drawing knows, such a task can require a great deal of patience as well.

As is the case in most jobs, technicians should have solid communication skills. They should be able to explain technical problems to a wide variety of people, from people who have no understanding of how something works, to people who actually design certain machines. It can be a delicate issue and actually more important than one might think; companies who specialize in service lose customers if technicians make them feel stupid or do not take the time to carefully explain what is going on.

## how do I become a computer and office machine technician?

After finishing his associate's degree in electronics at Devry Institute of Technology, Everett learned about the job opening at Nationwide through the school's placement office. He was originally hired to work on calculators and adding machines as well as to help out in the audio visual department. "Since I started," he explains, "I've been learning about new machines every week."

### education

#### High School

A high school diploma is the minimum requirement for getting a job as a computer and office machine technician. Traditional high school courses like mathematics, physics, and other laboratory-based sciences can provide a strong foundation for understanding basic mechanical and electronic principles. English and speech classes can help boost both written and verbal communication skills.

More specialized courses offered at the high school level, such as electronics, electricity, automotive/engine repair, or computer applications, are a very good source of practice in manual aptitude. Any other courses focusing on the use of flow charts and schematic reading are also beneficial. In addition, any experience with audiovisual equipment is a plus. Opportunities for such experience might be found in school theatrical productions or any other kind of multimedia presentation.

### Postsecondary Training

In smaller repair shops that do basic work on familiar machines, special training may not be required, given that the applicant has a good general knowledge of mechanics and electronics. For any more specialized positions, one or two years of courses in mechanics and electronics from a community college or vocational or technical school is recommended. Everett doesn't remember his classes at Devry being too difficult, but says those courses with which he was the least familiar required more studying time.

For those individuals interested in specializing in computer and peripheral repair, courses designed around computer technology, like microelectronics and computer design, should be selected. Some technical schools already offer specialized degrees in computer technology. It is important to note that computer repair positions usually require the completion of at least an associate's degree, whereas office machine repair positions might only require some formal education in electronics. The additional education required of computer technicians includes courses in elementary computer programming and the physics of heat and light.

Even with a degree in hand, new employees will receive a heavy dose of on-the-job training. From the employer's point of view, the degree proves that the new employee has the ability to do the work, but he or she still needs to be trained specifically on the machines most used by the company. Training programs vary greatly in duration and intensity depending on the employer and the nature of the position. Training courses may be self-study or held in organized classrooms. Generally, they will include some degree of hands-on instruction.

Keeping up with the technological times is extremely important for technicians. In fact, Everett advises anyone thinking about this career to be prepared for the stress involved in working with such rapidly changing technology. Technicians are expected to participate in seminars and workshops offered at regular intervals by the employer, a machine manufacturer, or an outside service company. Everett feels that, out of all of formal education, the workshops sponsored directly by machine manufacturers have been the most helpful to him in his job.

Another way technicians keep up is by reading detailed brochures and service manuals on new equipment. They also read a variety of magazines and newsletters on electronics and mechanical devices.

**fyi** There are currently about 141,000 people employed as computer and office machine technicians in the United States.

## certification or licensing

Certification as a computer and office machine technician is not mandatory to enter the field, but the International Society of Certified Electronics Technicians and the Electronics Technicians Association both offer a voluntary certification program. Individuals can take exams in order to be certified in fields such a s computer, industrial, and commercial equipment, and audio and radar system repair. Those technicians with less than four years' experience can take an associate's test, while those with more can become Certified Electronics Technicians (*see* "Sources of Additional Information").

Other voluntary certification exists for individuals who are specialized computer technicians. The Institute for Certification of Computer Professionals (ICCP) offers certification to computer professionals under the titles of Certified Computing Professional (CCP) and Associate Computer Professional (ACP) (*see* "Sources of Additional Information").

## scholarships and grants

There are no scholarships or grants offered uniquely to individuals wishing to pursue education for a career in computer and office machine repair. However, vocational and technical schools have financial aid offices that disburse, among other types of aid, corporate scholarships to qualified students. Interested individuals should contact the financial aid office of their prospective schools to find out about this type of opportunity. In addition, employers often have tuition reimbursement programs if educational courses benefit office operations. Check with the personnel or benefit office of current or prospective employers.

Professional organizations often have information regarding scholarship programs; *see* "Sources of Additional Information."

## internships and volunteerships

"Experience," says Everett, "is one of the most important qualifications for a job like mine." As is true for someone trying to break into any field, initial entry may be difficult. If you are still in high school, you may wish to get a summer or weekend job in a local repair shop. This experience will give you a feel for the variety of activities that go on there. Technical schools often have internship programs for students, offering off-quarter employment in various companies. Many of these internships turn into full-time jobs after graduation.

One way you can practice your skills is to offer repair services around the neighborhood, repairing VCRs, TVs, tape recorders, etc. In this way, you can begin to develop expertise at just those skills that will later make you an excellent computer and office machine technician.

## who will hire me?

The placement office of Devry helped Everett find his first job after graduation, and most vocational and technical schools manage similar offices. Placement offices work very closely with local companies that regularly recruit at those schools and so are generally successful in placing graduates in a job. To get an idea of how successful a school's graduates are at getting jobs, ask an admissions counselor for the school's placement statistics.

Approximately three of every five computer and office machine technicians work for wholesalers of computers and/or office machines and for independent repair shops. This includes service divisions of large computer companies that offer maintenance to the client with the purchase of a computer system; it also includes service companies that sell maintenance contracts to new equipment buyers or leasers.

Others, like Everett, work for large companies that have enough equipment to justify employing a full-time maintenance and repair staff. Corporations that typically have in-house repair departments are insurance companies, financial institutions (particularly those involved in any aspect of credit card processing), and banks. Large factories, whose production depends on large machinery, also have in-house technicians. Still other technicians work for retail companies. Different branches of the military, as well as federal, state, and local governments, provide opportunities for employment as well.

Most computer technicians work for computer manufacturers, often located in geographical clusters like California's Silicon Valley, or the "Valley up North" (that is, Washington state, where the headquarters of Microsoft, Inc. is found). Some computer repair specialists work for maintenance firms, and others work for corporations whose complex network of computers requires a full-time computer technician.

Office machine technician positions may be found everywhere, but are concentrated in cities or suburban office complexes. Individuals may look in the classified-ads of major metropolitan newspapers and related magazines for job opportunities. In addition, many professional organizations publish newsletters and magazines that include job lists (*see* "Sources of Additional Information").

## where can I go from here?

Everett has no plans to leave Nationwide. "They have great benefits, including a complete retirement package, plus I love my work and the people I work with." As long as he keeps working at his current performance level and stays active in training workshops, Everett will probably be a candidate when the position of *maintenance supervisor* opens up. "I would like that because I'd have more responsibilities and get to work on more complicated equipment."

After a period as maintenance supervisor, he might later become eligible for a managerial position. "It's a long way off, but I'm not sure I would want it anyway. Managers don't get to get their hands dirty too much; I would miss working on the machines." Managers get paid better, of course, so he's not ruling out the possibility just yet.

Most computer and office machine technicians start working on relatively simple machines and gradually become familiar with more and more complex machines as their experience increases. When they are trained in most of the equipment, they may be promoted to maintenance supervisor. Then, with demonstrated leadership and business skills, technicians may be promoted to managerial positions. Promotion to management often requires further education in business or engineering, and the job often entails more "office" work than repair work.

Another advancement route a technician can take is to become a sales representative for the company whose product he or she has had the most experience with. Technicians develop expert hands-on knowledge of particular

### Advancement Possibilities

**Maintenance supervisors** assign repairers to handle specific jobs when repair requests come into the office. They are in charge of a group of employees, fulfilling such administrative duties as verifying that repair records are kept in good order by each repairer and keeping track of hours spent on different machines. They help other repairers troubleshoot problems and might also have expertise in working on more technologically complicated and expensive machines.

**Maintenance managers** perform many of the same duties as supervisors but are usually in charge of a greater number of employees. They might also be responsible for deciding when a machine should be replaced, which new machines should be purchased, and how the office should be laid out in order to ensure ease of repair. In addition, they perform normal managerial functions such as preparing departmental budgets, interviewing, and hiring.

**Computer and/or office machine sales representatives** may get their start in computer and office machine repair because such experience provides them with a very solid understanding of particular machines. Their knowledge, as well as experience in explaining technical problems in easy-to-understand terms, gives them a head start in the active "selling" of a product that salespeople must develop.

machines and are thus often in a better position than anyone to advise potential buyers about important purchasing decisions.

Some technicians might aspire to owning their own business, which is also an option for advancement. Entrepreneurship is always risky and the responsibilities are great, but so are the potential rewards. Unless they have already determined a market niche, technicians usually find it necessary to branch out when they open their own repair shop in order to service a wide range of computers and office machines.

## what are some related jobs?

Although he wants to stay where he is, Everett says he would be interested in working in the automotive industry if he had to change jobs. "They are putting more and more electronics in new cars these days, and it seems like it would be fun."

The U.S. Department of Labor classifies office machine technicians under the headings *Business and Commercial Machine Servicers* (DOT) and *Engineering, Craft Technology: Mechanical Work* (GOE). Also listed under these headings are people who repair air-conditioning and refrigeration equipment, aircraft engines, automobiles, diesel engines, farm equipment, industrial machinery, lawn and garden equipment, motorboat and motorcycle engines, photographic equipment, scales, and statistical machines.

People who are interested in computer and office machine repair might also be interested in careers as television and radio technicians, photographic equipment repair technicians, electronic home entertainment equipment technicians, telecommunications equipment technicians, and instrument technicians.

Individuals interested in a career in computer repair who want to continue their education might also consider other computer-related jobs such as computer operators, programmers, or systems analysts.

| Related Jobs |
| --- |
| Aircraft mechanics |
| Appliance and power tool technicians |
| Automobile, motorcycle, and bicycle mechanics |
| Biomedical equipment technicians |
| Communications equipment technicians |
| Electrical technicians |
| Electronics engineering technicians |
| Farm equipment mechanics |
| Industrial machinery mechanics |
| Photographic equipment technicians |
| Service sales representatives |
| Telephone technician s |
| Vending machine technicians |

## what are the salary ranges?

Computer and office machine technicians in entry-level jobs earn an average of between $20,000 and $24,000 a year. Technicians already specialized in computer repair may start at slightly higher salaries in some companies depending on education and any prior experience.

As technicians gain valuable years of experience, their average yearly salary rises to about $30,100. Technicians who become experts in complex machines or computers may make even more. The highest-paid technicians earn as much as $50,000 a year after many years of experience.

In general, computer technicians make the highest salaries in this field. The rule of thumb about salaries for technicians is this: the more complicated a machine, the higher the salary. Self-employed technicians of even simple office machines do, however, have a potential for excellent earnings. In addition, some technicians may work for a company that pays commission for sales of machines and supplies and thus may be able to supplement their income in that way.

## what is the job outlook?

According to the U.S. Department of Labor, there will be no increase in demand for office machine technicians through 2006, although demand for computer technicians will increase as much as 30 percent in the same time period. The key for individuals interested in this career: keep up with technology and be flexible.

There will be demand in large corporations with in-house repair departments for well-rounded technicians, those who can maintain and repair both computers and office machines. Computer companies and service contractors will look for people with strong computer backgrounds, whether in education or professional experience. Interested individuals should make a habit of keeping up-to-date with technological advances.

The computer industry has proven to be a very volatile market; when sales are up, so are the number of employees hired by computer companies. But when sales are down, layoffs are numerous. If technicians are planning specializations in computer repair, they should be qualified in several different kinds of equipment. For example, instead of becoming an expert in maintaining and repairing only laser printers, or a particular brand of laser printers, a technician might also want to learn about terminal screens, color photocopiers, CD-ROMs, and modems as well. This flexibility will give technicians the

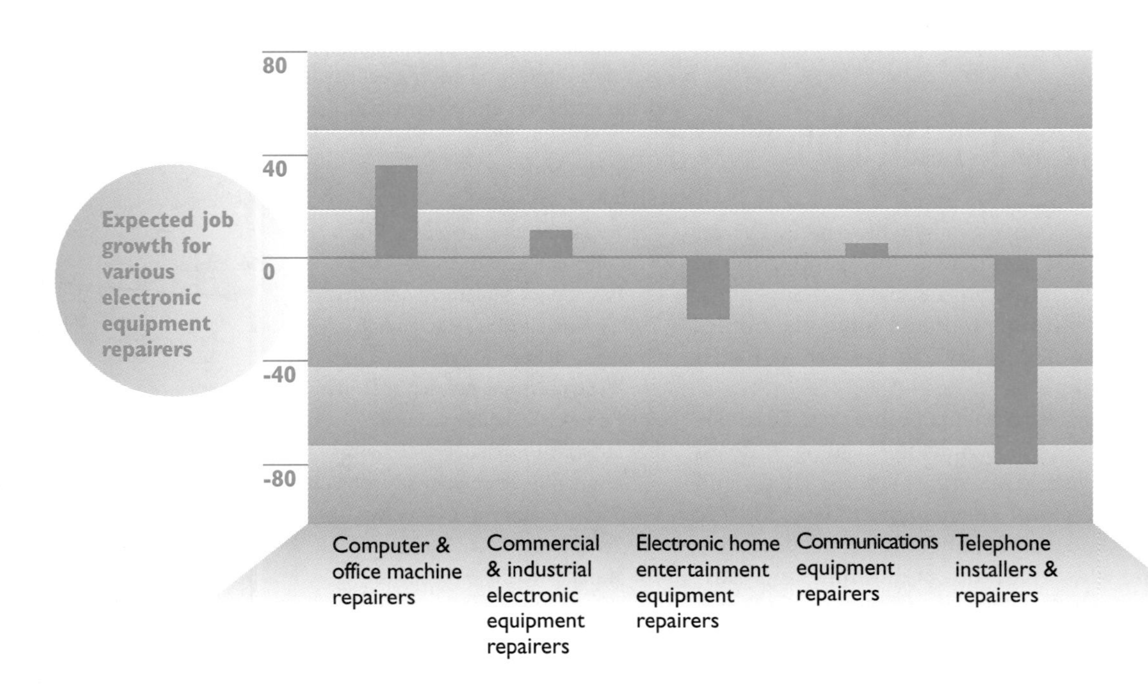

## in-depth

### Office Machines Over Time

The history of office machines can be traced as far back as ancient Babylonia and the invention of the abacus, a manual calculating device that is considered an ancestor of the computer.

Today's office machines stem from inventions of mechanical calculators in the seventeenth century, when French mathematician and philosopher Blaise Pascal developed the first such tool. Pascal's digital machine could perform addition and subtraction. American inventor William Burroughs developed the first truly practical adding machine in 1894.

The typewriter's history dates to the nineteenth century, although at this time many cumbersome typing machines were as big as pianos, and others resembled clocks. By the 1870s the Remington company was producing much more practical machines. Thomas Edison invented the first electrically operated typewriter in 1872, and by the 1930s, such machines were being used in offices.

The computer is the most recently engineered office machine. The first experimental versions of modern computers were built during the 1940s. Some early computers used as many as fifty thousand vacuum tubes, were very large, and required huge amounts of electricity.

Technical improvements made during the 1950s led to the first commercial computers. Their manufacturers leased them to users instead of selling them. By the late 1950s and early 1960s, the transistor was developed. It gradually took the place of vacuum tubes and helped make computers smaller, more reliable, and less expensive.

In the late 1960s the introduction of integrated circuitry made possible the development of minicomputers, which were smaller and cheaper but just as powerful as their predecessors.

The second phase of the computer revolution began in the early 1970s when the microprocessor became the heart of the modern computer. The discovery of how to store information on a tiny silicon chip (approximately .03 by .03 inches in size) rapidly caused new developments. The silicon chip opened up many new uses for computers. In addition, its development greatly reduced the computer's cost and increased the number of computer applications. This in turn increased the need for computer service technicians—a need that is still growing!

edge when jobs in computer companies are difficult to find.

Many opportunities in computer repair will continue to open up as long as computer sales remain steady or increase. Many computer companies offer service contracts with new purchases, and they need technicians to fulfill these contractual obligations.

In general, jobs in computer and office repair are relatively stable. When business is good, companies tend to buy newer machines, which need to be installed and properly maintained. However, this might also lower demand for technicians since new equipment is relatively free of problems. When business is down, businesses refrain from new purchases, relying more on repair of old equipment. No matter what, computers and office machines must be kept in working order; as long as there are computers and office machines there will be a need for the technicians who repair them.

## how do I learn more?

### sources of additional information

Following are organizations that provide information on computer and office machine technician careers, accredited schools, and employers.

**Electronic Industries Association**
2500 Wilson Boulevard
Arlington, VA 22201
703-907-7500

**Electronics Technicians Association International**
602 North Jackson Street
Greencastle, IN 46135
317-653-8262

**IEEE Computer Society**
1730 Massachusetts Avenue, NW
Washington, DC 20036-1992
202-371-0101

**International Society of Certified Electronics Technicians**
2708 West Berry
Fort Worth, TX 76109
817-921-9101

# bibliography

Following is a sampling of materials relating to the professional concerns and development of computer and office machine technicians.

## Books

Kanter, Elliott S. *Opportunities in Computer Maintenance Careers.* Lincolnwood, IL: VGM Career Horizons, 1995.

## Periodicals

*IEEE Micro.* Monthly. Concentrates on the most recent developments in computer technology. IEEE Computer Society, 10662 Los Vaqueros Circle, Box 3014, Los Alamitos, CA 90720-1314, 714-821-8380.

*Computerworld.* Weekly. Provides news reports as well as detailed feature articles on various types of computers used int he corporate community. Computerworld, Inc., 551 Old Connecticut Path, Box 9171, Framingham, MA 01701-9171, 508-879-0700.

# computer network specialist

**Definition**

Computer network specialists administer computer networks so that they operate smoothly and consistently for high efficiency and productivity.

**Alternative job titles**

Computer network administrators
Computer network security specialists
Data communications analysts
Data recovery planners
Network programmers
System administrators

**Salary range:**

$37,000 to $59,250 to $81,500

**Educational requirements**

High school diploma

**Certification or licensing**

Recommended

**Outlook**

Faster than the average

GOE 11.10.05

DOT 033

O*NET 22127

## High School Subjects

Business
Computer science
Mathematics

## Personal Interests

Computers
Figuring out how things work
Fixing things

the passwords pretty soon, instead of waiting for the scheduled date. Several people have been having problems lately," Randy Willy's supervisor says as he passes Randy's desk.

"I'll get right on it, Sam," Randy says, wondering where he will find the time for that, too. It's only 11:30 AM and Randy has already fielded seven phone calls from people in the office with computer problems. It's his job to see to it that each and every one of their problems is solved quickly. Without use of their computers, his colleagues can't work. If they don't work, his company loses money, plain and simple.

Today, in addition to troubleshooting their questions, he has fifteen new user accounts to open and a meeting at 1:30 PM to explain to the systems people what happened to the network when a power outage cut communications yesterday afternoon. On top of it all, his department is in the middle of a major recabling project and he is supposed to be gathering all the required information so that it happens without a hitch.

"Looks like I'll be here late tonight," he thinks as he goes back to work. Randy doesn't mind much; he is used to working overtime when big projects come along.

## what does a computer network specialist do?

Businesses and organizations choose to network, or connect, their computers for many different reasons. One important reason is so that numerous computer users can simultaneously access the same hardware, software, and other computer equipment like printers, modems, and fax machines. By networking such equipment, the business avoids having to purchase individual products for each user. For example, if a small business has four different computer terminals, all terminals might be able to share a single printer instead of having four of them. In addition, if the company maintains a database that is accessed by multiple users, it has to set up the system in such a way as to make the database, with its constant changes, available to everyone.

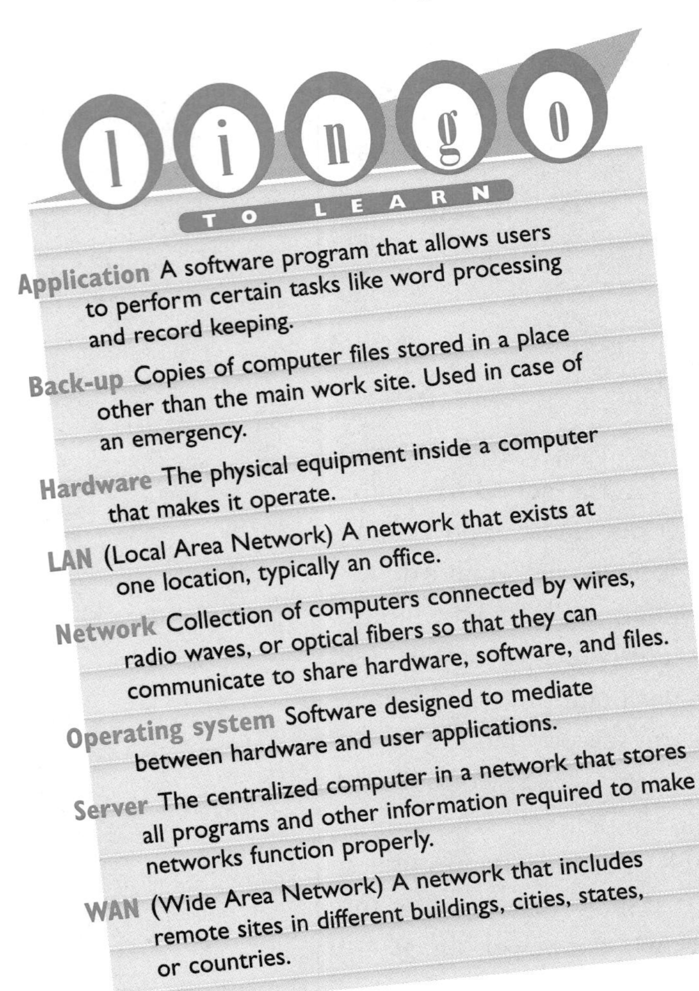
lingo TO LEARN

**Application** A software program that allows users to perform certain tasks like word processing and record keeping.

**Back-up** Copies of computer files stored in a place other than the main work site. Used in case of an emergency.

**Hardware** The physical equipment inside a computer that makes it operate.

**LAN** (Local Area Network) A network that exists at one location, typically an office.

**Network** Collection of computers connected by wires, radio waves, or optical fibers so that they can communicate to share hardware, software, and files.

**Operating system** Software designed to mediate between hardware and user applications.

**Server** The centralized computer in a network that stores all programs and other information required to make networks function properly.

**WAN** (Wide Area Network) A network that includes remote sites in different buildings, cities, states, or countries.

These networks can range in size from two computer terminals to hundreds and can operate with one or more of several different network servers or none at all. Each network is different and tailored to the needs of the business or large corporate department. And each network is invariably messy—problems come up often, employees have difficulties learning the system, passwords and file names are regularly changed, software is updated, back-up files must be made, communications lines are broken and must be reestablished—the list can go on and on. *Computer network specialists* must know the network well enough to be able to handle all of these situations.

The specific job duties of a computer network specialist vary greatly depending on the size and structure of the company or department and on the kinds of network systems used. For example, in larger companies the positions may be extremely well defined; they may employ one person for computer securities, one for network administration, one for data recovery, etc. In smaller companies or those departments in which networks are just being introduced, however, employers may hire one or two people to do everything. Keep this in mind as you read the following descriptions of various kinds of computer network specialists.

*Computer network administrators* are responsible for adding and deleting files to the network server. The server is a centralized computer that stores, among other things, software applications used on a daily or regular basis by the network users. The files updated might include those for database, electronic mail, or word processing applications. Network administrators also handle printing jobs. They must tell the server where the printer is located and establish a printing queue, or line, designating which print jobs have priority. They might also tell the server to hold or change certain files once they reach the printing queue.

Network administrators might also be responsible for setting up user access. Since highly confidential and personal information can be stored on the network server, employees generally have access to only certain applications and files; the network administrator assigns each user or group of users access to the appropriate files and often makes up passwords to be used by each employee. The passwords protect the system from both internal and external computer spying.

*Network security specialists* often work in companies that make extensive use of computer networks. They are responsible for regulating user access to various computer files, monitoring file use, and updating security files. If an employee forgets his or her password, the network security specialist changes it to grant access. Network security specialists ensure that security measures are running accurately by tracking nonauthorized entries, reporting unauthorized users to management, and modifying files to correct the errors. They maintain employee information in the security files; they add new employees to the list, delete former employees, and make official name changes.

*Data recovery operators* are responsible for developing, implementing, and testing off-site systems that will continue to work in case of emergencies at the main office, such as power outages, fires, or floods. This task is very important because many businesses cannot operate at all without their computer networks; some businesses, like insurance companies and banks, depend even more on their computer systems during emergencies. Data recovery operators must determine which hardware and software should be stored at the emergency site, which applications receive operating priority in case of an emergency, where to locate the emergency site, and which employees would be needed to run emergency operations. After testing the system they develop, they report the results to supervisors and management. Like network security specialists, they work primarily in larger companies.

*Network control operators* are in charge of all network communications, which usually operate over telephone lines or fiber optic cables. If a user cannot get the computer to send his or her file to the appropriate place, he or she calls the network control operator for help. The control operator then explains to the employee the step-by-step procedure he or she should have followed to send the file. If the error is not an employee mistake, the network control operator attempts to diagnose the problem by making sure information stored in computer files is accurate, verifying that modems are functioning properly, or running noise tests on the phone lines with special equipment. Sometimes, the control operator must ask for outside help from the company who sold or manufactured the malfunctioning system.

Network control operators also keep records of the daily number of communications transactions made, the number and type of problems arising, and the actions taken to solve them. They might also train staff to use the communications systems efficiently and help coordinate or install communication lines.

> When I develop solutions, I get to use my technical expertise in a creative way.

## what is it like to be a computer network specialist?

An employee of Nationwide Insurance in Columbus, Ohio, Randy Willy has always been into computers. He continues to be excited by what people and businesses can accomplish because of advances in computer technology and is very satisfied by his job. "I like explaining how the system works to my colleagues who are scared of computers at first. In general, they catch on quick and that helps make our department a lot more efficient."

Like most office employees, Randy usually works a five-day, forty-hour week in comfortable office surroundings. When major installation or other projects are undertaken by his department, however, he has to put in a lot of overtime, receiving either overtime wages or extra vacation days. During the last major project, in fact, Randy was teased by his co-workers for being at work too much. "They were convinced I had no life outside the office," he chuckles.

Working for a big corporation like Nationwide, Randy receives a complete benefits package and paid vacation. In addition, any courses or certification programs he completes that are linked to his job responsibilities are paid for by the company. "Tuition reimbursement is a great benefit, especially since computer classes are usually pretty expensive."

Randy likes to arrive at the office a little early to have some time to himself. During this time, he might work on his more routine tasks like changing security passwords (which he does every forty days), checking into how to solve minor network problems, or backing-up the whole system. He might also devote some time to the bigger projects, like converting an old system to a newer one, recabling the entire network with fiber optics, or catching up on changes in the company-wide system that affect how his network should operate.

When his co-workers encounter problems, he has to be ready to drop everything to help work them out. "It's my job to keep things running smoothly," he explains. "It doesn't matter what else I'm doing or how important it is." Sometimes, he feels like he's being pulled in all directions at once. "It takes a lot of concentration to be able to go from one situation to another and back again. I've always got a lot of things going on at once. It's already pretty stressful, but if I were any less organized, I'd be crazy by now."

Randy is responsible for setting up new workstations, or computer terminals, which involves a lot of detail work. He programs the network server to recognize new terminals and to route programs and information to and from them by assigning them various addresses. The process can be very complicated because an address recognized by his local server or particular software program might not be recognized by a wider server used by a bigger part of the company. In this case, he must assign addresses that each system can read correctly and keep a careful record of all of them. One of the bigger projects he is working on is researching new ways to streamline addressing, particularly since the users of his network must have access to many other company networks.

Large companies often have their own electrical generators because uninterrupted electrical service is extremely important to computer systems. If there is a power surge or outage, even for fractions of a second, it can mean the loss of millions of pieces of information. But even these systems fail sometimes and when Nationwide's power went down, Randy had his hands full. The bridge that linked his network to the company's other systems was cut, and he had to help the systems people reestablish the connection. Then, he had to figure out exactly what was lost and what needed to be restored. Most importantly, he had to calm his co-workers' frustration and resolve their confusion by explaining to them step-by-step what they had to do. His patience and communication skills proved invaluable.

Randy spends a lot of time working on his computer and is thus susceptible to eye strain and carpal tunnel syndrome, a condition resulting from repetitive motions that can make it painful to type on a computer keyboard. To prevent long-term eye damage, Randy has made a habit of briefly looking away from the screen at regular intervals and taking short breaks from it altogether every few hours. "You get used to it, but a lot of people's vision does deteriorate. I've been lucky enough not to have problems with my wrists, though some of my co-workers do."

**To be a successful computer network specialist, you should**

- Be organized and possess logic skills
- Understand basic computer principles and be willing to learn more complex skills as technology advances
- Enjoy challenges and solving problems
- Be able to patiently communicate complicated material in simple terms
- Be able to work well under pressure

## have I got what it takes to be a computer network specialist?

Computer network specialists work with a lot of detail; every type of computer, every software application, every network system has a particular set of codes that must be used to get things done. Since users contact them when they have problems, network specialists must have a running knowledge of all the codes, plus an understanding of how they all link together. This also requires strong logical thinking skills. "It's not easy to keep all that information in your head. Being able to pull back from it and visualize the bigger picture can be even more complicated," says Randy. To succeed as a network specialist, you must be very good at organizing and analyzing large amounts of detail at both the micro and macro levels. Often, network technicians are called on to solve multiple problems at once. "You've got to be a problem solver who loves challenges, because there's never a dull moment around here."

Working as a network specialist also involves a certain amount of pressure. "When your system's down and people can't work and you can't figure out what the problem is, you're stressed," explains Randy. Performing well under pressure involves the ability to think things through thoroughly without panicking. "You have to be able to concentrate and not lose your cool."

Effective communication skills are also important for network specialists, who must explain complex information to both technical experts and people who have little understanding of computers at all. "You've got to talk to them at their level and above all, be patient." Since the computer industry has generated a lot of jargon, it can be difficult to simplify explanations. "You get so far in it you just assume everyone knows what LAN, WAN, and AS/400s are."

Computer network specialists usually start with a basic understanding of broad computer principles, although there is always room for improvement and training. In fact, one of the things that worries Randy the most about his job is not being able to keep up with all the new developments in technology. "Continuing education is very important; there's always an interesting class or workshop to take, always a stack of magazines to read. Sometimes I think I could study twenty-four hours a day and still not know everything." And that's when network specialists have to remember that it is impossible to know everything. You must be able to realize your limitations without being discouraged from the whole process. "You just learn as much as you can without making yourself crazy," says Randy.

Given the problem-solving nature of his work, Randy finds it difficult to leave his job at the office. "I'll be out to dinner, thinking about ways to improve the system, and I'll think, wait a minute! I'm not supposed to be working now!" But despite the stress, Randy finds his job extremely rewarding. "When I develop solutions, I get to use my technical expertise in a creative way." And there is nothing more thrilling for him than to see one of his projects up and running satisfactorily.

## how do I become a computer network specialist?

When the manager of corporate printing at Nationwide Insurance decided his department needed a network specialist, he asked Randy to train as one, knowing Randy's interest in computers and experience in form design and analysis. Randy was enthusiastic and first completed Novell's *Computer Network Administrator (CNA)* certification. He is now working on becoming a certified *Computer Network Engineer (CNE)*. Eighty percent of Nationwide runs on Novell systems, yet there are currently no CNEs employed. "I'm making a great career move," Randy explains. "I'll be highly sought after when I'm finished." He took computer programming courses in high school which helped him develop logical thinking skills, but Randy also has a very creative side and uses it a lot in problem solving.

### education

#### High School

A high school diploma is mandatory for individuals getting started in the computer field. Formerly there were advancement possibilities for long-term employees without diplomas, but competition is so intense in the computer industry today that a diploma is considered the bare minimum requirement for new hires.

High school courses in computer science provide a solid understanding of basic computing principles. Math courses such as algebra and geometry help develop logical and analytical thinking skills, while those in psychology and English improve verbal and written communication skills. Any business courses will increase knowledge about the ways in which management decisions are made, often important for network administration and engineering choices. Joining computer clubs and playing on your home computer are important, too. The more experience you have, the better off you'll be. Summer jobs at local businesses that have computers can provide opportunities to learn about various network systems and issues.

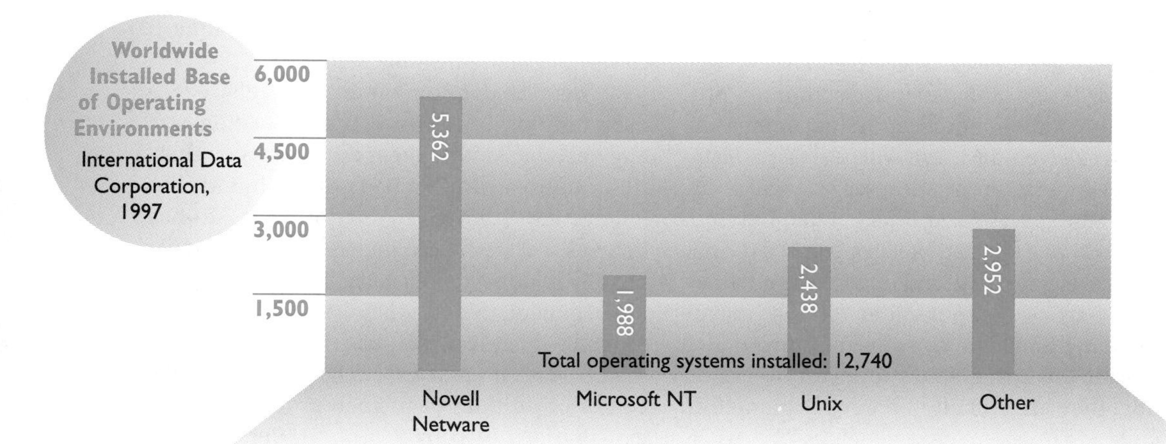

## in-depth

So, are you feeling confused about all this network stuff? Can't tell your WAN from your LAN or your intranet from your extranet? Join the club! With the high-tech industry changing by the minute, it's hard to keep up on the latest developments. Hey, it's hard enough just to figure out a basic word processing program, much less the differences between Microsoft NT and Novell Netware. Let's try to clear up some of that fog so you can have a rudimentary grasp of some of the common terms.

**LANs** are local area networks and usually refer to networks that reside in one building. LANs usually connect workstations and personal computers so they can communicate and share peripherals, such as printers. LANs can transmit data much faster than data can be sent over a telephone line, but they are limited in distance and in the number of computers they can support.

LANs that are connected to one another via telephone lines or radio waves are called WANs, or wide area networks. The Internet is an example of a WAN, so you can imagine how wide-ranging the capabilities of WANs are.

**Intranet** refers to a company's internal web server (and you thought it was just "Internet" spelled incorrectly). Intranets are accessible only by authorized users within an organization or corporation and are used to share information, such as with a network. Intranets work with private network systems and are growing in popularity because they are economical to build and maintain. Think of an intranet as a company's own little Internet.

Extranets are intranets that allow some accessibility to authorized outsiders. They come in handy when, say, a business wishes to share information or provide access to outside clients or business partners. The outsiders must have valid usernames and passwords to access the extranet. Extranets, like intranets, are enjoying popularity, and many customer service-based businesses are taking advantage of them. For example, Federal Express Corporation allows clients to access its package tracking system through an extranet.

Firewalls protect private networks from unauthorized usage. They prevent hackers and other Internet users from breaking into intranets by evaluating messages and rejecting those that do not meet security criteria. Firewalls use a variety of techniques to filter out unwanted messages. Some employ security measures when an outside connection is made, while others check all incoming and outgoing messages.

### Postsecondary Training

Although there is a way to advance in the computer industry without a college degree, an individual wishing to break into the field must give careful consideration to long-term career goals. More complex jobs like computer design, systems design and analysis, computer programming, and computer management are usually reserved for those with at least a bachelor's degree in some computer or engineering major. More specialized research and development positions may require an advanced degree.

College graduates tend to do the more theoretical part of computer networking, maybe developing hardware or software from scratch or deciding which systems to invest in. Importantly, they are usually the ones promoted to middle and upper management jobs.

Randy started working at his company right out of high school. Since they offered to pay for his Novell certification (which can run several thousand dollars for the training materials and testing), he is not interested in pursuing a college degree for the time being.

Several major companies offer postsecondary training in network administration and engineering. Among them are Microsoft, Unix, IBM, and Novell. Certified administrators, like Randy, are trained to perform day-to-day network tasks, while engineers are trained more broadly to do system installation, configuration, and maintenance.

Individuals can complete the required training courses either at a local licensed education center or on their own. Each provide detailed explanations of course content, personalized feedback on homework assignments, and some hands-on experience. "The program is pretty rigorous. They throw so much information at you at once that even if you take the classes, you still end up having to do a lot of studying at home." The assignments Randy found particularly helpful were ones that required him to create flow charts, set up new users on the classroom system, and get a good grasp of computer vocabulary. "The program makes you buckle down and study," he says, "and that's good because there's no other way to learn it all."

## certification or licensing

There are a number of certification programs, and Microsoft and Novell are the most popular. If an individual is already employed, he or she should probably become certified by a program designed to complement the system his or her company is already using, unless the goal is to branch off into a different career field. Novell's CNA and CNE and Microsoft's MSLE certifications are granted when an individual successfully completes a series of tests which are administered separately from the training classes.

Randy, who completed the Novell CNA certification process, says, "Study hard!" The tests can be difficult because they are timed, and if the time lapses, the test taker automatically fails. "Novell takes their certification very seriously. If you're a certified Novell professional, employers know you know your stuff." The same goes for Microsoft's Windows NT certification.

While certification may provide you with a competitive advantage over fellow job seekers, not all companies require certification. This is due in large part to the rapidly changing world of networking. You may receive certification one day and discover a new version of the operating system is being released the next. In addition, more and more systems will combine features from different companies and sources, so being strictly, say, a Microsoft NT specialist won't get you very far.

## scholarships and grants

For those choosing to pursue commercial certification in networking, check out tuition reimbursement plans offered by many large employers. Eligibility for these plans varies with the companies and positions held. In small- and medium-sized companies, employers might be persuaded to pay tuition costs if certification will have a direct impact on the company's operations.

Individuals pursuing a formal degree should contact their school's financial aid office for scholarship, grant, and student loan opportunities. In addition, many commercial businesses and professional organizations have fellowship programs for students with particular interests.

## internships and volunteerships

Since many people complete commercial certifications while working full time, there are not many internship opportunities available. Those people who would like to put their knowledge to work on a volunteer basis should contact local or national charities of interest. Most charity offices run on computers and such valuable help is often very sought after and appreciated.

College students should contact their college placement office for help identifying and applying to internship programs. Many large companies offer internships or may be persuaded to take you on—you just need to convince them of your value.

## who will hire me?

Computer network specialists work in companies that rely on computer networks to do business. As insurance companies, banks, and other financial institutions automate more and more of their services, they will count more heavily on computer networking, creating many job opportunities for individuals trained in these areas. Randy was offered his current position as network administrator in an insurance company because his experience as a forms analyst gave him a lot of background knowledge about how systems work.

Since most companies are moving toward networking, positions are multiplying, and demand for network specialists is currently very high. Some companies might decide to promote and train specialists from within, so employees wishing to work in networking should keep their eyes open on in-house job-opening lists. At this point, individuals already certified are likely to be hired rather quickly, especially if they have experience and knowledge of multiple systems.

Federal and state governments are also a good place to look for jobs in computer networking. Since many governmental offices have to manage huge amounts of information on many different networks, their overall need for network specialists is high. In addition, these positions are often on the top of the list to be filled even during hiring slowdowns or freezes.

There are several professional newsletters and magazines offering job opening lists in the area of computer networking, most of which can be found on the Internet, another indispensable job search tool. Job openings are also listed in all major newspapers' classified sections and in the back of computer trade magazines like *PC Computing, PC World,* and *MacWorld* (*see* "Bibliography").

In addition, many professional associations, including the Association for Computing Machinery, the Network Professional Association, and the Association of Information Technology Professionals provide job lists.

As with many areas of business, network specialists might find out about good job openings from their colleagues in the industry. As individuals work their way through education requirements for network certification, they might ask instructors and other education center personnel about where to apply.

College students should work through their college placement offices, since they have established recruiting programs with many employers. Other than companies specializing in the service industries like insurance companies and banks, graduates might look at Hewlett-Packard Co., Microsoft Corporation, and Novell Inc., since they are remaining strong despite recent computer market fluctuations. Competition for jobs here is very stiff, however.

## where can I go from here?

"I really want to get into multimedia," says Randy. "I'm excited by the possibility of really exploiting my creativity while putting my technical expertise to use." Working in multimedia is a long-term goal for Randy. In the meantime, he is going to concentrate on his CNE and on learning as much about computers as he can.

Advancement for computer network specialists varies greatly, depending on education, experience, and personal interest. If an individual begins as a network administrator, he or she might next work toward becoming a network engineer, like Randy. Then, he or she might progress to the level of enterprise network engineer, a position which includes responsibility of all computer systems within a single, medium- or large-sized company.

But a network specialist might wish to become more specialized and train as a network security specialist, data recovery planner, or data communications operator. If a network specialist wants to get into computer programming or systems analysis and design, he or she will probably need to pursue a college degree in order to get the theoretical knowledge necessary.

Another possible career path to take is information systems management. Larger companies need someone in charge of all their computer systems and services. Specialists who have strong, effective communication skills, the ability to motivate and organize team projects, as well as technical expertise of the computer systems involved in their work make exceptional candidates for such management positions. Network managers work closely with network specialists and systems analysts and are often responsible for selecting which computer equipment to buy.

### Advancement Possibilities

**Computer network engineers** set up computer networks, often from scratch. They interview employers and employees to determine their needs, analyze and prioritize those needs, select appropriate hardware and software, make any necessary changes, supervise installation and initial operations of the system, and sometimes provide training to network users.

**Enterprise computer network engineers** perform many of the same duties listed above but at a higher level. For example, they may work on company-wide networks or systems instead of networks serving only one department or several departments in a large company.

**Multimedia specialists** design ways to make graphics, audio, and video work well together, producing multimedia games, presentations, and other types of programs.

## what are some related jobs?

The U.S. Department of Labor classifies computer network specialists under the headings *Occupations in Data Communications and Networks* and *Occupations in Computer Systems and Technical Support* (DOT) and *Clerical Machine Operation: Computer Operation* (GOE). Also listed under these headings are people who research, test, and recommend communication hardware and software, people who offer technical support to computer users, and people who supervise groups of computer professionals.

| Related Jobs |
|---|
| Computer network engineers |
| Computer programmers |
| Computer scientists |
| Software engineers |
| Computer security coordinators |
| Data communications analysts |
| Computer systems hardware analysts |
| Network control operators |
| Computer systems analysts |

## what are the salary ranges?

Salaries for computer network specialists vary significantly depending on education level, experience, and the scope of job duties. If the demand for good network specialists continues to rise faster than people can become adequately trained, salaries will increase as well. According to a 1998 salary survey by Robert Half International, Inc., a staffing agency specializing in information technology, finance, and accounting, network administrators earn in the range of $37,000 to $58,000. LAN/WAN specialists make anywhere from $47,000 to $64,500.

A similar survey by Dowden & Co., a compensation research firm, indicates that the Southwest is the lowest-paying region for network administrators, where the median income is $53,300. The Northeast, on the other hand, is the highest-paying area for network administrators, where the median income is $66,000 and the high more than $81,500.

## what is the job outlook?

Job opportunities for computer network specialists are expected to grow faster than average through 2005. There are several reasons for this. Many companies who used to rely on very large-sized computers (mainframes) are now finding it better to develop a series of networks made of smaller computers that can communicate with each other and achieve the same results. The companies are already beginning and will continue to search for well-qualified people to help administer and engineer networking projects.

In addition, many service industries that used to rely more heavily on paperwork for record keeping now prefer to automate and keep records on computer databases. Insurance companies, for example, are looking to eliminate all paper forms from the insurance process. Instead, they want to have on-line forms that can be filled out by the client on the computer. This would allow them to avoid delays and expenses caused by the post office and paper processing. Computerized form procedures would be handled through a network which, in turn, would need administrators and engineers to run it. The economic and productive advantages of networking currently make it such that companies will continue to invest in network development, even in times of economic difficulty. This means that computer network jobs should be relatively easy to find for the next several years.

The growing trend toward networking is occurring particularly in insurance companies, banks, and other financial institutions, although any business with more than one computer may be heading toward networking. As you begin to prepare for careers in computer networking, you should pay attention to the current economic climate to ascertain which industries may be more financially stable.

The federal government is also very involved in the process of setting up networks among different offices and departments. This, combined with the fact that the government is handling ever-growing amounts of information and pays competitively in mid-level jobs, makes it a prime target in the future job hunting process.

If you develop expertise in a specific area, like computer security, you can look to larger companies and computer companies for jobs corresponding better to your exact qualifications. It is important to remain well trained in several areas, however, because companies may choose to hire one or two people to do everything instead of numerous specialists when business is down.

Many computer companies are shifting to a service-based paradigm, offering comprehensive service to clients. This means that big computer companies will be hiring more and more technical support staff who can help clients install their networks, train client staff, and answer client questions. Computer network specialists might be well qualified for some of these positions.

Individuals with a college degree should consider software engineering and systems analysis, as these will be areas of big job growth through 2005. In addition, companies that develop emerging technologies such as multimedia, virtual reality, and pen-operating systems will also experience wide growth and will therefore hire more computer professionals in the near future.

# how do I learn more?

## sources of additional information

Following are organizations that provide information on computer network specialist careers, schools, and employers.

**American Society of Information Science**
8720 Georgia Avenue, Suite 501
Silver Spring, MD 20910-3602
301-495-0900
http://www.asis.org

**Association for Computing Machinery**
1515 Broadway
New York, NY 10036
212-869-7400
http://www.acm.org

**Association of Information Technology Professionals**
315 South Northwest Highway, Suite 200
Park Ridge, IL 60068
800-224-9371

**IEEE Communications Society**
305 East 47th Street
New York, NY 10017
212-705-7865
http://www.comsoc.org

**Network Professional Association**
401 North Michigan Avenue
Chicago, IL 60611
888-379-0910
http://www.npa.org

## bibliography

Following is a sampling of materials relating to the professional concerns and development of computer network technicians.

### Books

Peterson's Guides, *Peterson's Job Opportunities for Engineering, Science, and Computer Graduates.* Princeton, NJ: Peterson's Guides, Inc.

Krol. Ed. *The Whole Internet: User's Guide & Catalog.* This book covers the basics and gives new users a step-by-step manual for retrieving information. Cambridge, MA: O'Reilly & Associates, 1994.

### Periodicals

*Association for Computing Machinery Journal*. Quarterly. Features scholarly articles in the field of computer science. Association for Computing Machinery, 1515 Broadway, 17th Floor, New York, NY 10036-5701, 212-869-7440.

*Data Communications*. Eighteen times per year. Directed at professionals involved in computer network integration and implementation. McGraw-Hill, 1221 Avenue of the Americas, New York, NY 10020, 212-512-2000.

*IEEE Network*. Bimonthly. Focuses on computer networking systems.Institute of Electrical and Electronics Engineers, 345 East 47th Street, New York, NY 10017-2394, 732-981-0060.

*MacWorld*. Monthly. Trade magazine devoted to providing information on Macintosh computer systems. Macworld Communications, 501 2nd Street, Suite 500, San Francisco, CA 94107, 415-243-0505.

*PC Computing*. Monthly. Computer trade magazine. Ziff-Davis Publishing Company, 50 Beale Street, 14th Floor, San Francisco, CA 94105-1813, 415-578-7000.

*PC World*. Monthly. Trade magazine devoted to users of IBM and IBM-compatible personal computers. DJ Communications, Inc., One Exeter Plaza, Boston, MA 02116, 617-534-1200.

# computer programmer

**Definition**

Computer programmers write the instructions (also called programs or software) that tell computers what to do.

**Alternative job titles**

Software engineers

**Salary range**

$19,500 to $40,000 to $65,000

**Educational requirements**

High school diploma

**Certification or licensing**

Voluntary

**Outlook**

Faster than the average

GOE 11.01.01

DOT 030

O*NET 25105

## High School Subjects

Business
Computer science
Mathematics
Physics

## Personal Interests

Computers
Figuring out how things work

rolls, soars, and then dives, plummeting toward the earth in a tailspin before the pilot suddenly pulls up on the controls and the jet swiftly changes direction. And then the program crashes. Frustrated, the flight instructor shuts down her computer and calls the company who sold her the software for the computerized flight simulation program.

"Greg Gasson at Synergy Microsystems. What seems to be going wrong?" Greg wrote the flight simulation program. He listens carefully and jots down some notes on a scratch pad. "Okay, I'll pull the program up here on my computer and see if I can duplicate your problem. Then we should be able to tell where the glitch is and get the software working for you." Greg quickly types some information into his computer, and the monitor displays a flight simulation computer graphic with blinking numbers and gauges.

# what does a computer programmer do?

*Computer programmers* work in a variety of business, industrial, professional, and governmental settings to create the detailed instructions that tell a computer what to do. These instructions are called programs or software.

Because computers cannot think, computer programmers must know how to arrange instructions in an order and language that the computer can follow. The programmer's first step is to think about the task and how to instruct the computer to perform it. At this stage, programmers usually create a flow chart to illustrate each step the computer will have to follow to get the desired results. Programmers must then translate this flow chart into a coded language that computers can follow. There are many programming languages, including traditional languages such as COBOL, FORTRAN, and C, object-oriented languages such as Java, C++, and Visual Basic; fourth- and fifth-generation languages; graphic user interface (GUI); and more. Programmers input the coded steps of the language into the computer, thereby creating a program.

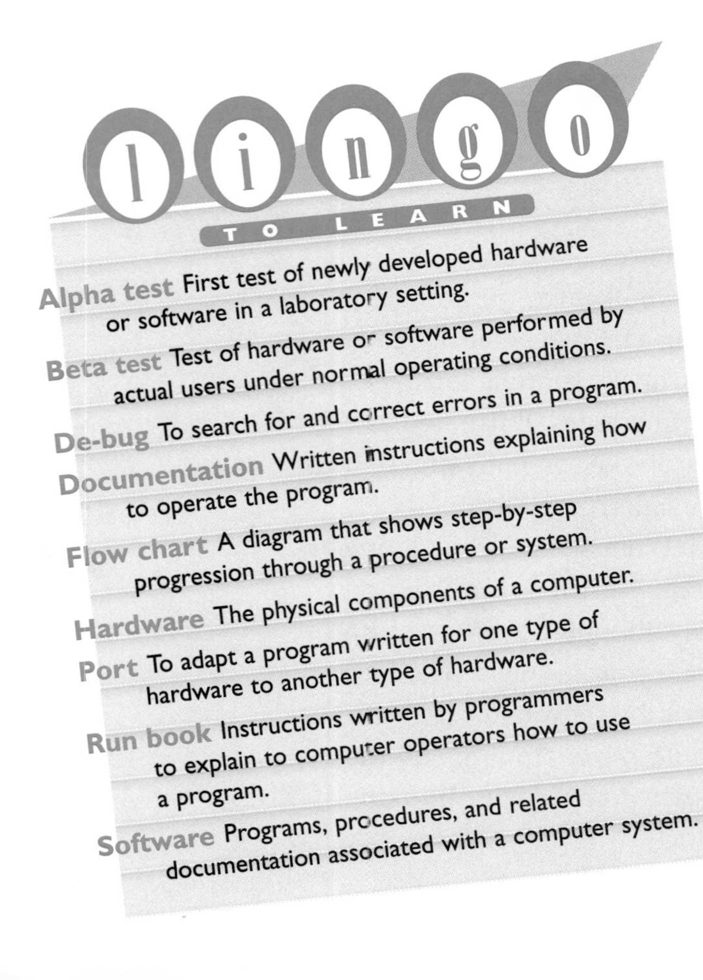


After the programmers have saved the program on disk, they must test it using sample data to see if it runs correctly. If the task is not being performed as intended, the programmers must examine the program and make changes in it until it provides the desired outcome. This process is called debugging.

Once a program is debugged and running as expected, the programmers may help write an instruction manual explaining how to operate the program. Because the end user may have limited technical knowledge, the programmer must be able to communicate in simple, clear language.

There are two main types of computer programmers: applications programmers and systems programmers. *Applications programmers* create or revise software for specific jobs or tasks, such as software to help a business office process payments or to help the Navy monitor the course of submarines. As new tasks arise at an organization, applications programmers are asked to develop software that will perform the desired job. Applications programmers often specialize in one of two fields: business and commercial applications programmer or engineering and scientific applications programmer. *Business and commercial applications programmers* write software to help a business run more smoothly. Often, their programs will involve accounting, billing, payroll, inventory, and database procedures. *Engineering and scientific programmers* create programs that are used for scientific or mathematical purposes. For example, their programs may instruct computers to analyze medical data or assist in air traffic control.

*Systems programmers* control and maintain the overall operation of the organization's computer system, including the central processing unit and peripheral equipment, such as printers, terminals, and disk drives. They instruct the computers on how to accept and store information and how to communicate with other equipment via cables, known as a network. Systems programmers often assist applications programmers with troubleshooting and problem solving because of their in-depth knowledge of the entire system.

Whether working in systems or applications, a computer programmer is the link between the computer needs of an organization and the capabilities of the machines.

## what is it like to be a computer programmer?

> You have to look at a whole project and then break it into components in order to be able to tackle it.

Most computer programmers work in relaxed environments with casual dress codes. While visions of barefoot, soda-guzzling programmers working all night aren't exactly off the mark, many work relatively stable hours (with shoes on). Greg Gasson has worked in the computer programming field for seven years. He has spent the last four years as an applications programmer at Synergy Microsystems in Tucson, Arizona, a company that specializes in developing flight simulation software. Most of Synergy's software is sold directly to the military or to companies that have contracts to sell to the U.S. Department of Defense. Because most of the work he does involves only himself and the computer, Synergy allows Greg to set his own work hours. He typically chooses to work from about 10:30 AM to 7:00 or 8:00 PM, but he adds, "If there's a big project going on, I'm liable to work sixty-hour weeks." Greg works Monday through Friday and rarely has to go to the office on weekends.

Greg usually begins his day by checking the E-mail messages on his computer and responding to them as necessary. "Sometimes customers need help with a product we've sold them and I'll have to solve the problem and sort of walk them through it. I get a couple calls like that a week." After Greg responds to his messages, he devotes the rest of his day to developing software. "As programmers, we know what projects are coming up and a lot of times we get to choose what we want to work on," says Greg. Other times, programmers are assigned software projects by a manager or by a *chief programmer* who is responsible for overseeing the work of the programming staff.

When Greg started work at Synergy, most of his projects involved programming for computer graphics. "Now, though," he says, "I'm writing programs that talk to, or interact with, the hardware inside the computer. I think I still prefer graphics, but this is also interesting." Often, Greg is asked to work with a program that was written by another company for a different type of hardware. "In this case, my job is to make that program work on *our* hardware," Greg says. "This is called *porting*. Right now I'm working on porting a real-time operating system for a flight simulator."

Once Greg knows what project he will be working on, he begins to think about how to make the computer do what his company wants it to do. He considers what procedures the computer could take to do the task, and he writes them down in the logical sequence of a flow chart. Then Greg begins typing codes that the computer will understand into the computer, telling the computer the steps, in order, that it should perform. "You have to keep testing things out, making sure the steps will work in the way you have planned," he says.

Another important part of computer programming is writing documentation to go along with the software. Greg explains, "When you create a program you have to write comments every few lines on what everything does, what every step in the program is telling the computer to do. Good documentation is really important so that other programmers who will be working on your software can see right away what's going on. Also, it's even helpful for me on my own programs. When I go back and look at code I've written a few months ago, it reexplains to me exactly what I was having the computer do at each point."

## have I got what it takes to be a computer programmer?

It is clear from the description of his workday that Greg spends a lot of time working alone on the computer. He knows what jobs he has to get done, and he works toward these goals with little supervision or assistance. "You do have to be self-motivated and well organized," says Greg. "If you're not good at working on your own, making your own schedules and deciding for yourself how you're going to approach things, then you probably won't like being a programmer. Most good programmers I meet are really independent thinkers." As telecommuting, or working from home or another remote location, grows in popularity, self-motivation will be more critical than ever.

Computer programming also requires a great deal of patience, persistence, and concentration. When creating software, programmers must focus on determining the correct steps and instructions for the computer to follow. This takes a great deal of thought, planning, and immense creativity. Programmers must test and retest steps of the program, making sure they are working before proceeding to the next steps. Finally, programmers must have the patience and problem-solving skills to debug a program, going through the codes and carefully examining them for error, entering test data, and running the program again. For someone with little patience, this type of work might quickly become frustrating.

"Programmers also have to be able to think logically," says Greg. "You need a good method of breaking things down. You have to look at a whole project and then break it into components in order to be able to tackle it. Technically, a large computer program is divided into a lot of subroutines. One part does one function, another does another."

Being accurate and detail oriented is also essential to computer programming. Because computers only understand specific instruction codes, even the smallest error will make an entire program unusable. Finally, since computer programmers act as the link between the capabilities of the computers and the needs of the computer operators, programmers have to be able to communicate with both people and machines. They have to be able to listen to an abstract description of a potential computer project and turn this description into precise instruction code that the computer can follow.

**To be a successful computer programmer, you should**

- Be able to think logically
- Be creative
- Be well organized, precise, and detail oriented
- Have a long attention span
- Like to work independently
- Enjoy solving problems

# how do I become a computer programmer?

As a computer programmer, Greg has the opportunity to combine several of his interests. "In high school I enjoyed science and math," says Greg, "and I did best in subjects and on tests that involved logical thinking." He adds that he's always had an interest in machines and electronics, and as a teenager, his hobbies included building his own electric guitars, aviation, and sky diving. "All of these things helped me realize I have an aptitude for programming," he says.

## education

### High School

If you are interested in becoming a computer programmer, you must complete high school. You should take college-preparatory courses in mathematics, which is the basis for computer programming and will help you develop the logical sequencing abilities necessary in the field. High-level courses in physics will also help you. "Really, studying all the hard sciences is important," says Greg. "If you end up working in science applications, it will help a lot if you have a background in chemistry, biology, all those subjects."

If your high school offers courses in computers and has computer laboratories on campus, you should take as many of these courses as possible. In this way, you can become well-acquainted with the functions of computers before entering college and the job market. High school courses in business are especially important if you are interested in becoming a business and commercial applications programmer or systems programmer for a business or corporation. Finally, Greg adds, "Take typing and learn to do it so you're quick and accurate. I never thought I'd need typing, but I do it every day now."

### Postsecondary Training

As popularity and demand both skyrocket, most employers now require college degrees. In fact, according to a study in the *Occupational Outlook Handbook*, nearly 60 percent of computer programmers held at least a bachelor's degree in 1996, and only 10 percent had high school diplomas or less. Although it is possible for those with exceptional backgrounds and work experience but no college degree to find employment, your chances will be much stronger with a degree, and pursuing a college education is highly recommended.

Many junior colleges and technical colleges provide training in computer programming, operations, and data processing, providing a jumping-off point for transfer into a four-year college, job, or internship. Many employers do not specify a major field for college study for potential programmers. Greg has a bachelor of science degree in computational mathematics from Arizona State University. Currently, with the increase in the number of college and university computer science departments, people hired as programmers often have degrees in this field. However, this is not always the case. Programmers interested in business and commercial applications might major in business, finance, accounting, or marketing. Those interested in engineering and scientific programming could consider a major in engineering, mathematics, or physics. Greg says, "Often the programmers I meet who are really outstanding in their fields don't have a degree in computer science, but instead studied the science they are working in. They have degrees in engineering, astronomy, chemistry, things like that."

Many employers also send new programmers to special training courses before the programmer is allowed to begin working with company computers. In this situation, expenses are usually paid by the employer, and training time can take anywhere from two weeks to two months. Some computer programming jobs do require graduate degrees.

## certification or licensing

Criteria and qualifications for being hired as a computer programmer varies between employers. Some employers may require work experience, others may require degrees from two- or four-year colleges, still others may look for a combination of the two. Most people who work as computer programmers are not certified in this field. It is possible, however, to become certified as a computer programmer, and certification could help in attracting the attention of potential employers. The Institute for Certification of Computing Professionals (ICCP) offers a series of exams that, once passed, will provide the programmer with certification at one of two levels: associate computing professional (ACP) or certified computing professional (CCP). The level at which programmers are certified depends upon the programmers' work and educational backgrounds and the exams taken. While ICCP certification is not required for most programming jobs, it may provide you with a competitive advantage over job seekers.

**Unlimited possibilities...** Computer programmers may find work in a wide variety of settings, including medical laboratories, software companies, universities, banks, small businesses, large corporations, nonprofit organizations, research companies, robot manufacturers, commercial aviation companies, and the military.

## internships and volunteerships

Today, most homes and even the smallest businesses have computers, providing ample opportunity for you to gain hands-on experience with computers and programming. At home, try surfing the Internet for information about computer-related careers, education, and scholarships. You may want to join a local or school computer club, and you should pick up some books or magazines about programming. You might be able to find summer or after-school jobs or internships working with a company computer, helping to input or process data and becoming familiar with the ways a computer is used in a business setting. You may be able to work at small shops such as book and pet stores helping to maintain inventory records on computer; in a library using computerized library catalogs and loan information; or on a school or local newspaper that uses computers for writing, graphics, and production. Some young people's business organizations such as Junior Achievement may also offer high school students an opportunity to work with computers. While computer-oriented jobs for most high school students will not involve programming, these jobs will help you gain exposure to and experience with a variety of computer capabilities.

## who will hire me?

While he was still in college, Greg heard about a job opening at Williams Air Force Base in Phoenix, Arizona. "They were looking for a college student who was interested in doing some entry-level programming. I've always liked aviation, so I gave them a call." Greg went to Williams for an interview and answered questions about his interests, the college courses he was taking, and his grades. He was hired for the position. At this job, Greg was asked to help adapt an F-16 flight simulator program to a new computer. "Then, in my spare time during lunch, I got permission to play with the computer graphics programs. That's really how I got into the field. Before the Williams job, I had taken only a couple computer classes. Pretty much everything I know about computers, I learned on the job." Greg adds, though, that his degree in computational mathematics has been very important to his career. "Nearly everything I do involves math," he says.

While he was programming on the flight simulator at Williams, Greg worked in close contact with Synergy Microsystems in Tucson, because Synergy specializes in these types of programs. Synergy was impressed with Greg's work at Williams, and after he graduated from college, they offered him a job.

The first step in finding a job in computer programming is for programmers to isolate the type of work they are interested in doing. Because computers are such a large part of today's world, programmers can find work at small or large businesses, in science, health, or the military, or even, like Greg, at a company that creates software to sell to other companies. Programmers should think about whether they are interested in systems or applications programming and whether they are more interested in the business and commercial applications of computers or the engineering and scientific aspects of the field.

Once programmers have selected an area of interest, they may follow numerous approaches to job hunting. Often, the newspaper classified section advertises job fairs and openings for computer programmers, but many now turn to the Internet for the job search. Thousands of companies and organizations advertise job opportunities on their web sites, and there are many web sites devoted specifically to helping you land a job. It is even possible to post a resume or apply for open positions directly online. Programmers with a specific interest, such as flight simulation or medical technology, might want to find companies in this field and contact them directly with a resume and cover letter. Systems programmers usually work for larger corporations or companies with extensive computer systems, and application efforts should be directed toward these. In addition, some agencies specialize in placing computer professionals in both temporary and permanent positions. These agencies work to match programmers to companies needing computer assistance. Computer consulting and independent contracting is also a growing field. Rather than hiring computer programmers as permanent employees, many companies are turning to computer contractors who are hired to come into the company and perform computer work on an as-needed basis. These consulting firms are usually found on the Internet or in the yellow pages under "Computers-Consultants."

## where can I go from here?

Greg enjoys his job and plans to stay in it for several more years. "For me, advancement will just mean working on more and more complex programs and getting more money." He adds that some programmers at Synergy go into management and deal directly with the clients. They then tell the programmers what types of software the clients want to buy. "I don't want to do that, though," says Greg. "I want to keep working with the machines."

Some companies offer a programmer with extensive programming background the job of *chief programmer*. This job entails overseeing the work of the programming staff, assigning projects, and consulting with management to determine deadlines and special needs and concerns. Programmers could also advance into the position of *systems analyst*, computer specialists employed by many large companies to examine the computer needs of the company and then create methods to improve the system.

Computer programmers may get involved in software or hardware sales, helping clients to select appropriate computer equipment and providing them with instructions on its proper use. They may open their own businesses to sell computer equipment or to offer consulting services to companies needing computer help. Computer programmers are also hired as computer science instructors at high school, community college, and university levels. With the prevalence of computers in today's society, skilled computer programmers have a wide variety of opportunities for advancement in their field.

## what are some related jobs?

Trained computer programmers can work at a variety of jobs that involve knowledge in technology and computer operations. Because of the expanding role of technology in the modern world, skilled programmers are an asset to nearly every business or corporation and scientific or engineering organization.

**Advancement Possibilities**

**Chief computer programmers** plan, schedule, and direct preparation of programs to process data and solve problems by use of computers, often consulting with managerial and systems analysis personnel.

**Systems analysts** analyze user requirements, procedures, and problems to automate processing or to improve existing computer systems. They confer with personnel of organizational units involved to analyze current operational procedures and identify problems.

**Programmer analysts** plan, develop, test, and document computer programs, applying knowledge of programming techniques and computer systems.

The U.S. Department of Labor classifies computer programmers under the headings *Occupations in Systems Analysis and Programming* (DOT) and *Mathematics and Statistics: Data Processing and Design* (GOE). Also under these headings are people who manage data processing equipment and work as technicians in mathematics and computer-related fields.

| Related Jobs |
|---|
| Software technicians |
| Systems analysts |
| Records managers |
| Information scientists |
| Engineering analysts |
| Operations research analysts |
| Mathematical technicians |
| Software engineers |
| Electronic data processing managers |
| Detail programmers |

## what are the salary ranges?

According the the 1998–99 *Occupational Outlook Handbook*, computer programmers can earn anywhere from $19,500 to more than $65,000 per year, depending on their level of experience and the type of programming they perform. For example, a computer programmer for NASA space shuttles will probably have a higher salary than a programmer for a small accounting firm. The median salary for full-time computer programmers is $40,100 per year, with the lowest 10 percent earning under $22,700 per year and the highest 10 percent earning over $65,200 per year. On average, systems programmers earn more money than applications programmers, and programmers in the Northeast and West usually have higher salaries than those in the South and Midwest.

## what is the job outlook?

There are more than 568,000 computer programmers employed in the United States. As the use of computers increases, there should continue to be plenty of employment opportunities for programmers. Programmers interested in database management, computer networking, and artificial intelligence are currently in high demand, and programmers who have a background in the field in which they program will have an advantage in finding a job. For example, it would be advantageous for programmers interested in working for a bank to have banking knowledge and for civil engineering programmers to have worked or studied in this field.

Computer programming fields that look exceptionally strong for the future include data processing firms, software companies such as Microsoft that create software to be sold to other companies and individuals, and companies that provide computer consulting or operation services to other organizations who do not employ their own computer personnel. The downside of the growth in consulting services is that many small companies are laying off their in-house computer programmers and opting for computer assistance from consultants on an as-needed basis. For the companies, this enables them to cut costs by reducing employee salaries and benefits.

Because of the increasing use of computers in business, programmers with expertise in business and commercial application programming are expected to be in especially high demand, as are programmers who are able to work with multimedia systems.

To maximize chances of finding employment as a computer programmer, students should complete a college degree and get as much experience as possible in their field of interest before applying for permanent positions.

# how do I learn more?

## sources of additional information

Following are organizations that provide information on computer programming careers, accredited schools and scholarships, and possible employers.

**American Society of Information Science**
8720 Georgia Avenue, Suite 501
Silver Spring, MD 20910-3602
301-495-0900
http://www.asis.org

**Association for Computing Machinery**
1515 Broadway
New York, NY 10036
212-869-7440
http://www.acm.org

**Association of Information Technology Professionals**
315 South Northwest Highway, Suite 200
Park Ridge, IL 60068
800-224-9371
http://www.aitp.org

**IEEE Communications Society**
305 East 47th Street
New York, NY 10017
http://www.comsoc.org

**Institute for Certification of Computing Professionals**
2200 East Devon Avenue, Suite 247
Des Plaines, IL 60018-4503
847-299-4227
http://www.iccp.org

# bibliography

Following is a sampling of materials relating to the professional concerns and development of computer programmers.

### Books

A large variety of computer books on programming can be found at most book stores. Because of the fast-changing nature of this area and the importance of reading up-to-date material, no specific books are listed here.

### Periodicals

*Computer Magazine*. Monthly. Discusses current developments in computer applications and research. IEEE Computer Society, 10662 Los Vaqueros Circle, PO Box 3014, Los Alamitos, CA 90720-1264, 714-821-8330.

*Computerworld*. Weekly. Newspaper covering a wide range of computer information. Computerworld, Inc., 551 Old Connecticut Path, Box 9171, Framingham, MA 01701-9171, 508-879-0700.

*Datamation*. Two issues per month. Trade magazine on computers. Cahners Publishing Co., 275 Washington Street, Newton, MA 02158-1630, 617-964-3030.

*SIGDOC Newsletter*. Quarterly. Examines and reviews papers from systems analysts, programmers, etc. Special Interest Group on Systems Documentation, Association for Computing Machinery, 1515 Broadway, 17th floor, New York, NY 10036, 212-869-7440.

# computer systems/ programmer analyst

**Definition**

Systems analysts plan and develop new computer systems or update existing systems to meet changing business needs.

**Alternative job title**

Programmer analysts

Systems analysts

**Salary range**

$24,800 to $46,300 to $76,200

**Educational requirements**

Bachelor's degree

**Certification or licensing**

Voluntary

**Outlook**

Much faster than the average

GOE
11.01.01

DOT
033*

O*NET
25102

## High School Subjects

Computer science
Mathematics

## Personal Interests

Computers
Figuring out how things work

**It's 11:30 AM** and Tim Sutter has only been in the office for a few hours. Already he's attended two staff meetings, his beeper has gone off four times, and he has fielded numerous calls from confused people asking for help with sick systems.

He's frustrated and ready to pull his hair out. Trying to explain highly technical concepts in plain English to an entire accounting department has taken all morning.

Solving problems is Tim's job. He is a computer programmer analyst for a large financial institution, and this is all part of a day's work.

## what does a computer systems/ programmer analyst do?

In the rapidly expanding world of technology, definitions, titles, and job duties change almost daily. Traditional job descriptions may no longer apply to the current job market. There are no uniform job titles. Both *systems analysts* and *programmer analysts* have similar, often interchangeable roles.

Computer systems/programmer analysts define specific business, scientific, or engineering problems and design solutions for them by using computers. They may plan and create entirely new programs or find better ways to use existing systems. They also install, modify, and maintain functioning systems.

Companies depend on computer systems/programmer analysts to make work easier, more streamlined, and more efficient. Essentially, their job is to save the company money.

**ASCII (American Standard Code for Information Exchange)** Numerical code used by personal computers.

**Database** A collection of information stored on the computer.

**Debugging** Identifying and correcting errors in software.

**GUI (goo-ey; Graphical User Interface)** A system that uses symbols (icons) seen on-screen to represent available functions.

**Network** Several computers that are electronically connected to share data and programs.

**LAN (Local Area Network)** A network that exists at one location, typically an office.

**Spreadsheet** A program that performs mathematical operations; used mainly for accounting and other record keeping.

**Wan (Wide Area Network)** A network that includes remote sites in different buildings, cities, states, or countries.

Analysts work a lot like detectives. When presented with a problem, they investigate. They analyze hardware and software and ask the people who actually use the system—the users—what they want the program to do. An analyst cannot create a good, workable program without the help of the people who need it. Because of this, good communication skills are important. "Knowing how to ask the right questions is one of the most important parts of my job," Tim emphasizes. "Users don't know the technological jargon. They just want to turn on their computers, point, and click."

Once the needs of management and users are established and system goals determined, the analyst gets to work on the program. Highly analytical and logical activities follow, and the analyst uses tools such as structural analysis, mathematics, data modeling, and cost accounting to determine which computers, hardware, software, and peripherals will be needed to reach the system goals. Analysts must carefully weigh the pros and cons of additional features and consider trade-offs between increased cost and extra efficiency.

Computer systems/programmer analysts then present reports and proposals to management. This is where attention to detail, organizational skills, and strong communication abilities come into play.

Once the system upgrades are approved, equipment is purchased and installed. This is where certain distinctions between a systems analyst and programmer analyst arise. Traditionally, a systems analyst does everything up to the actual programming of the computer. The computer programmer then writes the code and plugs everything into the system.

More and more, however, programmer analysts combine both of these functions, rolling two jobs into one. As an increasing number of businesses use computer technology, the ability to program becomes an important part of the job. In addition, tools such as CASE (Computer Aided Software Engineering) make programming easier by providing scripts of common commands.

Once the system is in place and the users trained, the analyst provides basic maintenance. Any problems or questions about the program? The analyst is once again called upon to clear this up. Often he or she must debug the system, cleaning up any mistakes or flaws in the program.

An analyst is also responsible for the security of the system, making sure that the data can't be accessed by anyone not authorized to use it. The analyst makes sure the system runs smoothly and accurately from start to finish.

## what is it like to be a computer systems/programmer analyst?

During a typical day, Tim spends more time fielding questions and talking to users than anything else. Often people outside the office come to him for help, tapping into his expertise, not to mention his personal time. "It chews up a lot of time. Most of it is not related to a current project," he says. Tim is asked questions such as, "Why won't this software work?" or, "How do I create a report on this program?" An analyst is often thought of as the "answer person."

When Tim comes into the office, he turns on his computer but spends little time there. "I rarely spend eight hours a day at my desk," and he says he likes it this way. The opportunity to work with other people and not just computers is a plus for him. "The day I just sit in front of my terminal is the day I leave."

Tim usually attends at least two meetings a day, sometimes more. Typically, a morning meeting with management is held to discuss the status and progress of a current project. Concerns and issues are addressed, and individual tasks are assigned.

After a program is written, the computer systems/programmer analyst often must produce written documentation. "You must explain the systems you develop so that anyone can run [or edit] them, even long after you're gone," Tim says. Some mornings Tim might spend hours documenting his work in this way. This is time consuming and his least favorite part of the job.

Depending on the employer, some analysts may be involved with setting up computer networks. This involves connecting computers or peripherals to work with each other. Analysts must design the hardware and software so that equipment from various manufacturers works together smoothly and efficiently.

Sometimes Tim works straight through lunch, eating at his desk. "I do things like document spacing and layout. Stuff I can do with one hand," eating a sandwich with the other, he says. He may then spend the afternoon debugging programs and working out problems and glitches in something he's designed. Problems can be simple fixes such as printer queue errors or more complex problems that may require days of work.

Later in the afternoon, Tim will probably attend yet another meeting with his managers, following up on earlier discussions, getting approval on projects, and discussing problems that have arisen or need to be fixed.

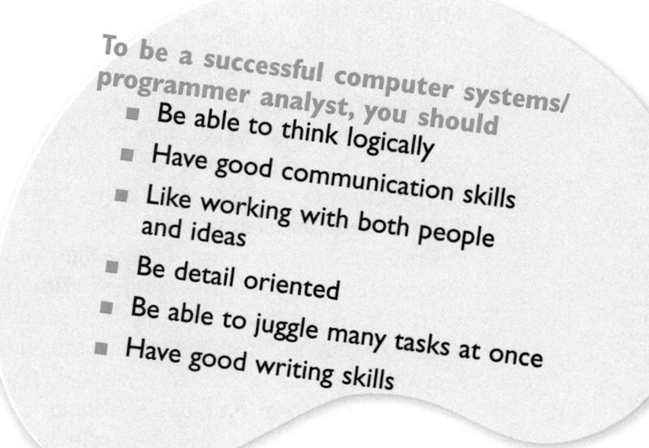


A computer systems/programmer analyst normally works an average forty-hour week, mainly in an office environment. Because of the prevalence of home computers and the sophistication of communications equipment, however, it's possible to work from home, and Tim often finds himself working at home after regular office hours. Forty-hour work weeks may be the average, he says, but to get ahead, a dedicated analyst will put in fifty or more hours a week.

Sometimes bringing work home is a drag, Tim says. "After a busy day at work, I come home and think, `Do I really want to turn on the computer?'" But working at home also offers Tim opportunities to take chances he might not normally take in an office environment. "Since I'm technically not on the clock, I can be riskier in developing programs. I use new methods and come up with unique solutions. It's a great way to learn something new and use things from work in different ways," he says.

## have I got what it takes to be a computer systems/programmer analyst?

Analysts often work long hours and deal with a wide variety of people. They work both alone and in groups. Patience and attention to detail are important qualities, as well as logical thinking and an ability to translate highly technical and complex concepts into simple language.

If you prefer to sit at a computer by yourself all day, every day, computer systems/programmer analyst isn't the job for you. An outgoing personality, good communication and listening skills, and being able to work well with others are as much a part of the job as computer knowledge. An analyst must be able to talk easily to both technical personnel, such as programmers and management, as well as to staff who might not have any knowledge of computers. One of Tim's least favorite roles, however, is that of go-between for the managers and users. "Everyone is not going to understand all the implications of the project," he says, and sometimes he gets caught in the middle. As analyst, Tim must approach his work from several different points of view so that he can clearly define and address everyone's needs.

Self-motivation and being able to juggle many different tasks at once are critical to the job. Tim thinks of himself as a jack of all trades. "I live and die by my own actions," he says. "It's my responsibility to do the research and solve the problems. I push myself." Tim enjoys the high visibility that his job brings and considers it one of the benefits.

Employers stress the importance of a well-rounded education to prepare for a job as a computer systems/programmer analyst. A degree in computer science is still the way to go, but analysts with business, management and communications skills are increasingly being sought after as well.

Ongoing education is also part of the job. "Not only are you doing work at home for the company, but you're constantly learning along the way," Tim says. For example, he frequently attends classes and seminars, learning about new systems and technology. Programs are usually offered through universities or the manufacturers of computer hardware and software systems and are usually paid for by the employer.

Tim also reads several trade magazines to familiarize himself with new and advanced technologies. These publications are usually on the cutting edge of new systems. Tim also finds user groups to be helpful. "Tech-heads get together, solve each other's problems, and come up with new stuff."

Technology changes at lightning speed and so must a good analyst. Staying on top of trends is very important. "If you fall behind, you're through," Tim stresses. Employers will always be looking for newer, faster, and more cost-efficient ways to do business. The analyst's job is to provide this service.

## how do I become a computer systems/programmer analyst?

Even in high school, Tim had a basic interest in computers but hoped to apply those skills to a career in business. He wanted the opportunity to work with other people, not just machines. A job as a computer systems/programmer analyst combines both.

"High school kids today have a better working knowledge of computers," Tim says. Being comfortable with computer programs and different kinds of software is a good start.

## education

### High School

A high school diploma is necessary for a future as a computer systems/programmer analyst. Computer use is more common in high schools today than when Tim attended. He stresses that a knowledge of IBM compatible computer programs as well as Macintosh systems is very important.

If possible, enroll in computer classes and labs to gain hands-on experience. Joining user groups or computer clubs won't hurt either.

High school courses in the sciences are also necessary. Many computer systems/programmer analysts work in industries directly involved in the scientific fields. Math classes are good for understanding the programming languages taught in college. Mathematical ability is also useful when writing and developing computer programs, and a good math background helps develop skills in logical thinking.

Also useful for developing analytical thinking skills are strategy games such as chess. Play with friends or against a computer. And to make computer usage as natural as brushing your teeth, play on your computer as frequently as possible—surf the World Wide Web, learn various software programs, and try to pick up some basic programming or networking skills.

Finally, don't ignore the basics, Tim warns. "Typing is the best class I ever took! If you can't type, just forget it."

### Postsecondary Training

There is no universally accepted way of preparing for a job as a computer systems/programmer analyst, but having a four-year degree from an accredited college, university, or technical institute is a must. Though not mandatory for analysts, many successful computer professionals hold not only an undergraduate degree, but a master's or doctorate as well.

A degree in computer science was once mandatory, but employers are now looking for applicants with more varied backgrounds. A large company will usually want a computer systems/programmer analyst to have a background in business management or a related field for work in a business environment. A background in the physical sciences, applied mathematics, or engineering is preferred by scientific organizations.

Tim has a degree in management information systems (MIS) with a minor in computer science. "You must have business experience if that's where you want to work," he says. "You need to understand where your computer skills are being applied. Users don't understand bit or byte, but they do know what they want. An analyst must know both."

Regardless of college major, employers look for people who know programming languages and have a good knowledge of computer systems and technologies.

## certification or licensing

Getting certified as a computer professional is not a requirement for a job as a computer systems/programmer analyst, but might give a jobseeker a competitive edge over other applicants.

The Institute for Certification of Computing Professionals offers the designation certified computing professional (CCP). Applicants must have at least four years of experience and pass a core examination. They then must pass exams in two specialty areas. The Quality Assurance Institute awards the title of certified quality analyst (CQA) and certified software test engineer (CSTE), to those who meet education and experience requirements, pass an exam, and endorse a specific code of ethics. (See "Sources of Additional Information" for contact information.)

## scholarships and grants

Following is a partial listing of the many organizations offering scholarships and grants for students pursuing a career in computer science. Some offer small amounts of money toward a degree, while others pay the full tuition and provide internships. Write early for information on eligibility, application requirements, and deadlines.

**Microsoft National Technical Scholarship**
**Microsoft National Women's Technical Scholarship**
Microsoft Corporation
One Microsoft Way
Redmond, WA 98052
scholar@microsoft.com
Attn: Mary Blain

**DeVry Institutes (Scholarship Program)**
One Tower Lane
Oakbrook Terrace, IL 60181
800-73DEVRY, ext. 2089
http://www.devry.com

**NAACP Willems Scholarship**
4805 Mount Hope Drive
Baltimore, MD 21215
410-358-8900
Attn: Director of Education

## internships and volunteerships

An important, but not mandatory, part of preparing for a career as a computer systems/programmer analyst is participating in an internship program. Internships provide practical hands-on experience for students. Universities and technical institutes usually have programs that help arrange internships for their students.

Tim spent eight weeks one summer during college working as an intern in Egypt. "It was a great experience, and it gave me an opportunity to see how things work in the real world." He worked for the Suez Canal Authority and took part in a study to find out if it was efficient for the Egyptian government to regularly dredge sand out of the canal.

The experience allowed Tim to work on a program from start to finish. He saw all the different aspects of his future job. Spending a summer in Egypt and having his expenses paid was a nice bonus.

While all internships may not be in such exotic locales, employers look for practical experience in job applicants. Most internships offer little or no pay for long hours and hard work, but the experience gained is priceless. Many students make valuable contacts this way, and some are hired as regular employees when the internship is finished.

## who will hire me?

As more businesses, organizations, and federal agencies expand their computer systems, jobs for computer systems/programmer analysts open up in a variety of areas. The largest number of analysts is found in computer and data processing firms, but opportunities in many other fields exist as well.

Government agencies, manufacturers of computer-related equipment, insurance agencies, universities, banks, and private businesses all employ the skills of computer systems/programmer analysts.

Most analyst jobs are concentrated in or near larger cities. Many employers actively recruit applicants on college campuses, interviewing students while they're still in school. Most simply advertise in the want ads of big city newspapers.

Tim got his first job right out of college by answering an ad for a position with a small accounting firm. He felt working in a smaller agency would allow for greater participation and give him valuable job experience. After two years he took a position with a larger company.

Following a trend in the overall workforce, growing numbers of computer systems/programmer analysts are employed on a temporary basis as short-term consultants. A company needing the services of an analyst to install or create a new computer program may not be able to justify hiring a full-time analyst but will turn to a computer consulting agency or contract a computer systems\programmer analyst to do the work. These contracts usually last for several months and sometimes up to two years.

Trade magazines often have job listings and are good sources for job seekers. They are also a great place to learn about changing technologies and trends in the industry. Reading several different publications may help to identify where the jobs are.

The Internet is also an excellent tool for finding job opportunities. Most major companies post job openings, and many web sites are dedicated solely to helping you land a position.

A computer systems/programmer analyst is not an entry-level job. Many analysts start in programming or data processing. Occasionally, some are promoted from another area altogether. For example, an *auditor* in an accounting department might become a systems analyst specializing in accounting systems development.

## where can I go from here?

Tim plans to stay with his current employer for two more years then branch out on his own. He hopes to one day have his own business and work as a private consultant. The prospect of working for himself has a lot of appeal. He'll make his own rules and hire out his services for higher wages than he can make on salary. "Hopefully, I can sell my services back to my old employer, do the same job, but get paid more as a consultant," he says. Many computer systems/programmer analysts have similar goals.

As a greater diversity of businesses become dependent on computers, the need for consultants in the field grows. For example, a small book store chain may not have the need for a full-time systems analyst but will hire one as specific projects arise, such as creating an automated inventory and purchasing system, for example.

The career path of a company computer systems/programmer analyst usually leads to management. After a few years' experience, those who show leadership ability, good communication skills, and diverse business and

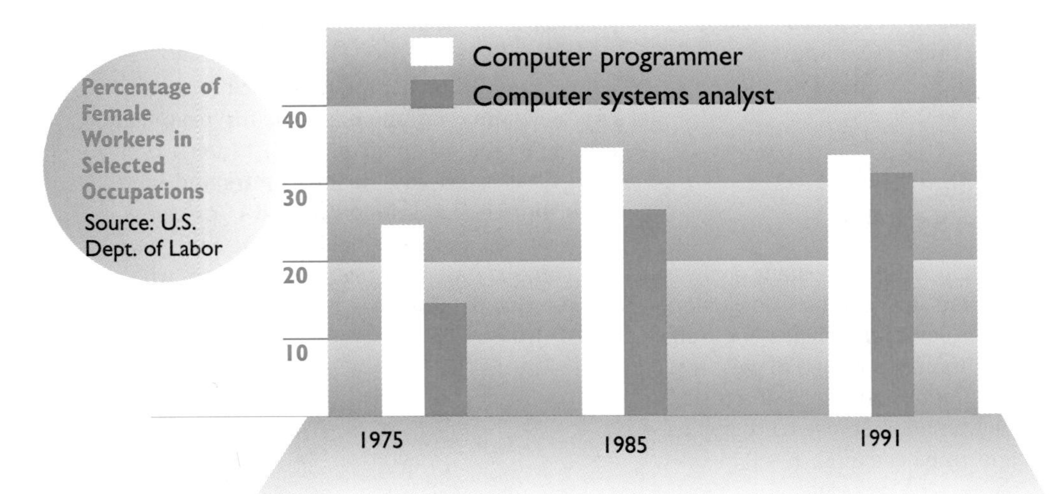

### Advancement Possibilities

**Senior systems/programmer analysts,** also known as **lead programmer/analysts,** are in charge of an entire project, coordinating and overseeing the work efforts of all the analysts working on a team.

**Managers of information systems**, also know as **project managers,** are overseers, operating as supervisors for all the projects the analysts are involved with.

**Database design analysts** work as liaisons between the computer systems/programmer analysts and the company. They evaluate project requests, their costs, and time limitations. They review and restructure the database to work with programs developed by the computer systems/ programmer analyst.

technical knowledge are promoted along these lines. If Tim decides to stay with his current employer, for example, there's a lot of room for advancement. He could be promoted from programmer analyst to *senior* or *lead programmer analyst,* and from there to *project manager.* He then might move up through management, eventually becoming a *vice president.*

## what are some related jobs?

A computer systems/programmer analyst can work anywhere there are computers in a wide variety of industries from programming and creating new software to product sales. The U.S. Department of Labor classifies systems analyst under the headings *Occupations in Systems Analysis and Programming* (DOT) and *Data Processing Design* (GOE).

| Related Jobs |
| --- |
| Engineering analysts |
| Information scientists |
| Operations-research analysts |
| Business programmers |
| Engineering and scientific programmers |
| Information system programmers |
| Process control programmers |
| Mathematical statisticians |

## what are the salary ranges?

According to the 1998–99 Occupational Outlook Handbook, systems analysts earned a median income of $46,300 in 1996. The lowest 10 percent made less than $24,800 while the highest 10 percent earned more than $76,200.

A 1998 salary survey by Robert Half International Inc., a staffing agency specializing in information technology, accounting, and finance, indicates that systems analysts at large companies earned anywhere from $47,000 to $60,000. Programmer analyst incomes ranged from $40,000 to $52,500.

The federal government in 1997 paid a starting salary of $19,520 a year to a recent graduate with a bachelor's degree. For an entrant with a superior academic record, the starting pay was $24,180.

Working for a large company or the federal government brings other nonsalaried benefits as well, such as health insurance, retirement plans, and paid vacations. Private consultants do not get these benefits and must provide their own, in addition to the extra work of running their consulting business.

## what is the job outlook?

The job of computer systems/programmer analyst is one of the fastest growing occupations. Demand for these skilled professionals is expected to increase much faster than average through the year 2006.

As smaller businesses utilize the efficiency of computers in the workplace, and technology becomes more affordable to private users, an even greater demand for systems and programmer analysts will arise.

There is an ever-increasing emphasis on personal computers. Businesses are moving away from the larger and extremely expensive mainframes that have traditionally been used by large companies. They are going toward a network of many smaller yet powerful computers that share the workload. As this trend continues, analysts will be indispensable in an office environment. Someone must program, organize, and link all these individual systems together.

As these intricate computer systems improve and grow, the analyst will be needed to continually upgrade and weed out errors in the programs. Maintenance and debugging will provide steady work as long as computers are part of our daily lives.

The role of computer systems/programmer analyst is an upwardly mobile one. Tens of thousands of jobs will be opening up as analysts move into managerial positions. Those who go into business for themselves or to private consulting firms will keep this field open for new analysts.

In 1996, computer systems analysts held 506,000 jobs in the United States. They mostly worked in large industries such as computer and data processing firms, government agencies, insurance companies, universities, and banks. While the private consultants made up a small percentage of these workers, the numbers are expected to grow though the beginning of the next century.

# how do I learn more?

## sources of additional information

Following are organizations that provide information on computer systems/programmer analyst careers and certification, accredited schools, and employers.

**Association for Computing Machinery**
1515 Broadway
New York, NY 10036
212-869-7440
http://www.acm.org

**Association of Information Technology Professionals**
315 South Northwest Highway, Suite 200
Park Ridge, IL 60068
800-224-9371
http://www.aitp.org

**IEEE Communications Society**
305 East 47th Street
New York, NY 10017
http://www.comsoc.org

**Institute for Certification of Computing Professionals**
2200 East Devon Avenue, Suite 247
Des Plaines, IL 60018-4503
847-299-4227
http://www.iccp.org

**Quality Assurance Institute**
7575 Dr. Philips Boulevard, Suite 350
Orlando, FL 32819
407-363-111

## bibliography

Following is a sampling of materials relating to the professional concerns and development of computer systems/programmer analysts.

### Periodicals

*Computer Magazine.* Monthly. Discusses current developments in computer applications and research. IEEE Computer Society, 10662 Los Vaqueros Circle, PO Box 3014, Los Alamitos, CA 90720-1264, 714-821-8330.

*Computerworld.* Weekly. Newspaper covering a wide range of computer information. Computerworld, Inc., 551 Old Connecticut Path, Box 9171, Framingham, MA 01701-9171, 508-879-0700.

*Datamation.* Two issues per month. Trade magazine on computers. Cahners Publishing Co., 275 Washington Street, Newton, MA 02158-1630, 617-558-4424.

*SIGDOC Newsletter.* Quarterly. Examines and reviews papers from systems analysts, programmers, etc. Special Interest Group for Systems Documentation, Association on Computing Machinery, 1515 Broadway, 17th floor, New York, NY 10036, 212-869-7440.

# computer-aided design technician

**Definition**

CAD technicians use computer-based systems to produce or revise technical illustrations needed in the design and development of machines, products, buildings, manufacturing processes, and other work.

**Alternative job titles**

CAD specialists

CADD technicians

Computer designer/drafters

**Salary range**

$13,500 to $25,000 to $42,000+

**Educational requirements**

High school diploma; one-year certificate or two-year associate's degree

**Certification or licensing**

Certification is voluntary; licensing may be required for certain projects

**Outlook**

Faster than the average

GOE 05.03.02*

DOT 003*

O*NET 22514A

## High School Subjects

Art
Computer science
Mathematics
Physics
Shop (Trade/Vo-tech education)

## Personal Interests

Building things
Computers
Drawing
Figuring out how things work
Programming

**A new building** is going up in the city, and an architectural firm is bidding for the job. Armed with a list of "wants" and "needs" from the customer, the designer sits down at the CAD system. She loads it with special software that knows all about architecture, and starts entering information about the project: cost limits; available materials; lot size. The CAD system, a souped-up PC, provides the processing power to crunch the numbers. Soon, the computer screen begins to display a series of suggestions about how the building could be designed—giving the architectural firm a starting point for developing their plan.

After the firm wins the bid, the design takes shape. The CAD system is used to develop a set of electronic working plans, the modern equivalent of the hand-drawn architectural blueprint. It will also be used to develop the plumbing, electrical, and heating/venting/air conditioning (HVAC) systems needed for the building; make a list of furniture required; check building codes; and provide other important information.

## what does a CAD technician do?

Design and drafting are two steps needed to put engineering ideas on paper. The designer develops the concept; the drafter puts the concept into technical illustrations, which can be read by the people who actually make the product.

Until about twenty years ago, most designing and drafting were done by hand—with a pen or pencil and paper, on a drafting board. To make a circle, you used a compass. To draw straight lines and the correct angles, you used a straight-edge, slide rule, or other tools. With every change required before a design was right, it was "back to the drawing board" to get out your eraser, sharpen up your pencil, and revise the drawing.

Everybody did it this way, whether the design was simple or complex: automobiles; hammers; printed circuit boards; utility piping; highways; buildings.

Some design and drafting are still done by hand, but the days of the "pencil drafter" are pretty much over. Computer-aided design and drafting (CAD or CADD) systems are the tool of choice in industries that require detailed drawings, because they greatly speed and simplify the designer's and drafter's work. CAD systems do more than just let the operator "draw" the technical illustration on the screen. They add the speed and power of computer processing, plus software with technical information that eases the designer's or drafter's tasks. CAD systems make complex mathematical calculations, spot problems, offer advice, and provide a wide range of other assistance.

*CAD technicians* operate the CAD systems. They may do designing, or drafting, or both. The most basic CAD technician work is similar to data entry: simply inputting the drafting instructions given by the industrial designer or engineer. With more training and experience, the technician may also design. Exactly what the CAD technician does depends on the industry, the company, and the CAD technician's know-how. CAD technicians usually specialize in one industry, such as aeronautics or automobile manufacturing, or on one part of design, such as new product development, structural mechanics, or piping.

Like everything else in the computer world, CAD systems have evolved a lot over the last fifteen or twenty years. The first ones were large, mainframe-based, and cost up to hundreds of thousands of dollars. Commands could only be entered at the keyboard.

Starting about ten years ago, lower-cost PC-based CAD systems began to be introduced, permitting more companies to afford the technology. Now, almost all CAD systems are PC-based. "Mac is used, too, but mainly for educational purposes," says John Jacobs, coordinator of the drafting technology program at Belleville Area College, Belleville, Illinois. "There's not much technical software for the Mac yet."

Most CAD systems look a lot like home or office computer systems. They may have bigger and better color displays, though, and more input devices—such as a light pen, which is touched to the screen to indicate a command, or a programmable "puck," which looks like a mouse. Output devices include printers and

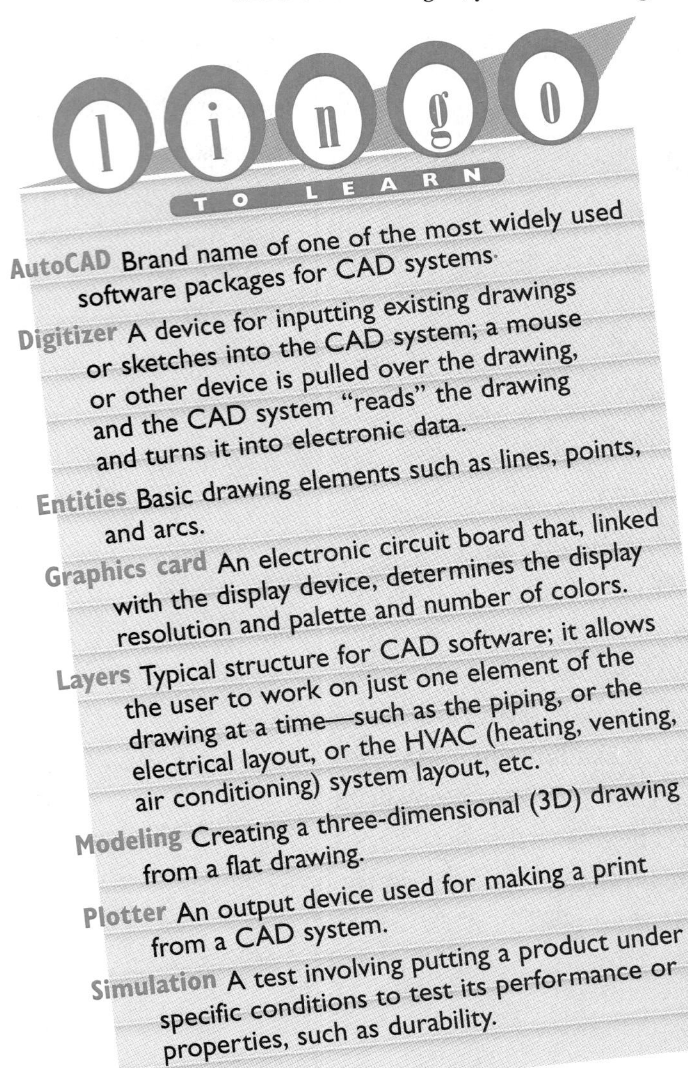


**AutoCAD** Brand name of one of the most widely used software packages for CAD systems.

**Digitizer** A device for inputting existing drawings or sketches into the CAD system; a mouse or other device is pulled over the drawing, and the CAD system "reads" the drawing and turns it into electronic data.

**Entities** Basic drawing elements such as lines, points, and arcs.

**Graphics card** An electronic circuit board that, linked with the display device, determines the display resolution and palette and number of colors.

**Layers** Typical structure for CAD software; it allows the user to work on just one element of the drawing at a time—such as the piping, or the electrical layout, or the HVAC (heating, venting, air conditioning) system layout, etc.

**Modeling** Creating a three-dimensional (3D) drawing from a flat drawing.

**Plotter** An output device used for making a print from a CAD system.

**Simulation** A test involving putting a product under specific conditions to test its performance or properties, such as durability.

# in-depth

**CAD, CAM, CAE, and CIM**

Closely related to computer-aided design and drafting (CAD or CADD) are computer aided manufacturing (CAM) and computer assisted engineering (CAE). CAM uses computers to determine which manufacturing processes and equipment are needed to make the product. It also monitors and controls the automated manufacturing of the product, guiding the factory's robots, automated measuring machines, computer-controlled machine tools, and other automated systems. CAD and CAM are linked in some operations, with information flowing back and forth via the computers.

Computer assisted engineering (CAE) is yet another possible step. It includes using computers to test and analyze various properties (such as weight, strength, durability, and other features) of a product or other design under specific conditions (simulation). Those doing CAE usually are four-year degreed engineers.

Computer integrated manufacturing (CIM) can wrap everything—design, manufacturing, assembly, sales, and other steps—into one combined process.

plotters that make oversized prints (up to six feet wide). CAD workstations also may be networked so drawings can be passed via computer among engineers, technicians, and supervisors.

As for software, different packages meet the needs of different industries and projects—from chemical, automotive, civil, structural, electrical, and other engineering, to clothing and furniture design. The software may provide simple drawing aids all the way up to sophisticated modeling tools that turn flat drawings into three-dimensional (3D) images.

Increasingly, the CAD technician who does more than just input others' instructions is most valued by employers. "It's like the word processing pools of years ago," says Peter Marks of Design Insight, a California design consulting firm that counts Ford and other large companies among its customers. "Back then, it probably seemed like a good idea to get a two-year degree to be a Wang system secretary. But those jobs just don't exist anymore. In the same way, it's just not a good idea for a CAD technician to be someone who only knows how to sit there and input. The best combination today is a blend of two things: some demonstrated aptitude on the CAD system, and expertise gained over time in a specific area, like plastic molding. The CAD technician who can bring specific knowledge to the job is going to be the 'value added' employee."

## what is it like to be a CAD technician?

A good example of a "value-added" CAD technician is James Stec, the lead designer for the design/drafting department of Morton International. James brings extensive design know-how to his work, which includes providing flow sheets and plant layouts as part of supporting twenty-two engineers in three of Morton's primary divisions: Chemicals, Salt, and Coatings/Adhesives. Based in Morton's downtown Chicago corporate offices, James also oversees the work of three other CAD technicians.

Given the wide range of products being produced, "you have to have a fair grasp of many areas," notes James, who has twenty-five years of experience as a designer/drafter. "We don't have to understand the concepts behind the products, say, the way the chemical engineers do. But we do help to design and develop the systems required to produce those products. And all of the divisions are constantly coming up with something new."

Technical illustrations produced by the department might be required when a new product is to be produced and a plant layout must therefore be changed. They might describe where equipment is to be placed, the structural framing/platforming required to support the equipment, how the production process will move from one piece of equipment to another, utility and process piping and electrical specifications, or other needs. As part of the flow sheet work, the technicians might check to be sure that they meet manufacturing and building codes and standards, such as those from OSHA, ASME, and NFP; they also might "tag" the piping on the flow sheet, identifying each kind of pipe by its universal specification number. Special work has included getting involved in putting up sleds for the crash dummy testing that's part of the work of yet another Morton division, which produces air bags.

If there is a question whether a plant building's foundation will be able to adequately support a new piece of equipment, "we usually send it out-of-house," James says. "Otherwise, utility piping and process piping [plans] we usually do in-house. Also the electrical, usually working through the details with an electrical engineer."

In addition to flow sheet work, James and his department also generate charts, graphs, and mechanical designs; run prints of the various flow sheets for themselves and the engineers; and maintain a formal document control system for keeping track of the work that goes through the department. "The documents have to be accessible for future reference," James says, such as when an addition to the same building or process is planned. Right now document control is handled manually, but it will eventually be computerized. The department also maintains Morton's engineering library.

Designing with CAD means picturing something and translating it into the computer. "You have to be creative," says James. "Be able to visualize what you want to do. Think, 'I can try it this way, rotate it, and it will look like this.' . . . You have to be able to think and chew gum at the same time."

At Morton, the CAD technicians are welcomed—and expected—to contribute to design ideas. "Otherwise, you're just a drafter, who always has to be told what to do," James says. "To just sit there and copy—no way. I always encourage suggestions. Granted, a lot of it is experience. But a person must be aggressive, and they must be able to talk to other people. CAD technicians also have to learn how to take rejection and handle constant change. You can't just say, 'I'll do it once, and it's done'—that hardly ever happens. There are always changes to be made."

Morton's design/drafting department is one large room with partitioned areas for the technicians, each of whom has an AutoCAD 12 system. "Our system looks like a PC you'd see in a house or office, except that our graphic monitor is bigger, twenty-one or twenty-two inches, and we have fourteen-inch command monitors," James says. "We use a sixteen-button programmable 'puck,' which looks like a mouse and can be programmed with up to thirty-two commands. We always work in color: different components of a flow sheet are a different color; for example, the process piping might be blue, the electrical systems red, and so forth. The technicians work under regular office lighting. "At first, there was some problem with glare," James notes. "But now we have monitors that cut down on the glare, and it seems to be ok."

**To be a successful CAD technician, you should**

- Be interested in the mechanical design or structure of things
- Like math, technical drawing, computer graphics, and engineering ideas
- Have an eye for detail and a passion for getting the "little things" right
- Like to work at the computer for long periods of time
- Enjoy sharing your work with others and discussing design points

## have I got what it takes to be a CAD technician?

James started out as a pencil drafter many years ago and has been a witness to the evolution from manual to CAD design/drafting. "I picked up a lot of CAD information on the job, and by reading books and attending workshops," he says. Does he see an advantage in having manual-drafting knowledge? "Definitely," he says. "I think sometimes you have a better feel for what you're putting into the CAD system."

Although CAD systems have increased productivity and greatly streamlined work, they don't let a CAD technician sit back and go on "auto-pilot"—and they have limits. "For some work, CAD is great—especially repetitive work; for example, you can copy or mirror an image easily," James says. "Conceptualizing can still be painstaking; designing piping or layout, for example, can still be time-consuming on the CAD system. As you're thinking up ideas, you have to zoom in and out, erase the areas that need to be changed. By hand, I'm putting it down on the paper as I'm thinking."

"Also, people always say revisions are faster with a CAD," he adds. "That's true to a point, but there's still time involved—you still have to go in and erase, and redraw."

CAD technicians spend long stretches of time sitting in front of the computer, working alone. "They have to be able to work by themselves, after being given some initial guidance," says Jacobs of Belleville Area College. "You can't be the kind of person who likes to keep hopping up to talk to other people."

Technical drawings range from the simple to the extremely complex, but all must be exactly right. Therefore, CAD technicians must be patient, methodical, and have a passion for getting the "little things" right. "They have to be meticulous as far as details," Jacobs says. If you're fascinated by things like architectural blueprints, diagrams of machines or engines, even maps, this might be the line of work for you.

With increasing responsibilities, CAD technicians must be good problem solvers, think logically and analytically, and help improve the designs they work on. Good communication skills are vital, because CAD technicians often interact with engineers or designers, fellow technicians, and supervisors. Though it's not necessary to be a programmer or other computer whiz, you clearly can't be "computer-shy" in this job.

FYI

**Design vs. Drafting**

The designer develops the concept. Automobile designers, for example, ask questions like, should it be x feet long, or longer? How much leg room will it have? How much trunk space? If we use this type of suspension system, how much impact can the car absorb? What should the car's exterior look like? How about the engine?

The drafter makes a technical illustration of what the design should look like, working from the instructions of the designer. Depending on the project, the first set of illustrations might present only the basic design. Later versions might get quite specific, showing materials to use, exact specifications to build to, the identity numbers of standard parts such as pipes to select, and so forth. The last set of illustrations may be used by manufacturing or construction people to actually produce the product.

## how do I become a CAD technician?

The minimum requirement for becoming a CAD technician is a high school diploma. However, the days when you could find a good job without postsecondary schooling may be past. "Today, high school alone doesn't provide enough training," Jacobs says. "Many more people on the market have associate's degrees. Companies prefer to hire them over someone who needs a lot of on-the-job training, because it saves them time."

In many ways, designing and drafting combine both artistic and technical skills. "When I got out of high school, I was going to go into programming," recalls James. "But I also liked art. I ended up going in the direction of designing and drafting.

However, at the same time, it's not necessary to like or be good at art. "It's not 'art'; it's technical drawing," Jacobs says. "You don't need to have special aptitude in art to be a designer or drafter."

James graduated from a one-year drafting/design program at a trade school twenty-five years ago. Since then, he's picked up a lot of on-the-job training, especially in computer design and drafting. "But today you'll see a lot of people coming into the field with a two-year associate's degree," he says. "I'd steer someone to college or a technical school. It also depends on the company; requirements vary."

### education

#### High School

Some jobs are still available for those with a high school diploma, but not as many as in the past. The positions also may be the most basic; additional training will help the CAD technician qualify to take on greater responsibilities.

Whatever route you choose, getting a good background in technical subjects in high school will pay off later. "You'll be working with engineers, so take math and the physical sciences," advises Jacobs. Algebra, geometry, and trigonometry, plus physics, are ideal.

"Take math and science," James agrees. "And drawing or drafting, if it's available. Also English—good communication skills are very important." Other helpful classes include machine shop and electronics.

Feeling comfortable around computers is a must. "It's the way the industry's gone," James says. Today he rarely does the work by hand, "maybe once or twice a year," he notes.

Jacobs makes the case for getting a good basic education in high school, with plans to go on for more schooling. "High school can't cover all the disciplines that we can" in a two-year associate's degree program," he argues. "Students are best off fulfilling their general education requirements in high school, and then going on to a two-year program."

### Postsecondary Training

One- and two-year programs in CAD design/drafting and technology are available at technical schools and community colleges. Belleville Area College, for example, offers a two-year associate's degree or a one-year certificate. "But ninety-nine percent of our graduates opt for the associate's degree," Jacobs says. "It gives them a competitive advantage." Though they will specialize once they're out of school, students who go through Belleville's program get a background in CAD work in eight basic areas: basic drafting, machine drawing, architecture, civil (with an emphasis on highways), process piping, electrical, electrical instrumentation, HVAC, and plumbing.

Different schools emphasize different things in their curricula, though, so it's a good idea to do a little exploring to find the program that best suits your interests. There can be classes in electronics, for example, for things like designing and drafting printed circuit boards. Some programs focus on specific areas, like architectural engineering technology. A broader two-year program might include courses in drafting and basic engineering topics such as hydraulics, pneumatics, and electronics; courses in computer programming, systems, and equipment; product design; industrial and architectural drafting; and computer peripheral equipment and storage. Some may also require the student to complete courses in technical writing, communications, social sciences, and the humanities.

Good sources of information about programs are local technical schools and community colleges, as well as trade associations such as the Society of Manufacturing Engineers (SME). The American Design and Drafting Association (ADDA) has developed a list of thirty schools, available from the association, with design/drafting curricula that meet ADDA standards (*see* "Sources of Additional Information").

### Advancement Possibilities

**Chief design drafters (utilities)** oversee architectural, electrical, and structural drafters in drawing designs of indoor and outdoor facilities and structures of electrical or gas power plants and substations.

**Controls designers** design and draft systems of electrical, hydraulic, and pneumatic controls for machines and equipment, such as arc welders, robots, conveyors, and programmable controllers.

**Data processing managers** oversee the gathering, storage, and retrieval of some type of computer data.

**Electronics design engineers** design and develop electronic components, equipment, systems, and products, applying knowledge of electronic theory, design, and engineering. May use CAE and CAD systems to formulate and test electronic designs.

**Lead designers** head CAD design/drafting department or operation, responsible for supervising and training other CAD designers/drafters.

**Integrated circuit layout designers** design the layout for integrated circuits (IC) according to engineering specifications using CAD systems and utilizing knowledge of electronics, drafting, and IC design rules.

**Printed circuit designers** design and draft layout for printed circuit boards (PCBs) according to engineering specifications, utilizing knowledge of electronics, drafting, and PCB design.

After a two-year associate's degree, any further degree will be in another field. "There is no four-year degree for drafting," Jacobs says. Four-year programs of interest might include engineering, marketing, computer programming, or other areas. "Another possible path is technical education," Jacobs says.

Once on the job, it's important to keep up with changes in the field. James, for example, attends workshops from time to time, especially on computer topics. "That's a good idea for anyone," he says. "There's constant software upgrades and other changes, and keeping up can save you a lot of time." He is a member of ADDA and an AutoCAD users' group. He also reads CAD magazines like *CADence* and *CADalyst* as well as industrial trade periodicals like *Plant Engineering* and *Powder & Bulk Handling*.

## certification or licensing

Certification for CAD technicians is voluntary. Certification in drafting is available from ADDA, which invites members and nonmembers regardless of formal training or experience to participate in its Drafter Certification Program. The certification process includes taking a ninety-minute test of basic drafting skills.

Licensing requirements vary. According to Rachel Howard, executive director of ADDA, licensing may be required for specific projects—such as in a construction project, when the client (such as a hotel) requires it.

## internships and volunteerships

Drafting-related jobs can sometimes be found through internships, and many future employers will look favorably on applicants with this kind of experience. Jobs relating to other engineering fields, such as electronics or mechanics, may be available, and they offer the student an opportunity to become familiar with the kind of workplace in which he or she may later be employed as a technician.

## who will hire me?

Before joining Morton fifteen years ago, James's first job was with a manufacturer of spray guns, spray exhaust systems, and related equipment. He worked there for ten years, starting out doing primarily mechanical design/drafting. "But I've always done a lot of plant layout, too," he says.

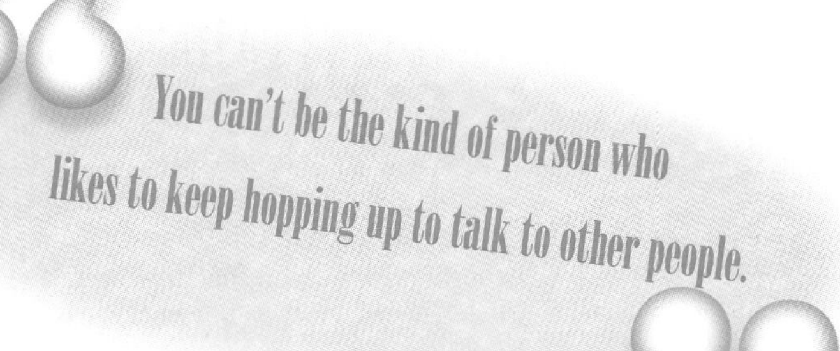


When he hires CAD technicians for the design/drafting department at Morton, James himself seeks creativity and a well-rounded design/drafting background. "It's harder to find people with well-rounded skills; they tend to be stronger in one area, like piping, and less strong in others," he notes. He encourages those considering the field to think about what area they'd like to work in, such as architectural, as they plan their career.

Drafting is so prevalent in so many areas that it would be impossible to list all of the opportunities here. However, a basic list of industries in which CAD technicians might work include the electrical, electronic, automotive, aeronautic, civil, mechanical, oil and gas, furniture, and construction fields.

Specialties are plentiful and interesting. *Aeronautical drafting*, for example, involves technical drawings for airplanes, missiles, and related equipment like launch mechanisms. A *commercial drafter* might create the technical illustrations for a store layout. *Furniture detailers* create the drawings used for the manufacture of chairs, sofas, and other furniture. *Oil and gas drafters* prepare technical plans and drawings for layout, construction, and operation of oil fields, refineries, and pipeline systems. There are also *technical illustrators* for textbooks and other printed materials who create drawings to show things such as assembly, installation, operation, maintenance, and repair of machines, tools, and other equipment.

Types of firms that use CAD technicians include construction companies, architectural firms, machinery manufacturers, engineering firms, electronics firms, electrical manufacturers, transportation manufacturers, communications manufacturers, consultants, high-technology companies, utility companies, and government agencies.

## where can I go from here?

The next step up from CAD technician might be designer, lead designer, or supervisor; opportunities will vary, depending on the company. Beyond designer/drafter positions, one possibility is to go back to school, earn a four-year degree, and become an engineer. You might stay within the same field: If you were working as an electrical or electronics CAD technician, for example, you might go on to become an electrical engineer. Engineering programs sometimes require a class or two in CAD, but most engineers are not extensively trained in this technology.

James's title at Morton is lead designer. In addition to his own work, he supervises the work of the other CAD technicians, generally starting them off on smaller projects when they first join the company, and gradually increasing their responsibility.

CAD technicians may first handle routine assignments, such as copying drawings or making minor revisions, and work up to helping to design and build equipment. Others get involved in concept and design right away. Titles for CAD technician positions include CAD specialist, computer designer, computer drafter, CAD designer, and CAD drafter.

Rather than becoming engineers, some technicians who continue their education and earn a bachelor's degree may become data processing managers or systems or manufacturing analysts. Other routes for advancement include becoming a sales representative for a design firm or for a company selling CAD, CAM, or CAE software, manufacturing services, or equipment. It also may be possible to become an independent contractor for companies using or manufacturing CAD or CAM equipment.

## what are some related jobs?

The U.S. Department of Labor classifies CAD technicians under the headings *Electrical/ Electronics Engineering Occupations, Civil Engineering Occupations,* (DOT) and *Engineering Technology: Drafting* (GOE). Students should check for the position of drafter under any of a wide range of industries, because most drafters today use CAD at least part of the time. These drafter listings include aeronautical, architectural, automotive design, automotive design layout, cartographic, castings, chief utility, commercial, detail, engineering, heating and ventilating, mechanical, oil and gas, and plumbing drafters. The DOT also has a separate category for drafters not listed elsewhere (017), such as furniture detailers.

| **Related Jobs** |
|---|
| Aeronautical drafters |
| Automotive design drafters |
| Auto-design detailers |
| Automotive design layout drafters |
| CAM technicians |
| Civil drafters |
| Civil engineering technicians |
| Commercial drafters |
| Detail drafters |
| Electrical drafters |
| Electronic drafters |
| Furniture detailers |
| Oil and gas drafters |
| Plumbing drafters |
| Structural drafters |
| Technical illustrators |

## what are the salary ranges?

According to Rachel Howard at ADDA, salaries vary by industry, by the part of the company in which the CAD operator works, whether it's a union shop or not, and other factors. Starting salaries for graduates of two-year technical programs typically fall in the range of $13,500 to $24,000. With increased training and experience, technicians can earn $25,000 to $45,000 a year. Some technicians with special skills, extensive experience, or added responsibilities may earn more.

Benefits usually include insurance, paid vacations, pension plans, and sometimes stock-purchase plans.

## what is the job outlook?

Overall, demand is growing for job applicants with a two-year associate's degree. As for specific jobs, opportunities are better in some industries than others—the healthier the industry, the healthier the prospects for the CAD technician. Manufacturers tend to like CAD because it helps boost productivity. There also is continuing interest in computer-aided manufacturing (CAM), with which CAD can be linked (*see* the chapter "Computer-Aided Manufacturing Technician").

While some say industrial designers and engineers will one day handle all design and drafting steps themselves, right now that doesn't seem too likely. "Engineers already have enough to do on their end," as James puts it. Currently, most engineering programs have just basic CAD training, if any. For now, technicians can make themselves most valuable to their employers by developing as much expertise in their industry as possible and contributing ideas, in addition to keeping up their CAD operator skills.

According to Jacobs at Belleville Area College, demand right now for that school's graduates seems to be best in civil, electrical, and mechanical areas. "Civil includes a lot of different areas, but especially highways," Jacobs says. Demand for the latter is up right now in many parts of the country, including Illinois, that need to rebuild their aging infrastructure, he notes. "When the tax dollars are there, these kinds of projects get under way," James says.

In addition to following your interests, prospective CAD technicians should take a hard look at their market and see what types of local companies use CAD and are hiring. "A lot of students like architecture when they first start the program, but that profession is pretty full right now," Jacobs says. He notes that some areas, like Belleville, don't even require drafted plans for building new houses. "By the time they graduate, students know they have to go where the jobs are."

**Sample list of ADDA-certified drafting programs:**

- architectural/CADD
- architectural drafting
- architectural engineering technology
- CADD
- computer aided design drafting technology
- design drafter
- design and drafting technology
- drafter
- drafting/CAD technology
- drafting technology
- drafting technology industrial management
- electro-mechanical/CADD
- engineering designer
- engineering design technology
- engineering drafting technician
- mechanical design engineering technology
- mechanical design technology
- mechanical drafting
- mechanical drafting and design technology
- mechanical technology

# how do I learn more?

## sources of additional information

Following are organizations that provide information on CAD careers, accredited schools, and employers.

**American Design and Drafting Association (ADDA)**
4709 Levada Terrace
Rockville, MD 20853-2261
301-460-6875

**Institute of Electrical Engineers**
1828 L Street, NW, #1202
Washington, DC 20036-5104
202-785-0017

## bibliography

Following is a sampling of materials relating to the professional concerns and development of CAD technicians.

### Books


Bone, Jan. *Opportunities in CAD/CAM Careers.* Lincolnwood, IL: VGM Career Horizons, 1994.

Turbide, David A. *Computers in Manufacturing.* New York: Industrial Press, Inc., 1991.

Valliere, David. *Computer-Aided Design in Manufacturing.* Englewood Cliffs, NJ: Prentice Hall, 1990.

## Periodicals

*CADalyst.* Monthly. Features the most recent advances in auto CAD systems and CAD applications. Advanstar Communications, 859 Willamette Street, Eugene, OR 97401, 503-343-1200.

*CADence.* Monthly. Directed at users of Autocad in the fields of construction and architecture. Miller Freeman, Inc., 600 Harrison Street, San Francisco, CA 94107, 415-905-2200.

*Computer Aided Design Report.* Monthly. Newsletter focusing on the application of computers by engineers in manufacturing design. CAD/CAM Publishing, Inc., 1010 Turquoise Street, Suite 320, San Diego, CA 92109-1268, 619-488-0533.

*Design Drafting News.* Monthly. Newsletter published for association members. American Design Drafting Association, Box 799, Rockville, MD 20848-0799, 301-460-6875.

# computer-aided manufacturing technician

**Definition**

Computer-aided manufacturing technicians operate computer-controlled machines, equipment, and systems in the manufacture of products.

**Alternative job titles**

CNC programmers

CNC setup/operators

Manufacturing engineers

Manufacturing technicians

**Salary range**

$18,000 to $30,000 to $40,000

**Educational requirements**

High school diploma; two-year technical training program; apprenticeship

**Certification**

Recommended

**Outlook**

Little change or more slowly than the average

GOE
05.05.09

DOT
609

O*NET
22514A

## High School Subjects

Computer science
Mathematics
Physics
Shop (Trade/Vo-tech education)

## Personal Interests

Building things
Computers
Figuring out how things work

**The call came** close to midnight. Tom Stryczek was reaching for his coat. Instead he reached for the phone.

"Machine Shop Three," he answered.

"Tom? Manipulator's down. Cylinder's cracked." It was one of the crew at the press, where 40,000-ton rods of hot, fresh steel were first forged into shape.

"Got it", Tom said. "Eighteen-degree arc inside, right?"

He flipped on the computer and called up the NC program. From a cabinet he pulled the cylinder's blueprints. But he knew this machine. He knew all the machines at A. Finkl & Sons, Chicago's oldest steelworks. He had spent time on each of them over the years he'd been working there. Tom swept the mouse across the screen, writing the program to guide the machining of the new cylinder. It had to match the specs exactly. There was his reputation, there was Finkl's reputation. But there was more than this. Without that cylinder, work would stop. Men would be sent home. It was up to Tom now.

He was already running the simulation, testing his program. He calculated the hours it would take to machine the part.

Tom cradled the phone to his ear. "You'll have it before the morning."

## what does a CAM technician do?

*Computer-aided manufacturing (CAM) technicians* are typically employed in two different phases of the manufacturing process. Some CAM technicians are involved in the initial design and setup of the process, organizing the materials, parts, and equipment to be used, as well as routing the various steps a product will take as it is manufactured. Other CAM technicians are more directly involved in running, maintaining, and repairing computer-controlled equipment. In smaller facilities, a CAM technician may be responsible for both phases of the CAM process. More experienced CAM technicians may also assist in the design, building, and testing of prototype products, machines, and robots.

CAM technicians' titles vary according to their actual roles but include *CNC (computer numerical control) programmers* and *numerical control tool programmers, CNC setup/operators,* and *manufacturing engineering technicians.* Generally, CAM technicians work under the supervision of engineers and designers, who initiate product designs and determine the types of materials needed, the specifications of different aspects of the product, and what types of machines must be used to manufacture the product. The engineers and designers draft blueprints of the product from which the CAM technician will work. The CAM technician acts as a link between the design phase and the production phase of the manufacturing process.

When the CAM technician receives the blueprint either on paper or from a file in the computer system, he uses specially designed computer software that allows him to translate the specifications on the blueprint to computer programs that will guide the various machines needed to perform each of the cuts, drillings, punchouts, and other work required. The technician enters specifications by inserting numbers and codes into the software program (more recent systems also make use of pull-down menus and a graphical interface as seen in many popular personal computer programs). With a graphical interface, the technician is able to enter the specifications and then run simulations that allow him to "see" how each machine will perform the required work. In this way, the technician is able to refine the program until he is certain that the product will exactly meet its specifications. Using computer simulations to test the machine programs, rather than using the actual parts and materials, is useful in avoiding costly mistakes.

After the program has been created, it is transferred, either by downloading directly or on punch tape, magnetic tape, or disk, to a central controller or to controllers attached to the different machines. These machines may include lathes, punch presses, stamping presses, milling and grinding machines, and lasers and may be operated individually, or linked together to perform as a group. CAM technicians who specialize in this part of the manufacturing process are often called *CNC programmers* or *numerical control tool programmers.* Their work may involve smaller production runs, or even

**Automation** The automatic operation or control of systems, equipment, or processes.

**Computer-aided manufacturing (CAM)** The use of computers to design, initiate, direct, control, and link the various areas of the manufacturing process.

**Computer-integrated manufacturing (CIM)** The system used to link together the manufacturing systems with design systems and business functions within a manufacturing facility.

**Computer numerical control (CNC)** Numerical control technology adapted to serve an individual machine tool through a dedicated control panel incorporated into the machine tool.

**Lathe** A machine that shapes metal or wood by high-speed rotation against a fixed cutting edge.

**Numerical control (NC)** Allows machine tools and other equipment to be operated using automated equipment by inserting codes, numbers, and other data into computer software programs.

**Specification** A detailed description and/or diagram of materials, measurements, and other qualities necessary for something to be manufactured, built, or installed.

producing individual parts. They must be familiar with all of the machines involved in the manufacture of a product, and with the computer-controllers that guide their movements.

Some CAM technicians work directly with these machines. Generally called CNC setup/operators, they are responsible for monitoring the machine's movements, and for the maintenance and repair of the machines. Because the machines operate at very high speeds, the technician makes certain that they are properly cooled and lubricated. Technicians may program the machine using a dedicated controller that is attached to the machine. They are also able to start and stop a program when that is necessary, such as when the part being machined must be repositioned. At times, they must also make changes to the program to accommodate for variations in the raw materials. For example, when one block of steel among several blocks to be lathed is longer or wider than the others, the technician can reprogram the machine to recognize these differences. These changes can be made on the controller attached to the machine. In this way, each part leaving the lathe will be exactly the same.

CAM technicians with the proper education, experience, and skills may oversee the entire manufacturing process. Called manufacturing engineers or manufacturing technicians, they are generally involved in large production runs, that is, in guiding many of the same products through the different steps of the manufacturing process. These technicians also begin with blueprints, programming each machine necessary to produce the product. They then use special software to plot the most efficient path the product will take along the automated production line.

The CAM technician enters a variety of data in the form of numbers and codes. These data include information on the type and amounts of materials to be used, types of machines needed, the programs to operate the machines, the amounts of lubricants and other coolant materials the machines will need, and the machines' availability and other conditions on the shop floor. The computer software interprets the data and directs the different aspects of the manufacturing process.

To achieve the greatest efficiency, information from other areas of the manufacturing facility are incorporated as well. The CAM technician includes the company management's requirements for the number and speed of production, and links the production process to the company's inventory, that is, the availability of parts and materials. Comprehensive manufacturing systems such as this are called computer-integrated manufacturing (CIM) systems. CIM systems link all of the manufacturing and business functions needed to produce a product, including the design and drafting phases, and the business-related areas of accounting, purchasing, and forecasting.

> "You have to be satisfied with yourself. Someone's not going to come along and pat you on the back. Making a good part is what you're supposed to do."

When all of these variables are accounted for, the computer software can initiate and direct the production process. The use of computers also allows changes to be made easily, speeding up or slowing down production, controlling the amounts of each product being produced, or allowing for new designs, or for variations and modifications in the design of products. The CAM technician uses data generated throughout the process to provide reports to management personnel. These data are essential for keeping the automated manufacturing equipment and the manufacturing process running smoothly and efficiently.

## what is it like to be a CAM technician?

Tom Stryczek is a machine shop foreman at A. Finkl & Sons, the oldest steelworks in Chicago, and he has been responsible for CNC programming since it was instituted there twelve years ago. "I'd been working there part-time while going to college when they converted some of the machines to CNC," Tom says. "Because I knew a little about computer programming from courses I'd taken, they offered me the full-time job as programmer. Since then, I've trained most of the operators on it, too. But I pretty much trained myself, using the manuals, some trial and error. Now I can write a program almost without thinking about it."

Tom's knowledge of the equipment he uses is an important part of his success on the job. "I do more here than only programming, which may be unusual. At other places, a CNC tool programmer may do only that. Here, I supervise the machinists, and I also run the machines myself, sometimes. And I've probably worked in every area of A. Finkl & Sons, from smelting the ore to administration. I know the process really well, I know the machines. That helps me do my job."

Typically, Tom arrives at work at 6:30 AM and leaves at 3 PM. "Although a lot of the time I'm here until five. And I'll sometimes work six or seven days a week. When there's work, we work. Sometimes we'll even go twenty-four hours. But I was raised to work hard at what you do, to give your best at all times. And I'm always aware of Finkl & Sons' reputation for being the best. I want to make sure it stays that way."

When Tom receives an order for a part, he studies the blueprints, performs the necessary calculations, then creates the program on the computer. "It's become a lot more user-friendly over the years. At first, I'd have to type in all the instructions with the keyboard. Now I'm able to use a mouse," Tom says, as he pulls down menus and chooses features that create a graphical representation on the computer screen of the part he's working on. "Most of the parts we produce at Finkl are what we call rough cuts, where we'll trim a block of raw steel to the initial shape a client needs. Then the client will perform the more intricate cuts. So usually it takes me five or ten minutes to write the simpler programs. Some parts, I'll spend forty-five minutes or an hour on. But it's always important that I'm precise, exact in what I do."

After he has written the program, Tom delivers it to the machinists, or loads the controller himself. "We still use paper tape, rather than magnetic or diskette, or downloading directly. That's partly because at a steelworks there's always a lot of dust to deal with. Because the controller reads the tape with light beams, there's less of a risk that something will go wrong. And we store the program in the computer, so we can always reproduce the tape if another order for the same part comes in."

**To be a successful CAM technician, you should**

- Be able to think logically and have good analytical skills
- Be methodical, accurate, and detail-oriented
- Be able to work independently and as part of a team
- Have solid mathematics, computer, and machine shop skills
- Have strong communication skills

Once the program has been installed, and the steel to be worked has been placed on the machine, the machinists perform adjustments to the program on a CNC panel to account for variations in the raw product. "We'll get ten pieces, some of which may be longer or thicker than the others," Tom says, "But by the time they're delivered to the client, every part will have exactly the same measurements."

Lenny Payne is a manufacturing engineer at Rockford Power Train in Rockford, Illinois. The title of "manufacturing engineer" should not, however, be confused with an engineer holding a four-year bachelor's degree in engineering. Such engineers are mostly involved in the design and development phases of a product, rather than in its actual production. Lenny, on the other hand, is a good example of a CAM technician who is responsible for guiding a product through the entire production process. "My duties include the CNC programming of the newly designed part or product, and writing the process plans to guide the part through the shop," Lenny says. "I set up the computer systems, and I troubleshoot problems on the shop floor. Like if one of the operators is having a problem getting the right tooling set up on a machine, I'll go out on the floor to help him."

Lenny, like most CAM technicians, works in an office located near the shop floor. Because most of his work is done on computer, his office is insulated from the noise and dust of the machines. "I know a lot of young people think manufacturing work is dirty and cumbersome, but it's really not like that," Lenny says, "Most of today's factories are clean and safe." Windows allow the CAM technician to keep an eye on the activity on the shop floor. When problems arise, the CAM technician must be ready to assist the shop personnel in solving the problems. "And I have to be quick," Lenny says, "Because every minute the machine is not producing parts, money is being lost."

Lenny typically arrives to work at 7 AM. Usually, he will find blueprints waiting for him. "It's my job to turn the blueprints into computer programs. Sometimes I'll have fifteen different designs to program before the end of a week. And it can take most of the morning just to program one design. I have to figure out exactly what it will take to take the part from raw steel, through the machinery, and to the final finished product." Using CNC software, Lenny programs the machining work needed for each aspect of the part. He then writes the computer process plan that will guide the part through the factory.

"The computer system lets me follow the part from the raw stock in inventory to the finished product going out the door to the customer," Lenny says. "The computer represents on screen what's happening on the shop floor. The hard part is done by the computer; what we do is feed in the information the computer needs."

Lenny's workday is typically a busy one. "Sitting in front of a computer is not all I do. And sometimes it's so hectic around here, I can't believe it when the day is over it's gone by so fast."

Raymon Avery is also involved in computer-aided manufacturing. At twenty years old, he is just beginning his career as an apprentice at Ingersoll Milling Machine in Illinois. "I entered the program right out of high school. I go to school in the morning; I'm going for an associate's degree in Automated Manufacturing Technology. By four o'clock, I'm at work at Ingersoll, where I'll work until 9 PM." As part of his apprenticeship, Raymon is involved in actually running each of the machines and producing the parts. "For the first couple of months, all I was allowed to do was watch. But depending on the operator, I'd sometimes be allowed to run the machine. After that, I've trained on the machines, like NC lathes, and manual lathes too, which is fun. But I'd rather be in the office programming," Raymon says, "That's why I'm going to school."

Raymon's apprenticeship training requires him to not only work in the machine shop but also to learn how to assemble the machines. In school, Raymon learns the principles of CNC programming, and he is also learning the basics of computer-aided design. "I love CNC programming," Raymon says, "And I also get to use the mainframe for designing parts. You want to get all the skills you need."

## have I got what it takes to be a CAM technician?

Tom, Lenny, and Raymon all agree that a strong attention to detail is essential for being a good CAM technician. "You can't glaze over the surface," Lenny says. "Everything has to be exactly right, because the machine will only do what you tell it to do. And if you tell it the wrong thing, then that's what it'll do." As Raymon says, "If you program a part wrong, it can cost the company a lot of money. You have to make sure you're doing your job accurately."

All three men take a lot of pleasure and pride in their work. "It's very satisfying to take a part from a piece of paper, giving the instructions on how to make the part, and seeing the part finished," Lenny says. "I love what I do. And you have to be satisfied with yourself around here, because someone's not going to come along and pat you on the back. Making a good part is your job. It's what you're expected to do." For Raymon, the different aspects of his apprenticeship training are very rewarding. "I think manufacturing is challenging. You learn something new everyday. Knowing things like why a part fits where it does. Calling up and creating a part and sending it where you want it to go. Choosing the right codes for the CNC program. You have to figure out how to get the part out right."

Math and computer skills are an important aspect of the job. "To create a drawing," Raymon says, "you need math. A good math background really is key." Lenny agrees: "Every day I'm using geometry and trigonometry, and basic adding and subtracting. But you don't have to be a scientist or a computer genius to do this job."

A strong knowledge of blueprint reading and drafting is important. "I can look at a blueprint of a part and already know how that part will look once it's produced," Tom says. "I see the part before it's even made." Tom enjoys his work. "I like seeing things being made, knowing that I've been responsible for it, seeing the finished product. I like knowing that a piece of steel I've worked on may be part of the car I drive, or it might have been a part needed to build my car. A lot of jobs don't offer that."

Lenny, who is not part of a union, says: "The hardest part of the job can be dealing with the union people. Sometimes the union shop workers resent your presence in the factory. They think that shop people belong in the shop and office people belong in an office. And there are a lot of different personalities on the shop floor. To do this job, you have to have people skills."

Combining school and on-the-job training is a difficult part of Raymon's apprenticeship program. "The stress can get to me sometimes," he admits, "You really have to be level-headed. If you make one mistake, it could cost you big! I'm always afraid that'll happen to me. If you ruin a part, if you scrap one out, you're losing $300 to $400." And learning CNC programming can be another source of stress. "You're not sure if a program is going to run or not. Still, I love my CNC programming classes. It's what I really want to do."

Working with others is another source of satisfaction for Raymon. "There's a lot of teamwork. Everyone helps me out." By the time Raymon arrives home from a long day of school and work, he's usually very tired. "Sure, being an apprentice and going to school can be stressful," he says, "But it's interesting and I really enjoy it. I wouldn't be doing this if I didn't like it."

## how do I become a CAM technician?

Tom Stryczek graduated with a bachelor's degree in marketing and advertising. "But I'd already been working at A. Finkl & Sons," he says. "And I'd taken computer programming courses. In high school, I'd taken a lot of shop courses, too. I knew how to read blueprints, mechanical drawing, how to take dimensions and measure tolerances. So when they asked me to become their programmer, I was really prepared."

Lenny Payne has worked in manufacturing for more than twenty years. He formerly worked as a production controller in the industrial engineering department of a machine shop. There, he received on-the-job training. But when Lenny was laid off from this job, he found it difficult to find other work as a programmer/production controller, despite his years of experience, because he lacked a formal education. That is why he returned to school for an associate's degree in Manufacturing Technology. As Lenny says: "It's hard to find someone to employ you without an associate's degree. And once you start school, your technical classes will let you know if you'll enjoy the work."

Raymon Avery, on the other hand, enrolled in his four-year apprenticeship program right out of high school. This means he spends his morning in classes for his associate's degree, and the afternoons and evenings at work in on-the-job training. He entered the field in part because of his family's background in manufacturing. "My father is an inspector, so I think that played a part in my choice. Also, I worked in my cousin's machine shop when I was sixteen."

Raymon recommends an apprenticeship program for students interested in this career. "If you're a junior or a senior in high school, try to get an apprenticeship. People won't hire you without training, unless they want to take the time to train you themselves. And you have a better chance to make more money than someone without that experience and without a degree."

## education

### High School

A high school diploma is required for technician positions in larger manufacturing companies. Smaller companies may not require a diploma, and may offer on-the-job training. However, a formal education will greatly increase a prospective CAM technician's chances for finding work.

High school students should focus on classes in algebra, geometry, trigonometry, physics, computers, and any shop or technical training classes offered through the school. Other classes that will be helpful for students entering this career include classes in mechanical drawing and blueprint reading. Because CAM technicians must be able to read specifications and write reports and other materials used in the manufacturing process, students should acquire good English skills, including grammar, spelling, composition, and reading comprehension.

Membership in high school clubs and other extracurricular activities, such as computer and electronics clubs, and participating in science fairs will provide good experience, as well as a background in and an understanding of some of the requirements of the field. Hobbies that include electronics, drafting, computers, mechanical equipment, and model building are ways for the prospective CAM technician to develop many needed skills. Reading books and articles on manufacturing and related areas will help familiarize the student with technical aspects of the field. Part-time and summer jobs in the manufacturing industry will also introduce high school students to the field.

**Advancement Possibilities**

**Manufacturing engineers** plan, direct, and coordinate manufacturing processes in industrial plants.

**Production engineers** plan and direct production procedures in industrial plants.

**Quality control engineers** plan and direct activities concerned with development, application, and maintenance of quality standards for industrial processes, materials, and products.

## in-depth

### Machine Tools

Machine tools are power-driven devices that shape metal by removing pieces such as chips, flakes, or grains. They produce the parts used in the construction of machines of all types. These parts range from nuts and bolts to engine blocks, from piston rings to railway wheels. Thousands of identical parts can be produced on a single machine.

Some machine tools are used to finish metal pieces produced by other methods, such as casting or forging; others are used to produce various shapes from bars or sheets of metal. Machine tools range in size from large immovable models to small types that would fit on a card table. Most are driven by electric motors.

There are seven basic types of machine tools: lathes, milling machines, planers and shapers, drilling and boring machines, grinding machines, sawing machines, and broaching machines. Many machine tools are combinations of two or more of these general types. Some shape metal by other means as well, such as by hammering or shearing. Some are designed to work on many different kinds of parts, while others are designed to produce one specific kind of part. A machining center is a computer-controlled machine tool that can perform a variety of machining operations on one piece of metal.

Lathes shape metal by rotating it against a cutting tool. Milling machines shape metal by making cuts in it with a rotating many-edged tool. Planers and shapers are used chiefly to finish flat surfaces. Drilling machines are used to cut or enlarge holes. Grinding machines are usually used to refine the surface of an object that has been shaped and rough-finished by another method. Sawing machines are used to cut off pieces of metal. Broaching machines are used to shape existing holes or to contour outer surfaces.

An automated machine tool is one that is controlled by another machine, rather than by a person. The controlling machine, in turn, is operated electronically by a magnetic tape similar to the kind used with tape recorders. The tape carries signals that operate mechanisms that control the machine tool.

One method of making the controlling tape is with a special machine tool operated by a machinist. His or her actions are recorded as electronic signals on the tape. Numerical control is another method, which uses computers to translate the dimensions of a piece of work into coded instructions on the tape. An automated machine using numerical control is far more accurate than one run by a person. It can produce simple shapes faster and more economically. It can machine in one solid piece a complex shape that, with conventional methods, must be produced in several pieces.

### Postsecondary Training

Many community colleges and technical trade schools offer associate's degree programs in manufacturing technology. An associate's degree will make a big difference in finding employment in this field. In such a program, the student can expect to study blueprint reading, shop math, CNC machine setup/operation, fundamentals of CNC programming, computer integrated manufacturing (CIM), and reading mechanical and industrial drawings. These programs will also require the completion of general education courses, including composition, business writing, humanities, mathematics, and speech.

Apprenticeship programs such as Raymon Avery's provide another path into the field. Such programs combine the formal education of an associate's degree program with on-the-job training. Apprenticeship programs are usually sponsored by a manufacturing corporation, and will often include financial aid for the apprentice's education. In Raymon's case, his company compensates him based on the grades he achieves. "If I get B's, they'll pay 70 percent of my tuition and books," Raymon says, "But if I get A's, they pay 80 percent. It's a great program."

Apprentices also receive certificates for each course they pass, and this certification can be useful while looking for work. Still, Raymon says, "You'll be much more competitive in the job market if you have an associate's degree."

Students interested in apprenticeship programs should speak to their high school guidance counselors or contact local community colleges and technical trade schools for more information on what may be available in their area.

A career in CAM technology may require continuing education even after the technician enters the work force because of advances in technology, new equipment, and new processes.

## scholarships and grants

The Society of Manufacturing Engineers offers scholarships for students considering a career in computer-aided manufacturing (*see* "Sources of Additional Information"). Students should write early for information on eligibility, application requirements, and deadlines. High school guidance offices and the admissions department at community colleges and trade schools may be able to provide more information on available grants and scholarships in the student's area.

## labor unions

Some CAM technicians may be asked or required to join a labor union, depending on where they work. At a Big Three automaker, a technician will join the United Auto Workers union. Other unions include the United Steelworkers, United Paperworkers, Teamsters, Sheetmetal Workers, and Metal Processors unions, among others.

# who will hire me?

Lenny Payne was first hired as a temporary employee while completing his associate's degree. A friend who works for the same company recommended him for the position. Lenny worked hard, and soon after he was hired on for full-time, permanent employment.

Tom Stryczek was promoted into his programming position after working at A. Finkl & Sons for several years.

Raymon Avery applied for the apprenticeship program through his local community college, where he found a listing of employment opportunities for high school and community college graduates.

Computer-aided manufacturing technicians are needed in many areas of industry, including architecture, manufacturing, high-technology industries, and government agencies. The Big Three automakers, Ford, Chrysler, and General Motors, are major employers of CAM technicians. The increasing use and application of high technology in the manufacturing process has created a demand for skilled technicians in nearly every area of manufacturing and industry. In addition, a CAM technician can work anywhere that computers are used.

One source for entering the field is through a school's job placement office. Companies employing CAM technicians often visit campuses to recruit students nearing graduation.

Other sources for employment leads include newspapers, employment agencies, and trade associations. Many associations publish their own newsletters and trade journals, and these are an excellent source for locating job openings nationwide (*see* "Bibliography").

# where can I go from here?

CAM technicians who do well at their jobs can expect promotions to greater responsibilities. They may work on the design staff, or in more demanding areas of manufacturing. Experience with the manufacturing systems will allow them to take on the important responsibilities of troubleshooting problems. A CAM technician may also be promoted to supervisory jobs, or become involved in training, where they may teach courses in the workplace or at local schools.

"After starting as NC programmer, I was promoted to foreman, then eventually to line foreman—first on the night shift, and now in day shift," Tom says. "I'm in charge of the machine shops, and the machinists, with a lot of supervisory duties."

There are twelve CAM technicians employed at Lenny's workplace. "I hope that my associate's degree will make me eligible for a management or supervisor's position," Lenny says. "With my degree, I have a better chance of advancement in my field." Raymon intends to work for a while after he gets his degree. "My two years at community college can transfer to a four-year engineering program," says Raymon. "If I decide to, I always have the option of returning to school."

Advancement may also mean that the CAM technician's duties become less and less routine. They may find themselves actually designing and building equipment. Technicians who elect to continue their education can advance to other areas of the manufacturing industry, becoming data processing managers, engineers and designers, and systems or manufacturing analysts. They may also begin a sales career as a representative of companies designing, building, selling, or servicing manufacturing technology.

## what are some related jobs?

There are many jobs related to CAM technology, including the variety of manufacturing positions using CAM technology, such as CNC programmers, numerical control tool programmers, and CNC set up/operators. The U.S. Department of Labor classifies numerical control machine operators and set up operators under the headings *Metal Machining Occupations, Not Elsewhere Classified* (DOT) and *Production Technology: Machine Set-up and Operation* (GOE). Also listed under these headings are occupations involved with shaping metal (and sometimes wooden) parts or products by removing excess material from stock or objects, such as gears, gauges, guns, pins, and screws.

Jobs involving computer programming in general, as well as data processing technicians, drafting and design technicians, computer-aided design technicians, and electrical and electronics engineers, are also related to CAM technology.

| **Related Jobs** |
|---|
| Graphics programmers |
| Industrial designers |
| Industrial engineering technicians |
| Robotics technicians |

## what are the salary ranges?

CAM technicians are normally paid by the hour, and receive overtime pay for work over forty hours a week. Wages vary greatly depending on education, experience, years with the same company, job title, and geographic region. A CNC programmer with five or more years of experience can earn up to $40,000 a year. An apprentice working forty hours a week, however, may make only $14,000 a year. Most employees will receive some benefits, but these can also vary widely, according to the size of the company and union membership.

## what is the job outlook?

More and more manufacturers are turning to computer-automated systems in an effort to remain competitive, and profitable, in the world economy. Many manufacturers report difficulties finding skilled employees. This translates into good job potential for educated, trained, and experienced workers.

In recent years, also, the economy has been doing very well. Manufacturers are posting record profits, and the housing market has been booming. Growth in the automobile and housing industries means growth in many other areas of manufacturing and industry. Builders of machinery and suppliers of parts to these industries benefit because they will receive more orders for their products. The boom in the housing market means that the suppliers of machines and parts necessary for construction will benefit from increased demands for their products. New houses and buildings also signify a potentially higher demand for the furniture, appliances, and other household and office fixtures needed to furnish them. All of this increased demand means an increased demand for skilled workers to manufacture these products.

On the other hand, the increasing adaptation of technology to manufacturing will mean fewer jobs for unskilled and low-skilled workers. Employers will continue to seek employees who will help them maintain high-levels of quality and a competitive position in their industry. Unskilled workers will increasingly be replaced by skilled workers. Several automakers, for example, have announced plans to replace as much as half of their unskilled work force in the next few years.

One further implication of the trend among manufacturers to strengthen their competitive edge is a resulting increase in competition among skilled workers. More college graduates and people with higher degrees may be employed even in lower level positions. Companies will attempt to hire only the best workers, and many employers will hire on the basis of competitive tests. For these reasons, a prospective CAM technician should concentrate on gaining not only experience but also a strong education.

# how do I learn more?

## sources of additional information

Following are organizations that provide information on CAM technician careers, accredited schools, and employers.

**American Design Drafting Association**
4709 Levada Terrace
Rockville, MD 20853-2261
301-460-6875

**Institute of Electrical Engineers**
1828 L Street, NW, #1202
Washington, DC 20036-5104
202-785-0017

**Robotic Industries Association**
PO Box 3724
900 Victors Way
Ann Arbor, MI 48106
313-994-6088

**The Precision Machined Products Association**
6700 West Snowville Road
Brecksville, OH 44141
440-526-0300

## bibliography

Following is a sampling of materials relating to the professional concerns and development of computer-aided manufacturing technicians.

### Books

Bone, Jan. *Opportunities in CAD/CAM Careers.* Lincolnwood, IL: VGM Career Horizons, 1994.

### Periodicals

*CAM-I Cameos.* Bimonthly. Newsletter featuring reports on computer-aided manufacturing and the field of robotics. Computer Aided Manufacturing International, 3301 Airport Freeway, No. 324, Bedford, TX 76021-6032..

*Computer Aided Design Report.* Monthly. Newsletter focusing on the application of computers by engineers in manufacturing design. CAD/CAM Publishing, Inc., 1010 Turquoise Street, Suite 320, San Diego, CA 92109-1268, 619-488-0533.

# conservation technician

**Definition**

Conservation technicians restore artistic and historic objects, such as pottery, statuary, and tapestries, to their original or natural appearance.

**Alternative job titles**

Armorer technicians
Ceramic restorers
Lace and textiles restorers
Paintings restorers
Paper and prints restorers

**Salary range**

$18,000 to $35,000 to $60,000

**Educational requirements**

Some postsecondary training

**Certification or licensing**

None

**Outlook**

About as fast as the average

GOE 01.06.02

DOT 102

O*NET 24302B*

## High School Subjects

Art
Chemistry
English (writing/literature)
History
Shop (Trade/Vo-tech education)

## Personal Interests

Drawing
Fixing things
Painting
Reading/Books
Science

valuable, and falling apart. Karen Clark focused on the task at hand, using thread to bind the pieces back together, but the job seemed to keep getting bigger as she worked. "For every row I stitched together, that area became stronger, so a nearby area would start to deteriorate very quickly," she recalls.

Then the problem was compounded as an adhesive on the back of the rug suddenly came through to the front. Karen spent hour after hour patiently scraping away the adhesive, restoring the old rug to its original beauty. As usual, the project was a learning experience and a test of her skills and creativity.

"Nothing comes in here quite the same. You may get five rugs, and they may be similar, but each one was made in a different way. They come from all over the world," she notes. The rug might have been brought back from a European vacation, or it might have been in the home of a local family for generations.

Like most of the items Karen helps stabilize and restore, the rug was unique, but she didn't allow herself to worry about damaging it as she worked. "Whatever you do is reversible," she kept reminding herself. "You can always fix it."

## what does a conservation technician do?

Without proper care, vintage clothing, pottery, rare books, and other valuable items begin to deteriorate and could be lost forever. Many of the treasures in private collections and in institutions such as museums, art galleries, and libraries have been stabilized and repaired by *conservators* and the *conservation technicians* who assist them. A technician focuses primarily on stabilizing and restoring objects but might also construct part of a display for it, such as a backboard for a feather headdress or a frame for a painting.

Research, artistic skills, and scientific procedures all play a part in the conservation of artwork and objects of historical or cultural value. Conservation technicians gather information to learn how an item originally appeared. They sometimes conduct chemical and physical tests to determine its age and composition. Then they develop a plan for cleaning, stabilizing, and restoring the object.

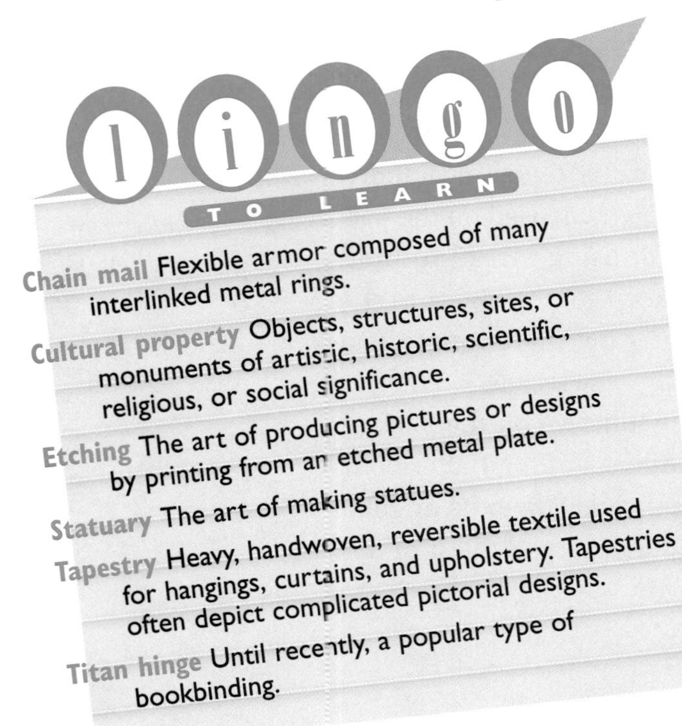


Different types of items require different care. A textile might be dry cleaned, but a suit of armor might be cleaned with chemical solvents. When making repairs, conservation technicians carefully sew, solder, or glue broken pieces back into place. They may also repaint objects, using materials of the same chemical composition and color as the original.

Conservation professionals frequently specialize in one type of object. *Armorer technicians* clean, repair, and determine the authenticity of medieval armor and arms. *Ceramic restorers* work with objects made of glass, porcelain, china, clay, or other ceramic materials. *Lace and textiles restorers* use their knowledge of weaving and sewing to clean and repair tapestries, clothing, lace, flags, and other fabric items. *Paintings restorers* clean, stabilize, and repair paintings. *Paper and prints restorers* work with paper objects of historical and artistic value, including books, documents, maps, and prints.

Conservation technicians usually work under the supervision of conservators. Other professionals in the field include *conservation scientists*, who develop varnishes and other materials used by conservators; *conservation educators*, who teach the principles, methods, and technical aspects of the profession; and *preparators*, who ensure that objects are displayed in a way that will protect them from deterioration.

## what is it like to be a conservation technician?

Karen Clark is learning to be a conservator through a sort of informal apprenticeship at Jessica Hack Textile Restorations, a privately owned studio in New Orleans, Louisiana. The business stabilizes and restores textiles and related items for museums and private clients.

Karen works with several other professionals, each bringing a specialization to the team: a master tailor, a master weaver, an accomplished quilter. Karen is most interested in lace and wedding veils, but she has restored bridal gowns, vintage Mardi Gras costumes, antique flags, hats, shoes, and even decorated boxes from the seventeenth century. "We work on whatever comes in," she says.

Karen usually starts by vacuuming the object, and she might prepare it for washing or dry cleaning. She performs any necessary repairs, which might require that thread or fabric be dyed to match the old fabric exactly. Then the item is pinned and sewed together; in some cases Karen uses an adhesive. She might cover a board with fabric to serve as a backing against which the item can be displayed, and if the object is an article of clothing, she might prepare a harness to hang it inside the display case.

She commonly spends several days on one project. Throughout the process, she documents her work with photographs and written reports for her own information and for the use of other conservators who might work with the object in the future.

Detailed reports are one of a conservation professional's primary duties, according to C. R. Jones, a curator-turned-conservator who has been in the profession since 1977. Another primary concern is monitoring the environment in which the object is stored or displayed, to ensure that the temperature, humidity, and other conditions will not damage the item.

C. R. is the conservator of the Farmers' Museum and the New York State Historical Association in Cooperstown, New York. He sometimes has help from volunteers, but usually he performs all the conservation duties for both organizations. His specialties are paintings, flat paper, and objects made of metal, wood, stone, or ceramics.

"One of the first things you have to do is figure out the anatomy of the object," he says. For that, he uses X-rays, ultraviolet rays, infrared scanners, microscopes, and other equipment.

He might do a test to see if a certain cleaner would harm the object, or he might examine a painting's canvas to determine if it needs support. Then he would take photographs, write a report, write a treatment proposal, clean the piece, and repair it. He might apply varnish to protect a painting's surface or place it on a new stretcher to pull it flat, then frame it and hang it in the museum.

Another conservation technician who works with paper, Jim Boydstun, spends his days examining and repairing the seven million books in the library at the University of California–Berkeley. Jim works at a desk in a small alcove or at a workbench nearby, repairing books damaged by rough handling from library patrons or deteriorating with age. Old, yellow, brittle paper and loose bindings are Jim's area of expertise. He is part of a team of workers who use scissors, shears, tweezers, erasers, glue, thread, and a special folder made of cattle bone to repair pages and bindings.

He repairs books with torn pages by gluing ripped edges back together or gluing a new corner on a page. Books that have become wet or whose pages are falling out of their bindings are more difficult to repair. Wet pages must be dried, and they may warp in the process and need further repair. Bad bindings must be torn out and sewed again. The job can be repetitive, but Jim remarks that with the large number of books in need of repair, "I'll never run out of work."

**To be a successful conservation technician, you should**

- Be extremely patient
- Have an appreciation for art and history
- Have manual dexterity and good hand-eye coordination for using tools
- Be able to stand or sit for long periods of time.

## have I got what it takes to be a conservation technician?

Conservation technicians need manual dexterity to handle tools and complete exacting repairs, such as making stitches so tiny that they can't be seen. Patience is essential for completing projects that often require hour after hour of intense concentration. Because the objects they handle are usually irreplaceable, technicians must exercise good judgment and refrain from making rash decisions that could further damage an item. The field requires a sense of dedication to art, history, and culture, coupled with the ability to do research.

Karen says textile restoration is a job for "someone who likes fabrics, likes looking at things like rugs and historic dresses." She has always been fascinated with the clothing in movies about the 1800s, for example. She adds, "You have to have a good sense of color and know about the way fabric is woven."

Many conservation technicians have training in art, which allows them to better understand how a book or art object was created. C. R. Jones says some talent with art materials is necessary to restore a painting, for instance, but since the goal is to repair someone else's work instead of producing a new painting, it's not necessary for the technician to be a creative artist. He enjoys the variety of his job and the opportunity to work with artifacts, but he dislikes working within a tight budget and having "too many things to work on and too little time to do it."

In contrast, Jim Boydstun likes his job because it doesn't put him under much pressure. "I really wanted to work in a stress-free environment," he says. "Mending books is a healthy occupation."

## how do I become a conservation technician?

Karen became interested in conservation four years ago, about twenty years after graduating from high school. She holds a college degree in special education, but so far she has taken no postsecondary courses to prepare for her new career in conservation. Instead, she's learning the old-fashioned way, by studying under a knowledgeable, established conservator.

Apprenticeship used to be the standard way to become a conservation professional, and it is still an accepted method of entry. Recently, however, graduate training, combined with on-the-job experience such as an internship, has become the preferred route.

## education

### High School

To prepare for a career in textile conservation, Karen recommends a well-rounded education that includes creative, artistic courses and practical, scientific classes. "You'll be using both your creative mind and your analytical mind," she notes. She also advises taking some business courses, in case you end up freelancing or owning a studio. If she ever had to leave the studio and work on her own, she says, "That (business administration) would be the roughest part. I'd have to write estimates. I'd have to write reports."

Classes in art are important for conservation technicians in almost any area of specialization, because they'll help you understand how objects were created and give you some skills, such as painting, needed to repair them. A knowledge of science is also critical, because conservation professionals need to know, for instance, how chemical agents interact. A familiarity with history will help you research the times in which an object was made. Studies in archaeology and anthropology are also recommended.

### Postsecondary Training

College programs in museum studies usually last from six months to two years and feature courses in art and the scientific procedures used to restore art objects. Classroom training is typically followed by an internship at a museum or library. It's possible to enter the trade via apprenticeship alone, but C. R. Jones says, "That would be difficult. You're going to have much better credentials if you have a college degree."

In a 1996 survey by the American Institute for Conservation of Historic and Artistic Works (AIC), more than 90 percent of conservation professionals who responded held at least a bachelor's degree. Nearly two out of three had completed a graduate degree program, and about a third held a graduate degree in an allied field. In addition, about two out of three had taken continuing education courses during their careers. Most respondents were conservators or department heads; only a small percentage were conservation technicians. Still, the survey shows that advancing in this field is generally easier if you hold a college degree.

C. R. Jones prepared for his career by obtaining a bachelor's degree in science, then earning a master's degree in museum studies and, ten years later, another master's degree in art conservation. "That's becoming pretty standard," he says. "If you want to break into the field (as a conservator or curator), you just about need a master's degree." He adds that some technicians hold only a bachelor's degree and may be able to work their way up from there, particularly if they are employed by the larger conservation laboratories.

**Advancement Possibilities**

**Curators** direct and coordinate activities of workers engaged in operating exhibiting institutions, such as museums, botanical gardens, arboretums, and art galleries.

**Art conservators** coordinate activities of subordinates engaged in examination, repair, and conservation of art objects.

**Museum registrars** maintain records of accession, condition, and location of objects in museum collections and oversee movement, packing, and shipping of objects to conform to insurance regulations.

# in-depth

**Painting Restoration**

C. R. Jones recalls one of his early restoration projects, a watercolor painting of Elizabeth Cooper, mother of the novelist James Fenimore Cooper. In years past, someone had attempted to protect the painting by coating it with shellac. That was an ill-conceived idea. When C. R. began work, the painting was murky and yellowed.

C. R. decided to try removing the shellac with solvent, because he knew that older watercolors, unlike some of their modern counterparts, are not soluble in alcohol. (If a watercolor touches certain liquids, the pigment will run, and the painting will be ruined.) C. R. tested the solvent on a small part of the painting, then immersed the entire piece and watched it carefully for any sign that the procedure might be causing damage. The idea worked nicely, and the shellac soon disappeared.

Next, C. R. made a mold of the ornamental front of the painting's antique frame, from which several pieces had been chipped. He used rock-hard water putty, a substance similar to plaster of Paris, to cast the missing pieces in the mold. Then he repaired parts of the frame that were embellished with a layer of gold leaf, which can be applied as a liquid or as thin sheets that are dropped onto wet varnish.

With the painting and frame repaired, C. R. cut an acid-free mat, a sort of inner frame made from thick paper that resembles cardboard. A mat is placed on top of a painting to keep it from touching the glass in the frame and sticking to it. Finally, the painting, mat, glass, and frame were assembled and hung where museum visitors could see Elizabeth Cooper's likeness, professionally restored.

## certification

No certification or licensing is available for conservation professionals. Technicians are hired on the basis of their education and experience. Participation in professional organizations is not required, but it's taken into consideration during the hiring process.

## scholarships and grants

Some museums and other organizations, such as the Smithsonian Institution, offer scholarships and grants for conservation technicians. Information is available through the American Institute for Conservation of Historic and Artistic Works and from colleges that offer training programs in the field.

## internships and volunteerships

To gain crucial on-the-job experience, you can volunteer or complete an internship or apprenticeship at a museum, bookbindery, or other facility. The AIC survey revealed that nearly a third of conservation professionals learned the trade through some sort of apprenticeship, and about half had completed an internship or postgraduate fellowship, which usually lasted one to two years.

Here are a few organizations that offer internships and volunteerships in conservation:

**Coordinator of Education and Training**
Conservation Analytical Laboratory
Museum Support Center, MRC 534
Smithsonian Institution
Washington, DC 20560
301-238-3700

**Office of Intern Programs**
National Museum of American Art
Room 270, MRC 210
Smithsonian Institution
Washington, DC 20560
202-357-2714

**Western Costume Company**
11041 Vanowen
North Hollywood, CA 91605
818-508-2148

## who will hire me?

Karen first met her mentor, Jessica Hack, when she brought an item to her to be restored. The experience piqued her interest, and she began working on small projects with the professionals at the studio. "I just found myself gravitating in this direction," she remembers. "Most of what I'd done led up to it. I'd sewed my own clothes for a long time."

Jim Boydstun also entered the trade through happenstance, fifteen years ago. "I kind of fell into it," he says. "A friend who owns a bookbindery needed some help. They taught me how to repair books, and I've been doing it ever since."

Most conservation technicians are employed by libraries, museums, historical societies, and state and federal agencies that have large, public collections of books, historical objects, or artwork. No matter where you hope to work in this field, networking is important. "I'd recommend that anybody who's interested should join a local historical society," says C. R. Jones. "You'll make good contacts, and you learn a lot, and it looks great on resumes." Joining a professional association is another good way to meet people in the field and learn of job openings, which are often published in the organization's newsletter. Some associations operate job banks that list openings and allow you to post your resume where it can be read by potential employers.

## where can I go from here?

Karen is happy at the textile restoration studio and would like to remain there indefinitely. If circumstances forced her to move on, she could apply for a job in a museum or open her own studio.

Jim Boydstun would also like to remain in his current position, but his job depends on the amount of state and federal funding the university receives. If the government decides to reduce spending, Jim's position could be cut. If funding is adequate, he could be promoted.

Many conservation technicians advance through a series of "grades" as they gain the required years of experience. A conservation technician can advance to conservator and take on supervisory duties within several years. With further education, some conservation technicians become administrators in an institution's collections department. Those with experience may also work as independent contractors, providing conservation services for libraries, museums, and individuals with private collections. Some appraise the value and determine the authenticity of art or artifacts for prospective buyers. Others become conservation educators.

## what are some related jobs?

With his knowledge of books and how they're made, Jim Boydstun could seek employment with a bookbinder or book manufacturer. C. R. Jones works with a team of four curators and was a curator, himself, before he decided to move into conservation. Curators are responsible for acquiring objects for an institution's collections, among other duties.

The U.S. Department of Labor classifies conservation technicians under the heading *Craft Arts: Arts and Crafts* (GOE). Also under this heading are people who make such things as models, props, jewelry, and patterns, as well as taxidermists, makeup artists, special effects specialists, and picture framers.

| Related Jobs |
|---|
| Paintings restorers |
| Lace and textiles restorers |
| Armorer technicians |
| Picture framers |
| Tattoo artists |
| Silversmiths |
| Stained glass glaziers |
| Wig dressers |
| Costumers |

## what are the salary ranges?

A conservation technician can expect to earn about as much as the average worker in America. Salaries vary, depending on the employer, geographic location, and other factors. "You have to really like this work," Karen comments. "It's not the world's best-paying job."

C. R. Jones says, "You will never get rich working for museums, so other considerations have to be the primary ones. In general, the salaries are adequate."

Starting salaries for entry-level positions average about $18,000 a year, according to the *Encyclopedia of Careers and Vocational*

*Guidance*. A recent survey by the Association of Art Museum Directors showed that associate conservators average about $37,000 annually; senior conservators average $40,000 to $50,000; and chief art museum directors average about $54,000.

According to a 1996 survey by the American Institute for Conservation of Historic and Artistic Works, many conservation professionals earn $25,000 to $50,000 annually. Most of the respondents to the survey were conservators or department heads, yet 24 percent earned less than $25,000 a year. Sixteen percent earned $50,000 to $80,000 annually; most of those workers were at least forty years old. Those with a graduate degree were earning about as much as those who had learned the trade via apprenticeship or who were self-taught. Those who were self-employed tended to earn less than most of the others. Benefits typically included paid vacations, health and life insurance, a retirement plan, and a profit-sharing plan.

## what is the job outlook?

Thanks to the nation's strong interest in cultural materials, the field of conservation and preservation is growing. New areas of specialization are emerging as workers combine their skills in conservation, curating, and registration. The outlook is particularly good for private conservation companies that perform contract work. Museums and other institutions will probably rely on private contractors more as they cut their own staffs in response to a drop in government funding through tax dollars.

## how do I learn more?

### sources of additional information

Following are organizations that provide information on conservation technician careers, accredited schools, and employers.

**American Association of Museums**
1575 Eye Street NW, Suite 400
Washington, DC 20005
202-289-1818
http://www.aam-us.org/

**American Institute for Conservation of Historic and Artistic Works**
1717 K Street, NW, Suite 301
Washington, DC 20006
202-452-9545
http://www.palimpsest.stanford.edu/aic
info@aol.com

**Association of Art Museum Directors**
41 East 65th Street
New York, NY 10021
212-249-4423
http://www.aamd.net
aamd.amn.org

**Costume Society of America**
55 Edgewater Drive
PO Box 73
Earleville, MD 21919
410-275-2329
http://www.costumesociety america.com
71554.3201@compuserve.com

**Intermuseum Conservation Association**
Allen Art Building
83 North Main
Oberlin, OH 44074
216-775-7331
http://www.oberline.edu/~ica

**Western Association for Art Conservation**
c/o Chris Stavroudis
1272 North Flores Street
Los Angeles, CA 90069

## bibliography

Following is a sampling of materials relating to the professional concerns and development of conservation technicians.

### Books

Oddy, Andrew, ed. *The Art of the Conservator.* Washington, DC: Smithsonian Institution Press, 1992.

*Science for Conservators.* Three volumes. Includes an introduction to materials, cleaning techniques, and adhesives. London: Routledge, 1992.

### Periodicals

*AIC News.* Bimonthly. Discusses the conservation of significant art works. American Institute for Conservation of Historic & Artistic Works, 1717 K Street NW, Suite 301, Washington, DC 20006, 202-452-9545.

*Antiques & Collectibles.* Monthly. Includes restoration information. Antiques & Collectibles, Inc., PO Box 33, Westbury, NY 11590, 516-334-9650.

*ARTnewsletter.* Biweekly. Reports on the world art market. ARTnews Associates, 48 West 38th Street, New York, NY 10018, 212-398-1690.

*Art & Archaeology Technical Abstracts.* Semiannual. Covers the treatment, analysis, and preservation of art. (IIC Abstracts) Getty Conservation Institute, 1200 Getty Center Drive, Suite 700, Los Angeles, CA 90049-1657, 310-440-7325.

# construction and building inspector

**Definition**

Construction and building inspectors examine new and renovated buildings, roads, bridges, dams, sewer and water systems, and other structures to ensure compliance with zoning requirements, building codes and ordinances, and contract specifications.

**Alternative job titles**

Building officials
Code officials
Fire safety inspectors
Home inspectors
Public works inspectors

**Salary range**

$19,400 to $32,300 to $57,500+

**Educational requirements**

High school diploma

**Certification or licensing**

Recommended

**Outlook**

Faster than the average

GOE 05.03.06

DOT 18.182

O*NET 21908A

## High School Subjects

Business
Computer science
English (writing/literature)
Mathematics
Shop (Trade/Vo-tech education)

## Personal Interests

Building things
Business management
Figuring out how things work
Fixing things
Helping people: protection

building inspector, Tom Wolff noticed problems almost as soon as he stepped inside the poorly maintained apartment complex. A leaky roof had allowed water to saturate some of the building materials, weakening them, and as a result, the ceiling was starting to collapse.

"We found stairways that were in disrepair, handrails that weren't there. It was virtually a deathtrap," he recalls.

It was five-thirty in the afternoon, and the building's numerous tenants were not pleased when Tom gave the order for them to vacate the premises immediately. The families were housed temporarily in other rental units their landlord owned. Meanwhile, Tom was arranging for the Red Cross to provide services that would help the tenants cope, filling out paperwork to document the incident, and working with the landlord to make sure the building's problems were corrected. Within a few days improvements had been implemented, and the tenants were allowed to move back into their apartments.

"It was just something that had to be done," Tom says. "All in all, it went well. The complex ended up being taken care of better. There were about forty-eight families that were better off because of it."

## what does a construction and building inspector do?

If a house is built on a defective foundation, it can collapse. If electrical wiring is not installed correctly, it can cause a catastrophic fire. If the concrete in a dam is not properly reinforced, the dam can burst and flood communities downstream. If a bridge near a fault line is not somewhat flexible, it can tear apart during an earthquake.

It is the job of *construction and building inspectors* to ensure that these and other structures are built, altered, and repaired properly, in accordance with building codes, ordinances, contract specifications, and zoning regulations. Some inspectors specialize in one area of expertise, such as mechanical components or plumbing, and others examine various aspects of homes and other structures.

Construction and building inspectors visit the site of new construction when work there is just beginning, make several additional visits as the project progresses, and perform an overall inspection after construction is complete. They also inspect older structures that are being altered or repaired, and sometimes they investigate reports of construction or renovation that is being done without the proper permits.

The job requires a thorough knowledge of building codes and other specifications, along with enough experience in construction to recognize potential problems before a disaster happens. Most inspections are visual, but the inspector sometimes tests the strength of concrete or uses metering equipment, survey instruments, tape measures, and other tools. Inspectors record their work in logs, document it with photographs, and compile reports. If necessary, they report violations or other causes of concern, which can stop work on the project until the problem is remedied.

*Building inspectors* watch for structural defects and safety problems in residential, commercial, and industrial buildings. Before construction begins, they examine the soil and footings where the foundation will be laid. Their inspections include an examination of fire sprinklers, smoke control systems, alarms, fire doors, and exits. Sometimes they calculate fire insurance rates, taking into consideration the type of building, its contents and fire protection system, and the types of buildings nearby.

Some building inspectors specialize in areas of expertise, such as reinforced concrete or structural steel. *Plan examiners* are specialists who study blueprints before the building is constructed; they ensure that the plans comply with building codes and that the building will be suitable for the site. *Plumbing inspectors* check the building's water and sewer pipes, fixtures, traps, drains, vents, and other components. *Electrical inspectors* ensure that lights, wiring, motors, generating equipment, sound and security systems, heaters, air conditioners, and appliances are installed and functioning properly. *Mechanical inspectors* check the installation of heaters, air conditioners, commercial kitchen appliances, butane and gasoline tanks, natural gas and oil pipes, and other appliances. Some also inspect ventilation equipment and boilers. *Elevator inspectors* examine elevators, escalators, lifts, hoists, ski lifts, moving sidewalks, inclined railways, and amusement rides.

### lingo TO LEARN

**Building code** Regulations that specify which materials and techniques can be used in various types of construction. The code also regulates other factors to address safety concerns, aesthetic values, and other considerations. For example, a building code might require that a garage be placed at least three feet from a street or the edge of the property, or it might require that a retaining wall be strong enough to withstand a mud slide during a heavy rainfall.

**Building permit** Formal permission, granted by a regulatory agency, to build or demolish structures such as houses, patios, garages, attics, stairways, and new roofs. The permit is granted after a professional determines whether the proposed project will comply with building codes, zoning requirements, and other regulations.

**Geotechnical review** An inspection that assesses the soil and other geological factors at a building site. The review is intended to detect potential problems, such as an unstable hillside that might slide if a house were built there.

*Home inspectors* examine houses that have just been built or that are up for sale; they are often hired by potential buyers to assess the home's condition. A home inspection includes almost every component of the building and its systems, from the foundation to the attic.

*Public works inspectors* examine structures built under the oversight of federal, state, and local governments, including dams, water and sewer systems, bridges, streets, and highways. They inspect the placement of forms for concrete; the pouring and mixing of concrete; asphalt paving; and projects that involve excavation, filling, or grading. Some specialize in reinforced concrete; structural steel; highways; ditches; or dredging for bridges, dams, or harbors. The job involves keeping records of materials used and the general progress of a project so contract payments can be calculated.

Construction and building inspectors sometimes work with *architects* who design buildings and other structures, but more often they deal with *trade foremen, construction superintendents, real estate salespeople*, and people purchasing real estate. They also work with *homeowners* who are required to obtain permits and schedule inspections for large projects and smaller undertakings, such as remodeling a patio.

**To be a successful construction and building inspector, you should**

- Have experience in construction
- Be detail oriented
- Have excellent communication skills
- Be in good physical condition
- Be organized and able to work independently
- Enjoy helping people
- Have self-confidence

## what is it like to be a construction and building inspector?

Tom Wolff is a senior building inspector for the Clark County Building Department in Las Vegas, Nevada. He arrives for work at seven in the morning and spends about half an hour answering telephone inquiries, then begins planning his day. "We get requests the day before for inspections. We do as many as fifteen or twenty stops in one day. We inspect, we compare, we advise, and we record," he says.

He looks at a building's structure, plumbing, electricity, heating, ventilation, and air conditioning systems. He compares the building to minimum code standards or minimum design standards. If he finds anything unacceptable, he advises the builders to contact their architects to remedy the problems, or if he's dealing with homeowners, he has them propose a solution and then tells them whether the proposal would correct the defect. Tom has to choose his words carefully and avoid proposing a solution himself, since he could be sued if anything went wrong because of his advice. Throughout the inspection process, he keeps records of what he has seen, said, and done.

Tom says that one of the best things about his job is that it gives him opportunities to share his experience in construction: "Personally, I enjoy helping people. It's really hard for me to bring bad news. If you get a kick out of giving people bad news, you probably shouldn't be an inspector. You need empathy."

Inspectors are in a somewhat awkward position, trying to enforce the law without offending people in the construction industry, which is a vital aspect of any community's economy. "You're not a cop," Tom explains. "You're a public servant. Treat your job accordingly. The construction industry is very important to local government entities. The government wants to make sure the construction industry is happy."

The government entities that perform building inspections are divided into various jurisdictions in Las Vegas. In Tom's jurisdiction seven inspectors cover about twenty-three square miles. Some of the inspectors specialize in one area, such as plumbing, but Tom does combination inspections of residential homes. In a typical day he might inspect a swimming pool, the framing of a roof that is being remodeled, the general structure of a new home under construction, and assorted other projects.

To inspect all these sites, Tom drives forty to seventy miles every day. He says he enjoys the variety of the work and being on the move instead of having to remain at a desk in an office all day: "I appreciate being outdoors, being on a different site from hour to hour."

Sometimes, though, Tom is frustrated by difficulties in communication, even though he is adept at technical writing. Talking to many people in many situations and writing clear, accurate reports is a major part of his job.

He also must continually read long, complex building codes, which can be a dry undertaking. There are four national codes, and inspectors must also be familiar with their jurisdiction's modifications to those codes, along with local zoning regulations. Tom says, however, that it's more important to know where to find the regulations than to attempt to memorize all of them.

Tom has to know his job well and keep accurate records, or he could be sued. Written records help maintain communication among the many people involved in construction and the county building department, and they can be used as evidence in case of a lawsuit. Although Tom has been to court only twice on civil matters, a construction and building inspector is always at risk of a lawsuit.

According to Gary Fuller, a home inspector who owns Building Tech, Inc., in Spokane, Washington, "There's a lot of liability. You have a chance of being sued every time you do an inspection." Gary carries liability insurance, but he says the key to avoiding lawsuits is to have enough experience in construction to recognize defects in a building. People who are about to purchase new or used homes hire him to check the buildings before they finalize their transactions, and they expect him to notice every flaw, from broken rafters to a leaky roof to a malfunctioning furnace. "You have to be like Superman with x-ray vision," he comments.

Usually, however, his clients appreciate the service he provides for them. As a member of the American Society of Home Inspectors, Gary examines components of each home as specified by ASHI. He reviews the building's general structure, foundation, basement, floors, walls, ceilings, windows, doors, attic, roof, visible insulation, plumbing, heating and air conditioning systems, and electrical wiring

He usually spends up to two hours on each inspection, taking notes as he works. Then he uses a computer to write a narrative report, but he says other inspectors merely make check marks on a printed list. Like most professionals in this trade, Gary has found that the ability to communicate and write clear, detailed reports is an important quality for an inspector.

### Advancement Possibilities

**Building officials** are the managers in charge of city and county building departments. They usually have a background in public administration or business administration.

**Plan examiners**, also known as **plan reviewers** or **plan checkers**, review architectural drawings and other documents that describe proposed structures, along with related specifications, structural design calculations, and soil reports. The examiner checks for compliance with building codes. Plan examiners typically work for government agencies, architectural and engineering firms, or private consulting companies.

**Supervisors** are experienced, knowledgeable inspectors who oversee the rest of an inspection staff. Frequently, they have completed studies in business management and personnel management.

## in-depth

**Projects that typically require a building permit**

The number of construction projects regulated by most cities might surprise you. For example, in many communities it is illegal to replace a bathroom sink or toilet, install a lawn sprinkler system, finish an attic or basement, put up a fence more than six feet high, or cut an opening for a new door in your own home without having it inspected by an official from the building department. A permit is required to abandon a cesspool, erect a small shed, enlarge a porch or roofed patio, pave a driveway, or alter the exterior of a building in an area zoned for certain environmental or scenic qualities.

Minor repairs and maintenance, such as replacing a concrete walkway or putting up shelves, do not require permits or inspections. Usually a homeowner can also paint a building, build a small patio or deck, or clear small areas of vegetation without a permit. In many cases an inspection is not necessary for larger projects if the homeowner does the work, but if a professional, friend, neighbor, or relative performs any of the work, a permit and inspection may be required.

A new construction project may require inspections by many specialists, such as officials from a city's planning, transportation, fire, water, forestry, and environment departments. Less complex projects, such as adding a roof over an existing outdoor deck, may require review by just one inspector from the building department.

## have I got what it takes to be a construction and building inspector?

A construction and building inspector should have experience in construction, have a good driving record, be in good physical shape, have good communication skills, be able to pay attention to details, and have a strong personality. Although there are no standard requirements to enter this occupation, an inspector should be a responsible individual with in-depth knowledge of the construction trades. Inexperience can lead to mistakes that can cost someone a staggering amount of money or even cause a person's death.

The trade is not considered hazardous, but most inspectors wear hard hats as a precaution. An inspector might need to climb ladders and walk across rooftops or perhaps trudge up numerous flights of stairs at building projects where elevators are not yet installed. Or they might occasionally find themselves squirming through the dirty, narrow, spider-infested crawl space under a house to check a foundation or crawling across the joists in a cramped, dusty, unfinished attic, inhaling insulation fibers and pesticides.

After the inspection a construction inspector needs to explain his or her findings clearly in reports and should expect to spend many hours answering questions in person, by telephone, and in letters. Because they often deliver bad news, they also need the emotional strength to stand firm on your reports, even when someone calls them a liar or threatens to sue.

On the other hand, an inspector knows that their work is to protect people. For example, they help ensure that a couple's new house will not be apt to burn down from an electrical short, and they might point out less dangerous problems, such as a malfunctioning septic tank or a leaking roof, that could require expensive repairs.

# how do I become a construction and building inspector?

Gary Fuller was a building contractor for twenty years before he became a home inspector in 1989. Like Gary, most construction and building inspectors have some previous experience in carpentry, plumbing, electrical work, or some other related occupation.

"Typically," Tom says, "they're working in one of the trades for several years and have gone through an apprenticeship. I would say certification and experience are probably the key to getting hired." To enter this profession, you can complete an apprenticeship program, study the building codes independently, or complete some college-level studies, then obtain certification from one of the professional associations.

## education

### High School

A construction and building inspector needs to have at least a high school diploma or GED. To prepare for a career in this field, you should take courses such as basic mathematics, computers, typing, shop, drafting, art, electronics, architectural drawing, and law. "We're dealing with law every day," Tom notes.

**Occupancy classification** The type of use for which a building has been approved by a regulatory organization. For example, a building classified as a residential rental property usually cannot be converted to an industrial use without a new inspection and new certificate of occupancy.

**Zoning** Regulations that designate the types of buildings and other structures, along with their uses, that will be allowed in particular areas of a community. Residential neighborhoods are usually separated from industrial areas and may also be separated from retail establishments.

In addition, a part-time or summer job in construction would give you valuable experience, help you decide whether this might be the best career for you, and put you in touch with people in the field. You might also consider becoming involved with a local chapter of one of the trade associations or the National Fire Protection Agency. "Students would be welcome to attend the meetings and to join," Tom says.

### Postsecondary Training

You don't have to go to college to become a construction or building inspector, but courses in subjects such as construction technology, building inspection, blueprint reading, mathematics, and public administration would give you a solid background for this career. Studies at college or vocational school can sometimes be substituted for on-the-job experience, and they will help you prepare for the certification examinations. Many employers are now requiring that their inspectors hold two-year, four-year, or advanced degrees in addition to practical experience. Finally, your promotion at some time in the future might depend on your educational background.

Nevertheless, Tom emphasizes that no amount of theoretical study can take the place of hands-on work in the field. Advanced education is good, he says, but it is not as important as experience and certification.

To retain his certification, Tom must participate in continuing education courses. Training of this type is available through trade associations, community colleges and vocational schools, or employers. Gary Fuller also stays up to date by attending continuing education seminars seven to nine times a year. He says he enjoys these learning experiences, where he studies the latest trends in areas such as heating, electrical, and plumbing systems.

## certification or licensing

In general, a license or certification is highly recommended but not necessarily required. Certification and licensing requirements vary widely among the states and individual employers. Many employers hire only licensed or certified professionals.

Certification for construction and building inspectors is available through three model code associations: The Building Officials and Code Administrators International (BOCA), the International Conference of Building Officials (ICBO), and the Southern Building Code Congress International (SBCCI). These three associations offer three different certification programs, but all are nationally recognized. Certification is available in various categories, such as certified building official, building inspector, electrical inspector, mechanical inspector, and plumbing inspector.

Tom expects to see the development of a national, standardized building code within three to six years. He predicts that hiring standards will vary with the employer, but certification will be standardized.

Home inspectors are certified through the American Council of Home Inspectors (ASHI). Gary Fuller is not required to hold such credentials in Washington State, but he voluntarily became certified.

## internships and volunteerships

Experience is probably the most important qualification for a construction and building inspector. Any hands-on work will provide valuable preparation for a career in this field. You could volunteer to help build houses with a service organization, such as Habitat for Humanity. Some home inspectors, particularly those who belong to the American Society of Home Inspectors, would be willing to let you accompany them as an observer on a few inspections.

A few internships for construction and building inspectors are available in the private sector. Some are designed to help construction workers who are undergoing a career change after an injury on the job.

More commonly, prospective inspectors are graduates of apprenticeship programs that provide them with on-the-job training. Many apprenticeship programs in the construction trades are administered by local offices of labor unions, such as the United Brotherhood of Carpenters or the Associated Builders and Contractors. For other apprenticeship opportunities, inquire at any Job Service office.

## labor unions

Tom belongs to the Service Employees International Union (SEIU). Other building inspectors employed by government agencies belong to the American Federation of State, City, and Municipal Employees (AFSCME).

### FYI

**Becoming a self-employed home inspector**

The American Society of Home Inspectors (ASHI) estimates that it takes at least $25,000 in working capital to launch an independent home inspection business. You would also need a dependable vehicle for driving to numerous job sites and a computer for writing reports and keeping track of clients and finances. Expect overhead costs (rent, utilities, and other expenses) to take about half your income.

Instead of opening an independent business, you could buy a franchise from an established home inspection organization. A franchise is a semi-independent, branch operation of a larger company. It requires an investment of your money, but it also features training and other support, such as advertising, from the parent company. A franchise is particularly attractive for new inspectors who have little experience in construction and business.

While preparing for a career as a home inspector, beware of inspection "schools" that can cost up to $10,000 but offer little real education. Some include classes that won't teach you much or that don't include the most important information a home inspector needs. Others waste most of your money on computer equipment and books that consist mainly of irrelevant material copied from other books. Before enrolling in such a training program, check with a local chapter of ASHI or discuss the program with a home inspector in your area.

## who will hire me?

Tom began his career as a building inspector in California in 1985. "I was working as a public works maintenance worker and was promoted from within," he recalls, adding that he was doing "everything from picking up trash in the park to trimming trees" before his promotion and subsequent three months of training.

Previously, the city had hired several inspectors, but they kept moving on when other opportunities arose. Tom was offered the position because he had shown loyalty by working for the city for some time; he stayed on as an inspector there for nine years. It was an unusual way to become an inspector. "I was in the right place at the right time," he says.

More commonly, candidates find openings in the public sector or government agencies by watching the classified advertisements in newspapers. Job openings are also listed in the classified advertising sections of Internet sites and newsletters of various trade organizations (*see* "Sources of Additional Information"). Several magazines, including *Western Cities, Jobs Available,* and the *International Conference of Building Officials' Building Standards*, feature advertisements for positions as an inspector.

For information about openings with government organizations, contact state or local employment services. You can also leave an "interest card"' with a government agency that interests you, and you will be contacted when an opening arises.

More than half of the construction and building inspectors in this country work for local governments, most frequently for city or county building departments, which may employ large staffs of specialized inspectors. These jobs are concentrated in cities and rapidly growing suburbs. Other inspectors work for state and federal government organizations, including the U.S. Army Corps of Engineers, the General Services Administration, the Department of Agriculture, the Department of the Interior, and the Department of Housing and Urban Development.

Nearly 20 percent of construction and building inspectors are employed by firms that provide architectural and engineering inspections for a fee or on contract. Many home inspectors are self-employed, but others work for large franchises.

## where can I go from here?

Tom stayed with his first inspector job for nine years, then moved to Nevada, because there was more opportunity there. He might become the supervisor of a staff of inspectors in Las Vegas, but he says he does not expect to advance further than that.

With enough experience and perhaps more education, he could become a building official in charge of the entire building department. He could also become a *plan checker,* one of the people who reviews blueprints and other documents before a building is constructed.

With credentials, a license, and enough money, a building inspector could follow Gary Fuller's example and launch a home inspection service. Getting started, he says, is "probably the toughest part of the business." To succeed as a self-employed home inspector, you would need to promote your business through avenues such as advertising, cultivating referrals, and doing a professional job to build a good reputation.

Gary remembers that, after his years as a building contractor, he expected less demanding duties as a home inspector: "I thought it would be like semi-retiring, but it wasn't." Even though he is self-employed, he finds that he must keep working full-time, because clients expect him to be available when they need his services. Still, he enjoys his new line of work and expects to continue in it indefinitely.

## what are some related jobs?

Construction and building inspectors are familiar with the principles of construction and associated legal issues. They are also proficient at communicating with various types of people, coordinating data, and identifying problems. These skills can be applied to several related professions, although some additional education may be required for certain fields. For example, an inspector could become an industrial engineering technician, an engineer, a surveyor, an architect, a drafter, an estimator, or a construction contractor or manager.

The U.S. Department of Labor classifies construction and building inspectors under the heading *Engineering Technology: Industrial and Safety* (GOE). Also under this heading are people who work in industrial engineering, flight engineering, and traffic engineering.

| Related Jobs |
| --- |
| Bricklayers |
| Carpenters |
| Cement masons |
| Cost estimators |
| Drafters |
| Electricians |
| Fire safety technicians |
| Industrial engineering technicians |
| Marble setters and tile setters |
| Plumbers and pipe fitters |
| Pollution control technicians |
| Traffic engineers |

## what are the salary ranges?

Tom estimates that construction and building inspectors in government jobs earn about $20,000 to $48,000 a year nationwide, and it's not unusual to start at $30,000 to $32,000. He says salaries are highest in larger jurisdictions and in the Southwest, where many buildings are currently being constructed. In Las Vegas a county building inspector is paid about $30,000 to $55,000 annually; and a building plans check specialist is paid about $41,000 to $63,000.

In addition to his salary, Tom receives paid vacation time, paid sick leave, and a pension plan. Like other government employees, he is exempt from paying Social Security taxes.

In contrast, a self-employed home inspector's earnings depend entirely on the individual. Expertise, professionalism, and the ability to attract a large number of clients are the deciding factors. Gary Fuller says, "I know inspectors who make $300,000 a year, down to $10,000 a year."

According to the U.S. Department of Labor, the median annual salary of construction and building inspectors was $32,300 in 1994. About half earned $25,200 to $43,800. The lowest 10 percent earned less than $19,400. The highest 10 percent earned more than $57,500.

According to Building Officials and Code Administrators International (BOCA), salaries in building code enforcement can range from $25,850 to $97,000 annually. Pay is generally higher in the West and lowest in the Southeast, and workers in large cities usually earn more than those in rural areas.

## what is the job outlook?

Tom expects to see standardized building codes and other changes within his profession during the next decade or so, but he says, "Building inspectors will be here forever. As long as they're building buildings, the career will be in demand."

The U.S. Department of Labor expects that the number of openings for construction and building inspectors will grow faster than the average for all occupations through the year 2006. More buildings and other structures are being inspected because of increasing concern for public safety and improvements in the quality of construction.

Although the field is expanding, most job openings will be created as inspectors transfer to other professions or leave the work force. There is a relatively high turnover in this field, because most inspectors are older workers who have years of experience but are nearing retirement age.

Workers with the most experience and education will have the best opportunities. Some college education, training in architecture or engineering, or certification as an inspector or plan examiner will give you a competitive edge. The ability to read and assess plans and blueprints is essential. There will be opportunities to work for architectural, engineering, and management firms that provide inspection services for government agencies, particularly at the state and federal level, as those agencies cut costs by contracting with the private sector.

Rapid growth is expected in the field of home inspections, because more real estate will be sold as the population expands. Home inspectors enjoy a degree of job stability even during recessions, since their services are required for maintenance and repair projects, not just for new construction. According to Gary Fuller, the field of home inspections is "growing by leaps and bounds. More people are becoming aware of inspection and finding a value in it."

## how do I learn more?

### sources of additional information

Following are organizations that provide information on construction and building inspector careers, accredited schools, and employers.

**American Society of Home Inspectors (ASHI)**
85 West Algonquin Road, Suite 360
Arlington Heights, IL 60005
312-372-7090
800-743-2744
http://www.ashi.com/

**Building Officials and**
**Code Administrators International (BOCA)**
4051 West Flossmoor Road
Country Club Hills, IL 60478-5795
708-799-2300
http://www.bocai.org/
info@bocai.org

**International Conference**
**of Building Officials (ICBO)**
5360 Workman Mill Road
Whittier, CA 90601-2298
562-699-0541
http://www.icbo.org/

**Southern Building Code**
**Congress International (SBCCI)**
900 Montclair Road
Birmingham, AL 35213-1206
205-591-1853
http://www.sbcci.org/
info@sbcci.org

## bibliography

Following is a sampling of materials relating to the professional concerns and development of construction and building inspectors.

### Books

Domel, August W. *Basic Engineering Calculations for Contractors*. Reference for contractors, project managers and other construction professionals. Complex engineering concepts and calculations are simplified with nontechnical formulas and explanations. New York: McGraw-Hill, 1997.

McConville, John G. *The 1996 International Construction Costs and Reference Data Yearbook*. Comprehensive, easy-to-use reference providing construction professionals with information essential to make accurate cost comparisons, assess and minimize risk, and select the right market for their services. New York: John Wiley & Sons, 1996.

Glover, P.V. *Building Surveys*. 3rd ed. Reference for the practitioner and student of architecture, building and surveying, covering the knowledge, techniques and equipment required to inspect and report accurately on all types and conditions of buildings. Portsmouth: Butterworth-Heinemann, 1996.

O'Brien, James J. *Construction Inspection Handbook: Total Quality Management*. 4th ed. A quality-control tool for construction inspectors who are looking toward managing a crew. New York: Chapman & Hall, 1997.

### Journals

*Builder Online*. Daily. An online tool for mastering the business. News, chats, product reviews, related links. http://www.builderonline.com.

*Building*. Monthly. A national news magazine focusing on issues of concern to the building development industry. Building Magazine/Canadian Interiors, 360 Dupont Street, Toronto, Ontario, Canada M5R 1V9.

*Construction Online*. Daily. Covers equipment, maintenance issues, education and training, and other topics. http://www.constmonthly.com.

*Metropolis*. Monthly. Covers the world of design with topics ranging from the sprawling urban environment to intimate living spaces to small objects of everyday use. Metropolis Magazine, 177 East 87th Street, New York, NY, 10128.

*Permanent Buildings and Foundations Online*. Daily. Covers building code regulations, public policy, conservation and restoration, and other key issues. http://www.pbf.com.

*Traditional Building*. Monthly. A magazine covering more than three hundred leading suppliers of historically styled products for restoration and renovation of older structures—as well as traditionally styled new buildings.Traditional Building Magazine, 69A Seventh Avenue, Brooklyn, NY, 11217.

# corrections officer

**Definition**

Corrections officers guard people who have been arrested and are awaiting trial or who have been tried, convicted, and sentenced to serve time in a penal institution.

**Alternative job titles**

Detentions officers

Jailers

**Salary range**

$19,100 to $28,300 to $41,730

**Educational requirements**

High school diploma

**Certification or licensing**

Not mandatory, but useful for certain job functions

**Outlook**

Much faster than average

GOE 04.01.01

DOT 372

O*NET 63047*

## High School Subjects

Government
Physical Education
Psychology
Sociology

## Personal Interests

Helping people: emotionally
Law
Sports
Student government

**An hour ago,** corrections officer David Pilgrim was chatting with a convicted murderer in the Oklahoma State Penitentiary about today's lineup of college football games. The man admitted a soft spot for his home state's team and asked if David would find out what the final score was.

That conversation was a pleasant diversion from the beginning of David's shift, when a fight broke out between two inmates in the showers. One inmate had cleverly crafted a knife out of his plastic light shade and took it into the shower with him. Fortunately, the man had dropped the knife the instant David stepped in. Now David has rolled up the shirt sleeves of his dark blue uniform and is wrist-deep in an enormous vat of green beans. "I take myself pretty seriously at times like this," he jokes with another officer, jabbing a thermometer into the beans to test that they're heated to the correct temperature. He dishes up a plate and hands the tray to one of the inmate workers, who will deliver it to the others. Once meals are served to all ninety-eight inmates on his floor, David will take a few minutes to eat his own dinner that he packed from home. Then it'll be time for him to walk the hallways for another routine head count.

## what does a corrections officer do?

*Corrections officers* are hired by federal, state, and local prisons and jails to maintain order according to the institution's policies, regulations, and procedures. They are concerned with the safekeeping of people who have been arrested and are awaiting trial or who have been tried, found guilty, and are serving time in a correctional institution.

Corrections officers keep watch over inmates around the clock—while they're eating, sleeping, exercising, bathing, and working. In order to prevent disturbances, corrections officers carefully observe the conduct and behavior of inmates. They watch for forbidden activities, as well as for poor adjustment to prison life. They try to settle disputes before violence can erupt. They may search the inmates or their living quarters for weapons or drugs and inspect locks, bars on windows and doors, and gates for any sign of tampering. They conduct regular head counts to make sure all inmates are accounted for. Some corrections officers are stationed on towers and at gates to prevent escapes. In the case of a major disturbance, a corrections officer may have to use a weapon or force. After such a violation or disturbance, corrections officers are responsible for filing detailed reports. Corrections officers cannot show favoritism and must report any inmate who breaks the rules.

Corrections officers assign work projects to the inmates, supervise them while they carry out their work, and teach them about unfamiliar tasks. Officers try to ensure inmates' health and safety by checking the cells for unsanitary conditions and fire hazards. They are in charge of screening visitors at the entrance and inspecting mail for prohibited items. Officers are also responsible for escorting inmates from one area of the prison to another and helping them get medical aid. Certain officers are charged with transporting inmates between courthouses, prisons, mental institutions, or other destinations.

Some officers specialize in guarding juvenile offenders who are being held at a police station or detention house pending a hearing. These officers often investigate the background of first offenders to check for a criminal history and to make a recommendation to the court. Lost or runaway children are also placed in the custody of these officers until their parents or guardians can be contacted. In small communities, corrections officers may also serve as deputy sheriffs or police officers.

The person in charge of supervising other corrections officers is often called the *head corrections officer*. This person assigns duties, directs the activities of groups of inmates, arranges for the release and transfer of inmates, and maintains overall security measures.

While psychologists and social workers work at the prison to counsel inmates, a secondary aspect of a corrections officer's job is to provide informal counseling. Officers may talk with inmates in order to help them adjust to prison life, prepare for return to civilian life, and avoid committing crimes in the future. On a more immediate level, they can help inmates arrange a visit to the library, get in touch with their families, suggest how to look for a job after being released from prison, or discuss personal problems. Corrections officers who have college degrees in psychology or criminology often take on these more rehabilitative responsibilities.

Corrections officers keep a daily record of their activities and make regular reports to their supervisors. These reports concern the behavior of the inmates, the quality and quantity of work they do, as well as any disturbances, rule violations, and unusual occurrences. Because

**Contraband** Any forbidden item that is in a prisoner's possession.

**House** An inmate will refer to his cell as his "house."

**Run** A hallway lined with inmates' cells.

**Shake-down** Corrections officers conduct a thorough search of an inmate's cell, looking for contraband.

**Yard-out** This means that it's time for the inmates to go to the yard for exercise. Corrections officers also say "chow-out" when it's time for the inmates to eat and "shower-out" when it's time for showers.

prison security has to be maintained at all times, corrections officers sometimes are expected to work nights, weekends, and holidays. Generally, a work week consists of five eight-hour days. Work takes place both indoors and outdoors, depending on the officer's assigned duties on a given day. Conditions range from a well-lit, well-ventilated area to a hot, noisy, and overcrowded one.

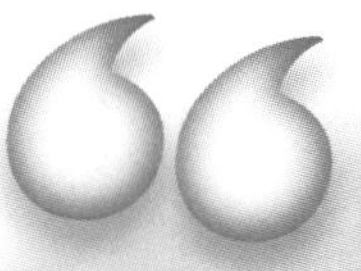

*The inmates like to hear what's going on in the outside world. They're always real interested in what I brought for lunch.*

## what is it like to be a corrections officer?

David Pilgrim is a corrections officer with the rank of sergeant at the Oklahoma State Penitentiary, a maximum-security facility housing fourteen hundred inmates. He's thirty-three years old and has worked at the prison for three years. During this time, he has been promoted twice and was recently awarded Outstanding Correctional Officer at his facility.

Much of his work is routine, but he has to be prepared for violence to erupt at any time. As a matter of fact, that's one of the things David will tell you right off the bat that he likes about his job. "It's the unpredictability," he says. "I like the notion that anything could happen when I go for a shift." His current shift runs from 2 PM to 10 PM, with Wednesdays and Thursdays off. He begins the workday by attending a briefing fifteen minutes before his shift begins. Here, he and the other corrections officers going on duty find out everything that's happened and learn which unit they'll be working that day, if they don't have a permanent assignment.

Once he gets to the unit, the first order of business is to take a count of inmates, which he'll repeat at two-hour intervals. "Doing the count and recording it in the log is routine, and following a routine is important here." Absolutely everything that happens during a shift, from the mundane to the extraordinary, must be documented in this log. His shift is likely to include escorting small groups of inmates to the shower or out to the yard for exercise. At mealtime, he and the other officers are responsible for dishing up the meals, which are delivered to inmates' cells by other inmates granted "worker" status.

Beyond those routine tasks, David says he deals with "whatever breaks out." Attempted suicides, knife fights, and feigned illness are all standard fare. "Last night," he describes, "a couple of them taped the doors up in their cell and jammed books in the locks because they were conducting a gang initiation." But he emphasizes that a good deal of his work is not so tense. "The inmates like to hear what's going on in the outside world. They're always real interested in what I brought for lunch, because the food in here, is . . . well . . . ," he trails off.

Talking to inmates and to fellow officers is one of the things he most enjoys. "I like being around people," he says. "Some folks like being assigned to the towers or to the control room, but that's torture to me. You're just sitting there alone waiting for your shift to go by." As a sergeant, he particularly likes being able to help and coach the new officers assigned to him for on-the-job training. "I'm very security-minded and I try to pass that on to the cadets," he says.

Unlike most of his peers at the prison, David has a bachelor's degree. He majored in wildlife conservation, with the expectation that he'd become a game warden. But he discovered that it was difficult to break into that field. At the prison, he's actually made greater use of his minor in occupational safety and health. This background has been extremely useful, he notes, because sanitation is a major concern among corrections officers. "You treat everyone as though they have HIV, because so many of the inmates do. We wear rubber gloves whenever we're in direct contact."

The Oklahoma State Penitentiary is under twenty-three-hour lockdown, which means that corrections officers must escort inmates everywhere that they go. "Sometimes it seems like a big daycare," jokes David. Officers there don't carry weapons on them and instead keep anything that could be used against them—like restraints, handcuffs, and weapons—in the control rooms. The officers rely on radios and an intercom system to communicate with one another when a need arises.

## have I got what it takes to be a corrections officer?

There's no denying that handling the inherent stress of this line of work takes a unique person. In a maximum-security facility, the environment is often noisy, crowded, poorly ventilated, and even dangerous. Corrections officers need the physical and emotional strength to handle the stress involved in working with criminals, some of whom may be violent. A corrections officer has to stay alert and aware of prisoners' actions and attitudes. This constant vigilance can be harder on some people. Work in a minimum-security prison is usually more comfortable, cleaner, and less stressful.

Officers need to use persuasion rather than brute force to get inmates to follow the rules. Certain inmates take a disproportionate amount of time and attention because they're either violent, mentally ill, or victims of abuse by other inmates. Officers have to carry out routine duties while being alert for the unpredictable outbursts. Sound judgment and the ability to think and act quickly are important qualities for corrections officers. "We have quite a few new cadets who freeze up when the first fight breaks out among inmates," David says. "It takes a while before I feel comfortable putting my life in the hands of someone who hasn't been tested." With experience and training, corrections officers are usually able to handle volatile situations without resorting to physical force.

**To be a successful corrections officer, you should**

- Not be easily intimidated or influenced by the inmates
- Have physical and emotional strength to handle violent or abusive prisoners
- Be able to use persuasion rather than brute force to get inmates to follow the rules
- Be able to stay alert and aware of prisoners' actions and attitudes
- Have sound judgment and the ability to think and act quickly
- Be able to communicate clearly both verbally and in writing

The ability to communicate clearly verbally and in writing is extremely important. "You've got to be able to get across to all walks of life," David says. Corrections officers have to write a number of reports, documenting routine procedures as well as any violations by the inmates. On a busy shift, David's eight-hour shift can easily extend to ten hours because of the reports that must be written.

An effective corrections officer is not easily intimidated or influenced by the inmates. There's a misperception, however, that corrections officers need to be tough guys. While it's true that a person needs some physical strength to perform the job, machismo only gets in the way. "You get types who think that with a badge on they can do and say anything," David says. "That attitude will only get you injured. If I'm in a unit without air conditioning on a hot day, the inmates get awfully irritated and may verbally abuse me when I walk by. There's no sense in doing anything but to keep on walking."

## how do I become a corrections officer

### education

#### High School

Entry requirements vary widely from state to state. A high school diploma, or its equivalent, is the minimum requirement for employment at most correctional institutions. While most high school classes are not directly relevant to corrections, health classes may offer an introduction to issues that will be covered thoroughly during formal training as a corrections officer, such as sanitation, universal precautions, and first aid. English classes that focus on writing are recommended as well for anyone interested in a career as a corrections officer. At most prisons, jails, and penitentiaries, reports are required to be well-written, with attention paid to sentence structure and spelling. David commented that officers who write sloppy reports are often the last to be promoted.

Spanish-language classes would be useful for future corrections officers who plan on working in a region of the country with a large Spanish-speaking population.

### Postsecondary Training

Some correctional institutions require that corrections officers possess a college degree. In certain states, officers need two years of college with an emphasis on criminal justice or behavioral science, or three years as a correctional, military police, or licensed peace officer. Generally, states that require more education offer higher entry-level salaries and have shorter-duration training academies. At federal institutions, applicants must have at least two years of college or two years of work or military experience.

The most relevant areas of study in college include psychology, criminal justice, police science, and criminology. Some correctional facilities offer internships to students who are earning their degrees in these areas.

## certification or licensing

The American Correctional Association and the American Jail Association provide guidelines for prison training programs. These programs generally introduce new corrections officers to the policies of their particular institution and prepare them for handling work situations. The training lasts several weeks and includes crisis intervention, contraband control, counseling, self-defense, and use of firearms.

Training ranges from special academies to informal, on-the-job training. The Federal Bureau of Prisons operates its training center in Glynco, Georgia, where new hires take part in a three-week training program. In Oklahoma, new hires work for a month to "try the career on" before going to the six-week training academy. After the training academy, a new officer will usually spend two to six months under the supervision of an experienced officer.

While there are numerous certifications available to corrections officers, these are optional in most states. Common certifications include self-defense, weapons use, urine analysis, shield and gun, shotgun/handgun, CPR, and cell extraction. Many officers also take advantage of additional training that is offered at their facility, such as suicide prevention, AIDS awareness, use of four-point restraints, and emergency preparedness. At most prisons, there is annual mandatory in-service training that focuses on policies and procedures.

## labor unions

Joining a union is not required but many correctional officers find it advantageous to join. Officers who work for state-run facilities often have the option of joining the union for all state employees. In return for weekly or monthly dues, members receive services intended to improve their working conditions. David says that he joined his union recently but has not become very involved yet. Currently, his union is focusing on rising concerns among Oklahoma correctional officers about the extreme shortage of staff, overtime pay, and safety issues.

## who will hire me?

There are facilities in virtually every part of the country that need corrections officers. You could work in a prison or jail that houses men, women, juveniles, or a combination. Roughly 60 percent of officers work in state-run facilities. The rest are employed at city and county jails, while the smallest percentage works for the federal government. A number of officers are hired by privately run correctional facilities. Depending where you live, you may also have the choice of maximum-, medium-, and minimum-security facilities.

To apply for a job, simply contact your state's department of corrections or the Federal Bureau of Prisons and request information about entrance requirements, training, and job opportunities. In addition, there are numerous journals that include a list of job openings. For example, three of the professional organizations listed at the end of this book—the American Correctional Association, the International Association of Correctional Officers, and the Federal Bureau of Prisons—have both publications and web sites with job listings. Other publications that post job openings are Corrections Compendium and Inside Corrections.

Most correctional institutions require candidates to be at least eighteen years old (sometimes twenty-one years old), have a high school diploma, and be a U.S. citizen with no criminal record. There are also health and physical strength requirements, and many states have minimum height, vision, and hearing standards. Other common requirements are a driver's license and a job record that shows you've been dependable.

## where can I go from here?

David feels that his facility offers good opportunities for advancement. He began as a cadet, was promoted to corrections officer I, then corporal, and now sergeant. He envisions himself moving into counseling or case management, and eventually unit management. He says that his college degree has probably helped differentiate him when he has applied for promotions, but that an excellent performance record is important, too.

Many officers take college courses in law enforcement or criminal justice to increase their chances of promotion. In some states, officers must serve two years in each position before they can be considered for a promotion. Training officer Wayne Ternes at the Montana State Prison says that his facility is seeing more college graduates applying for the position of corrections officer. "They see that it's not just a job to fill the time until something better comes along. It's an excellent career path." Wayne himself began as a corrections officer, and then worked in food service, all the while with his eye on training.

With additional education and training, experienced officers can also be promoted to supervisory or administrative positions such as head corrections officer, assistant warden, or prison director. Officers who want to continue to work directly with offenders can move into various other positions. For example, probation and parole officers monitor and counsel offenders, process their release from prison, and evaluate their progress in becoming productive members of society. Recreation leaders organize and instruct offenders in sports, games, and arts and crafts.

**Just Who Lands in a Federal Prison?**

If you go to work in the federal prisons, do you wonder about what kind of criminals you'll be dealing with? It might surprise you to learn that drug offenses far and away bring the most people to federal prison. Sixty percent of inmates in federal prisons committed crimes involving drugs. The list below shows the rest of the picture:

| | |
|---|---|
| Drug offenses | 60.2% |
| Robbery | 9.6% |
| Firearms, arson | 9.0% |
| Property offenses | 5.9% |
| Extortion, fraud | 5.6% |
| Immigration | 3.4% |
| Violent offenses | 2.6% |
| White collar | 0.7% |
| National security | 0.1% |

## what are some related jobs?

The U.S. Department of Labor classifies corrections officers with workers in detention occupations and in protective service occupations. Included in these categories are people who guard prisoners at institutions other than prisons and jails, such as mental institutions or courthouses; people who maintain law and order during courtroom proceedings, such as bailiffs; or people who prevent crime, such as security guards and police officers. Some of these related jobs include armored car guards, security guards, bodyguards, bailiffs, parole officers, patrol conductors, probation officers, police officers, immigration guards, fire fighters, and fish and game wardens.

| Related Jobs |
|---|
| Armored car guards |
| Bailiffs |
| Bodyguards |
| Fire fighters |
| Fish and game wardens |
| Immigration guards |
| Parole officers |
| Patrol conductors |
| Police officers |
| Probation officers |
| Security guards |

## what are the salary ranges?

According to a 1996 survey in *Corrections Compendium,* the national journal for corrections officers, salaries of corrections officers vary dramatically from institution to institution. Entry-level salaries at state-run facilities, for instance, range from $12,930 to $31,805. Salaries of experienced corrections officers at state facilities range from $17,300 to $41,730.

At the federal level, starting salaries were about $19,000 in 1994. Federal officers with several years of experience earned about $30,000, and supervisory officers at the federal level averaged around $55,000.

Most corrections officers participate in medical and dental insurance plans offered by their facility, and can get disability and life insurance at group rates. They also receive vacation and sick leave, as well as retirement benefits. Many correctional facilities, such as the Oklahoma Department of Corrections, offer full retirement benefits for corrections officers after twenty years of service.

## what is the job outlook?

Corrections officers can count on steady employment and good job security. Prison security has to be maintained at all times, making corrections officers unlikely candidates for layoffs, even when there are budget cuts. These jobs are rarely affected by changes in the economy or government spending. Due to a high turnover rate, a budget can usually be trimmed by simply not replacing the officers who leave voluntarily.

There are other factors pointing to the strong job outlook for corrections officers. The prison population has more than doubled in the past ten years, and new prisons are being built to house these inmates. This growth is expected to continue, with the trend in the United States being toward mandatory sentencing guidelines, longer sentences, and reduced parole. All of this translates into a strong need for more corrections officers. It is estimated that another 142,000 jobs will be created in the next fifteen years, an increase in employment of 61 percent. While entry-level jobs are plentiful, competition will continue to be stiff for the higher-paying supervisory jobs.

Certain technological developments—such as closed-circuit television, computer tracking systems, and automatic gates—do allow a single corrections officer to monitor a number of prisoners from a centralized location, but the impact of these technologies on overall staffing needs is minimal.

## how do I learn more?

### sources of additional information

Following are organizations that provide information on careers in corrections.

**The American Correctional Association**
4830 Forbes Boulevard
Lanham, MD 20706
301-918-1800

**The American Probation and Parole Association**
c/o Council of State Governments
Iron Works Pike
PO Box 11910
Lexington, KY 40578
606-244-8203
Email: appa@csg.org

**The International Association of Correctional Officers**
Box 53, 1333 South Wabash Avenue
Chicago, IL 60605
312-996-5401

**The American Jail Association**
2053 Day Road
Hagerstown, MD 21740-9795

**Advancement Possibilities**

**Head corrections officers** supervise other corrections officers. They assign duties, direct the activities of inmates, arrange for the release and transfer of inmates, and maintain overall security measures.

**Wardens**, sometimes known as **prison directors**, oversee all correctional staff. They are ultimately responsible for the safety of the prisoner population, as well as the security of the prison and its employees.

**Probation and parole officers** monitor and counsel offenders, process their release from prison, and evaluate their progress in becoming productive members of society.

**Federal Bureau of Prisons**
National Recruitment Office
320 First St. NW, Room 460
Washington, DC 20534

# bibliography

Following is a sampling of materials relating to the professional concerns and development of corrections officers.

## Books

Bales, Don L. *Correctional Officer Resource Guide.* A guide for the aspiring corrections officer, filled with information about where the jobs are and how to get them. Lanham: American Correctional Association, 1997.

Schroeder, Donald J., and Frank A. Lombardo. *How to Prepare for the Correction Officer Examination.* Hauppauge: Barrons Educational Series, 1996. A study guide to the exam that also offers insight into what a corrections officer can expect after passing it.

Steinberg, Eve P. *Correction Officer.* 11th ed. Information about test-taking practice. Includes sample exams, detailed explanations, and illustrations. New York: Arco Publishers, 1997.

## Periodicals

*Office of Correctional Education News.* Quarterly. Discusses and evaluates innovations in correctional education, and provides useful information on correctional research, history, and other federal and private resources available to practitioners serving incarcerated individuals. Office of Correctional Education, 600 Independence Avenue SW, MES 4529, Washington, DC 20202-7242..

# cytotechnologist

**Definition**

Cytotechnologists study cells. They assist in the collection of body cells, prepare slides, and examine cells using microscopes. Cytotechnologists search for cell abnormalities in order to aide in the diagnosis of disease. They concentrate on finding abnormalities that may signal the beginning and progression of cancerous growths.

**Salary range**

$26,000 to $35,000 to $50,000

**Educational requirements**

High school diploma; at least one year of professional instruction in cytotechnology after or included in a bachelor of science degree.

**Certification or licensing**

Highly recommended with mandatory licensing in some states

**Outlook**

About as fast as the average

**GOE**
02.04.02

**DOT**
078

**O*NET**
24308G

## High School Subjects

Biology
Chemistry
Computer science
English
(writing/literature)
Mathematics

## Personal Interests

Building things
Computers
Drawing
Helping people:
physical
health/medicine
Science

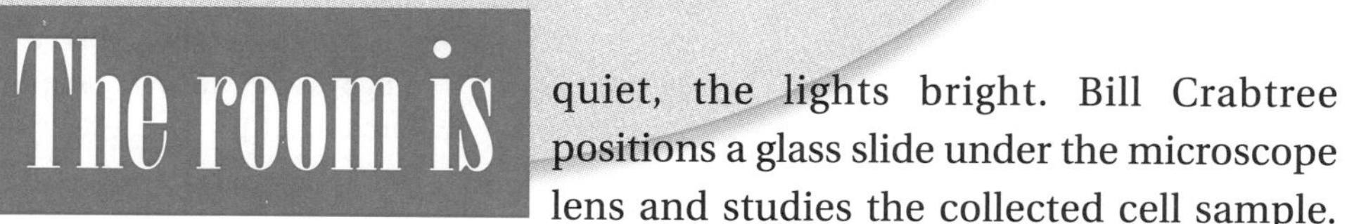

**The room is** quiet, the lights bright. Bill Crabtree positions a glass slide under the microscope lens and studies the collected cell sample. He's on the lookout for abnormal growth patterns. A cytotechnologist and director of the Indiana University School of Medicine's cytotechnology program, Bill will spend the majority of his day at the bench, peering through the microscope and picking out cell samples that appear to be cancerous. "We're cell detectives," says Bill. "We really affect people's lives. What we do can be a matter of life or death."

Like good detectives, cytotechnologists are careful and precise. They are laboratory specialists who search cell specimens, seeking out abnormalities. Some forms of cancer, a tumor growing on someone's liver for instance, can be seen with the naked eye, but other types of cancer are not so easily detected. Cytotechnologists are particularly effective at finding cancer of the cervix. Much of their work involves diagnosing Pap smears, cell samples that are taken during routine gynecological exams. Because of the detective skills cytotechnologists bring to their profession, the death rate from cancer of the cervix is 25 percent lower than what it was forty years ago.

"Cytotechnology is a challenging field," says Bill, "but it's a good one if you like laboratory work. Plus it's a field where you can really make a difference."

## what does a cytotechnologist do?

*Cytotechnologists* perform the majority of their work by looking through a microscope at prepared slides. They study cell growth patterns and check to see whether the specimens under the lens have normal patterns or abnormal patterns. Abnormal patterns can indicate the presence of disease. Cytotechnologists search for changes in cell color, shape, or size. A change in any one of these can be cause for concern. In any single slide there may be more than one hundred thousand cells; so cytotechnologists must be patient and thorough in order to make accurate evaluations.

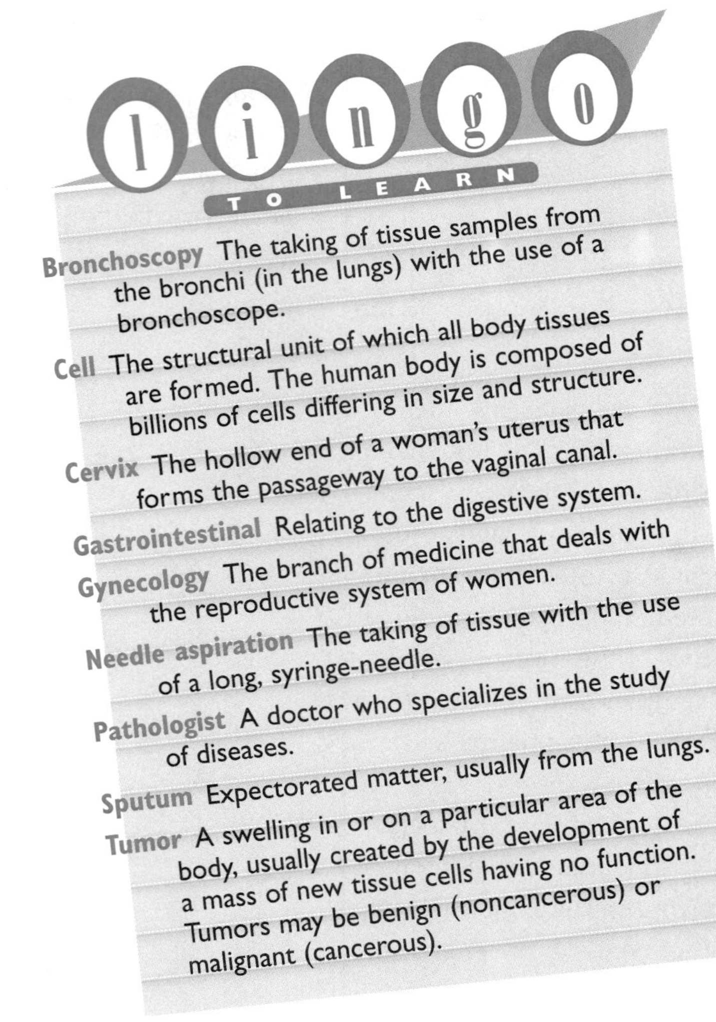


Cytotechnologists do more than peer through microscopes. When they are away from the laboratory, they may work at patients' bedsides assisting doctors in the direct collection of cell samples. The respiratory system, urinary system, and gastrointestinal tract are some of the body sites from which cells may be gathered. Cytotechnologists also assist physicians with bronchoscopies and with needle aspirations, a process that uses very fine needles to suction cells from many locations within the body. Needle aspirations sometimes replace invasive surgeries as a means of gathering microscopic matter for disease detection. Once cells are collected, cytotechnologists may prepare the slides so that the cell samples can be examined under the microscope. In some laboratories, medical technicians prepare slides rather than cytotechnologists.

Another part of the cytotechnologist's day is spent keeping records, filing reports, and consulting with co-workers and pathologists on cases. Cytotechnologists can issue diagnoses on Pap smears if the diagnosis is normal. However, if cell examination indicates any abnormalities, Pap smear results as well as other cytological results are sent on to supervising cytotechnologists or to pathologists for review.

Most of the time cytotechnologists work independently. They may share lab space with other personnel, but the primary job of a cytotechnologist is to look through the microscope and search for evidence of disease. Most cytotechnologists work for private firms that are hired by physicians to evaluate medical tests, but many cytotechnologists also work for hospitals or university research institutions.

## what is it like to be a cytotechnologist?

"Where I work," says Bill, "it's routine for cytotechnologists to spend 50 or 60 percent of their day at the microscope. Like all cytotechnologists, we see a lot of Pap smears, but at the Indiana University School of Medicine we also look at cell samples that are nongynecological too. We look at material collected from any kind of solid tumor, at abdominal fluids, thoracic fluids, urine samples, sputum, brushes and washes of the lungs and bronchial passages, lesions from the gastrointestinal tract, scrapings from the mouth or skin, and spinal fluid. Occasionally we can identify microbiological infections, bacteria, fungus and such, but principally we're diagnosing cancer or its precursors."

Cytotechnologists usually work a standard eight-hour day, five days a week. They also do a lot of work on computers. They enter and retrieve information using computers. Diagnostic results are entered into computers so that pathologists can directly access the information and make the results available to patients.

"Since I'm at a university medical school," says Bill, "part of my day is involved with the education program for cytotechnology students and part of my day is involved with research. Another part of my time is spent working on quality control and going through quality assurance procedures. I also file reports and keep records using the computer."

Susan Dingler also works at a teaching hospital. "Simply working in a laboratory can be repetitive," she says, "but here at the School of Cytotechnology at Henry Ford Hospital in Detroit, cytotechnologists rotate job duties. One week I might work in the preparation area, extracting cells and preparing slides. (We use various methods, depending on the source of the cells and the amount of sample available.) Another week I might assist with needle aspirations. Or I could be working in the CAT scan room. Then again, I could be involved with coordinating our education program."

"Everybody's cells and every single tumor is different," says Susan."There are similarities—otherwise we wouldn't be able to do our work—but because no two cells look exactly the same, I'm never bored when I'm looking through the microscope. Cytotechnology is an especially challenging field."

Bernadette Inclan works for a private company in Phoenix, Arizona. Unlike Bill and Susan, she does not teach students or perform research. She does not work for a university. As a quality insurance inspector, she oversees the work done in several private laboratories located across the Southwest and along the West Coast.

Because the diagnoses that cytotechnologists make can literally be matters of life and death, the profession is governed by a system of checks and balances. Bernadette and other supervisors help make sure that the medical tests cytotechnologists perform are accurate. For example, she does second screenings to assure quality control and to confirm test results on high-risk patients.

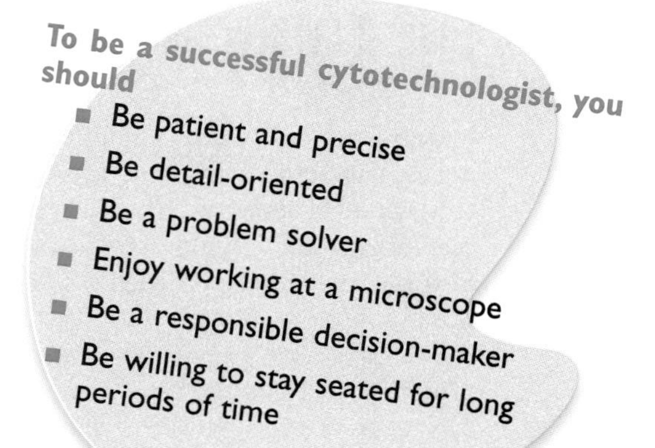


When specimens are hard to screen because the sample itself is very small or because the cells are obscured by too much blood, Bernadette is also called upon to give her expert opinion. In fact, any time any cytotechnologist in her company's laboratories marks a sample "Please Check" (meaning that the cytotechnologist is unable to make an unequivocal decision about the contents of the cell sample), Bernadette or one of the other supervisors double-checks the work.

Bernadette is also involved in managing people. "It's an exciting time for us,"she says. "My firm is merging with another health lab and so that means a lot of changes. I'll do a lot of traveling between company sites and help make sure that the merger goes well for our cytotechnology team. I'll also make sure that the work continues successfully in the lab."

Team work is an important feature of cytotechnology. "Sure, we're behind individual microscopes a lot of the time," says Bill. "But really we're rather unique in the health care field. Cytotechnologists work more closely with physicians, especially with pathologists, than do most laboratory workers. Plus we work together as cytotechnologists, consulting with each other and with our supervisors on unusual cases."

## have I got what it takes to be a cytotechnologist?

"The person who gets straight A's does not necessarily make the best cytotechnologist. Our job involves more than learning facts and memorizing information. You need to know how to apply what you learn when you look through the microscope. You need a knack for detail," Bill says. "Cytotechnologists are very art-oriented really." He goes on to explain that cytotechnologists must be good observers. Like artists they search for subtleties in color, shape, and size. "Cytotechnology is an art as well as a science," Bill says.

"If you like to work jigsaw puzzles, cytotechnology just might be the career for you," suggests Susan. "Like jigsaw puzzle fans, cytotechnologists enjoy comparing the shapes and sizes of small objects, scanning a lot of similar objects as they try to detect subtle differences. Both puzzles and microscope work require hard concentration, patience, and observation of acute detail."

Bernadette adds, "You must be meticulous and able to make your own decisions. The supervisor is there to back you up, but a lot of your original work will be done at the bench, working alone at the microscope. Cytotechnologists use worksheets and must follow the printed orders exactly. In addition, you must be able to sit for long periods of time without moving from the bench. It's not like working with a computer keyboard. You can't shift positions and place the microscope on your lap."

FYI

**Preparing a Slide**
To prepare a slide, cells are spread, or "fixed," in the center of narrow glass rectangles. Following this, colored dye is added to emphasize cell structure and make disease detection easier. Finally, using a smaller piece of glass, the specimens are covered and sealed in order to preserve them.

"One big advantage of working in cytotechnology," says Bernadette, "is that you can come to work, do your job, and then go home. It is literally impossible to take your work home with you. Oh, you may go home and mull over an interesting case sometimes, but that's all you can do. Plus, if you're ever unsure about a diagnosis, there's always another set of eyes there to help you out. Still, cytotechnology can get monotonous at times, especially if you're working in a huge lab and doing nothing but processing Pap smears."

Cytotechnology is a good field for someone who is less people-oriented, but who still enjoys working in the medical field. "I wanted a job with stability and lots of opportunities," says Bill, "where I could make good money and work at an interesting job in a laboratory."

"It's an especially challenging field," Susan says. "The best thing for me is that cytotechnologists are involved in patient diagnosis. We don't just handle specimens and pass on the results. We're the first to evaluate. We get to give our opinion. Cytotechnology appeals to people who want to be responsible and who want to be involved in something that will have a direct effect on patient care."

## how do I become a cytotechnologist?

Bill fell into the career of cytotechnology by accident. "I always enjoyed studying biology," he says. "I was interested in disease, but I knew that I didn't want to become a physician. Then I stumbled across a brochure that described laboratory careers, including cytotechnology. It sounded good." He attended school at the University of Tennessee and then had the chance to work on a large research project, the National Bladder Cancer Project. He's been in the field for eighteen years and now directs a program to train new cytotechnologists.

Susan entered the field of cytotechnology after studying medical technology for two years. She'd taken seven chemistry classes already and didn't look forward to taking any more. She enjoyed studying the biological sciences, however, and began searching for a new course of study that could take her in that direction. Cytotechnology fit what she was looking for, and, thirty-two years later, she still enjoys her work as a cytotechnologist.

## education

### High School

Biology, chemistry, and other science courses are necessary for students wishing to become cytotechnologists. Math, English, and computer literacy classes are also important. In addition, students should be sure to fulfill the entrance requirements of the college or university they plan to attend.

### Postsecondary Training

There are two routes that students may take to become cytotechnologists. One route involves obtaining a bachelor's degree in biology, life sciences, or a related field. Following this, students enter a one-year, postbaccalaureate certificate program sponsored by an accredited hospital or university.

The second route involves transferring into a cytotechnology program during the junior or senior year of college. Students on this track earn a bachelor of science degree in cytotechnology. In both cases, students earn a college degree and complete at least one year of training devoted to cytotechnology.

General college class work includes biology, microbiology, parasitology, cell biology, physiology, anatomy, zoology, histology, embryology, genetics, chemistry, computer science, and mathematics. Additional classes include cytochemistry, cytophysiology, diagnostic cytology, endocrinology, medical terminology, the study of inflammatory diseases, and the history of cytology. Students learn how to prepare slides, use microscopes, and follow safe laboratory procedures.

> "The person who gets straight A's does not necessarily make the best cytotechnologist."

## certification or licensing

Cytotechnology graduates (from either degree programs or certificate programs) may register for the certification examination given by the Board of Registry of the American Society of Clinical Pathologists. Most states require cytotechnologists to be certified, and most employers insist that new employees be certified. Usually it is a requirement for advancement in the field.

Many continuing education programs exist for professionals working in the field of cytotechnology. It is important that practicing cytotechnologists remain current with new ideas, techniques, and medical discoveries.

## scholarships and grants

Most colleges and universities offer general scholarships. Institutions with specific cytotechnology programs are most likely to have scholarships available for students interested in the field.

In exchange for a promise of two to three years of staff work at a private laboratory, some employers offer scholarships to students.

### Advancement Possibilities

**Teaching supervisors in medical technology** teach one or more phases of medical technology to students of medicine, medical technology, or nursing arts.

**Chief medical technologists** direct and coordinate activities of workers engaged in performing chemical, microscopic, and bacteriologic tests to obtain data for use in diagnosis and treatment of diseases.

**Cytology supervisors** supervise and coordinate activities of staff in cytology laboratories.

**Pathologists** are medical doctors who specialize in the study and diagnosis of diseases.

## internships and volunteerships

Colleges and universities, along with professional organizations, are sources of information on work-study projects and student internships. For more information, contact the program director at individual teaching institutions. A list of accredited cytotechnology programs may be obtained through the American Society of Cytopathology (*see* "Sources of Additional Information").

## who will hire me?

Like many veteran cytotechnologists, Susan now directs a teaching program. More than sixty hospitals and universities have programs in cytotechnology. Other cytotechnologists are involved in research. Some cytotechnologists work for federal and state governments and some work in private industry, nursing homes, public health facilities, or businesses. The majority of cytotechnologists work for either hospitals or for private laboratories.

Demand for cytotechnologists is high, and recruiters often visit universities and teaching hospitals in the months prior to graduation. Professional journals also list advertisements for employment (*see* "Bibliography").

## where can I go from here?

Cytotechnologists who work in larger labs may move up to supervisory positions. However, cytotechnologists seeking managerial or administrative positions in smaller labs may find limited opportunities for advancement. Another career move might be to enter the teaching field and direct classes or oversee research.

Some cytotechnologists join forces with medical directors and open their own laboratories. One creative cytotechnologist opened his own business by concentrating on his expertise at staining cells. He developed his own line of chemicals and is now a leader in the staining industry.

## what are some related jobs?

The U.S. Department of Labor classifies cytotechnologists under the headings *Occupations in Medical and Dental Technology* (DOT) and *Laboratory Technology: Life Sciences* (GOE). Also under these headings are medical technologists who perform laboratory tests and analyze data for diagnosis, treatment, and prevention of disease; orthotic assistants who help with the fabrication and fitting of orthopedic braces for patients with disabling conditions; and pheresis specialists who collect blood components and provide therapeutic treatments, such as replacement of plasma.

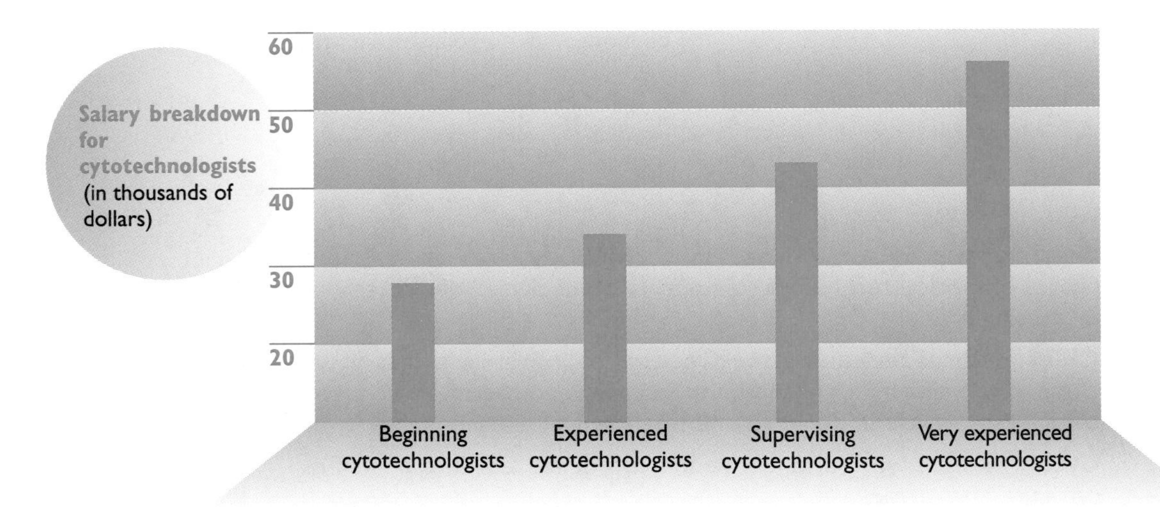

# in-depth

**Cytotechnology and the Pap Smear**

The field of cytotechnology is only a half century old. It began in the 1940s, more than ten years after Dr. George N. Papanicolaou (1883–1962), a Greek-American physician, developed a procedure for early diagnosis of cancer of the uterus in 1928. Dr. Papanicolaou collected cell samples by scraping the cervixes of female patients. He placed these cell samples on glass slides. The slides were then stained so that individual cell differences and abnormalities could be studied more easily. Using microscopes, medical laboratory workers then compared cells known to be healthy against those known to be diseased. As the value of the "Pap smear" (the term used for the test Dr. Papanicolaou developed) became more widely accepted, demand for laboratory personnel trained to "read" Pap smears grew, and the career of cytotechnologist was born. Over time, the field of cytotechnology expanded to include examination of other cell specimens besides gynecological samples.

| Related Jobs |
| --- |
| Biological aides |
| Food testers |
| Medical laboratory technicians |
| Public health microbiologists |
| Embalmers |
| Biomedical technicians |
| Microbiology technologists |
| Veterinary laboratory technicians |

## what are the salary ranges?

Because of the need for skilled cytotechnologists, salaries are increasing. Salaries depend on the education and experience of the employee and on the type of employer. Those employed by the federal government earn slightly less overall. Cytotechnologists working in private laboratories earn slightly more than those working in hospitals. Geographically, salaries are highest in the West.

Beginning staff level cytotechnologists earn an average salary of between $26,000 and $28,000 annually. More experienced staff cytotechnologists earn about $35,000 a year, and supervising or highly experienced cytotechnologists often earn more than $50,000 annually.

## what is the job outlook?

Competition to enter cytotechnology programs is keen and shortages still exist for qualified graduates. The demand for cytotechnologists is especially high in private industry. As more and more hospitals contract with private companies to perform work formerly done inside hospitals, more jobs will open up. Additional governmental regulations now limit the number of slides cytotechnologists may work with each day and this adds to the shortage of qualified personnel.

In the future, the demand for cytotechnologists may be slowed somewhat by advances in laboratory automation, but for now demand remains very high.

Vacancy rates for cytotechnologists are highest in the southern United States and in the mountain states of the West. Vacancy rates are lowest in areas closest to universities or teaching hospitals with cytotechnology programs, but shortages in this field exist in every geographical area.

# how do I learn more?

## sources of additional information

Following are organizations that provide information on cytotechnology careers, accredited schools, and employers.

**American Society of Clinical Pathologists**
2100 West Harrison
Chicago, IL 60612
312-738-1336

**American Society for Cytopathology**
400 West 9th Street, Suite 201
Wilmington, DE 19801
302-429-8802

**American Society for Cytotechnology**
920 Paverstone Drive
Raleigh, NC 27615
919-848-9911

**Division of Allied Health Education and Accreditation**
535 North Dearborn Street
Chicago, IL 600610
312-645-4624

## bibliography

Following is a sampling of materials relating to the professional concerns and development of cytotechnologists.

### Periodicals

*ASC Bulletin.* Bimonthly. American Society of Cytopathology, 400 West 9th Street, Suite 201, Wilmington, DE 19801, 302-429-8802.

*Cell.* Biweekly. Information about cell biology and other biological fields. Cell Press, 1050 Massachusetts Avenue, Cambridge, MA 02138, 617-661-7060.

*Cell Biochemistry and Biophysics.* Bimonthly. Humana Press, 999 Riverview Drive, Suite 208, Totowa, NJ 07512, 973-256-1699.

*Cell Transplantation.* Bimonthly. Analyzes how cell transplantation relates to curing human disease. Elsevier Science Inc., Box 945, New York, NY 10159-0945, 212-633-3730.

*Cellular Immunology.* Fourteen times per year. Investigates immunological cell activity. Academic Press, Inc., 525 B Street, Suite 1900 San Diego, CA 92101-4495, 619-230-1840.

*Cellular Physiology and Biochemistry.* Bimonthly. Research journal discussing cellular function. S. Karger AG Allschwilerstr, 10, PO Box CH-4009, Basel, Switzerland, 41-61-3061111.

*Cytotechnology.* Nine times per year. Kluwer Academic Publishers, PO Box 358, Accord Station, Hingham, MA 02018-0358, 617-871-6600.

# database specialist

**Definition**

Database specialists design, install, update, modify, and otherwise maintain computer databases and provide technical support and training.

**Alternative job titles**

Database administrators
Database design analysts
Database managers
Database programmers
Information systems managers

**Salary range**

$50,000 to $67,000 to $84,000

**Educational requirements**

Associate's degree

**Certification or licensing**

Voluntary

**Outlook**

Faster than the average

GOE 11.01.01

DOT 039

O*NET 25103A

## High School Subjects

Business
Computer science
Mathematics

## Personal Interests

Computers
Figuring out how things work

executing final revisions on the database system he has custom-designed for one of his company's major clients when the phone rings.

"This is Sandy, Danny. We thought of one more piece of information we would like to encode on each record. We want to cross-reference each member's profession with his or her place of employment."

"I thought we decided on encoding only profession since employers change so frequently," says Danny, on the edge of exasperation.

"Yeah, well," Sandy continues, "we've been talking and it might be nice to know both at any one time."

"We'll be a couple weeks behind," warns Danny. "It's going to mean a heavy load of reprogramming." But Sandy persists, "I'll bet you can get it done faster than that. I know what a great worker you are."

As Danny hangs up the phone, he is frustrated. "I wish they'd think of all this stuff before I begin. They have no idea how much work is generated by just the slightest change. Oh well, back to the drawing board."

## what does a database specialist do?

We are now firmly into an era referred to as the "information age." Because of the incredible strides in computer technology and the promise of continued advances, issues surrounding information management are at the forefront of the industry. What is the best way to store information in a computer? How should data be organized to allow easy access by multiple users? How can files be encoded so that only specified ones are retrieved by a given search? Everything from judicial court decisions to credit card customer accounts to professional organizations' mailing lists has to be stored in computers in an organized fashion so that the information can be accessed and updated easily, accurately, and efficiently.

The branch of computer science and technology that attempts most directly to answer these questions is database management. A database management system, also called a database engine, is the commercial software used to organize information. The term "database" refers to the actual collection of information stored by the database engine.

*Database specialists* design, install, update, modify, maintain, and repair computer databases. This range of duties involves numerous and varied tasks, usually too much work for one person to handle (unless he or she works for a small company). Typically, several individuals or a whole group of database engineers and technicians team up to make sure that all functions are covered. Specific assignments for each position in database technology and management differ according to the nature and scope of the company or department, as well as the education and work experience of the end user. Although job titles are flexible, most database specialists perform some combination of the responsibilities described in this section.

*Database administrators* rely on their knowledge of database management in order to code, test, and install new databases. They also coordinate changes made to already existing ones. Many administrators work closely with other computer professionals in order to determine the most logical and efficient way to operate databases in a given situation.

They might instead concentrate on reviewing descriptions of database changes and thinking through how those changes might affect the way information is stored, where it is stored, and the method used to access it. For example, if the management of a bank decides to use a new system for the assignment of personal identification numbers for automated teller machines, the database administrator has to analyze, among other things, how those changes will be encoded in the database system at large, how such codes will be entered into each customer's account number, and how and by whom they will be accessed since the information is highly confidential.

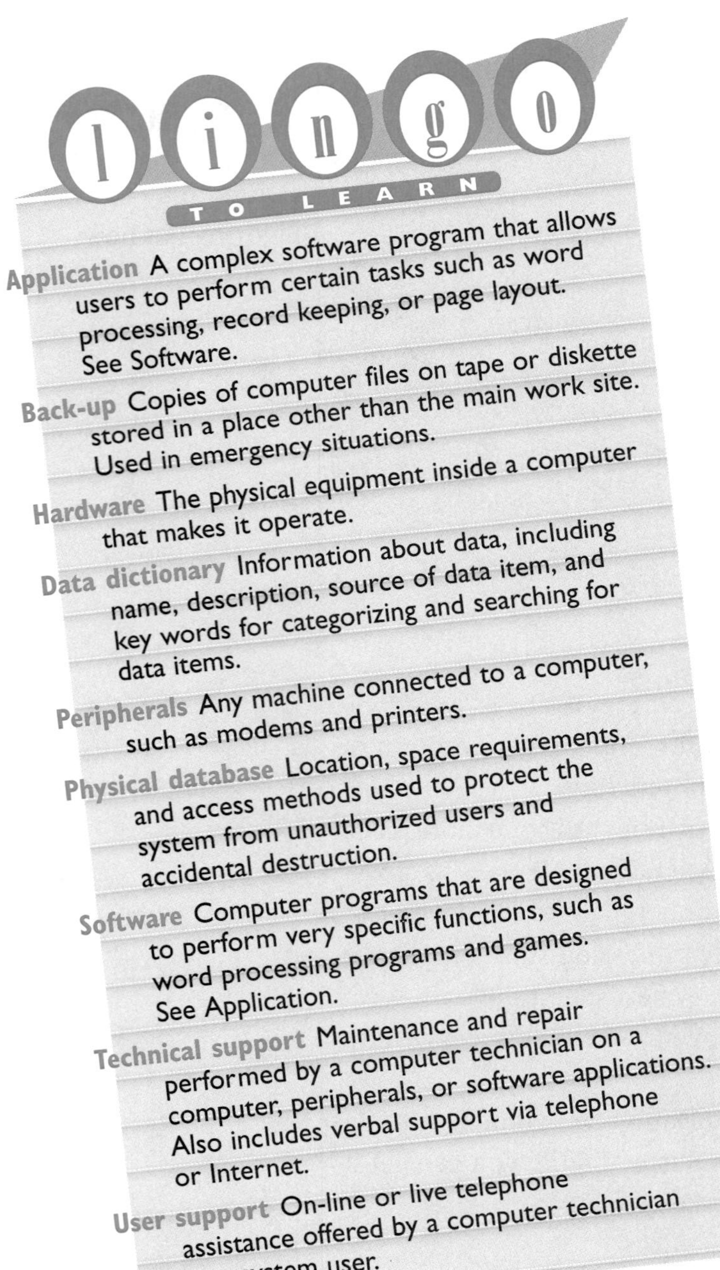
lingo TO LEARN

**Application** A complex software program that allows users to perform certain tasks such as word processing, record keeping, or page layout. See Software.

**Back-up** Copies of computer files on tape or diskette stored in a place other than the main work site. Used in emergency situations.

**Hardware** The physical equipment inside a computer that makes it operate.

**Data dictionary** Information about data, including name, description, source of data item, and key words for categorizing and searching for data items.

**Peripherals** Any machine connected to a computer, such as modems and printers.

**Physical database** Location, space requirements, and access methods used to protect the system from unauthorized users and accidental destruction.

**Software** Computer programs that are designed to perform very specific functions, such as word processing programs and games. See Application.

**Technical support** Maintenance and repair performed by a computer technician on a computer, peripherals, or software applications. Also includes verbal support via telephone or Internet.

**User support** On-line or live telephone assistance offered by a computer technician to a system user.

They then execute all necessary programming changes, such as specifying who can get access to what files when, testing and correcting errors, selecting a utility program to monitor database performance, and advising programmers and analysts about the need for further improvements. They may also be involved in training users on new procedures.

*Database design analysts* design databases and coordinate database development as part of a team. Database design includes very broad as well as very detailed work. First, design analysts review the database request, meeting at length with clients and colleagues. Next, they analyze the desired parameters of the database and decide whether or not existing database programs can be updated to include changes. Only very rarely do they design an entirely new database management system.

Design analysts then present to clients or colleagues a time and cost estimate of the implementation of the proposal. If it is approved, design analysts use database manuals to figure out how changes should be made. They revise the current data dictionary and write descriptions for programmers of how programs should access data. They also write descriptions of the logical and physical databases.

Those who specialize in adding, deleting, and modifying data items in data dictionaries are sometimes called *data dictionary administrators.*

## what is it like to be a database specialist?

Danny Robinson has worked at Pony-X-Press Printing in Columbus, Ohio for about twenty years. His professional involvement in database management happened almost by accident.

"Working in the graphic arts industry, I discovered our association clients needed services that databases could provide. I became interested in the challenge of using a data source for mailings, directory publication, membership records, and catalog production." For Danny, this need was enough—he got involved. His company is still small enough that he does many of the database tasks himself.

His typical day is quite hectic. "Clients tend not to realize how much careful, methodical, planned work goes into database design and programming; I end up working right up to the deadline in order to satisfy last-minute requests," he says. His work load is divided between three major areas: technical support to systems he has already developed and installed, database design for new clients, and database programming for a wide range of applications and conversions necessary for the customized needs of individual clients. "As for a daily schedule, it's up to me. I have to keep track of all my clients and their respective deadlines; no one looks over my shoulder."

"I end up working right up to the deadline in order to satisfy last-minute requests"

On any given day, Danny might have a planning meeting with a client. There, he discusses their database needs—whether they want, for example, a simple updateable mailing list or a more complex system that allows information to be encoded on each file and cross-referenced for many different purposes. Dealing with people not trained in database technology can be challenging in itself. Clients may know their data, but not how it could, or should, be structured in a database. Danny must listen to their requests and ideas and guide the clients through the process to give them a preliminary idea of the feasibility of the project. Client meetings are held both in his office and at the client's place of business.

Danny usually spends at least a few hours every day actually programming. This involves typing out, detail by detail, every single command in the huge programs. "It can be tedious and frustrating work," says Danny. In general, he is programming from his own designs and so has little problem clarifying design specifications. In larger companies, however, two different people might perform these tasks and might therefore have to meet to work through discrepancies.

Danny also programs a lot of custom conversions. The process involves revising standard database applications in order to make them meet very specific needs of a particular client.

He is also the technical support person in his company. When systems he has already designed and installed are giving clients difficulties, he has to make modifications to the programs or computer equipment in order to fix the problems. He might also assist users when they run into system malfunctions.

When design and programming are completed, Danny oversees installations. For some clients, this involves working at their business locations. He might also provide training to client users of the system.

Danny has a very large office with good lighting and that's well equipped. Since he is the main database employee in his company, he enjoys the luxury of having several computer terminals, modems, CD ROMs, and printers at his immediate disposal. Working with computers day in and day out can lead to some deterioration of vision and carpal tunnel syndrome, a painful condition in the wrist. While Danny has experienced some loss of vision, he has managed to avoid wrist problems.

In theory, database specialists maintain normal forty-hour work weeks, but Danny works a lot more than that. Since he receives a salary, he is not compensated for overtime hours. "Any extra time I put in helps me in the long run, though," says Danny, "because I am always learning new ways of doing new things. In an industry that is constantly changing, that's a definite plus."

Database specialists, like most computer professionals, work for the most part in companies that offer a full benefits package, including health insurance, sick leave, paid vacation, and educational allowance.

To be a successful database specialist, you should

- Be detail oriented
- Have strong analytical and problem-solving skills
- Like to figure out creative solutions to challenging problems
- Be able to communicate effectively with people on all levels of technical understanding
- Be able to analyze a problem from several angles

## have I got what it takes to be a database specialist?

Database specialists, whether involved with design, programming, or administration, must eventually be able to account for every bit of information used by the system. The ability to pay extremely close attention to detail is therefore an indispensable quality of a database specialist.

"Programming commands have to be exact or the computer simply won't understand it," explains Danny. In addition, client needs are frequently very detail-ridden, making use of long strings of numbers, for example, for identification and account naming purposes. And their requests tend to be complicated, requiring the technician to keep track of a lot of details at once.

To be a successful database specialist, you must be a strong problem solver. "Much of our work boils down to this: a client wants something; we have to figure out how to give it to him or her using computers. It's really the most basic problem-solving situation," says Danny. You need good analytical thinking skills and a flair for intellectual creativity in order to be able to invent new ways of doing things. "Sometimes you come up against a wall and have to be able to step back and examine the problem from a different perspective," he says.

For Danny, the challenges of problem solving are the most rewarding aspects of his job. He experiences a great sense of accomplishment when he is able to provide a computerized solution to a problem or make a system work better and more efficiently.

Solid communication skills, both verbal and written, are essential if you want to be a database specialist. Technicians deal all the time with people who have little technical knowledge. "It's one of the most frustrating things about my job—dealing with people who do not understand that defining a project scope is critical to its design." He has to be able to explain to them in nontechnical terms the nature of the situation.

Although database specialists are trained to develop the most efficient and logical solution to a problem, you should not become frustrated if time limitations or client preferences somehow impinge on that. "Sometimes I know my way is better or that I could improve the thing if the client just gave me two extra weeks, but that's not always the reality of business," Danny says.

# how do I become a database specialist?

After working fifteen years in graphic arts, Danny went back to school to earn his bachelor's degree in a computer-related field. "I got to the point," he explains, "where I just couldn't continue my professional growth without the academic background."

## education

### High School

A high school diploma is required for all database specialists. Among the traditional subjects, mathematics, including algebra and geometry, are particularly good at helping students develop strong analytical and problem-solving skills. Science classes, especially those like chemistry and physics that rely on investigative laboratory experiments, train students how to approach a problem and solve it by employing precise scientific methods.

English and speech classes help students develop solid communication skills. Though Danny was literally terrified of a public speaking course he took, he remembers it as one that showed him he could successfully give group presentations, a skill he uses to this day in proposal presentations to clients.

High school courses in computers, whether in basics, business applications, programming, or computer science, give prospective database specialists an excellent background. Business classes like accounting or statistics can be important for technicians involved in the decision-making process that surrounds project selection, as well as giving an insight into the basic needs and operations of businesses (and future clients!). Drafting and mechanical drawing classes allow students to practice skills that become important in flow chart and schematic drawing construction and analyzing.

### Postsecondary Training

Database technology is relatively new in many businesses around the country. In many instances, companies pay long-time employees to acquire training in database management through self-study, vendor educational workshops, and technical school classes. Danny participated in all of these before returning to school full time.

Self-study involves reading related trade books and magazines, investigating and learning about the computer products on the business and consumer market, talking with people in the field to find out what's out there, and practicing database design and programming. "I read everything I could get my hands on," says Danny. "Anything that had anything to do with computer technology. And I hacked around on my computer a lot."

Hardware and software vendor workshops provide employees of a client company with training on the systems purchased by their employers. "The workshops are run by people who know the product inside and out. They are helpful and know what they are doing," says Danny.

Depending on the scope and needs of the company and the prior experience of the employee, postsecondary training is not always necessary. Employees interested in pursuing database management who work in a company that could benefit from such on-staff expertise should investigate these types of opportunities. Employers may offer educational assistance or time off to facilitate study.

Newcomers to the workforce, however, should be formally educated in some area of computer technology. Many technical schools offer courses in database management as part of an associate's degree in a business-related computer applications program. An associate's degree demonstrates to a prospective employer that the applicant has the intellectual ability, general computer knowledge, and professional drive to become an outstanding database specialist.

Some people, like Danny, feel the need to earn a bachelor's degree in order to remain competitive in the job market. "I realized that my knowledge was so specific to company functions that I might not be qualified to get a job elsewhere," he explains. And in an ever-changing market, that's not good.

Danny completed a bachelor of science degree in computer and information systems. Some of the courses, like business management, were easy for him, since he had already been employed in the field for fifteen years. He found classes like micro and macro economics boring because he was not at all interested in them, nor did he see the pertinence of such disciplines for his occupation. "That's the thing about a university degree as opposed to a technical degree," he explains. "You have to take general education requirements that may not interest you."

Overall, though, Danny is glad that he went back to school. "I'm even thinking about a master's now." And indeed, the more formal education an individual has, the more likely he or she will progress to top-paying positions.

## certification or licensing

There is no standard certification for database specialists. Certain hardware and software vendors might offer training certificates when their workshops are completed successfully, but these carry relatively little weight with new employers.

Certification through commercial associations and companies are springing up in other computer specialties, such as network administration and engineering certification offered by Novell, Inc. and others. As database management continues to expand in importance, commercial organizations might begin development of similar certification programs. Interested individuals should keep an eye on advertisements in computer magazines, keep up with on-line discussion groups, and surf the Internet to find out about future possibilities.

## scholarships and grants

There are currently no scholarships or grants specifically designated for individuals wishing to pursue education in database management. As mentioned above, individuals already employed should speak with supervisors, managers, and the personnel office about opportunities for tuition reimbursement and scheduled time off for education.

Individuals seeking formal education should work closely with their school's financial aid department to keep on top of the latest scholarships, grants, and federally subsidized student loans. Some professional associations geared toward the computer industry offer student scholarships and grants.

## internships and volunteerships

There are not many internships available for individuals interested in database management. This is due in part to the fact that so many people train themselves in the field while already employed full time. As demand for database specialists grows, however, companies may be open to internship possibilities. It may be to your benefit to convince them.

As in any computer specialty, hands-on experience is vitally important. One way to put database knowledge to work is to offer volunteer services to a local charity, church, or small business. These organizations have a lot of information to keep track of, like membership and donor files, mailing lists, and monthly newsletters, that database automation is crucial to efficient operations. Joining local or school computer clubs and learning database software programs at home will also be to your benefit.

## who will hire me?

Database specialists work in any business or company that relies on computer databases for part of its business operation. Retail stores, catalog companies, insurance companies, on-line database or communications services, financial institutions, hospitals, government agencies, universities, public and private school systems, computer companies, and any service industry business are just a few of the types of organizations for whom database management is crucial.

In order to understand who typical employers of database specialists might be, just imagine a company like a mail- and phone-order catalog. A retail company such as this has hundreds of items for sale and all pertinent information about them is stored on the computer. Item prices, available stock, and general descriptions are readily accessible to sales representative and warehouse personnel alike. In addition, the status of an item changes frequently—prices change, stock is depleted, or the item is discontinued. Then, imagine the company's mailing lists; they may have customers numbering in the thousands and tens of thousands. Customers have identification and account numbers; their addresses change and they might only receive some, rather than all, catalog mailings. This is a lot of information to encode, store, and organize, and database specialists do it all. So, in any business where such a large quantity of information needs to organized, there is demand for database specialists.

The diversity of companies employing database specialists becomes clear when considering that Danny's job in printing services led him to the field. His company specializes in the printing of a client company's annual reports, member directories, and mailing lists. Every year, these projects would require an enormous amount of labor. Since relatively few changes occurred from one year to the next, Danny figured that inputting such data in databases would save both time and money. By training himself in how to achieve that, Danny created his own professional position.

If you're seeking a position in database management, you should consult local newspaper classified ads. You should also read the job lists in computer trade magazines, like

# in-depth

### Databases in Everyday Life

Many people do not realize the extent to which they come into contact with computer databases in their daily lives. In fact, databases are used all around us in order to make everyday interactions more efficient and accurate. Think about some typical Saturday afternoon errands to understand just how much this is true.

The first stop might be to return the movie videos rented the night before. Since the video store keeps a record of all its videos and clientele on a computer database, the clerk logs the return into the computer, and the database automatically credits the account, tallies appropriate fees, and marks the video as available for the next customer.

The second stop is the library. Here, you look up a book on the electronic card catalog, which itself is a huge database that might even be connected to even bigger library databases around the world. Some libraries have on-line access to databases offering information on about anything you can think of. If you want to read reviews of current movies, see what's good on TV, or go through recent publications to research your English paper, you can find it quickly and easily, thanks to databases.

The next stop is the music store. If you want to know if a favorite old song is out on compact disc, the clerk checks either directly on the computer database or consults a printed book, the hard-copy product of the commercial music database.

Virtually any purchase made involves contact with a database. At the grocery store, the clerk uses a laser scanner that reads the product code on your box of Pop Tarts, which has the product name and price information stored on a database that is accessed with the input of the code. At the department store, the clerk there uses a similar scanner to total up your new pair of jeans. Databases allow the businesses to do much more than check out customers, however; they also make it easier for store managers to do inventory and ordering (that's one less pair of jeans, 25 are left, order more when the total reaches 15), and for owners and business executives to do better financial analysis and market studies (Jane Smith, aged 18, bought one pair of jeans on October 23rd; last purchase 2 T Shirts, October 21, send sale advertisement when jeans are on sale again).

Databases of a much larger scale also affect our lives every day. Social security numbers, and the employment, benefits, and tax information stored with them, are organized in databases. The U.S. Census Bureau keeps all of its data on databases as well. The bureau compiles statistics from these files, which store hundreds of millions of pieces of information. The Internal Revenue Service maintains databases similar in scope, as do welfare and Medicaid offices, and all other government agencies.

Have you or your parents ever received junk mail? Lists of names and addresses for political and commercial mailings are kept on databases. With the use of special codes, companies can keep track of regional or even individual consumer preferences.

These are only a few of the databases we come into contact with every day. A complete list would be so long and intricate, we would need another database to organize it!

*PC Computing, PC World,* and *Mac World,* as well as check company and job search sites on the Internet.

The Association for Computing Machinery publishes a list of job openings in database management and maintains a resume file for prospective employers of member professionals seeking employment (*see* "Sources of Additional Information").

As with many areas of business, and especially the computer industry, database specialists might find out about good job openings from colleagues employed in the industry. Professional networking can also be done with vendors, workshop instructors, and on-line contacts. Students can find out about employment opportunities through their school's placement office.

## where can I go from here?

Danny is not quite sure what he would like to do in the future, but he wants to continue his formal education in order to be qualified once he finds the right position. "There are so many developments in the industry that it's really hard to say what option I might have next," he says. The important thing for advancement, then, is keeping up with technology. "As long as I am current, I can keep meeting the growing needs of our clients."

In a large computer and information services (CIS) department, a database specialist who starts as a database administrator might next get into database design analysis and programming. These jobs require more extensive knowledge of the database programs being used. Danny already performs many design and programming functions. Design analysts and programmers have to have a solid understanding of the entire scope of the project and some idea of how the larger goals will translate into program specifications. Some design analysts and programmers might have technicians under them, who do the very detailed work of filling in the programming gaps. Further education, in any of the forms described above, is usually required for these positions.

Computer professionals generally have the option of advancing in one of two ways. The first, the more technical route, includes working with computer projects of larger and larger scope. Instead of one department's small database, for example, a technician might go on to work on the overarching system of a company's databases.

The second route is to go into the managerial side of things, which requires at least some formal business education or demonstrated leadership ability.

Along the technical route, a programmer may be promoted to the position of systems analyst; along the managerial route, he or she may go into the position of computer and information services manager.

Another possibility for advancement is to become an independent database consultant. This option really combines both the technical and managerial paths since consultants operate as small business owners. Currently, another option for database specialists is to tackle the year 2000 problem. As we approach 2000, companies with older computer systems, as well as any business that depends on dates and databases, may run into some problems. Basically, most dates on computers are stored as two digits. For example, 1997 is stored as 97. When we reach the year 2000, the computer will store the date as 00. The computer may think

### Advancement Possibilities

**Database programmers** design database systems by converting commercial software programs to systems appropriate to the company's needs. They interview future database users and company management to determine the nature and scope of the project They conceive of and articulate in reports and flow charts the overall design of the database, including explanations of what it will accomplish, how information will be stored and accessed, and what input procedures will be. They do much of the actual programming themselves, but often pass detail work on to junior programmers or technicians.

**Information services managers** oversee a large corporate or governmental department that specializes in the installation, maintenance, and repair of information systems such as databases. They make decisions concerning the feasibility of new project proposals, offer technical support to interdepartmental users of the systems, and meet with other managers to investigate their computing needs. They perform administrative duties like making the budget, approving and rejecting major purchases, meeting with clients and vendors, and supervising departmental work.

**Independent database consultants** are hired by small and medium-sized companies who require temporary assistance in designing a customized database. Consultants are generally hired to do the entire project, from start to finish; many jobs are acquired through reputation in the local business community.

that 00 is earlier than 97, which is where the issues are. Imagine, for instance, that you open a bank account in 1998. When the year 2000 arrives, the computer database that has your account information may think you don't have an account, since it interprets 00 as coming before 98. This problem will mostly affect financial institutions with old systems. Database specialists with experience with older programming languages such as COBOL are suited for year 2000 work.

## what are some related jobs?

The U.S. Department of Labor classifies database specialists under the headings *Computer-Related Occupations, Not Elsewhere Classified* (DOT) and *Mathematics and Statistics: Data Processing Design* (GOE). Also under these headings are people who help users work efficiently on their computer systems; make changes in programs when they fail; install, maintain, and repair computer communications systems; plan information back-up files to be used in case of an emergency; and establish passwords and other security codes to reserve computer access to legitimate users.

People who test the performance of computer software applications, who help software engineering solve problems with computers, and who train users on new applications are also related to this job category.

| **Related Jobs** |
|---|
| Database consultants |
| Computer programming technicians |
| Computer programmers |
| Systems analysts |
| Microcomputer support specialists |
| Computer network administrators and engineers |
| Computer security coordinators |
| Data communications analysts |
| Data recovery planners |
| Quality assurance technicians |
| Software engineering technicians |
| Computer scientists |
| Software engineers |
| Technical support technicians |
| User support technicians |

## what are the salary ranges?

Salaries for database specialists vary significantly depending on education level, experience, and specific job responsibilities of the employee. Salaries also depend greatly on the nature and size of the employer. According to a 1998 salary survey conducted by Robert Half International, Inc., a staffing agency specializing in information technology, accounting, and finance, database analysts earned between $50,000 and $62,000, while database administrators ranged from $55,000 to $69,000. Database manager salaries ranged from $67,000 to $84,000.

A similar 1998 survey by Dowden & Co., a compensation research firm, identified the Southwest as the lowest-paying region for senior database administrators, with a median salary of $65,100. The highest-paying region for this category is the Northeast, where senior database administrators can earn a median income of $70,300. Year 2000 analysts can earn the most on the West Coast, where median incomes are $70,600. They can expect to make the least amount in the Southeast, where the median income is $66,700.

## what is the job outlook?

According to the 1998-99 *Occupational Outlook Handbook,* database administration is the fastest growing occupation today. The major reason for this is that businesses of the "Information Age" simply continue to generate more and more information that needs to be organized on computer databases. In addition to this, businesses that before kept their records in hard copy form, like doctors with their long shelves of patient files, are finding it more accurate and efficient to maintain computer-automated files. As always with technological advances, business demands will become more sophisticated as technology becomes more complex.

Prospective database specialists should watch the economic situation surrounding the largest employers of database specialists. Banks, insurance companies, and other major service corporations will need more database specialists; they promise to enjoy economic stability in the years to come. Even during periods of economic slump, computer automation projects such as database design, implementation, and maintenance will tend to have priority since they offer long-term financial savings and highly efficient productivity in return. Local, state, and particularly federal government agencies will also rely heavily on database technology in the years to come, since they are also primary players in the transition to file and record automation.

As aforementioned, demand for year 2000 specialists will remain strong through 2000. Some cleanup and maintenance work will be required after 2000, but demand will drop off. Currently, however, demand is much greater than supply, and year 2000 specialists command hefty salaries. Database specialists might also find increasing opportunities as user or technical support staff for major database software manufacturers. They might also be employed as workshop instructors who teach customers how best to use the programs. The trend in the computer industry is toward offering this kind of training and troubleshooting service, so possibilities are expected to increase.

# how do I learn more?

## sources of additional information

Following are organizations that provide information on database specialist careers, accredited schools, and employers.

**American Society of Information Science**
8720 Georgia Avenue, Suite 501
Silver Spring, MD 20910-3602
301-495-0900
http://www.asis.org

**Association for Computing Machinery**
1515 Broadway
New York, NY 10036
212-869-7440
http://www.acm.org

**Association of Information Technology Professionals**
315 South Northwest Highway, Suite 200
Park Ridge, IL 60068
800-224-9371
http://www.aitp.org

**IEEE Communications Society**
305 East 47th Street
New York, NY 10017
http://www.comsoc.org

## bibliography

Following is a sampling of materials relating to the professional concerns and development of database specialists.

### Periodicals

*ACM Transactions on Data Base Systems.* Quarterly. Journal directed at the database community. Association for Computing Machinery, 1515 Broadway, 17th Floor, New York, NY 10036-5701, 212-869-7440.

*Communications of the ACM.* Monthly. Covers research and development in the field of computer science and reports on computing practices, including database techniques. Association for Computing Machinery, 1515 Broadway, 17th Floor, New York, NY 10036-5701, 212-869-7440.

*Data Base Newsletter.* Bimonthly. Newsletter providing information on various topics related to the field of data administration. Data Base Research Group, 31 State Street, Suite 1150, Boston, MA 02109, 617-227-2583.

*Database Programming & Design.* Monthly. Devoted to individuals who work with databases and related fields. Miller Freeman, Inc., 600 Harrison Street, San Freancisco, CA 94107, 415-905-2200.

*Info DB.* Bimonthly. Newsletter covering important developments in the field and providing the latest information on technology and new products. DataBase Associates, PO Box 310, Morgan Hill, CA 95038-0310, 408-779-0436.

*Macworld, the Macintosh Magazine.* Monthly. Trade magazine devoted to providing information on Macintosh computer systems. Macworld Communications, 501 2nd Street, Suite 500, San Francisco, CA 94107, 415-243-0505.

*PC Computing.* Monthly. Computer trade magazine. Ziff-Davis Publishing Co., 50 Beale Street, 14th Floor, San Francisco, CA 94105-1813, 415-578-7000.

*PC World.* Monthly. Trade magazine directed at users of IBM and IBM compatible personal computers. DJ Communications, One Exeter Plaza, Boston, MA 02116, 617-534-1200.

*SIGMOD Record.* Quarterly. Newsletter featuring scholarly articles and news reports concerning research on database systems. Association for Computing Machinery, Special Interest Group on Management of Data, 1515 Bropadway, 17th Floor, New York, NY 10036, 212-869-7440.

# dental assistant

**Definition**

Dental assistants help dentists treat and examine patients. They clean, sterilize, and disinfect equipment and prepare dental instrument trays, and assist the dentist by passing the proper instruments, taking and processing X-rays, preparing materials for making impressions and restorations, and instructing the patient in oral health care. They also perform administrative and clerical tasks, such as making appointments, maintaining patient records, and handling billing.

**Salary range**

$17,264 to $23,088 to $24,544

**Educational requirements**

High school diploma; on-the-job training

**Certification or licensing**

Voluntary

**Outlook**

Faster than the average

GOE 10.03.02

DOT 079

O*NET 66002

## High School Subjects

Biology
Business
Chemistry
Health
Psychology

## Personal Interests

Business
Helping people: emotionally
Helping people: physical health/medicine
Science

**The boy is** screaming, his face red from the effort and streaked with tears. He hasn't even met the dentist, and he's already terrified of having his teeth examined. While his mother tries desperately to pull him out of the waiting room chair he's gripping with both hands, his older brother sits nearby, smiling smugly after having filled his brother's head with horror stories about the dentist.

Although Melissa Mosier's job as a dental assistant entails many duties such as cleaning and sterilizing the dental office and performing clerical tasks, the best part about being a dental assistant is calming nervous patients, she says. She walks over to the boy and after a few soothing words and her promise that he can tour the office first and then decide whether or not to see the dentist, the little boy takes Melissa's hand and together they walk through the office. By the time he meets the dentist, his face is dry of tears and he's smiling as he waves a pack of sugar-free gum at his older brother.

Dental assisting requires a lot of attention to detail and repetition throughout the day. Melissa must clean up after each patient, making sure the dental operatory is clean and sterilized for the next, she must wear proper protective equipment for herself, and she must follow very specific industry guidelines for cleaning the dentist's instruments. Speed and efficiency are of utmost importance. The most valuable quality for patients who are often afraid to visit the dentist, however, is the dental assistant's ability to be personable.

## what does a dental assistant do?

Dental assistants perform a variety of duties in the dental office, including helping the dentist examine and treat patients, and performing office and laboratory duties.

Although individual states regulate what assistants are allowed to do, generally, they help patients get comfortable before dental treatment and assist during treatment, by operating suction machinery that keeps the mouth clear of blood and saliva so the dentist can see himself or herself work. Dental assistants also take and develop X-rays, ask medical history questions before an examination, and take patients' blood pressure, pulse, and temperature. Many dental assistants perform office duties, such as scheduling appointments, answering telephones, handling billing, and working with vendors to replenish office and clinical supplies. Dental laboratory duties may include making impressions of a patient's teeth so the dentist can use the models to study the patient's condition and monitor progress.

A dental assistant's tasks can be divided into many categories. An *administrative assistant* acts as a receptionist, handles appointments, manages patient records, and may be responsible for inventory control, and handling correspondence and bookkeeping. A *chairside assistant* works directly with the dentist by seating patients, preparing the instrument tray, operating the suction devices while the dentist works, performing X-rays, and educating patients. A *coordinating assistant* may work where needed, such as processing X-ray film and performing laboratory procedures. Dental assistants often act as business managers who perform all nonclinical responsibilities such as hiring auxiliary help, scheduling and terminating employees, and overseeing accounting, supply ordering, and records management.

**Amalgam** An alloy containing silver, mercury, and other metals used as a dental filling material.

**Gypsum** Powdered materials used to make models, dies, and denture molds that help the dentist diagnose patients and develop a treatment plan.

**Operatory** Patient treatment room which contains an electronically controlled reclining chair and a dental unit containing an overhead light, a small sink, a saliva ejector, an instrument tray, and air hoses.

**Periapical films** An X-ray that helps detect those suspected diseases that show no physical symptoms.

**Saliva ejector** A small suction pump used to keep the patient's mouth dry and free of blood or saliva during treatment.

## what is it like to be a dental assistant?

Dental offices typically are clean, modern, quiet, and pleasant. They are also well lighted and well ventilated. In small offices, dental assistants may work solely with dentists, and in larger offices and clinics they may work with dentists, other assistants, dental hygienists, and laboratory technicians.

Although dental assistants may sit at desks to do office work, they spend a large part of the day beside the dentist's chair where they can reach instruments and materials. About one-third of dental assistants work forty-hour work weeks, sometimes including Saturday hours. About one-half work between thirty-one and thirty-eight hours a week. The remainder work less, but some part-time workers work in more than one dental office. Some offices offer better benefit packages than others, such as paid vacations and insurance coverage.

Taking X-rays poses some physical danger if handled incorrectly because regular doses of radiation can be harmful. However, all dental offices must have lead shielding and safety procedures that minimize the risk of exposure to radioactivity.

## have I got what it takes to be a dental assistant?

"My favorite part of my job is making people comfortable,"says Melissa, whose typical day involves a lot of rushing between operatories to keep patients on schedule while making them comfortable. She also must get along well with people and be able to sympathize with their fear of dental work, despite advances in pain control.

Sometimes patients are so afraid of getting dental work that they must be taken to the hospital where they can be put under general anesthesia. "When they get put under, we can do massive restorative work." Melissa says she likes going to the hospital because everything is clean and neat. There's even a dental chair in the operating room. "Nurses help me set up by showing me where everything is," she says. "Of course, when the patient isn't awake, we get our work done much faster. And when we're done, we can just leave because the hospital staff cleans up everything."

**To be a successful dental assistant, you should**

- Be cheerful; you're the patient's first impression of the dentist
- Have compassion and understanding for people with fears about dental procedures
- Be patient, calm, and flexible
- Be able to anticipate the dentist's needs
- Be able to sit still and remain alert for several hours through longer procedures

## how do I become a dental assistant?

Melissa Mosier's interest in a dental assistant's career wasn't the result of long-range deliberate planning, she says. She wasn't quite sure what she wanted to do when she graduated from high school. Having taken all college-preparatory courses in high school, she attended a state university for a year and decided she wanted to come home and work for a while. While she was working in a nursing home, a friend told her about a dental assisting program where she could get on-the-job training.

Melissa was always a good student and had taken courses necessary for a health career, such as chemistry. She had the discipline necessary to work while going to school, and most of all, she liked to help people.

In choosing a dental assisting career, Melissa lucked into a field where assistants are in high demand, and the degree of educational preparation largely depends on the individual's willingness to seek higher education. Dental assistants commonly are trained on the job or at community colleges, vocational schools, technical schools, or universities. In the latter case, they receive a certificate for completing course work lasting one to two years, depending on the program. However, since most dental assistant positions are entry-level, many assistants learn their skills on the job. Most positions usually require little or no experience and no education beyond high school. Aspiring dental assistants can also receive training while serving as members of the armed forces.

### education

#### High School

Students considering a career as a dental assistant should take science courses, such as biology and health, and obtain office skills such as typing and bookkeeping. With an increasing number of small businesses using computers, one or more computer courses are also recommended. In addition, some dental assisting programs require students to pass physical and dental examinations and have good high school grades.

#### Postsecondary Training

Students who attend two-year college programs receive associate's degrees, while those who attend trade and technical school programs earn a certificate or diploma after one year .

Graduating from an accredited school ensures that dental assisting graduates learn all necessary information required to successfully practice dental hygiene. There are approximately 250 accredited programs in the United States that are approved by the American Dental Association's Commission on Dental Accredita-

**Advancement Possibilities**

**Dental hygienists** perform many of the same skills as dental assistants, but they also take more courses and certification exams in order to assume some of the responsibilities of a dentist.

**Assistants to specialists** study the procedures of a dental specialist, such as an orthodontist or pediatric dentist, to assist in complicated or specialized techniques or oral surgeries.

**Dental office managers** schedule appointments, maintain business records, and depending on the size of the dental office and its staff, manage staff schedules. In addition, they are responsible for maintaining and ordering the proper clerical and dental supplies.

tion (CDA). The commission is responsible for setting standards for dental assisting programs.

## certification or licensing

Some states require dental assistants to become licensed or registered, but even for assistants practicing in states that don't require it, certification helps boost knowledge and earning power. Graduates from CDA-approved schools are automatically eligible to take examinations required to become certified. The certification process evaluates a dental assistant's knowledge. After Melissa earns her certificate, she plans to take the exam that will test her on such things as dental terminology, proper materials, and how to mix materials to achieve their desired function. "For example, with impression materials that come in tubes, you have to make it smooth without bubbles and you have to know how to do it at a certain speed," Melissa says.

Dental assistants who have not graduated from an accredited school or who have been trained on the job may take the certifying exam after they have worked full-time for two years as a dental assistant.

**FYI**

Dental implants are a new alternative to dentures. An implant consists of a small post that protrudes from the gum tissue and is anchored to the jaw. Prosthetic teeth are attached to the post. Dental assistants will need to be familiar with the fitting and care of implants.

## who will hire me?

"I got a job in my first year because assistants and hygienists are in such high demand," Melissa says. Dentists who practice in the area where Melissa lives frequently call her school or post signs there announcing job openings. Since the dental community is rather small in most cities and towns, many positions are learned about by word of mouth. High school and college guidance counselors, family dentists, dental schools, dental employment agencies, and dental associations are several ways to learn about job openings. Also, many dental assisting training programs offer job placement assistance.

Dental assisting associations also help students gain a foothold into the field and help members cultivate knowledge and skills while developing a local and national network of friends and colleagues. The American Dental Assistants Association keeps assistants up to date on all aspects of their profession by offering home study courses, monitoring local and national legislation that affects dental assistants, and publishing a newsletter identifying clinical and practice trends in dental assisting. The Dental Assisting National Board administers the certifying exam for dental assistants (*see* "Sources of Additional Information").

Most dentists work in private practices, so that's where an aspiring dental assistant will most likely end up working. An office may have a single dentist or may be a group practice with several dentists, assistants, and hygienists. Other places to work include dental schools, hospitals, public health departments, and U.S. Veterans and Public Health Service hospitals.

> My favorite part of my job is making people comfortable.

## where can I go from here?

Although dental hygiene is a natural step for an assistant looking to advance in the dental field, Melissa says she's not "school oriented," so she will not take that route. However, she may become an office manager, whose duties run the gamut from basic secretarial tasks like coordinating schedules and answering phones to ordering all supplies. Melissa is on her way to more responsibility because in addition to her clinical duties of assisting patients and taking X-rays, she now orders supplies.

Dental assistants advance in their careers by moving to larger dental practices where they can take on more responsibility. An assistant's ability to command higher pay is tied to the prestige of the dentist for whom the assistant works. By upgrading skills, continuing education, and achieving national certification, dental assistants may achieve higher pay in small offices. Specialists in the dental field, who typically earn higher salaries than general dentists because of their specialized knowledge, often offer higher salaries to their assistants.

Besides improving skills and adding more responsibilities, a dental assistant may choose to advance in the dental profession by studying to become a dental hygienist. This route requires taking more courses and taking state and national licensing exams designed for hygienists. Dental assistants considering moving into dental hygiene should carefully consider what dental assisting program courses will be accepted toward a hygiene degree. Many schools don't accept credits earned in dental assisting courses.

Dental assistants also may use their dental knowledge to obtain sales jobs at dental product companies or work for placement services or insurance companies.

## what are some related jobs?

The U.S. Department of Labor classifies dental assistants under the headings *Occupations in Medicine and Health, Not Elsewhere Classified* and *Child and Adult Care: Patient Care* (GOE). Such health support jobs include medical assistants, physical therapy assistants, occupational therapy assistants, pharmacy assistants, and veterinary technicians. Other related jobs include dental sales representatives, insurance personnel, and business office assistants.

| **Related Jobs** |
|---|
| Administrative assistants |
| Ambulance attendants |
| Business office assistants |
| Chairside assistants |
| Medical assistants |
| Optometric assistants |
| Physical therapy aides |
| Psychiatric aides |
| Surgical technicians |

## what are the salary ranges?

Although wages for assistants have not kept up with inflation, salaries naturally increase with experience. Working for a specialist, such as a pediatric dentist or orthodontist, often results in higher pay. According to the Bureau of Labor Statistics, dental assistants working full-time earned an average salary of $18,772 a year in 1996. Experienced dental assistants who worked thirty-two hours a week or more in a private practice averaged $21,112 a year. Part-time wages are slightly higher.

## what is the job outlook?

The employment outlook for dental assistants looks bright. Employment for dental assistants is expected to grow faster than all occupations through 2006, according to the U.S. Department of Labor. About 60,000 more jobs are expected to open in dental assisting by then. As the median age of the U.S. population rises, and people become more aware that they can keep all of their teeth and be healthy, more people will seek dental services for preventive care and cosmetic improvements. Moreover, younger dentists who earned their dental degrees in the '70s and '80s are more likely than other dentists to hire one or more assistants. Because these dentists have become established enough in their careers, they are also financially able to hire more assistants. As dentists increase their clinical knowledge of innovative techniques, such as implantology and periodontal therapy, they will delegate more routine tasks to assistants so they can make the best use of their time and increase profits. Job openings also will be created through attrition as older assistants retire and others assume family responsibilities, return to school, or transfer to other occupations.

# how do I learn more?

## sources of additional information

Following are organizations that provide information on dental assisting careers, accredited schools and scholarships, and possible employers.

**American Association of Dental Examiners**
211 East Chicago Avenue, Suite 760
Chicago, IL 60611
312-440-7464

**American Association of Dental Schools**
1625 Massachusetts Avenue, NW
Washington, DC 20036
202-667-9433
aads@aads.jhu.edu

**American Dental Assistants Association**
203 North LaSalle Street, Suite 1320
Chicago, IL 60601
312-541-1320

**American Dental Association**
211 East Chicago Avenue
Chicago, IL 60611
312-440-2500
http://www.ada.org/prac/careers/dc.menu.html

**Dental Assisting National Board**
216 East Ontario Street
Chicago, IL 60611
312-642-3368

**National Association of Dental Assistants**
900 South Washington Street, Suite G13
Falls Church, VA 22046
703-237-8616

## bibliography

Following is a sampling of materials relating to the professional concerns and development of dental assistants.

### Books

Ehrlich, Ann B. and Hazel O. Torres. *Essentials of Dental Assisting*. 2nd ed. Philadelphia: W. B. Saunders, 1996.

### Periodicals

*ADA News*. Biweekly. General information relating to the dentistry field. American Dental Association, 211 East Chicago Avenue, Chicago, IL 60611, 312-440-2867.

*Dental Assistant Journal*. Bimonthly. Covers technical and theoretical aspects of the profession. American Dental Assistants Association, 203 North LaSalle Street, Suite 1320, Chicago, IL 60601-1225, 312-541-1550.

*Explorer.* Monthly. General newsletter for dental assistants. National Association of Dental Assistants, 900 South Washington Street, Suite G13, Falls Church, VA 22046, 703-237-8616.

# dental hygienist

**Definition**

Dental hygienists provide preventative dental care by performing clinical tasks such as cleaning and scaling teeth to remove tartar and plaque, instructing patients in proper oral care, taking X rays, administering anesthesia, and assisting dentists.

**Salary range**

$19,000 to $24,000 to $39,000

**Educational requirements**

High school diploma; associate's degree or bachelor's degree from an accredited dental hygiene program

**Certification or licensing**

Mandatory

**Outlook**

Faster than the average

GOE
10.02.02

DOT
078

O*NET
32908

## High School Subjects

Biology
Chemistry
English
(writing/literature)
Mathematics

## Personal Interests

Helping people:
physical
health/medicine
Science
Teaching

**Hectic** isn't the word for Mary Hafner Myers' morning, more like controlled chaos. After getting her kids dressed and dropping them off at the babysitter's, she has to dash off to the office to be in time for a 9:00 AM appointment.

Fortunately, Mary is one organized woman. She won't have to set up her dental instrument trays this morning because she normally does that at the end of each day, after she has cleaned and sterilized all of the instruments required to clean and scale debris from her patients' teeth. Once in the office, she takes off her jogging pants and walking shoes and slips into freshly laundered "scrubs," her uniform as a dental hygienist, and then pulls on rubber soled shoes comfortable enough to stand in all day. Mary then washes her hands and dons rubber gloves for her and her patients' protection. She slips a clear plastic shield over her head to block any saliva or blood that may splatter while peering inside her patients' mouths. Mary's daily goal in a fast-paced environment of seeing eight or nine patients a day is to be relaxed enough to put patients at ease and spend enough time educating them about proper oral care and good nutrition. After getting dressed, Mary glances down at her watch and smiles. She still has ten minutes to spare before her appointment with Mr. Vincenzo. She decides to go out to the waiting room and chat with him.

## what does a dental hygienist do?

*Dental hygienists* perform clinical tasks, serve as oral health educators in private dental offices, work in public health agencies, and promote good oral health by educating adults and children.

In clinical settings, hygienists help prevent gum diseases and cavities by removing deposits from teeth and applying sealants and fluoride to prevent tooth decay. They remove tartar, stains, and plaque from teeth, take X rays and other diagnostic tests, place and remove temporary fillings, take health histories, remove sutures, polish amalgam restorations, and examine head, neck, and oral regions for disease.

Their main responsibility is to perform "oral prophylaxis," a process of cleaning teeth by using sharp dental instruments, such as scalers and prophy angles. With these special instruments they remove stains and calcium deposits, polish teeth, and massage gums. They teach patients proper home dental care, such as choosing the right toothbrush or how to use dental floss. Their instruments include hand and rotary instruments to clean teeth, syringes with needles to administer local anesthetic (such as Novocaine), teeth models to explain home care procedures, and X ray machines to take pictures of the oral cavity that the dentist uses to detect signs of decay or oral disease.

A hygienist also provides nutritional counseling and screens patients for oral cancer and high blood pressure. More extensive dental procedures are done by dentists. In some states, a dental assistant is permitted to perform some of the same tasks as a dental hygienist, such as taking X rays, but a hygienist has more extensive training and has received a license to perform the job.

Like all dental professionals, hygienists must be aware of federal, state, and local laws that govern hygiene practice. In particular, hygienists must know the types of infection control and protective gear that, by law, must be worn in the dental office to protect workers from infection. For example, dental hygienists must wear gloves, protective eyewear, and a mask during examinations. As with most health care workers, hygienists must be immunized against contagious diseases, such as hepatitis.

Dental hygienists also are required by their state and encouraged by professional associations to continue learning about trends in dental care, procedures, and regulations by taking continuing education courses. These courses may be held at large dental society meetings, colleges and universities, or in a more intimate setting such as a nearby dental office. These meetings also foster comraderie among fellow dental professionals, which is important in a field where the majority of people work in offices with small staffs.

In some private dental offices, a dental hygienist may perform office duties, such as answering phones, ordering dental supplies, keeping patient records, scheduling appointments, and processing dental insurance claims. Hygienists also visit local schools to perform oral prophylaxis on students and to teach them how to properly brush and floss their teeth. While most hygienists in clinical practice work in private dental offices, others may work in hospitals, correctional facilities, health maintenance organizations, and school systems.

Hygienists also carry out administrative, educational, and research responsibilities in private and public settings. They hold administrative positions in education, public health, hospitals, and professional associations. They sell dental products and supplies and evaluate dental insurance claims as consultants to

**Curette** A spoon-shaped blade on a long handle used for extensive tartar removal on teeth and below the gum line.

**Mouth mirror** A small round mirror on a long handle that allows the hygienist to see hard-to-reach areas of a patient's mouth.

**Probe** A tapered, rodlike blade on a long handle. The probe is inserted under the the gum line to measure gum depth, an indicator of gum disease.

**Scaler** A sickle, chiseled, or hoe-shaped blade instrument on a long handle used to remove "tartar" from the tooth surface. A scaler is also used to smooth the tooth surface so it will resist reaccumulation of deposits that cling to rough surfaces.

## in-depth

**Endodontics**

Dental hygienists work in every dentist's office, from the orthodontist to the endodontist.

Endodontists treat diseased inner tooth structures, such as the nerve, pulp, and root canal. Every tooth has the same basic structure. The outer covering of the tooth exposed above the gum line is called the *enamel*, the hardest substance in the human body. Beneath the enamel is another layer of hard material, *dentin*, that forms the bulk of the tooth. *Cementum*, a bonelike substance, covers the root of the tooth. Finally, a generous space within the dentin contains the *pulp*, which extends from just beneath the *crown*, or top of the tooth, to down through the root. The pulp contains blood vessels that supply the tooth and the lymphatic system.

Not so long ago, if a tooth was diseased, that was it; yank—no more tooth. Things have changed. Modern preventative endodontic techniques now make it possible to save many teeth that would have been extracted once decay spread into the pulp canal. These specialized procedures include *root canal therapy*, *pulp capping*, and *pulpotomy*.

In *root canal therapy*, the endodontist first examines the pulp to determine the extent of infection. Using rotary drills and other instruments, the pulp is removed and the empty *canal* surrounding the root is sterilized and filled with *gutta-percha* (a tough plastic substance) or silver, or a combination of the two.

*Pulp capping* consists of building a cap over the exposed pulp with layers of calcium hyroxide paste and zinc oxide. These layers are then topped with a firm dental cement.

*Pulpotomy* involves removing the pulp within the crown while leaving intact the pulp within the root canal. A pulp-capping procedure is used to seal and restore the crown of the tooth.

insurance companies. Dental hygienists may teach in dental hygiene schools, present seminars, conduct clinical research, write grant proposals, and publish scientific papers.

## what is it like to be a dental hygienist?

Mary Hafner Myers, a registered dental hygienist with a bachelor of science degree (R.D.H., B.S.), works full time, about thirty hours in a four-day work week. On average, she sees five or six patients in the morning, and she takes an hour lunch. During the remaining three hours, she sees the rest of her patients.

Most patients are on a regular recall schedule, such as every six months, to get a checkup, cleaning, and to have their health history updated. "I clean their teeth and go over special areas they need to pay attention to, such as plaque or tartar accumulating in certain areas," Mary says. "I spend a lot more time with new patients, and I show kids their X rays. They're always excited about that."

Most hygienists work in private dental offices. Dental hygienists must be flexible to accommodate varying patient schedules, which includes working evenings or Saturdays. Approximately 50 percent of all hygienists work full-time, about thirty-five to forty hours a week. Full-timers and part-timers may work in more than one office because dentists typically only hire hygienists to work two or three days a week. Many piece together part-time positions at several different dental offices and substitute for fellow hygienists who take days off. About 88 percent of hygienists see eight or more patients daily, and 68 percent work only in one dental practice.

Flexibility is key for hygienists. Before Mary got married, she worked in many dental offices as a substitute for the staff hygienist. Fortunate to have an annual eight-week vacation at her present job, Mary can easily fill up her schedule subbing at nearby dental offices whenever she wants the extra work and money. Usually, she prefers to enjoy her vacation. "If I didn't want to take a week off, there are at least four dentists I could call to fill a schedule for me," Mary says. "I can make good money doing that because as a sub they aren't paying my benefits, so the hourly rate is a little higher."

Work conditions for the dental hygienist in a private office, school, or government facility are pleasant with well-lit, modern, and adequately equipped facilities. The hygienist usually sits while working. State and federal regulations require that hygienists wear masks, protective eyewear, and gloves as well as follow proper sterilizing techniques on equipment and instruments to guard against passing infection or disease, such as hepatitis or AIDS.

Mary wears scrubs in the dental office because they are designed to protect her from transmitting or getting an infection from patients, particularly those who have infectious diseases. She doesn't wear any jewelry, such as earrings or her wedding band. She used to launder her work clothes at home, but new government infection control procedures for health care workers require her to change into her street clothes and leave her work clothes at work. Some dentists have a washer and dryer on site to launder clothes according to government guidelines. Mary's dentist pays a laundry service to pick up all his employees' uniforms and launder them. "I used to wear short sleeves, but now I'm covered from head to toe," Mary says.

As a licensed health care worker, Mary knows all the laws governing infection control, and she knows how to properly clean and sterilize her instruments. The dentist pays for all of her instruments and the machinery necessary to keep them clean and safe. He buys all the barrier items, such as plastic to cover chairs and trays. Whenever Mary needs a new instrument, she just asks her employer and he orders it because he wants her to do the best job she can.

Government hygienists work hours regulated by the particular agency. For a salaried dental hygienist in a private office, a paid two- or three-week vacation is common. Part-time or commissioned dental hygienists in private offices usually have no paid vacation. Benefits will vary, however, according to the hygienist's agreement with the employer.

When Mary was pregnant with her two children, the dentist accommodated her schedule and always understands when family emergencies arise. "My dentist was super with both of my pregnancies. I could take off as much time as I needed," she says. "I also like that he's very much for continuing education, and I have full-time benefits, health insurance, and pretty much everything that I need."

**To be a successful dental hygienist, you should**

- Enjoy working with people of all ages
- Be patient, flexible, and calm in stressful situations
- Be articulate and organized
- Work well as part of a team
- Be willing to follow strict safety and health guidelines

## have I got what it takes to be a dental hygienist?

"There's a lot of blood in dentistry, but it's not the life-and-death kind that you find in medical careers," Mary says. Dental hygienists operate in a more controlled atmosphere, and that appeals to her.

However, being a hygienist requires grace under pressure, Mary says. Although patient schedules are often full with one patient after another, hygienists must possess the manual dexterity to properly clean and scale below the patient's gum line, answer the patient's questions, and make the patient feel so relaxed that he feels he's all that matters. Fortunately, Mary has a dental assistant who helps her in the afternoons to usher in patients, clean up after every appointment, adjust the chair, and make patients feel comfortable.

"Hygienists must have patience to work on really tight schedules, be organized, and able to work on a team," Mary says. Dental hygiene school prepared her for real practice because in that strict atmosphere, she learned discipline. Mary had to find her own patients and learn to talk to strangers. Skipping a class was unheard of. "You have to really feel bad not to go to work," Mary says. "Patients will be there waiting for you and they depend on you. School definitely prepared me for that."

As an oral care educator, hygienists must be capable of communicating with patients because if they don't understand, they won't come back, Mary says. She also tells patients what she's doing and what instrument she's using because if they haven't been to the dentist in a while, sharp instruments can startle them if they don't understand their purpose.

# how do I become a dental hygienist?

Mary was naturally drawn to college prep courses in high school because challenging herself provided personal satisfaction. She knew she wanted to do something related to health and education, and the comraderie among dental workers at her family dentist's office convinced her to seek a career in dental hygiene.

**fyi** Dental hygienists clean teeth of plaque and calculus deposits to prevent gum damage, or *periodontal disease*. Untreated teeth cause gums to become inflamed and infection to spread to the roots of the teeth. Regular cleanings help prevent the disease.

## education

### High School

Dental hygiene programs vary, but all require students to have a high school diploma or general educational development (GED), also known as general equivalency degree. Recommended high school courses include mathematics, chemistry, biology, and English. College entrance test scores are needed, and some dental hygiene programs require prerequisite college courses in chemistry, English, speech, psychology, and sociology.

Math and science courses proved to be most helpful as Mary first worked to earn an associate's degree in dental hygiene science and later a bachelor's degree in public health, which gave her teaching experience on top of her clinical skills.

### Postsecondary Training

Dental hygiene education takes a minimum of two years in an accredited two-year program that offers a certificate or associate's degree, or an accredited four-year program offering a bachelor's degree. A master's degree may be an option for those seeking opportunities in education, research, or administration. Thirty percent of hygienists have a bachelor's or master's degree.

The dental hygiene curriculum generally consists of 1,948 clock hours of instruction, including 600 hours of clinical experience that involves working on patients. Courses include chemistry, anatomy, physiology, biochemistry, dental anatomy, radiology, pain control, dental materials, and pharmacology. Dental hygiene science courses include oral health education and preventive counseling, patient management, clinical dental hygiene, and ethics.

Mary earned an associate's degree and immediately went to work for a private dental office almost fifty miles from her home. She completed her studies toward a bachelor's degree in dental hygiene so she could teach and increase her earning power.

"Dental hygiene school was tough because I worked full-time to pay for school," Mary says. "The instructors are very strict, and you have to be very attentive and disciplined. There is so much responsibility because you have to set up your own schedule, find your own patients."

## certification or licensing

A state or clinical examination is required for licensing, as is a written national dental examination that is required by all states and the District of Columbia. Upon passing required exams, a dental hygienist becomes a Registered Dental Hygienist (R.D.H.). Other designations include Licensed Dental Hygienist (L.D.H.), and Graduate of Dental Hygiene (G.D.H.). If a hygienist moves to another state, he or she must pass that state's licensing exam and requirements. In Alabama, for example, hygienists may forego college and obtain on-the-job training in a dentist office that has been approved by the state. This is called preceptorship.

Mary's dentist supports her need to update her education and pays for her continuing education courses. To maintain her license, she must take a certain number of continuing education courses. She takes others to keep up to date and because she finds them quite interesting.

## who will hire me?

Once hygienists pass national board exams and licensing exams for their particular state, they must decide where to work, such as a private dental office, a school system, or a public health agency. Hospitals, industry, and the armed forces also employ a small number of dental hygienists. Graduating students have little difficulty finding a satisfactory position. Most dental hygiene schools maintain placement services, and dentists make announcements at local dental hygiene meetings. Often, temporary services match hygienists with dentists.

Upon earning an associate's degree, Mary had no trouble finding a job. As a matter of fact, the job found her. A dentist from a small community fifty miles from her home asked her school for a list of new graduates, and she was invited for an interview. She was hired to work three days a week, and the dentist paid her an extra dollar for each hour worked to compensate for travel expenses.

"I liked it because it was a small office and it just seemed like a real homey atmosphere," Mary says. "I thought even if I just worked there one year, it would be fun to get out of my general area and meet new people. Who would have thought I'd end up living there?"

## where can I go from here?

Opportunities for advancement, other than salary increases and benefits that accompany experience in the field, usually require post-graduate study and training. Educational advancement may lead to a position as an administrator, teacher, or director in a dental health program or in a more advanced field of practice. Only a small number of dental hygienists have continued their education to become practicing dentists.

With her bachelor's degree, Mary is qualified to teach college-level dental hygiene courses, which she has done. She plans to earn a master's degree in a different field, although she hasn't quite figured out what field that will be. "I'll probably branch out more into teaching," Mary says.

## what are some related jobs?

The U.S. Department of Labor classifies dental hygienists under the headings *Occupations in Medical and Dental Technology* (DOT) and *Nursing, Therapy, and Specialized Teaching Services: Therapy and Rehabilitation* (GOE). Workers in other occupations who support health practitioners in an office setting include dental assistants, ophthalmic medical assistants, and podiatric assistants.

Working as a dental assistant may serve as a stepping stone to a hygiene career and be a source of insight and information about the field. As a dental assistant, an individual may closely observe the work of dental hygienists, assess the personal aptitude required for this work, and discuss any questions with other hygienists before enrolling in dental hygiene school (*see* chapter "Dental Assistant").

### Advancement Possibilities

**Dental school instructors** teach college-level dental hygiene courses. This includes assigning and grading papers and administering exams.

**Directors of public dental programs** administer and carry out the policies of agencies whose mission is to educate the public about the importance of dental care and maintenance as well as provide free or affordable dental care.

**Dentists** try to maintain the dental health of their clients through preventative and restorative practices, such as cleaning or replacing teeth, filling cavities, and performing extractions. With additional years of schooling, a hygienist could advance to become a endodontist, periodontist, prosthodontist, pedodontist, oral pathologist, or oral surgeon.

| Related jobs |
|---|
| Dental assistants |
| Medical assistants |
| Ophthalmic medical assistants |
| Podiatric assistants |

> There's a lot of blood in dentistry, but it's not the life-and-death kind that you find in medical careers.

## what are the salary ranges?

A dental hygienists' income is influenced by such factors as education, experience, geography, and type of employer. Most dental hygienists who work in private dental offices are salaried employees, though some are paid a commission for work performed, or a combination thereof.

According to the American Dental Association, experienced dental hygienists who work full-time in a private practice earn an average annual salary of $39,000. Beginning hygienists earn between $19,000 and $24,000 a year. Salaries are typically higher in large metropolitan areas, as compared to small cities and towns.

Dental hygienists working in research, education, or administration may earn higher salaries. Another factor affecting earning power is the hygienist's level of responsibility. In addition, an increased demand for dental care and higher wages has provided incentives for hygienists to work in the field longer or to return to the field.

Since 1983, average hourly wages for hygienists have climbed by 52 percent compared to an average 31.1 percent increase against inflation.

## what is the job outlook?

Job opportunities in dental hygiene are growing faster than overall job growth. About sixty-four thousand new dental hygiene positions are expected to be created between 1996 and 2006. The demand for dental hygienists is expected to grow as younger generations who grew up receiving better dental care, keep their teeth longer. For example, 66.7 percent of dentists employed a hygienist in 1990 compared to 53 percent in 1983, according to the American Dental Association.

Older dentists, who are less likely to hire one hygienist, let alone more than one, will retire, and younger dentists will hire one or more hygienists to perform preventative care so they can have more time to perform more profitable, medically complex procedures. Population growth, increased public awareness of proper oral home care, and the availability of dental insurance should result in more dental hygiene jobs. Moreover, as the population ages, there will be a special demand for hygienists to work with older people, especially those who live in nursing homes.

Because of increased awareness about caring for animals in captivity, hygienists are also among a small number of dental professionals who volunteer to help care for animals' teeth and perform annual examinations. Dental professionals are not licensed to treat animals, though, and must work under the supervision of veterinarians.

## how do I learn more?

### sources of additional information

Following are organizations that provide information on dental hygienist careers, accredited schools and scholarships, and possible employment.

**American Association of Dental Examiners**
211 East Chicago Avenue, Suite 844
Chicago, IL 60611
312-440-7464

**American Association of Dental Schools**
1625 Massachusetts Avenue, NW
Washington, DC 20036
202-667-9433
aads@aads.jhu.edu

**American Dental Association**
211 East Chicago Avenue, Suite 1814
Chicago, IL 60611
312-440-2500
http://www.ada.org

**American Dental Hygienists' Association**
444 North Michigan Avenue, Suite 3400
Chicago, IL 60611
312-440-8929
http://www.adha.org

**National Dental Hygienists' Association**
5506 Connecticut Avenue, NW, Suite 24-25
Washington, DC 20015
202-244-7555

## bibliography

Following is a sampling of materials relating to the professional concerns and development of dental hygienists.

### Books

Cowan, Fred F. *Dental Pharmacology.* 2nd ed. Covers pharmacology as related to dental hygienists. Philadelphia: Lea & Febiger, 1992.

Kendall, Bonnie L. *Opportunities in Dental Care Careers.* Discusses the job of dental hygienist. Lincolnwood, IL: VGM Career Horizons, 1991.

### Periodicals

*ADA News.* Biweekly. General information relating to dentistry. American Dental Assn., 211 East Chicago Avenue, Chacago, IL 60611, 312-440-2867.

*Dental Teamwork.* Bimonthly. Articles and reports directed toward the entire dental staff. American Dental Assn., 211 East Chicago Avenue, Suite 840, Chicago, IL 60611-2616, 312-440-2500.

*Dentistry Today.* Ten times per year. General information. Dentistry Today, Inc., 26 Park Street, Montclair, NJ 07042, 201-783-3190.

*Journal of the American Dental Association.* Monthly. Current news in the dentistry industry. American Dental Assn., 211 East Chicago Avenue, Chicago, IL 60611, 312-440-2500.

*Journal of Dental Hygiene.* Nine times per year. American Dental Hygienists Association, 444 North Michigan Avenue, Suite 3400, Chicago, IL 60611, 312-440-8900.

### Specialties

**Endodontists** Treat diseased inner tooth structures, such as nerve, pulp, and root canal.

**Periodontists** Treat diseased gums and other tissues that support the teeth.

**Prosthodontists** Create artificial teeth, dentures or implants, to precise measurements.

**Pedodontists** Treat children's dental problems.

**Oral pathologists** Examine and diagnose tumors and lesions of the mouth.

**Oral surgeons** Perform difficult extractions, set jaw fractures, and remove tumors from the gums or jaw.

**Orthodontists** Use braces and other devices to correct irregularities in the development of teeth and jaws.

# dental laboratory technician

**Definition**

Dental laboratory technicians make and repair dental appliances and replacements for missing, damaged, or poorly positioned teeth according to dentists' written prescriptions. Appliances include dentures, inlays, bridges, crowns, and braces made with materials such as plastic, ceramics, and metals.

**Alternative job titles**

Crown and bridge specialists

Dental ceramists

Dental technicians

Orthodontic technicians

**Salary range**

$16,848 to $32,656 to $50,000+

**Educational requirements**

High school diploma

**Certification or licensing**

Voluntary

**Outlook**

Little change or more slowly than the average

GOE
05.05.11

DOT
712

O*NET
89921

## High School Subjects

Anatomy and Physiology

Art

Chemistry

Shop

## Personal Interests

Building things

Sculpting

Everything, if you're a dental laboratory technician. Gerson Shapiro owns a dental laboratory, and every day he and his staff of twenty-two create and repair a whole slew of dental appliances so they can keep people smiling. Gerson, as the owner, focuses on the big picture, such as managing employees, insuring quality control, planning complicated cases, and communicating with and keeping his dentist-customers happy.

But it's the details that have made Gerson a success in the business of fabricating dental appliances to help dentists improve their patients' oral health and outlook. Since high school, Gerson's daily routine has revolved around sitting at a workbench equipped with a Bunsen burner and tools, such as wax spatulas and wax carvers and grinding and polishing equipment.

As a dental technician, he fills a dentist's prescription for dentures, braces, bridges, crowns, and inlays. It takes a lot of time, patience, and artistic skills to create durable, lifelike teeth. Gives you something to think about next time you brush your teeth, doesn't it?

## what does a dental laboratory technician do?

*Dental laboratory technicians* work in the trenches, filling dentists' prescriptions for crowns, dentures, bridges, braces, and other dental prosthetics and devices. A hands-on job, technicians spend several hours a day working at a bench making and perfecting these dental appliances. Dentists send the laboratory impressions, or molds, of the patient's teeth or mouth. Using knowledge of oral anatomy and restoration, the technician then creates a model of the mouth or teeth by pouring plaster into the mold from the dentist. After the plaster sets, the technician places the model on an articulator, a device that mimics the movement of the jaw opening and closing. After studying the model and determining the best position so the upper and lower teeth will fit together when the mouth is closed (known as occlusion), the technician builds up wax over the model using wax spatulas and carvers. A lathe equipped with polishing wheels is used to clean and buff the model.

Once the wax model is complete, the dental laboratory technician pours a mold and casts a metal framework. The technician uses small, hand-held tools to prepare the metal surface so it will bond with porcelain. Porcelain layers are applied layer by layer to accurately create the shape and color of a real tooth. The dental appliance is then baked in a porcelain furnace so the porcelain will adhere to the metal framework. The technician touches up the appliance by shaping, grinding, and removing excess porcelain or adding more porcelain, and voila—a precise replica of the patient's tooth or teeth.

While some dental laboratory technicians perform all stages of the work, others specialize as they gain more experience in the field. There are five areas of specialization: full dentures, partial dentures, orthodontic appliances, crowns and bridges, and ceramics.

Technicians specializing in complete dentures not only create dentures, but also repair them. When repairing dentures, technicians cast plaster models of replacement parts and match the new tooth's color and shape to the adjacent teeth. They cast reproductions of gums, fill cracks in dentures, and rebuild linings using acrylics and plastics. They may also bend and solder wire made of gold, platinum, and other metals. Occasionally, technicians fabricate wire using a centrifugal casting machine.

Partial dentures restore missing teeth for patients who have some teeth remaining on the jaw. Technicians use the same techniques as those used to make full dentures, but partials require metal clasps to secure them to remaining teeth. The clasps also facilitate removal of the partials for cleaning.

*Orthodontic technicians* bend wire into intricate shapes and solder them into complex positions to make and repair frames and retainers for positioning teeth. In other words, the technicians make braces, something most teenagers are familiar with. The tasks include shaping, grinding, polishing, carving, and assembling metal and plastic appliances.

*Crown and bridge specialists* restore the missing parts of a natural tooth. Crowns and bridges are made of plastics and metal and are sometimes called fixed bridgework because they are permanently cemented to the natural part of the tooth. Crown and bridge specialists are adept at melting and casting metals and must also wax and polish the finished product.

### lingo TO LEARN

**Artificial crown** A restoration that reproduces the entire surface of the natural crown of a tooth.

**Articulator** A device that mimics the movement of the mouth; used to make dental laboratory models.

**Bridge** An artificial appliance that replaces lost teeth and is held in place by attachments to nearby natural teeth.

**Denture** An artificial or prosthetic replacement for natural teeth and adjacent tissues.

**Implant** A foreign object, such as a metal root, set into the jaw bone to support an artificial tooth or set of teeth.

**Inlay** A tooth filling shaped and cemented into place.

**Occlusion** How the upper and lower teeth fit together when the mouth is closed.

**Prosthodontics** Dentistry involving artificial replacements and devices.

Dental laboratory technicians who specialize in porcelain are known as *dental ceramists* and are involved primarily with cosmetic dentistry. They make natural-looking replacements to fit over natural teeth or to replace missing ones, including crowns, bridges, and tooth facings. Ceramists apply multiple layers of mineral powders or acrylic resins to a metal framework then fuse the materials in an oven. This process is repeated until the product is exactly as specified. Ceramists must possess natural creative abilities and understand all phases of dental technology. They are, therefore, generally the highest paid of dental laboratory technicians.

**To be a successful dental laboratory technician, you should**

- Have manual dexterity and mechanical aptitude
- Be artistically inclined and like to do things with your hands
- Be patient
- Be able to sit in one place for long hours
- Take orders well and understand instruction

## what is it like to be a dental laboratory technician?

On a typical day when Gerson makes an implant prosthesis, he situates himself at his workbench to begin a process that will take several days to complete. He attaches a model of the patient's mouth to an articulator and fashions a wax pattern of teeth in preparation for a metal casting.

He puts the wax pattern into a mold and heats the mold to get rid of the wax pattern. A metal alloy, the substance that will make the metal casting, the basis of the prosthesis, is cast into the mold. The metal fills the void that had been filled by the wax pattern. When the cast has filled the space left by the melted wax, it is allowed to cool and become a solid mass. This process takes place in a furnace with a special exhaust fan that sucks out fumes. "It takes about two hours, but we put it in the oven overnight, and a timer turns on the oven so that we can cast the metal when we start work in the morning," Gerson says. "You can't let the metal get too hot because it will burn out the ingredient that makes the metal, but you also have to make it hot enough so it flows into the cast. When it cools, we break it out of the mold and finish the metal. We grind it and take out the scratches and polish it, like you would a ring."

The next step is adding "teeth" to the metal appliance. Gerson uses a porcelain paste that matches the color of the patient's real teeth and forms the teeth using his bare hands, instruments, and knowledge of tooth anatomy. He also uses a color guide and a description from the dentist to help him match the teeth perfectly. "It's like an art. You have to pick the right colors. It's the same porcelain used to make a bathtub, but it's finer and translucent. It's almost the quality of a fine China dish."

Forming teeth takes expert artistic skill, Gerson says. He molds the porcelain to tooth shapes and then puts them into a furnace to go through a heating and drying process. This only takes about five to ten minutes.

"You have to file and shape it," he says. "We use dental hand pieces and different types of grinding stones and diamond stones, very similar to what a dentist uses. Then we have to stain and glaze the porcelain so it has a sheen to it."

When complete, the implant prosthesis will either be attached with tiny screws to the metal roots that have been implanted into the patient's jaw bone or attached to a connective bar. The prosthesis is then polished, checked, and sent out to the dentist.

Each procedure takes a day or two, and the dentist has the patient try on the prosthesis to make sure he or she can speak, eat, and swallow properly. Gerson makes as many adjustments as needed to achieve the right fit for the patient.

Dental lab technicians generally work five days a week, eight hours a day with required overtime if a special order needs to be filled. "You can't be a clock watcher," Gerson says. "When your normal assignment is done, try another assignment. Most laboratory owners would be pleased to have someone like that."

Wages tend to be low when a person first starts working as a lab technician because of the high potential to easily damage a product, Gerson says. Dental laboratory technicians must prove themselves over time by mastering small tasks and progressively accepting more responsibility. Trainees generally start by performing simple tasks, such as mixing plaster and pouring molds. Making dental appliances well requires years of hands-on experience; whether trained on the job or graduated from a two-year applied science or arts program, it generally takes three to four years before a trainee is formally considered a full-fledged technician. Experience is reflected in salaries. "We have technicians that make $30,000 to $50,000 or more a year, but that doesn't happen overnight," Gerson says.

## have I got what it takes to be a dental laboratory technician?

Lab technicians must be creative and artistic, says Gerson, who went into dental laboratory technology because he wanted to do something health-related and liked working with his hands. Precision, patience, and dexterity are the key skills required if you want to be a successful dental laboratory technician. You must understand and carry out verbal and written instructions according to the dentist's prescription. You must have good color vision to be able to distinguish between innumerable shades of "pearly white," and you must have the ability to perform delicate and intricate tasks with your fingers. A friendly personality helps a lot, Gerson adds, and a good command of dental terminology is essential.

Lab technicians must also be able to sit for several hours while producing an appliance. "This can get very boring," Gerson admits. If you like to sit in one place and do routine work, you will have a better chance at success, he says. If you are considering dental laboratory technology, you should also like to learn new things because the only way to improve technique and skill is to tackle more challenging assignments in school or during on-the-job training.

## how do I become a dental laboratory technician?

Gerson attended a vocational-technical high school, and that was where he began learning about dental laboratory technology. In the mornings, he would take traditional high school courses such as English and math, and in the afternoons, Gerson learned dental anatomy, dental chemistry, biology, and fabrication of dental appliances, such as dentures, partial dentures, and crowns and bridges. He also learned how to make a model from a patient's impression taken by a dentist. "We had to make a positive of the patient's mouth from the negative sent over by the dentist," Gerson explains. "The dentist would tell the lab what his plans were, and we'd fill the prescription."

## education

### High School

Dental laboratory technicians must have a high school diploma. Science and math courses are essential to future success as a lab technician. Chemistry, biology, anatomy, shop, mechanical drawing, and ceramics are all useful courses you should take in preparation for a career as a lab technician.

Metallurgy courses, which teach the science of metals and working with them, are also helpful. Metals have certain properties that make them less effective if they are not prepared properly, Gerson explains. If certain metals are overheated, the dentist wouldn't get medically acceptable results, and the metal would be harmful to patients.

Any other courses or activities that would allow you to gain skills practiced by dental laboratory technicians should be taken advantage of. If you can learn how to solder or mold or learn about the chemistry of plastics, seize the opportunity.

Part-time or summer jobs as laboratory helpers may be available. You may also want to visit a local dentist or dental laboratory to see firsthand what the work entails.

### Postsecondary Training

There are primarily two ways to get training to be a dental laboratory technician: on-the-job training or earning an associate's degree or certificate in a two-year applied science or arts training program at a community college, technical college, vocational school, or dental school. Many dental lab technicians have been successfully trained on the job under the supervision of veteran technicians. It generally takes three to four years of on-the-job training to achieve the same level of knowledge as someone trained formally in school. Trainees start with simple tasks and work their way up to more complex tasks.

The American Dental Association's Commission on Dental Education has approved more than thirty-five dental laboratory technology programs. Programs consist of classroom instruction in chemistry, dental materials science, oral anatomy, fabrication procedures, ethics, metallurgy, and more. Students gain hands-on experience at school or at an associated dental laboratory.

### Advancement Possibilities

**Dental laboratory supervisors** supervise and coordinate activities of workers engaged in fabrication, assemblage, and repair of full or partial dentures, crowns, bridges, inlays, and orthodontic appliances.

**Dentists** are medical doctors who treat diseases, injuries, and malformations of teeth and gums and related oral structures.

**Dental and medical equipment and supplies sales representatives** sell medical and dental equipment and supplies, except drugs and medicines, to doctors, dentists, hospitals, medical schools, and retail establishments.

Although classroom training provides a good introduction into the field, dental laboratory technician trainees must undergo additional training and practice to acquaint them with their lab's specific procedures. Students of dental laboratory programs, regardless of whether they have graduated, are considered good candidates for training because of their previous exposure in class.

The military is a third, less common route for students to obtain dental laboratory technology training. After high school, Gerson joined the U.S. Air Force where he was able to continue his training and do more concentrated work. "Because of my previous education, I was able to get a good position," Gerson says. "In high school, I wasn't really able to work on my own. Students had to have complete supervision." In the Air Force, Gerson mastered the art and science of fabricating crowns and bridges, got advanced training in inlays, and learned to understand contour and how to properly use materials. He also learned how to fabricate various connecting devices and precision attachments.

"The more knowledge you have, the more the dentist will consult with you because they have very limited training in laboratory technology," Gerson says. "In dental schools of the past, dentists spent a great deal of time learning dental lab technology. Now they learn more oral medicine and don't have time to be dental lab technology experts."

## certification or licensing

Certification is an option but not a requirement for dental laboratory technicians. Certification signifies that a technician has attained a high level of academic and practical achievement in dental technology. It also may be an advantage when applying for a job.

Technicians who have five years' experience or an associate's degree with two years' experience are eligible for certification. The certification exams are administered by the National Board for Certification, a division of the National Association of Dental Laboratories (NADL), and include written and practical (an exam where you have to prove that you can actually do the work) examinations for five laboratory specialties: crowns and bridges, ceramics, complete dentures, partial dentures, and orthodontic appliances. Technicians who successfully pass certification tests earn the right to put the initials C.D.T. after their names, signifying that they are Certified Dental Technicians. Courses taken during dental technology training may count toward certification.

Every year, certified dental technicians must take a certain number of continuing education courses to maintain their status and to further master skills. Gerson, who has served as the president of NADL, says he originally sought certification for "my own self-respect." Continuing education courses are easy to attend as they are usually held at nearby schools or a large dental laboratory, Gerson says. Local dental societies offer many of these dental laboratory technology courses.

One of the biggest benefits of taking continuing education courses is meeting with other dental laboratory technicians to share tips and discuss the pluses and minuses of different products, Gerson says.

fyi

**Metal-free Restorations—Wave of the Future?**

Composite materials are making headway into dental laboratory technology, providing a metal and porcelain-free alternative to dentists and patients. Ideal for crowns, bridges, and inlays, these materials are made of fiber-reinforced composites that include resins and polymers. These composites are cured by light and heat.

Some other advantages? Many patients are allergic to various metals, and fiber-reinforced composites can solve the problem. They're also easy to repair, simpler for dentist to work with and affix in patients' mouths, and offer improved wearability to the patient (the composites rub against adjacent teeth more smoothly and with less erosion). They are also more flexible than porcelain prostheses, leading to fewer fractures.

## scholarships and grants

The American Dental Association Endowment and Assistance Fund, Inc. offers the Allied Dental Health Scholarship for students studying to become dental laboratory technicians. For requirements, details, and deadline information, contact the American Dental Association (ADA). (*See* "Sources of Additional Information.")

Scholarship and grant opportunities may also be available through one of the accredited dental laboratory technology programs or dental schools. Check with your financial aid office or counselor.

Other sources for possible scholarship opportunities include state and local dental societies or large commercial dental laboratories. The ADA may be able to point you in the right direction.

## internships and volunteerships

While some internship opportunities may exist, the common route for aspiring dental laboratory technicians is to start as trainees. Trainees, regardless of whether or not they graduated from a dental laboratory technology program, begin with simple tasks such as mixing plaster and pouring it into molds. Trainees with no formal training in the field may spend three to four years mastering the techniques of an accomplished technician. Graduates of dental laboratory technology programs may progress more rapidly out of the trainee ranks and become fully qualified technicians in a few years.

# who will hire me?

Although Gerson worked in dental laboratories throughout high school as part of his diploma requirements, his first official job was in the Air Force where he was fortunate to get a good position because of his prior education. The Air Force also sent him to advanced courses, which he used to his advantage to get a job when he left the service.

New graduates of dental laboratory technology programs should make use of school placement offices to find a job, or they may apply directly to dental laboratories and dentists' offices, which sometimes have on-site labs. Private and state employment agencies may be helpful in finding employment. Networking at dental laboratory association meetings is also a good way to find out about job opportunities.

Government opportunities may be found by applying at regional offices of the U.S. Veterans Administration or at the Office of Personnel Management. Many Department of Veterans Affairs hospitals provide dental services and may have the need for technicians.

Experienced technicians may contact dental supply houses and salespeople they meet on the job for potential leads. Sales representatives often know about staffing needs because they are in constant contact with dentists and dental laboratories.

New technicians usually start off with small jobs performing several routine tasks, such as making and trimming models, making minor denture repairs, or polishing dentures. Those working in large commercial laboratories may be assigned to various departments. The average dental laboratory, however, employs fewer than five technicians.

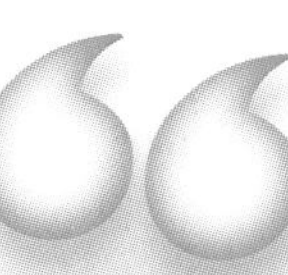

> We have technicians that make $30,000 to $50,000 or more a year, but that doesn't happen overnight.

## where can I go from here?

Gerson originally wanted to be a dentist, but eventually changed his mind and decided to become a laboratory tech instead. He spent several years working in a dental laboratory, working his way up to be a manager. "A lot of that had to do with my experience in the service and working in clinical environments with dentists," Gerson says. He spent six years managing a laboratory for three prominent big-city dentists. He had to know more about supervising people, scheduling cases, and communicating directly and effectively with dentists. "As you prove your ability, you can advance," says Gerson, noting that some people are more content sitting and doing one thing all day than being responsible for the running of a laboratory.

As dental technicians improve their skills and techniques, they can work on more complex or special assignments. They also can become supervisors or managers like Gerson, which involves resolving problems with prescriptions, establishing costs and delivery arrangements, and training new workers. Lab supervisors also inspect work and order materials and supplies. Managers may be assigned to a specific department, covering single areas such as dentures, partial dentures, or porcelain.

Technicians also may become teachers in dental laboratory programs or become technical representatives or salespeople for dental product companies. Outstanding technicians also may teach continuing education courses.

After Gerson built a reputation for being accurate and personable, he was able to start his own dental laboratory. According to the 1998–99 *Occupational Outlook Handbook,* one in seven technicians is self-employed. Dentists also call Gerson when they want to try a new technique and need to have a laboratory technician help them plan the case. For example, Gerson worked on some of the first cases of modern tooth implants with a group of university-based dentists. Implantology is one of the new techniques used by dentists to replace patients' teeth in a more permanent manner than dentures. "It's one of the most dynamic things going on in the dental field," Gerson says. Planning implant cases requires technicians to make study models and discuss plans for tooth placement with the dentist. Technicians also give dentists an indication of how much the laboratory work will contribute to the overall cost of the procedure.

## what are some related jobs?

The U.S. Department of Labor classifies dental laboratory technicians under the headings *Occupations in Fabrication and Repair of Surgical, Medical, and Dental Instruments and Supplies* (DOT) and *Craft Technology: Scientific, Medical, and Technical Equipment Fabrication and Repair* (GOE). Also listed under these headings are opticians, ophthalmic laboratory technicians, artificial plastic eye makers, instrument makers, taxi meter repairers, watch repairers, glass blowers, biomedical equipment repairers, arch-support technicians, orthotics technicians, prosthetists (artificial limbs), and camera repairers. Jobs that require similar skills also include biological technicians, dental assistants, dental hygienists, jewelers and jewelry repairers, and museum exhibit technicians.

| Related Jobs |
| --- |
| Opticians |
| Prosthetists |
| Jewelers |
| Orthotics technicians |
| Camera repairers |
| Watch repairers |
| Dental assistants |
| Dental hygienists |

## what are the salary ranges?

Wages usually are low for beginning technicians because they have to fully develop skills and techniques to obtain higher pay. Earnings also vary depending upon the laboratory size, geographic location, and responsibilities of the technician.

According to the 1995 Economic Conditions Survey of the Commercial Dental Laboratory, conducted by the National Association of Dental Laboratories (NADL), the national average wage of a trainee in 1994 was $6.14 an hour. A graduate of a two-year program could expect to earn an average of $8.10 an hour. Those with more than ten years of experience earned an average of $15.70 an hour. Technicians specializing in ceramics are the best paid, making $13.90 an hour once they are skilled and need no supervision.

The 1998–99 *Occupational Outlook Handbook* indicates that technicians' earnings increase sharply with experience, and that the average annual salary for all dental laboratory workers in 1995 was $23,723. Skilled technicians with artistic abilities can earn more than $50,000 a year.

## what is the job outlook?

According to the 1998–99 *Occupational Outlook Handbook,* opportunities for dental laboratory technicians should be favorable, although employment of technicians is projected to grow more slowly than the average through the year 2006. This is attributed to the improvement in the overall dental health of the population. People are taking better care of their teeth and are suffering from fewer cavities because of fluoridated drinking water. Full dentures are therefore often unnecessary.

According to the NADL, demand for highly skilled dental laboratory technicians will increase as restorative and cosmetic dentistry become more sophisticated and more popular. The NADL also notes that the job outlook should be positive for skilled technicians because fewer people are entering the field. This decrease may be due to the low entry-level wages and the fact that many are unaware of the field of dental technology.

## how do I learn more?

### sources of additional information

The following organizations offer listings of schools and information on careers, scholarships, and certification programs.

**American Dental Association**
211 East Chicago Avenue
Chicago, IL 60611
800-621-8099
http://www.ada.org

**National Association of Dental Laboratories**
8201 Greensboro Drive, Suite 300
McLean, VA 22102
703-610-9035
http://www.nadl.org

### bibliography

Following is a sampling of materials relating to the professional concerns and development of dental laboratory technicians.

#### Periodicals

*ADA News.* Biweekly. General information relating to the dentistry field. American Dental Assn., 211 East Chicago Avenue, Chicago, IL 60611, 312-440-2867.

*American Academy of Esthetic Dentistry Newsletter.* Semiannual. American Academy of Esthetic Dentistry, 500 North Michigan Avenue, Suite 1400, Chicago, IL 60611, 312-661-1700.

*American Journal of Orthodontics and Dentofacial Orthopedics.* Monthly. Analyzes orthopedic techniques. Mosby, 11830 Westline Industrial Drive, St. Louis, MO 63146, 314-872-8370.

*Dental Teamwork.* Bimonthly. Articles and reports directed toward the entire dental staff. American Dental Assn., 211 East Chicago Avenue, Suite 840, Chicago, IL 60611-2616, 312-440-2500.

*Dentistry Today.* Ten times per year. General information. Dentistry Today, Inc., 26 Park Street, Montclair, NJ 07042, 201-783-3190.

*Journal of Oral Implantology.* Quarterly. Directed toward oral implantology personnel, including laboratory technicians. International Association for Dental Research, 1111 14th Street NW, Suite 1000, Washington, DC 20005, 202-898-1050.

*Journal of Prosthetic Dentistry.* Monthly. discusses all aspects of restorative dentistry. Mosby, 11830 Westline Industrial Drive, St. Louis, MO 63146, 314-872-8370.

# desktop publishing specialist

**Definition**

Desktop publishing specialists use computers to prepare files for printing. They take files that others have created and manipulate the images and text so they will print properly.

**Alternative job titles**

Digital prepress operators
Digital production operators
Electronic prepress operators
Imagesetters
Preflight technicians
Prepress workers

**Salary range**

$18,000 to $30,000 to $83,000

**Educational requirements**

High school diploma; some postsecondary training highly recommended

**Certification or licensing**

Voluntary

**Outlook**

Faster than the average

GOE
01.06.01

DOT
651

O*NET
34002E

## High School Subjects

Art
Computer science

## Personal Interests

Computers
Drawing
Figuring out how things work
Photography

with a hand-drawn sketch on butcher paper. The sketch somehow had to end up on a paper cup for instant hot cereal, and it was up to Kathy Richardson to get the drawing into her computer and manipulate it to produce a file ready to print. Kathy needed to figure out how to maintain the integrity of her client's artwork—the colors, lines, and details—while also getting the design to wrap around the cup without becoming distorted. Kathy looked for ideas on the Internet, spoke with printers, and consulted members of professional organizations, but she came up empty handed. She couldn't very well tell the designer she was stumped. "Technology is supposed to solve all of their problems," Kathy says, "and they want to hear solutions."

Kathy finally found a product promoted as a 3-D special effects tool for web designers. Although her project had nothing to do with special effects or the World Wide Web, she applied it to the design, and it worked. Her client, the graphic designer, was happy because she had kept his design intact. The designer's client, the cereal company, was pleased because the designer delivered exactly what had been requested. The hot cereal cups now line the shelves at grocery stores. Kathy modestly states, "I got the job done."

## what does a desktop publishing specialist do?

*Desktop publishing specialists* work on computers, converting and preparing files for printing presses or other media. Much of desktop publishing fits into the prepress category, and desktop publishing specialists typeset, or arrange and transform, text and graphics into finished products or products ready to be printed. Typesetting and page layout work entails selecting font types and sizes, arranging column widths, checking for proper spacing between letters, words, and columns, placing graphics and pictures, and more. Desktop publishing specialists also deal with technical issues of files, such as resolution problems, colors that need to be corrected, and software difficulties.

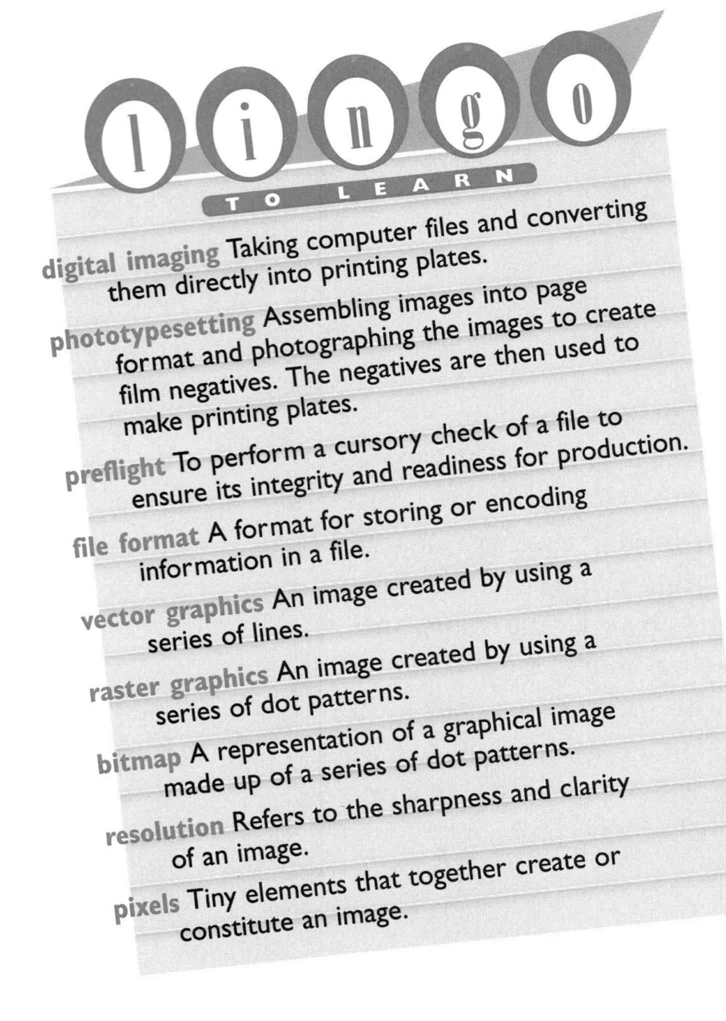


Desktop publishing specialists who work for service bureaus handle the technical issues of graphic designers and provide prepress services, including film output. *Graphic designers* use their creativity and artistic skills to create designs, often from scratch. Some may use computer software programs, while others draw with pencil and paper. They provide the desktop publishing specialists with their designs, and the desktop publishing specialists must convert these designs to the format requested by the designers. A designer may come in with a hand-drawn sketch, a printout of a design, or a file on a diskette, and he or she may want the design to be ready for publication on the World Wide Web, in a high-quality brochure, or in a newspaper. Each format presents different issues, and the desktop publishing specialist must be familiar with the processes and solutions for each. Service bureaus also provide services such as color scanning, laminating, image manipulation, or poster production.

Desktop publishing specialists at commercial printing houses generally focus less on the technical issues of designs and more on the printing end and prepress operations, although many commercial printers now have in-house service bureaus. Desktop publishing specialists take disks from customers, check the files for problems, then print the files to film or directly onto printing plates. The process of converting files on disks to film or printing plates is known as digital imaging. The job of the desktop publishing specialist is to ensure that the images on the film or plates will print perfectly and accurately.

One specialization within desktop publishing at the commercial printing house is the *preflight technician*. After a customer brings in a disk, the preflight technician performs an initial check of the files to make sure the files are ready to go into production. This can entail checking the disk contents against a hard copy or printout supplied by the customer and making sure the fonts, colors, resolutions, and all other details are satisfactory. Once the check is complete, the process of printing the files to film can begin.

Desktop publishing specialists can also specialize in scanning. *Scanner operators* focus on color correction, color separation, and image manipulation. They use computerized equipment to output the film that will be used to print the final product. The computer handles the color separation process, which involves producing four-color separation negatives from a print. In printing, photographs must be printed from images consisting of millions of tiny dots. In order to create an accurate reproduction of an original color print, it's necessary to produce separation negatives that will be combined during the printing process. Each scan produces an image of the tiny dots representing one of four colors—cyan, yellow, magenta, and black. Separate printing plates are made for each of these scans and, using transparent color inks, they are printed one at a time. The final result combines all the colors and produces a replica of the original print or photograph. The scanner operator corrects color errors and enhances color where necessary. For instance, the original print may have uneven color or fading problems.

**To be a successful desktop publishing specialist, you should**

- Have a good eye and design sense
- Possess strong problem-solving skills and analytical thinking skills
- Be creative and curious
- Enjoy working with computers and software tools
- Have endurance to see things through to completion

## what is it like to be a desktop publishing specialist?

Kathy Richardson spends most of her day sitting at a Macintosh computer, troubleshooting and manipulating her designer clients' files so she can provide them with the end results they want. "Basically," says Kathy, "what you would say my position is is a problem solver for the graphic designer." Kathy works at Direct Imaging, a computer imaging service bureau in San Luis Obispo, California.

Kathy works primarily with five software packages: QuarkXPress, Adobe PageMaker, Macromedia FreeHand, Adobe Illustrator, and Adobe Photoshop. These are the basic tools used by desktop publishing specialists who work with Macintosh computers, and Kathy may use all of them on one project. Kathy understands the strengths and weaknesses of each program, which enables her to provide software solutions to her clients. Part of her job is also to respect the clients' designs. "What they end up getting is what delivers without compromising their design," Kathy states. "They want someone like me who respects their design, who doesn't trash it, who doesn't start redesigning it."

Because Kathy finds she is more productive after the shop is closed, she usually arrives at work around 10 AM. Direct Imaging has clients all over the world, so the first part of Kathy's day is spent checking messages and making East Coast and international calls. The bulk of Kathy's day, however, is spent on projects. Some may take one day to complete, while others may take a week or longer. She may ready a logo for letterhead, prepare a catalog for the printer, or work on a file that will be published on the World Wide Web.

Kathy works with two deadlines every day. "Everything revolves around FedEx and UPS," she asserts. The latter part of each afternoon is devoted to getting projects completed and ready for either FedEx or UPS. This can sometimes cause problems, especially when there are rush jobs. "I would say that practically all the jobs that are rush jobs are the jobs with the most problems."

Scott Gordon, president and prepress manager of Haagen Printing in Santa Barbara, California, spends much of his time at the Macintosh computer as well. His situation, however, is unique. "I wear the hat as prepress manager as well as the president," he explains, "so one minute I might be color correcting an image, and the next minute I might be in a meeting with a banker or a lawyer." Although a fair amount of prepress work is done on IBM-compatible PCs, Scott finds, "Out of 1,000 jobs, a handful are PCs. Ninety-nine percent of our production is done on a Macintosh."

Haagen Printing is a commercial printing firm that specializes in high-quality printed material, such as annual reports, custom printing jobs, and eight-color brochures. The prepress department must take the customers' files and ready them for the printing presses. This may include preflighting the files, troubleshooting certain types of problems or files, and color correcting and scanning. Each desktop publishing specialist in the prepress department has a propensity for one niche of the business, but "the ultimate goal is to get what's in the computer out onto a piece of film or out onto a plate or whatever you're trying to print to," Scott says.

Kathy enjoys the flexibility and variety her job offers her. Because new software is released on a constant basis, Kathy is always learning something new and discovering new solutions, which she finds fun. She also likes working with graphic designers and teaching them about the capabilities of the software, although this can sometimes work against her. "I try to educate them," Kathy says, "and in some cases I've ended up shooting myself in the foot because I've taught myself out of a job."

Scott also enjoys the variety and creativity of desktop publishing. "I get to make something new every day," Scott declares. "You're making film, and you're making proofs, and you're making pictures, and they're all different."

A calm temperament comes in handy for desktop publishing specialists. You have to be able to work under pressure and constant deadlines. Sometimes Kathy finds herself caught in the middle between graphic designers and printing firms, which can be difficult. The designers may blame Kathy for a problem with the design, and printers may blame her for issues with the output. Kathy takes this all in stride and strives to find solutions rather than dwelling on who is to blame.

## have I got what it takes to be a desktop publishing specialist?

If you want to be a desktop publishing specialist, you must be detail oriented, possess problem-solving skills, and have a sense of design or some artistic skills. A good eye and patience are critical, as well as endurance to see projects through to the finish. As Scott relates, "I often call our prepress department the emergency room. Every day new patients come in, and we open them up and fix them. Every job is basically just problems we solve. They don't print themselves, and it takes a lot of work to get them out right."

You should also have an aptitude for computers, the ability to type quickly and accurately, and a natural curiosity. Kathy says, "You have to like taking chances. You can't be afraid of the computer." Kathy recalls working with people who were wary of touching certain keys or clicking on buttons for fear that the computer would crash. Kathy boldly tried everything to gain a better understanding of the capabilities and limits of the computer and software tools.

A design background is helpful so you can comprehend the designer's approach. "You have to be technically inclined," Kathy states, "but you also have to have some kind of creative background to understand where these designs are coming from." This background can also assist you in troubleshooting problem files. "You're trying to get things to output," explains Scott, "and when you run into problems, you have to get inside the mind of the designer and understand how he built the file, or what he did there."

## how do I become a desktop publishing specialist?

Both Kathy and Scott entered the desktop publishing industry from the design side. Kathy's father was a printer, so she grew up in print shops. She studied graphic design in college but realized her strength lay in working with computers and software tools rather than creating art from scratch.

Scott worked as a commercial photographer and provided design services as well. "It just so happened that that was the era that desktop publishing was invented," he recalls. "It was the early eighties, and in 1984 the Macintosh hit the streets." Scott naturally developed prepress skills because he used the Macintosh as a tool for his photography and design business. Scott followed along as improvements to the Macintosh were made and new software tools were created.

## education

### High School

Classes that will help you develop desktop publishing skills include computer classes and design or art classes. Computer classes should include both hardware and software, since understanding how computers function will help you with troubleshooting and knowing the computers' limits. In photography classes you can learn about composition, color, and design elements. Typing, drafting, and print shop classes, if available, will also provide you with the opportunity to gain some indispensable skills.

Kathy also suggests enrolling in a chemistry class. "I learned how to follow a recipe and do what it said in real life. I learned how to apply what's in a manual to doing something. There's not that many classes where you can do that." Laboratory experiments taught Kathy how to follow instructions and pay attention to detail.

Working on the school newspaper or yearbook will train you on desktop publishing skills as well, including page layout, typesetting, composition, and working under a deadline. Learn the software as quickly and as well as you can, and others will turn to you for help and advice, Kathy says.

Endurance sports, such as cross-country running or long-distance swimming, will teach you the discipline to see projects through to the finish. Kathy emphasizes that "you have to have endurance development, and you don't develop that on a computer," which is why she advises participating in activities "that help you go for the long haul."

Joining computer clubs or volunteering at small organizations to produce newsletters or flyers are other activities that will be to your benefit. If you have a computer at home, use it and experiment with it. You may also be able to find part-time or summer employment with printing shops or companies that have in-house publishing or printing departments.

### Postsecondary Training

Although both Kathy and Scott graduated from four-year colleges, a college degree is not required for desktop publishing work. Kathy had planned to become a graphic designer and studied in the graphic communications department at California Polytechnic State University in San Luis Obispo, California. She ended up switching from the design emphasis to the management concentration halfway through her schooling when she realized she preferred to help designers rather than be one herself.

Kathy stresses that experience is the key to becoming a good desktop publishing specialist, and she gained experience by working in the field while attending school. It took her ten years to graduate, but, she says, "I was able to apply what I was learning while I was learning it. I don't think you should learn in a vacuum for this kind of field, because it changes so quickly that you can't expect to earn a living just by what you learn on campus."

Some two-year colleges and technical institutes offer programs in desktop publishing or related fields. A growing number of schools offer programs in technical and visual communications, which may include classes in desktop publishing, layout and design, and computer graphics. Four-year colleges also offer courses in technical communications and graphic design. There are many opportunities to take classes related to desktop publishing through extended education programs offered through universities and colleges. These classes can range from basic desktop publishing techniques to advanced courses in Adobe Photoshop or QuarkXPress and are often taught by professionals working in the industry.

## certification

Certification is not mandatory, and currently there is only one certification program offered in desktop publishing. The Association of Graphic Communications has an Electronic Publishing Certificate designed to set industry standards and measure the competency levels of desktop publishing specialists. The examination is divided into a written portion and a hands-on portion. During the practical portion of the examination, candidates receive files on a disk and must manipulate images and text, make color corrections, and perform whatever tasks are necessary to create the final product. Applicants are expected to be knowledgeable in print production, color separation, typography and font management, computer hardware and software, image manipulation, page layout, scanning and color correcting, prepress and preflighting, and output device capabilities.

### Advancement Possibilities

**Prepress managers** supervise prepress departments and delegate work to prepress operators or desktop publishing specialists. Managers also make purchasing decisions for the department.

**Printing sales representatives** work for printing firms and solicit work from clients and companies.

**Graphic designers** create designs and artwork based on clients' requests and needs.

**Software developers** work in research and development for software publishing firms. They design and test software tools and make recommendations to programmers.

## in-depth

### The Big Five Rundown

Desktop publishing specialists who work with Macintosh computers use the following five software programs extensively. Kathy Richardson has provided some insight into the strengths and weaknesses of each.

**Macromedia FreeHand** specializes in vector-based drawings and images. Vector-based images are more flexible than bitmap or raster graphics when it comes to sizing. FreeHand provides a slew of tools for the desktop publishing specialist, including 3-D capabilities, web publishing functions, and the ability to work with different page sizes in the same document.

**Adobe Illustrator** also specializes in vector-based drawings and images and has been the industry standard for years. If you can use Illustrator, you should have no problem learning FreeHand. Illustrator can be used to create graphical images that are then transferred to PageMaker or QuarkXPress.

**Adobe PageMaker** is a desktop publishing program that is ideal for documents such as newsletters, letterhead, business cards, advertisements, and envelopes. PageMaker is often the preferred format for a file before delivery to the printer.

**Adobe Photoshop** specializes in bitmap-based images and is the preferred tool for editing photographs and converting images. Photoshop provides versatility for web design as well. Bitmaps offer more flexibility than vectors when it comes to creating images.

**QuarkXPress** is another desktop publishing program and is ideal for longer documents such as books, catalogs, newspapers, and pamphlets. QuarkXPress is a bit harder to master than Adobe PageMaker but offers more flexibility and features once you've gotten the hang of it. QuarkXPress is the preferred format for the fully composed document before delivery to the printing shop.

Desktop publishing specialists who work with IBM-compatible computers shouldn't feel left out. The Big Five can be used on a PC as well as on a Macintosh. Some software programs specifically geared toward the PC market include the following:

**Corel Draw** is a vector- and bitmap-based program that has 3-D capabilities and powerful functions and tools for desktop publishing.

**Adobe FrameMaker** is ideal for book and textbook publishing because of its capabilities for managing large amounts of text. FrameMaker can also be used for web publishing and preparing documents in multiple languages.

The Printing Industries of America, Inc. (PIA) is in the process of developing industry standards in the prepress and press industries. PIA may eventually design a certification program in desktop publishing or electronic prepress operation.

## scholarships and grants

A number of professional organizations and schools offer scholarship and grant opportunities. The Graphic Arts Education and Research Foundation (GAERF) and the Education Council of the Graphic Arts Industry, Inc., both divisions of the Association for Suppliers of Printing and Publishing Technologies (NPES), can provide information on scholarship opportunities and research grants (*see* "Sources of Additional Information"). Other organizations that offer financial awards and information on scholarship opportunities include the Society for Technical Communication, the International Prepress Association, PIA, and the Graphic Arts Technical Foundation, which offers scholarships in graphic communications through the National Scholarship Trust Fund.

Colleges and universities that offer programs in desktop publishing and related fields may also grant scholarships. Contact your advisor or financial aid office for additional information and resources.

## internships and volunteerships

Internships and cooperative work experiences are common and highly recommended in the desktop publishing industry. Many major newspapers offer internship and apprenticeship opportunities in the pressroom. Most internships in the publishing or printing industry are nonpaid or do not pay well, but the experience and connections you gain will pay off in the long run. When Kathy was in college, USA Today had an internship program in desktop publishing and prepress operations that was highly coveted by students. Landing a well-respected internship will facilitate your career and provide you with credibility.

A good place to look for internship opportunities is through your school counselor or advisor, whether you are in high school or college. Professional associations often have information regarding internships, and you may wish to contact major newspapers, magazines, or publishing houses as well.

If you are unable to find an internship, there are plenty of organizations that would be happy to have volunteers adept at desktop publishing. Many small businesses and nonprofit organizations need help producing newsletters, brochures, letterhead, flyers, catalogs, and more. Volunteering is an excellent way to try new software and techniques and to gain experience troubleshooting and creating final products.

## who will hire me?

Kathy found her first job in the field through connections and says, "Most people find jobs by networking." Kathy worked primarily as an administrative assistant to the art director at a company that manufactures fitness attire. Her duties included clerical tasks as well as desktop publishing projects using PageMaker. Kathy discovered that her co-workers were not very skilled at using the software tools, so she was able to help them and demonstrate her value as an employee.

Scott was involved with desktop publishing before the term was even coined. "I just sort of grew up with it," he states. Scott developed prepress skills because he worked as a commercial photographer, media coordinator, and art director, and desktop publishing and design are natural extensions of those jobs. Scott had been a customer at Haagen Printing for years, and when the shop decided to switch to digital prepress operations, it contacted him and asked if he would help with the transition. He's been with Haagen ever since.

The primary employers of desktop publishing specialists are printing shops, service bureaus, newspaper plants, and large companies with in-house graphics or design staffs. Basically, any organization with a printing department will have a need for desktop publishing specialists. Printing shops handle both commercial and business printing. Commercial printing involves catalogs, brochures, and reports, while business printing encompasses products used by businesses, such as sales receipts or forms.

Jobs with the federal government are another option for desktop publishing specialists. The Government Accounting Office (GAO) and the Government Printing Office (GPO) publish a large amount of documents. The GPO even has a Digital Information Technology Support Group (DITS Group) that provides desktop and electronic publishing services to federal agencies.

Kathy suggests looking for a position as a production artist if you're just starting out. Production artists take the work of graphic designers and work on layout. Kathy explains that production artists are not designers or technical experts. "You're just a grunt," she says bluntly, "but you can learn, and if you know your stuff, you can help them when something goes wrong." Printing houses and design agencies are places to check for production artist opportunities.

Both Scott and Kathy agree that networking is the best way to find a job, so it might be a good idea to keep up on your membership dues for appropriate professional organizations and clubs. Most professional organizations offer career services and job listings to members. If you enroll in classes or school, talk to your instructors about job openings they've heard about.

The Internet is another job search tool that can come in handy, especially, says Scott, if you're willing to relocate. Newspaper classified advertisements and trade magazines are also sources for job leads.

## where can I go from here?

Kathy enjoys working with computers and pushing software tools to their limits, which is why she is interested in working as a software developer for a software publisher at some point in her career. She thinks it would be exciting to be on the cutting edge of technology, helping programmers develop software innovations. Kathy is also interested in moving into management. "I would like to work for a publishing house, to be the head of a department and make purchasing decisions." Kathy would also like to get more involved in teaching. Currently, she teaches about one software class per session through extended education programs.

Scott became a co-owner of Haagen Printing in June of 1997, so his future goals revolve around building the business and making it the best printing shop possible. Providing the highest quality and the best service are goals for Scott, and after that, maybe he'll retire and go sailing.

Desktop publishing specialists can move into middle management or sales positions within a printing firm. Prepress managers oversee prepress departments and supervise staff members. Prepress managers may be responsible for scheduling, staffing, and purchasing of equipment, including computer hardware and software. Sales representatives work for printing firms or publishing houses. Their job is to find new customers and expand business.

Scott and Kathy entered the desktop publishing realm with design backgrounds, and both feel it is important to have an understanding of design to be a successful desktop publishing specialist. It is possible, however, to transition into graphic design from prepress if the desktop publishing specialist has an aptitude for it. Graphic designers are artists, and not everyone possesses these skills.

## what are some related jobs?

Desktop publishing specialists can shift into web design or specialize further in electronic prepress operations. Web designers, however, use a different set of tools to accomplish their goals, so some additional training would most likely be necessary. Desktop publishing specialists with a considerable amount of experience may find work as art directors for large companies or advertising agencies.

The U.S. Department of Labor classifies desktop publishing specialists under the heading Printing Press Occupations (DOT) and Craft Arts: Graphic Arts and Related Crafts. Also included under these headings are those in lithographic occupations and people who use artistic skills in their work, such as ceramic artists, jewelers and jewelry repairers, special effects technicians, taxidermists, and conservators and conservation technicians.

| Related Jobs |
| --- |
| Lithographers |
| Conservators and conservation technicians |
| Taxidermists |
| Jewelers and jewelry repairers |
| Ceramic artists |
| Special effects technicians |

## what are the salary ranges?

There is limited salary information available for desktop publishing specialists, most likely because the job duties of desktop publishing specialists can vary and overlap with other jobs. Scott, who handles staffing in the prepress department at Haagen Printing, believes a good prepress operator can earn from $18 to $28 an hour. Kathy's experience has been that desktop publishing specialists in the San Luis Obispo area, which is a small region in the central coast area of California, can earn between $18,000 and $30,000 a year.

According to a salary survey conducted by PIA in 1997, the average wage of desktop publishing specialists in the prepress department ranged from $11.72 to $14.65 an hour, with the highest rate at $40 an hour. Entry-level desktop publishing specialists with little or no experience generally earn minimum wage. Electronic page makeup system operators earned an average of $13.62 to $16.96, and scanner operators ranged from $14.89 to $17.91.

According to the 1998–99 *Occupational Outlook Handbook*, full-time prepress workers in typesetting and composition earned a median wage of $421 a week in 1996. Wage rates vary depending on experience, training, region, and size of the company.

## what is the job outlook?

According to the 1998–99 *Occupational Outlook Handbook*, the job outlook is excellent for desktop publishing specialists. As technology advances, the ability to create and publish documents will become easier and faster, thus influencing more businesses to produce printed materials. Desktop publishing specialists will be needed to satisfy typesetting, page layout, and design demands. With new equipment, commercial printing shops will be able to shorten the turnaround time on projects and in turn can increase business and accept more jobs. For instance, digital printing presses allow printing shops to print directly to the digital press rather than printing to a piece of film, and then printing from the film to the press. Digital printing presses eliminate an entire step and should appeal to companies who need jobs completed quickly.

Prepress machine operators may notice a decline in employment opportunities as their work becomes more automated. Printing plants may also lose jobs to large companies with in-house printing and preparation capabilities. Desktop publishing specialists are best suited to fill these positions because of their skills with computers and electronic prepress operations.

According to a survey conducted by PIA in 1997, the printing industry is growing, which can be attributed partly to the growth experienced by the North American economy. The electronic prepress segment of the printing market enjoyed the most growth, with an average change from 1996 of 9.3 percent. Traditional prepress, on the other hand, suffered a decline of 5.7 percent.

PIA's survey also indicates that printing firms have been experiencing difficulties finding new, qualified employees. This is a good sign for desktop publishing specialists with skills and experience.

Both Kathy and Scott concur that the job outlook is positive. Scott says, "It's looking good for the next five years." He feels computers will increasingly mechanize printing processes, and the prepress operator's job will become easier. Kathy also feels this is a good time to be in desktop publishing. "It's growing so much," she asserts. She acknowledges that many prepress jobs are being phased out because many are learning to do their own layout and design on computers, but she says, "There will still always be the need."

## how do I learn more?

### sources of additional information

The following organizations provide information on schools, scholarship opportunities, careers in desktop publishing, and employment.

**Association of Graphic Communications**
330 Seventh Avenue, 9th Floor
New York, NY 10001
212-279-10001
http://www.agcomm.org

**Association for Suppliers of Printing and Publishing Technologies (NPES)**
Education Council of the Graphic Arts Industry, Inc.
Graphic Arts Education and Research Foundation
1899 Preston White Drive
Reston, VA 20191
703-264-7200
http://www.npes.org

**Graphic Arts Technical Foundation**
PO Box 1020
Sewickley, PA 15143
800-662-3916
http://www.gatf.org

**International Digital Imaging Association**
84 Park Avenue
Flemington, NJ 08822
908-359-3924
http://www.idia.org

**International Prepress Association**
7200 France Avenue, Suite 327
Edina, MN 55435
612-896-1908
http://www.ipa.org

**National Association of Printers and Lithographers**
780 Palisade Avenue
Teaneck, NJ 07666
201-342-0700
http://www.napl.org

**Printing Industries of America, Inc.**
100 Daingerfield Road
Alexandria, VA 22314-2888
703-519-8100
http://www.printing.org

**Society for Technical Communication**
901 North Stuart Street, Suite 904
Arlington, VA 22203-1854
703-522-4114
http://www.stc-va.org

# bibliography

Following is a sampling of materials relating to the professional concerns and development of desktop publishing specialists.

## Books

Giambruno, Mark. *3D Graphics and Animation: From Starting Up to Standing Out.* Teaches the basic concepts of 3D design, explains the market for 3D artwork, and gives advice on finding work in the 3D graphics industry. Indianapolis: New Riders Publishing, 1997.

Litwiller, Dan and Patrice-Anne Rutledge. *The Essential Publisher 97 Book: The Get-It-Done Tutorial.* Covers publishing tasks and design techniques; how to add and alter graphics and text; and work with Page Wizards, templates, boundaries, rulers, multiple and facing pages, and the Design Checker. Rocklin: Prima Publishing, 1997.

Parker, Roger C. *Desktop Publishing and Design for Dummies.* Rev. ed. Tutorial on creating desktop-published documents. Foster City: IDG Books Worldwide, 1995.

Toor, Marcelle Lapow. *The Desktop Designer's Illustration Handbook.* A guide to professional illustration and animations. New York: John Wiley & Sons Press, 1997.

Wempen, Faith. *10 Minute Guide to Office Pro 97 for Windows 95: Access, Excel, Powerpoint and Word.* A comprehensive, easy-to-follow guide to three key desktop publishing applications. Indianapolis: Que Education & Training, 1997.

## Periodicals

*3D Artist.* Monthly. Illustrated magazine covering the latest trends in graphics. Columbine, Inc., P.O. Box 4787, Santa Fe, NM 87502.

*DTP Journal.* Monthly. A features-oriented publication, covering desktop publishing trends and extensive software reviews. Desktop Publishers Journal, 462 Boston Street, Topsfield, MA, 01983-1232.

*PC Magazine Online.* Daily. Website featuring information, opinions, reviews, news, previews, and features. http://www.zdnet.com.

*Publish RGB: The online magazine for electronic publishing professionals.* Daily. A comprehensive online resource containing reviews of software and design products, industry trends, a list of conferences and events, and interactive technique "workshops." http://www.publish.com.

*Windows Magazine.* Monthly. Windows Magazine, P.O. Box 420211, Palm Coast, FL, 43132-0211.

# dialysis technician

**Definition**

Dialysis technicians set up and operate hemodialysis (artificial kidney) machines. These machines filter the blood of patients whose kidneys no longer function, removing waste materials and other impurities. Dialysis technicians also maintain and repair this equipment.

**Alternative job titles**

Hemodialysis technicians

Nephrology technicians

Renal dialysis technicians

**Salary range**

$15,000 to $35,000 to $40,000

**Educational requirements**

High school diploma; on-the-job training

**Certification or licensing**

Voluntary; mandatory in California and New Mexico

**Outlook**

Faster than the average

GOE
10.02.02

DOT
078

O*NET
32999E

## High School Subjects

Anatomy and Physiology
Chemistry
Biology
Mathematics

## Personal Interests

Fixing things
Helping people: physical health/medicine
Science

**Stan's job** is to keep people alive. As a dialysis technician at Bio-Medical Applications of Evanston, Illinois, he oversees the artificial kidney machines that purify the blood of patients whose kidneys no longer function normally. So that these patients will not die of uremic poisoning, Stan and his fellow patient-care technicians connect them to the machines that actually pump the blood from their bodies, cleanse it of impurities, and return the purified blood to their bodies. "It's a big responsibility," Stan says, "so I know I have to be a responsible person."

## what does a dialysis technician do?

The kidneys are vital organs; they remove the waste products of daily living that accumulate in the bloodstream and are normally eliminated from the body as urine. Many people, particularly those who are diabetic or suffer from undetected high blood pressure, develop a condition known as chronic renal failure (CRF) in which their kidneys no longer function properly. Before artificial kidney machines were developed in the 1940s, such patients would die of uremic poisoning as toxic products built up in their bloodstream.

The use of artificial kidney machines is called hemodialysis. In the process of hemodialysis, blood is pumped from the body through a dialyzer, in which it passes through tubes constructed of artificial membranes. The outer surfaces of these membranes are bathed with a solution called the dialysate; body waste chemicals pass from the blood through the membrane into the dialysate, but blood cells and other vital proteins do not. The cleansed blood is returned to the patient's body without the harmful waste products. The rate of waste removal depends on the extent of the patient's kidney failure, the concentration of waste products in the blood, and the nature and strength of the dialysate.

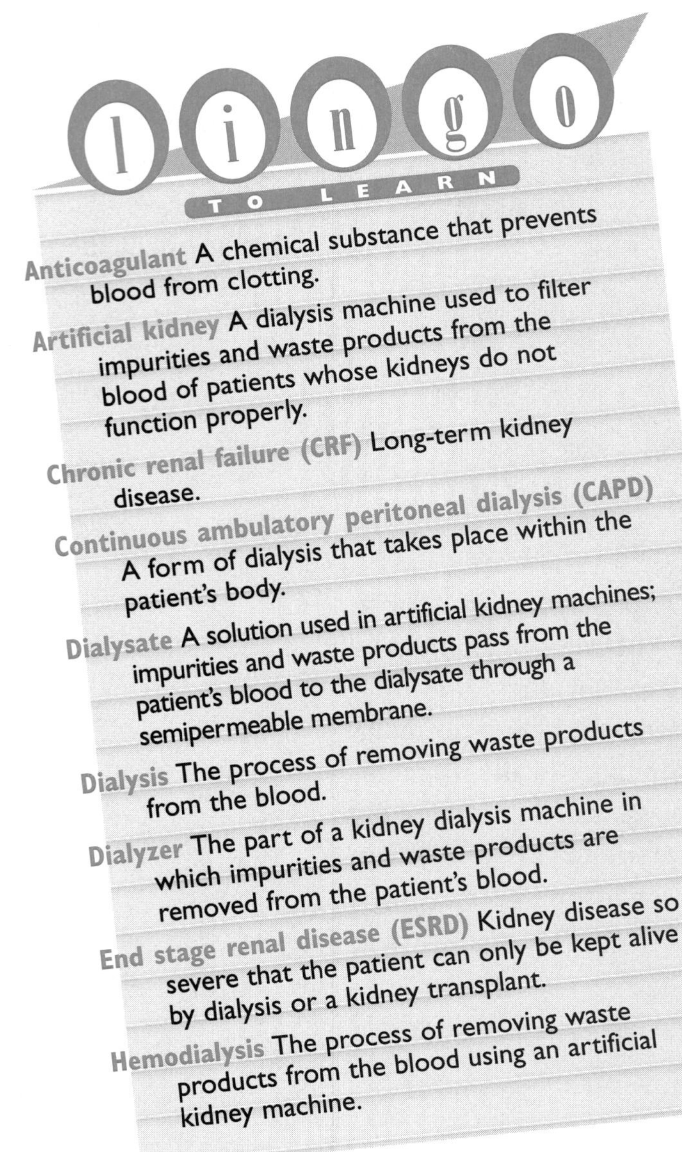

**lingo TO LEARN**

**Anticoagulant** A chemical substance that prevents blood from clotting.

**Artificial kidney** A dialysis machine used to filter impurities and waste products from the blood of patients whose kidneys do not function properly.

**Chronic renal failure (CRF)** Long-term kidney disease.

**Continuous ambulatory peritoneal dialysis (CAPD)** A form of dialysis that takes place within the patient's body.

**Dialysate** A solution used in artificial kidney machines; impurities and waste products pass from the patient's blood to the dialysate through a semipermeable membrane.

**Dialysis** The process of removing waste products from the blood.

**Dialyzer** The part of a kidney dialysis machine in which impurities and waste products are removed from the patient's blood.

**End stage renal disease (ESRD)** Kidney disease so severe that the patient can only be kept alive by dialysis or a kidney transplant.

**Hemodialysis** The process of removing waste products from the blood using an artificial kidney machine.

The National Association of Nephrology Technicians/Technologists recognizes three types of *dialysis technician*, although in some hospitals and dialysis centers the responsibilities may overlap. These are the *patient-care technician*, the *biomedical equipment technician*, and the *dialyzer reprocessing (reuse) technician*. Dialysis technicians always work under the supervision of medical personnel, usually nurses.

Patient-care technicians are responsible for setting up the dialysis machine and connecting it to the patient's body, for measuring the patient's vital signs (including weight, pulse, blood pressure, and temperature), and for monitoring the process of dialysis. They must be able to administer cardiopulmonary resuscitation (CPR) or other life-saving techniques if an emergency occurs during a dialysis session. In some states, including Illinois, technicians are not permitted to administer drugs to patients; this can only be done by nurses.

Biomedical equipment technicians are responsible for maintaining and repairing the dialysis machines (*see* chapter "Biomedical Equipment Technician). Reuse technicians care for the dialyzers—the apparatus through which the blood is filtered. Each one must be cleaned and bleached after use, then sterilized by filling it with formaldehyde overnight so that it is ready to be used again for the patient's next treatment. To prevent contamination, a dialyzer may only be reused with the same patient, so accurate records must be kept. Some dialysis units reuse plastic tubing as well; this too must be carefully sterilized.

The spread of hepatitis and the growing risk of HIV infection have necessitated extra precautions in the field of hemodialysis, as in all fields whose procedures involve possible contact with human blood. All patient-care personnel must observe universal precautions, which include the wearing of a protective apron, foot covers, gloves, and a full face shield.

While most hemodialysis takes place in a hospital or a free-standing dialysis center like BMA, the use of dialysis in the patient's home is becoming more common. In this case, technicians may travel to patients' homes to carry out the dialysis procedures or to instruct family members in assisting with the process.

Another form of dialysis is continuous ambulatory peritoneal dialysis (CAPD). In CAPD, the membrane used is the peritoneum (the lining of the abdomen), and the dialysis process takes place within, rather than outside, the patient's body. Dialysis technicians are not needed for this form of treatment.

**To be a successful dialysis technician, you should**

- Like working with people
- Be comfortable dealing with the chronically ill
- Be able to follow directions and procedures exactly
- Be familiar with the metric system and able to do calculations
- Keep cool in emergency situations

## what is it like to be a dialysis technician?

Each morning Stan arrives at the center at 6:15 AM. His first task is to prepare the machines, rinsing out the sterilizing solution that has filled the dialyzers overnight. Both he and another member of the dialysis team must sign off on the log certifying that each dialyzer is perfectly free of chemicals before he can connect it to the containers of dialyzing solution that will be used in filtering a patient's blood.

By now the patients are arriving, usually one for each of the four stations that Stan oversees. One at a time he checks each patient, recording blood pressure, temperature, and pulse rate. He also weighs each of them, comparing the result to the "dry" weight their doctor has established as ideal. If a patient is over this weight it means there is excess fluid in his or her body, and Stan will adjust the machine to remove the appropriate amount of fluid as it filters the blood.

Because the patients are "regulars," each of whom is treated two or three times a week on an ongoing basis, Stan knows them well and is familiar with the particular needs of each one and the required machine settings.

With the patient seated in a recliner, a nurse connects an intravenous (IV) device and administers heparin (an anticoagulant that will prevent the patient's blood from clotting). Stan then connects the blood pump to specially prepared blood vessels, usually in the patient's arm. Blood will be pumped out of the patient's body and through the coils of the dialyzer. These coils are bathed with the dialysate solution, into which the waste products and impurities will pass. The filtering process will take anywhere from two and a half to four hours, depending on the severity of the patient's kidney problem.

Blood pressure and temperature are continually monitored during this time; an alarm will sound, bringing Stan to the patient's side immediately, if any reading slips outside the "safe" range. Stan doesn't panic if this occurs, because he is well trained in emergency procedures and has confidence in his ability to use them, as well as in the other members of his team.

With all four of his patients "on the machine," Stan gets a bit of down time. He and his colleagues (three other technicians who oversee the other twelve machines) take turns going on their morning break.

By midmorning, the patients with the shortest dialysis time are finished with their treatment. Stan makes sure that all of the blood that is circulating through the dialyzer is returned to the patient's body and, if necessary, that saline (salt) solution is added to replace excess fluid lost during the treatment. He disconnects the blood pump and once again checks the patient's vital signs. Again, all measurements are written down and both Stan and another team member sign the log. Now a reuse technician will begin the twenty-four-hour sterilization process that will ready the dialyzer for the patient's next appointment.

The first half of the work day is over, and Stan begins to prepare the machines for his second group of patients. The morning's routine will be repeated, with a lunch break for each technician during the treatment period. At this center each technician works four ten-hour days a week. Another shift will be added soon to accommodate patients who need evening appointments. When that happens, the technicians will switch to five eight-hour days a week.

Although he acknowledges that there's a certain degree of monotony in following the same procedures day in and day out, Stan enjoys his work at the center. One of the best things about it is the wide variety of people—both patients and staff—he comes in contact with. Patients come from all walks of life and all levels of society. "It's like a mini UN," he says.

## have I got what it takes to be a dialysis technician?

The ability to talk easily with patients and their families is essential. Kidney patients, especially those who are just beginning dialysis, are confronting a major—and permanent—life change. The technician must be able to help them deal with the emotional as well as the physical effects of their condition. Good interpersonal skills are crucial not only in the technician-patient relationship but in working with other team members.

At the same time, a good technician can't afford to get so involved with a patient or patients that he or she loses objectivity. If necessary but uncomfortable parts of the routine are slighted, or if too much time is given to a favorite patient, the overall quality of care will suffer.

Stan says it's very important that the patients perceive him as a trustworthy reliable person, since they literally put their lives in his hands each time he treats them. He says knowing that even a little mistake can have deadly consequences calls for a great deal of maturity.

A good head for mathematics and familiarity with the metric system are required. Technicians must be able to calibrate machines and calculate the correct amounts and proportions of solutions to be used, as well as quickly determine any necessary changes if there are indications that a patient is not responding to the treatment appropriately.

Technicians keep logs and fill out daily reports; their signature on the log verifies the accuracy of its content.

Parry Stashkiw is Director of Nursing at BMA of Evanston. When asked what she looks for when she hires a technician, she stresses reliability, teamwork, and willingness to learn. A technician must be able to follow procedures exactly and to think quickly in an emergency. One of Stan's regular patients, Noah Goldman, points out that manual dexterity is also essential, since a technician must insert a good many needles into his or her patients!

## how do I become a dialysis technician?

Like most dialysis technicians, Stan learned dialysis techniques through on-the-job training at the first center that employed him. Although there is a movement toward providing more formal academic training in the field of renal dialysis, presently only a few community colleges, such as Malcolm X College in Chicago, offer formal academic training for dialysis technicians. By far the majority of technicians learn their skills in hospitals or dialysis centers. Most people entering the field have either some type of experience in a medical setting or college training in biology, chemistry, or health-related fields.

Stan attended college for two years; he was working in another field when a friend told him of a job opening in a dialysis center. Although he did not have any health care experience, his college work in science and mathematics qualified him to enter the training program. One of his colleagues at BMA entered the field after having cared for her husband in home dialysis for many years.

### education

#### High School

Interested high school students should study general science, chemistry, biology, mathematics, and communication. Volunteering in a hospital, nursing home, or other patient care facility can give students a taste of what it's like to interact with patients in a health care setting.

#### Postsecondary Training

Students interested in the requirements for becoming a dialysis technician may obtain job descriptions from the National Association of Nephrology Technicians/Technologists (NANT). Those whose interest lies specifically in the area of nursing may wish to contact the American Nephrology Nurses' Association (ANNA). (*See* "Sources of Additional Information" for contact information.) Until there are a greater number of organized and accredited training programs, those who are interested in this career must seek information about educational opportunities from local sources such as high school guidance centers, public libraries, and occupational counselors at technical or community colleges. Specific information is best obtained from dialysis centers, dialysis units of local hospitals, home health care agencies, medical societies, schools of nursing, or individual nephrologists (physicians who specialize in treating kidney disease).

Other ways to enter this field are through schools of nurse assisting, practical nursing, or nursing and programs for emergency medical technicians. In these programs students learn basic health care and elementary nursing. After that, they must gain the specific knowledge, skills, and experience required to become a dialysis technician. The length of time required for a person to progress through the dialysis training program and advance to higher levels of responsibility should be shorter if he or she first completes a related training program.

Most dialysis centers offer a regular program of in-service training for their employees. At BMA of Evanston, the monthly training events may be presented by a member of the nursing staff, an outside speaker, or one of the technicians. Topics range from patient-care skills to advances in machine technology.

## certification or licensing

In most states, dialysis technicians are not required to be registered, certified, or licensed. California and New Mexico require certification. In some states dialysis technicians are required to pass a test before they can work with patients.

The Board of Nephrology Examiners—Nursing and Technology (BONENT) offers a voluntary program of certification for nurses and technicians. The purposes of the program are to identify safe, competent practitioners, to promote excellence in the quality of care of kidney patients, and to encourage study and advance the science of nursing and technological fields in nephrology. Technicians who wish to become certified must be high school graduates. They must either have at least one year of experience and be currently working in a hemodialysis facility or have completed an accredited dialysis course.

The certification examination contains questions related to anatomy and physiology, principles of dialysis, treatment and technology related to the care of patients with end stage renal disease, and general medical knowledge. Certified technicians use the title CHT (Certified Hemodialysis Technician) after their names (*see* "Sources of Additional Information").

Recertification is required every four years. To be recertified, technicians must continue working in the field and present evidence of having completed career-related continuing education units.

## who will hire me?

Dialysis technicians are employed by most major hospitals and by free-standing dialysis units like BMA of Evanston, which is part of a nationwide chain of more than five hundred dialysis centers. Many of these are listed under "Clinics" in the local Yellow Pages. Health care chains that provide home dialysis, either independently or in conjunction with a clinic, also employ larger numbers of technicians.

## where can I go from here?

Dialysis technicians who have gained knowledge, skills, and experience advance to positions of greater responsibility within their units and can work more independently. The NANT guidelines encourage a distinction between technicians and technologists, with the latter having additional training and broader responsibilities. Not all dialysis units make this distinction.

A technician looking for career advancement in the patient-care sector may elect to enter nurses' training; many states require that supervisory personnel in this field be registered nurses. Social, psychological, and counseling services may appeal to others who find their greatest satisfaction in interacting with patients and their families.

Someone interested in advancement in the area of machine technology may elect to return to college and become a biomedical engineer. Technical support and equipment maintenance is of major importance, and biomedical/equipment technicians may go on to become management personnel in this field.

**Advancement Possibilities**

**Biomedical equipment technicians** repair, calibrate, and maintain medical equipment and instrumentation used in health care.

**Counseling psychologists** provide individual and group counseling services in universities, schools, clinics, rehabilitation centers, Veterans Administration hospitals, and industry to assist individuals in achieving more effective personal, social, educational, and vocational development and adjustment.

**Registered nurses** provide general medical care and treatment to patients in medical facilities, under the direction of physicians.

## what are some related jobs?

The U.S. Department of Labor classifies dialysis technician under the headings *Occupations in Medical and Dental Technology* (DOT) and *Therapy and Rehabilitation* (GOE).

| Related Jobs |
|---|
| Biomedical equipment technicians |
| Cardiopulmonary technologists |
| Dental hygienists |
| Nuclear medical technologists |
| Occupational therapy assistants |
| Physical therapist assistants |
| Radiologic technologists |

## what are the salary ranges?

Dialysis technicians can earn between $15,000 and $35,000 a year, depending on their job performance, responsibilities, and length of service. Some employers pay higher wages to certified technicians than to those who are not certified. Technicians who rise to management positions can earn from $35,000 to $40,000, and those who work for industrial engineering or consulting firms may do even better.

Technicians receive the customary benefits of vacation, sick leave or personal time, and health insurance. Many hospitals or health care centers not only offer in-service training but pay tuition and other education costs as an incentive to further self-development.

## what is the job outlook?

The number of patients receiving dialysis in the United States doubled in the ten years from 1978 to 1988, and it continues to grow. Individuals who become dialysis patients will need to continue treatment for the rest of their lives unless they are able to have a kidney transplant. Technicians make up the largest proportion of the dialysis team, since they can care for only a limited number of patients at a time (the ratio of patient-care technicians to nurses is generally about four to one). In addition, there is a shortage of trained dialysis technicians in most locales and a high turnover rate in the field.

Two factors that could slow the future demand for dialysis technicians are the growing use of peritoneal dialysis and an increase in the number of kidney transplants, since patients who have had a successful transplant no longer require dialysis. However, there are a limited number of kidneys available for transplant, and until researchers discover a cure for kidney disease, dialysis technicians will be needed to administer treatment

## how do I learn more?

### sources of additional information

Following are organizations that provide information on careers in renal technology, training programs, and employers.

**Board of Nephrology Examiners—Nursing and Technology (BONENT)**
PO Box 15945-282
Lenexa, KS 66285
913-541-9077

**American Nephrology Nurses Association (ANNA)**
East Holly Avenue, Box 56
Pitman, NJ 08071
609-256-2320

**National Association of Nephrology Technicians/Technologists (NANT)**
PO Box 2307
Dayton, OH 45401-2307
513-223-9765

### bibliography

Following is a sampling of materials relating to the professional concerns and development of dialysis technicians.

#### Periodicals

*Contemporary Dialysis & Nephrology*. Monthly. Reports on current news regarding legislative, scientific, and administrative issues in the field of renal care. Contemporary Dialysis, Inc., 6300 Variel Avenue, Suite I, Woodland Hills, CA 91367-2513, 818-704-5555.

*Dialysis and Transplantation*. Monthly. A professional journal covering new clinical techniques and technologies and relevant legislative issues. Creative Age Publications, 7628 Densmore Avenue, Van Nuys, CA 91406-2042, 818-782-7328.

# dietetic technician

**Definition**

Dietetic technicians usually work under the supervision of a dietitian on a food service or health care team. Their responsibilities include taking dietary histories, planning menus, supervising food production, monitoring food quality, and offering dietary counseling and education.

**Salary range**

$15,000 to $25,000 to $35,000

**Educational requirements**

High school diploma; two-year associate's degree

**Certification or licensing**

Voluntary

**Outlook**

About as fast as the average

GOE 05.05.17

DOT 077

O*NET 32523

## High School Subjects

- Biology
- Business
- Chemistry
- English (writing/literature)
- Family and Consumer Science
- Health
- Mathematics

## Personal Interests

- Business management
- Cooking
- Exercise/personal fitness
- Helping people: physical health/medicine
- Teaching

**Karen Lucas** stops in front of Room 203 in the Cardiovascular Services wing and taps lightly on the door, glancing down at the chart in her hand. After pausing a moment, she pushes the door open and smiles broadly at the elderly woman in the bed nearest her.

"Good afternoon, Mrs. Breeden. How are you feeling today?" She pulls a chair nearer the bed and sits down.

"Oh, pretty good, honey," the woman responds weakly.

"It looks like you get to go home tomorrow, right?" Karen asks, as she organizes the papers on her lap.

Mrs. Breeden's face brightens. "Dr. Whiting says I do. I'll be glad to sleep in my own bed."

Smiling, Karen nods. "Now, Mrs. Breeden, you know Dr. Whiting has put you on a salt-restricted diet for your heart. We don't want you to end up in here again."

"I always watch my salt, honey."

"Good! Good for you. You're a step ahead, then. I'm just going to give you this list of foods that are high in sodium to take home." Karen pauses to hand Mrs. Breeden a printed sheet. "The doctor says you should be getting no more than four grams a day. Now, what that means is . . ."

## what does a dietetic technician do?

*Dietetic technicians* work in a variety of different settings, such as hospitals, nursing homes, community programs and wellness centers, public health agencies, weight management clinics, schools, day care centers, correctional facilities, and food companies. They may work independently, or in partnership with a dietician, depending on their employer and what they do.

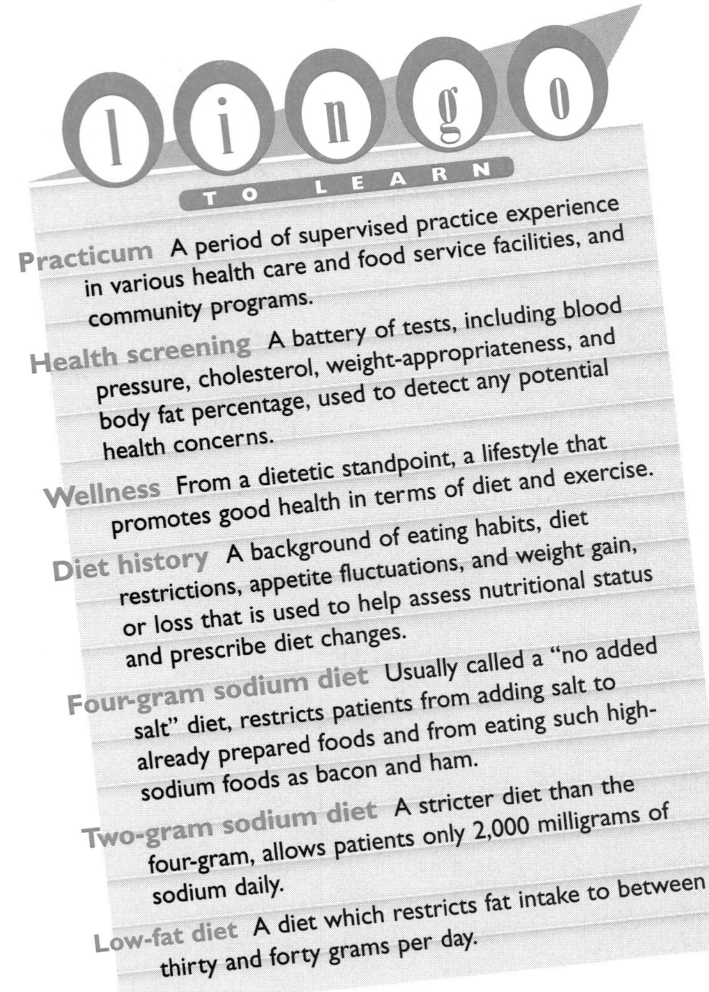


Dietetic technicians work in food service administration and clinical nutrition, which is the nutritional care of individuals. Technicians in smaller facilities may be involved in both areas of work, while in a larger facility they will probably have more specific duties in one area or the other.

Dietetic technicians working in food service administration will probably be involved in the management of other food service employees. They may develop job descriptions, plan work schedules, and help train staff members in methods of food production and equipment operation. They may also work directly in the kitchen, supervising actual food preparation, or in the cafeteria, supervising the workers who assemble and serve the food. In some cases, dietetic technicians are responsible for meeting standards in sanitation, housekeeping, and safety.

Another area of responsibility for the technician in food service administration can be diet and menu planning. Technicians may help modify existing recipes, or create new menus to meet the particular needs of individuals or institutions. They may also be in charge of monitoring the quality of the food and service. Finally, some dietetic technicians are involved in purchasing supplies and equipment, keeping track of the inventory, supervising food storage, and budgeting for cost control measures.

Dietetic technicians working in clinical nutrition at hospitals and nursing homes, for example, may observe and interview patients to obtain diet histories and food preferences. Using this information, they can work with a dietician to determine each patient's nutritional status and dietary needs, and to develop diets that meet these needs. They also may counsel and educate the patients and their families on good nutrition, food selection and preparation, and healthy eating habits. Some dietetic technicians may make follow-up contacts with these patients to monitor their progress and offer them further help.

If employed by a community program, such as a public health department, clinic, youth center, or home health agency, technicians may have many of the same counseling duties as the technicians in patient care facilities. They may provide health screenings and dietary education for low-income families, elderly persons, parents of small children, or any groups of people who might have special questions about nutrition and health care. In some cases, they make follow-up visits to their clients' homes to check on their progress, or make permanent arrangements for continuing care for the needy, such as hot meals for the housebound, or school lunch programs for children.

Dietetic technicians in a community program may also be in charge of developing and coordinating community education efforts. Technicians may help create brochures and teaching materials, or plan classes in nutrition, weight loss, and other health-related topics. Some diet techs may actually teach or co-teach the classes. They may also work with other community groups, corporations, or schools to promote an interest in health and nutrition.

*"You need to be interested in some aspect of nutrition. I don't think you'd be in this field if you weren't."*

## what is it like to be a dietetic technician?

Karen Lucas and Joan Shaw are both dietetic technicians working in the area of clinical nutrition at the same hospital, and their duties are almost exactly the same. Because one of them must be there every day, during the hours meals are served, they do not work together.

"We have ten-hour days," Joan says, "and we alternate weekends. It works out so that we work eight days and have six days off in a two-week period." Although this particular schedule is not the most common for dietetic technicians, the forty-hour work week is standard.

For Karen and Joan, the typical day at work begins at 6:30 AM. "The first thing that must be done is to check the breakfast tray line for completeness and accuracy, because at 6:45, the first wave of breakfast trays goes upstairs to the patients." According to Joan, it usually takes about an hour to check all the trays.

The diet tech on duty then has the responsibility of collecting all of the patient menus for the next day. Each day, the patients are given menus that contain all of their food options are instructed to mark their choices on the menu form. Every morning, Karen or Joan goes to each floor to collect all menus that have already been completed. After taking the completed forms back to the department to be processed, she checks a computer printout to see which patients still need to make up menus.

"We try to get everyone to fill out a menu," Karen says, "so we go to each patient's room. If they haven't already done it, we try to help them make the choices." Because patients may be sleeping, out of the room, being bathed, or receiving doctors' visits, seeing every one of these patients may take a number of trips, and often lasts until nearly lunch time.

Afternoons are mostly reserved for nutrition assessment and education, according to Karen and Joan. After conferring with one of the three dieticians to determine which patients need attention, and receiving diet orders for these patients, the dietetic technician checks each individual's chart to find out if he or she needs nutrition education. Not everyone on a special diet needs the training. "Sometimes they already know what they're supposed to be eating," Joan says. "Maybe they've had the counseling before, or have been on the diet for a long time already."

If the patient does need dietary counseling, Joan or Karen makes a personal visit to explain the prescribed diet and to go over the restricted foods. "We talk through it with them," Karen says, "and we have printed material, also, for them to take home with them."

There are different diets that the technicians must understand and be able to explain, such as two- and four-gram sodium, bland, low-fat, low-residue or low-fiber, and no sugar. "We don't usually see the diabetic patients," Karen says. According to her, the dieticians do the counseling for those patients, because the education is much more extensive and time consuming. Many of the patients Joan and Karen see are on low-salt and low-fat diets because of heart problems.

Although most of the days follow the same routine, Karen's duties do vary slightly on one day each week. On these days, she works for the Wellness Center, which is an offshoot of the hospital. "I do on-site health screenings for corporations and organizations," she says. The Wellness Center also offers on-site classes in better nutrition, stopping smoking, weight loss, and other health-related issues. Karen helps teach some of these classes. "It is a part of the hospital," she says, "but it serves the whole community." She enjoys her time at Wellness, as she calls it, because it adds variety to her job.

## have I got what it takes to be a dietetic technician?

The duties of individual dietetic technicians vary widely from person to person, depending both on workplace and area of service. Likewise, the personal qualities necessary to excel depend somewhat on the responsibilities of the particular job.

Karen and Joan agree that one of the most important skills is being able to communicate and deal well with people. "You have to really like people," Joan says. "We have to deal with patients all the time. A large part of our day is spent with them." Although the majority of patients she sees are very friendly and cooperative, she acknowledges that some of them are difficult to communicate with. "Sometimes, you walk into a room and the patient won't talk to you," she says. "Sometimes because of medication, or they're groggy, or they're depressed. Occasionally, but not often, they are unpleasant and argumentative." Because the technician has to deal with all sorts of people in all different circumstances, compassion and a desire to serve others are important. "The personal interaction can be a positive point about the job," Karen says. "You really feel like you're helping, and that's very satisfying."

Interpersonal skills are significant for the dietetic technicians who work in food service administration, as well as for those who, like Karen and Joan, work in nutrition and counseling. Food service administration may involve managing, scheduling, training, and evaluating other employees. Communication skills and an aptitude for dealing with people are essential for this area.

For the dietetic technicians who work in food service administration, it is also extremely important to be well organized, efficient, and able to deal with stress. Both Karen and Joan worked previously in food service supervision, and they say it's very different than the jobs they now perform. "Supervising is stress times ten," Karen says. "I was always swamped with work. Staffing the department, making sure the employees are there, and finding replacements for the ones who cancel at the last minute can be especially taxing." On the other hand, she says, she did enjoy the challenge of the job, and the duties she now performs, while much less stressful, can sometimes be a bit too routine.

Finally, the dietetic technicians say that an interest in nutrition and health care is important to being a success in this job. Joan has a longtime interest in cooking and home economics, while Karen has always been interested in health and fitness. "I think you need to be interested in some aspect of nutrition," Joan says. "I don't think you'd be in this field if you weren't.

**To be a successful dietetic technician, you should**

- Have a desire to serve people
- Be a good communicator
- Have an interest in health care and nutrition
- Be able to follow instructions well
- Be a good planner and organizer

## how do I become a dietetic technician?

Karen and Joan took different paths to becoming dietetic technicians. Joan has a four-year bachelor of science degree in home economics, and was a high school teacher before deciding to move into dietetics. Karen, on the other hand, enrolled in the two-year dietetic technician program at her college, after learning that such a degree was offered. "I'd always been interested in nutrition and wellness, and I wanted a two-year degree, so it seemed like a natural choice," she says.

### education

#### High School

To become a dietetic technician, you'll need at least a two-year college degree. The American Dietetic Association (ADA) suggests that students who are considering a technician career should emphasize science courses in their high school studies. Biology, anatomy, and chemistry will provide a very important background for success both in college classes and in the course of a career. High school math and business courses will also be good training, since the college requirements for this degree include some accounting and purchasing classes. Finally, the ADA recommends sociology and psychology classes, to broaden the student's understanding of people, as well as English to improve the student's communication skills.

The interested high school student might also want to check in his or her area to see if there are summer or part-time jobs, or volunteer opportunities available in the field of dietetics. Actually working in the field can provide valuable experience, as well as insight into what the jobs are like.

### Postsecondary Training

In order to become a dietetic technician, the interested student must enroll in one of the more than seventy colleges offering the ADA's approved program. The program is a combination of classroom training and a set number of hours of supervised practical experience, usually called a practicum. The classes generally include a number of science courses, such as biology, anatomy, and chemistry, and some business and administrative courses, such as accounting and institutional administration. General education classes, such as English and psychology, are also part of the curriculum. The technical, specialized dietetic training may include classes in food preparation, therapeutic diets, meal management, community nutrition, quantity food purchasing, and nutritional management of disease.

Over the two-year period, the student also gets a certain amount of supervised practical experience in various health care and food service facilities. The type of field experience a student dietetic technician might receive includes practicums in clinical nutrition; food service planning, purchasing, equipment use, sanitation, and training; and management. Students may be assigned to a patient care facility for practicum, where they help in preparing schedules, ordering food, cooking, or instructing patients. If they are assigned to a community agency, they might go on home visits with a nutritionist, help teach individuals, or assist in demonstrations and classes.

Karen especially enjoyed the practicums. "You really get a feel for what the job is like in practicum," she says. She doesn't remember the classes being extremely difficult, although she did have to study. Overall, she felt like her college program left her well prepared to enter the dietetics field.

## certification or licensing

After successfully completing the associate's degree, you are eligible to take the ADA's registration examination for dietetic technicians. The exam is given in October each year and consists of 240 multiple choice questions broken down into four subject areas.

FYI

The concept of dietetics is by no means a new one. Guidelines about food have existed for thousands of years in religious beliefs, and ancient and folk traditions.

Technicians who have passed the registration exam are known as Dietetic Technicians, Registered, and are allowed to use the initials DTR after their names. They are also eligible to become members of the American Dietetic Association.

It is not a requirement that dietetic technicians be licensed or certified. However, many do choose certification. As a newly graduated dietetic technician, it is a good credential to have, since it indicates that you have met a certain standard of competence. While it may not be necessary to be registered in order to get a job, it might provide a competitive edge in some cases.

Registered dietetic technicians are required to earn fifty hours of continuing education every five years to maintain the credential. They can do this by attending hospital programs, symposiums, or college classes. They can also perform approved self-studies or self-assessment modules for credit.

## scholarships and grants

Students in the dietetics field have a number of possibilities for financial aid to help pay for their schooling. In addition to federal grants, the ADA also offers aid, in the form of scholarships, to encourage eligible students to enter the field. Students in the first year of a dietetic technician program may apply (*see* "Sources of Additional Information").

Finally, qualified students might be able to obtain a grant or scholarship from corporations, community or civic groups, religious organizations, or directly from the college or university they plan to attend.

## who will hire me?

When Karen graduated from college, she sent her resume out to several hospitals, nursing homes, and food service companies. Her first dietetics job was as manager of a corporate cafeteria for a vending and food service company. After four years in that position, she decided she'd like to work in a clinical setting, because of her interest in therapeutic nutrition. She applied at a hospital near her home, and was hired as the assistant director of the food service department.

The majority of dietetic technicians are employed by hospitals. Technicians who work in long-term patient care facilities, such as nursing homes, make up the second largest group. A smaller percentage of dietetic technicians work in community health care programs or outpatient clinics. Finally, some dietetic technicians work in settings that are not directly related to health care, such as schools, colleges, hotels, or, as in Karen's case, employee cafeterias.

Graduates of a dietetic technician program who are looking for a first job should check the placement office of their school. Also, since they have spent a considerable amount of time in various dietetic workplaces to complete practicums, they may have excellent contacts in the field, which can serve as job leads. Applying directly to the personnel offices of all area hospitals, nursing homes, and public health programs is another possible route to finding a position. Finally, the classified ads of local newspapers, private and public employment agencies, and job listings in health care journals are all potential employment sources.

In some locations, the labor market for dietetic technicians may be flooded, particularly in areas near schools that offer the training program. In such cases, the new job-seeker might have more luck by broadening his or her search to include less competitive areas, and by looking for creative job opportunities. Karen advises that the dietetic technician who is looking for a job should keep an open mind about the possibilities. "Anywhere a dietician is, a dietetic technician could investigate," she says. Joan agrees. "Look for situations where there's too much work for one dietician, but not enough for two," she counsels.

**Gender statistics for dietetic technicians**
According to a 1993 ADA survey of its members

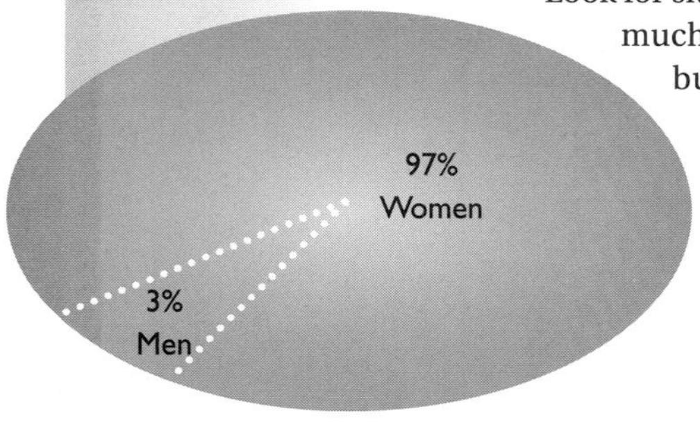

97% Women

3% Men

## where can I go from here?

Karen is currently in the process of finishing her bachelor's degree in business administration. She hopes that the extra schooling in administration, combined with her training and experience in dietetics, will help her obtain a managerial position. "I'd like to get more into the Wellness Center type of work," she says. "I'd like a supervisory job in something like that."

Beginning technician positions are usually closely supervised, but after spending some time on the job, the dietetic technician may be able to take on more responsibilities. Many technicians, after proving their abilities, are allowed to perform some of the same functions as entry-level dieticians, such as diagnosing nutrition problems, prescribing diets, helping develop educational materials, and being involved in the financial management of the department. With the expanded range of duties, technicians may then earn higher pay, while either keeping the same title, or officially changing positions. For example, a dietetic technician could be promoted to kitchen manager.

A very common means of advancement in the field of dietetics involves further schooling. The dietetic technician who wants to attain a higher position, such as dietician, may decide to pursue his or her bachelor's degree. A major in dietetics, nutrition, food science, or food service systems management, plus a year of internship, are the requirements for becoming a dietician. Although earnings vary widely with employer and amount of experience, the salary range for dieticians is $28,000 to $40,000.

Further advancement possibilities for experienced dieticians include assistant, associate, or director of a dietetic department. With a graduate degree, the dietician could move into research or an advanced clinical position.

## what are some related jobs?

Dietetic technicians use the principles of nutrition, working under the supervision of a dietician, in a variety of ways. They must have an understanding of food science, menu planning, and dietary needs, and may have, as well, a special aptitude for making food look and taste good.

**Advancement Possibilities**

**Clinical dietitians,** also known as **therapeutic dietitians,** plan menus and oversee preparation of meals for patients in hospitals or nursing homes, consult with doctors to determine diet needs and restrictions, and instruct patients and families in nutrition and diet planning.

**Community dietitians** coordinate food services for public health care organizations, evaluate nutritional care, instruct individuals and families in diet and food selection and provide follow-up, and conduct community dietary studies.

**Administrative dietitians,** also known as **dietetic department directors,** or **chief dietitians,** direct food service and nutritional care department of institutions, establish policies and procedures, hire and supervise staff members, and are responsible for menu planning, meal preparation, purchasing, sanitation, and finances.

The U.S. Department of Labor classifies dietetic technicians under the headings *Dietitians* (DOT) and *Craft Technology: Food Preparation*. All types of dieticians are, likewise, classified in these categories. Also under these headings are people who prepare, test, decorate, and analyze various foods and beverages, or who supervise others in performing these activities.

| Related jobs |
|---|
| Food products tasters |
| Cake decorators |
| Food and beverage analysts |
| Chefs |
| Cooks |
| Cook apprentices |
| Passenger vessel chefs |
| Food service managers |
| Home economists |
| Clinical dietitians |
| Research dietitians |
| Community dietitians |
| Consultant dietitians |

## what are the salary ranges?

Salaries for dietetic technicians have been increasing gradually for the past few years and are expected to continue to rise. Beginning dietetic technicians might expect to earn between $15,000 and $20,000 a year. Those with ten to fifteen years of experience would likely be making an annual salary of $20,000 to $25,000, and might look forward to eventually making $30,000 to $35,000.

Aside from the amount of experience a dietetic technician has, a significant factor affecting salary level is the area of work he or she has chosen. For those employed in clinical nutrition, which involves patient assessment and counseling, the overall median salary is $22,350. For those who work in food and nutrition management, which involves the supervision of food service employees and the overseeing of food production, the median is slightly higher at $25,255.

Most technicians are offered a benefits package by their employers, usually including health insurance, paid vacations and holidays, and meals during working hours.

## what is the job outlook?

Although the *Occupational Outlook Handbook* indicates that the outlook for dieticians is expected to be average, it appears as though opportunities for dietetic technicians may be better than average. Because the position is a fairly new one, dating back only to the early 1970s, the demand for technicians has previously been unsteady and patchy. Now, however, as employers are becoming more aware of the advantages of hiring dietetic technicians, the need is increasing yearly and is expected to continue to expand.

One major reason for the positive job forecast is that hiring dietetic technicians is cost-effective for employers. Many functions that dieticians used to perform can be done easily by dietetic technicians, leaving the dieticians free to concentrate on the work that only they can do. Since dieticians are expensive to hire, it makes sense for employers to supplement their nutrition or food service team with technicians who can do many of the same tasks but do not earn as high a wage. This method of reducing expenses may become even more popular, with the increasing public and governmental concern over health care costs.

The emphasis on nutrition and health in today's society is another reason for the positive outlook for dietetic technicians. In future years, more health services, some of them involving nutrition, diet training, and monitoring, will be used. The population is growing, and with it, the percentage of older people who have the greatest health care demands. The increasing need for health care services translates into an increasing need for workers.

# how do I learn more?

## sources of additional information

The following organization provides information on dietetic technician careers, accredited schools and scholarships, and possible employers.

**The American Dietetic Association**
216 West Jackson Boulevard, Suite 800
Chicago, IL 60606-6995
800-877-1600
http://www.eatright.org

## bibliography

Following is a sampling of materials relating to the professional concerns and development of dietetic technicians.

### Books

Caldwell, Carol Coles. *Opportunities in Nutrition Careers.* Lincolnwood, IL: VGM Career Horizons, 1991.

Kane, Michael T., ed. *Role Delineation for Registered Dieticians and Entry-Level Dietetic Technicians.* Chicago, IL: American Dietetic Association, 1990.

### Periodicals

*Directory of Dietetic Programs.* Annual. Lists ADA-accredited internship programs and ADA-approved dietetic technician programs. American Dietetic Association, 216 West Jackson Boulevard, Suite 800, Chicago, IL 60606-6995, 312-899-0040.

*Journal of Nutrition.* Monthly. Features reports on current research in the field of nutrition. American Institute of Nutrition, 9650 Rockville Pike, Bethesda, MD 20814, 301-530-7027.

*Journal of the American Dietetic Association.* Monthly. Addresses professional dietitians and nutritionists and reports on research and developments in the field of nutrition therapy. American Dietetic Association, 216 West Jackson Boulevard, Suite 800, Chicago, IL 60606-6995, 312-899-0040.

*The Official Journal of the American Naturapathic Medical Association and the American Association of Nutritional Consultants.* Bimonthly. Directed at members of the association, including dietetic technicians. American Association of Nutritional Consultants, 880 Canarios Court, Suite 210, Chula Vista, CA 91910, 888-828-2262.

*Topics in Clinical Nutrition.* Quarterly. Provides clinical information for dietitians and nutritionists. Aspen Publisher, Inc., 200 Orchard Ridge Drive, Suite 200, Gaithersburg, MD 20878, 800-638-8437.

# dispensing optician

**Definition**

Dispensing opticians are health professionals who fit eyeglasses and contact lenses using prescriptions written by eye doctors.

**Salary range**

$18,000 to $27,900 to $35,000

**Educational requirements**

High school diploma; two-year training program or apprenticeship program with eyewear company

**Certification or licensing**

Required in twenty-one states; recommended otherwise

**Outlook**

As fast as the average

GOE 05.10.01

DOT 299

O*NET 32514

## High School Subjects

Biology
Mathematics
Physics
Shop (Trade/Vo-tech education)
Speech

## Personal Interests

Fixing things
Helping people: physical health/medicine

By 3:30 on a Tuesday afternoon, Carla Hawkins has been at work for several hours. She's had to send back a young man's contact lenses because he couldn't adjust to them. She spent an hour helping an elderly woman who didn't want glasses in the first place to find a pair of frames she liked, and that was no easy task. A shipment of new display frames arrived damaged, and one of her co-workers called in sick. Carla hopes that the rest of her day will be easy. Instead she sees a little girl, around eight years old, nearly dragged in by her mother.

"We need glasses," the mother says when Carla approaches them with a smile. She hands Carla a prescription. The little girl is pouting.

"Not too excited to get glasses, huh?" Carla says, leaning down. The little girl doesn't answer. "Come over here with me," Carla says, and leads her to a stool where she can sit. Carla and the mother sit, too.

"You might feel like the only one," she says, "but I see kids your age and much younger every day who come in here and don't want glasses, feel mad or disappointed, or just that they're not going to look good, and I've had nearly everyone leave pretty happy with what they've chosen." The little girl is starting to show some interest.

"Maybe you're afraid you can't play ball as well with glasses or that you're going to stick out, but I will show you some frames you'll like. We can even make you some sunglasses."

"None of my friends have glasses," the little girl says finally.

"They will sometime," Carla says. "I promise you, it's going to feel good to see properly. Tell me about what you do after school."

A little while later they've found some frames the little girl likes, and she and Carla are looking for a chain for them to hang from around her neck. "I think it was the idea of the chain that won her over," her mother says to Carla as her daughter tries one on in the mirror.

## what does a dispensing optician do?

With 60 percent of the population needing corrective eyewear, *dispensing opticians* are in demand. Carla is one of many people throughout this country as well as throughout the world, who help customers find glasses or contact lenses that fit their needs and their lifestyles. Customers count on dispensing opticians to guide them through their eyewear process, and even the most experienced customer needs to be sure that her glasses or contacts fit and correct her vision.

**Refractometry** The process by which visual accuity is measured.

**Visual accuity** The measurement of how well an individual is able to see.

**Diopter** Measurer of visual deficiency.

**Fluoresciens** Die used to diagnose whether a lens is fitting the eye properly.

**Hyperopia** Scientific term for farsightedness, or inability to see close up.

**Presbyopia** Latin for "old eyes"; the process of vision deterioration that occurs in everyone once they pass the age of forty.

**Myopia** Scientific term for nearsightedness, or inability to see far away.

Glasses were widely used in the 1500s when printed matter for reading first became readily available, and the use of corrective lenses continued over the next several hundred years. From the start, dispensing opticians have been necessary to ensure the accuracy of the corrective eyewear prescribed. The optician needs to be sure that eyeglasses or contacts are made according to the optometrist's prescription specifications. They need to be certain of the placement of the lenses in relation to the pupils of the eyes. Dispensing opticians are valuable in helping customers select lenses or frames appropriate for their lifestyle. They also prepare work orders for the optical laboratory workers.

Dispensing opticians work around people all day long. Once a customer has obtained a prescription from the optometrist or opthalmologist, he or she gives the prescription to an optician and begins the selection process. The optician asks the customer about what he or she does. Will the glasses be for everything? Just for reading? Will they be needed while playing sports or engaging in hobbies? What types of jobs does the customer have? What kinds of hobbies? The optician can tell from the prescription whether the customer will require thick or thin lenses in the glasses, and can therefore advise certain frames. They offer suggestions as needed to the customer regarding the style and color of the frames.

Dispensing opticians are responsible for making certain that glasses fit the customer's face once the glasses have returned from the lab. While the process of finding and fitting glasses used to take some time, it has been turned into a one- or two-hour process thanks to the optical superstores with in-store labs.

Regardless of the time period, however, the optician fits the glasses for the customer when they return from the lab. They use small tools and instruments to make minor adjustments to the frames.

When fitting contact lenses, opticians must be precise and skilled. The optician measures the curvature of the cornea and prepares very specific directions for the optical mechanic who will make the lenses. They must be very careful and patient when placing the lenses on the customer's eyes and when teaching the customer to insert and remove the lenses.

# what is it like to be a dispensing optician?

To be a successful dispensing optician, you should

- Genuinely enjoy working with people
- Have steady hands
- Possess good eye-hand coordination
- Have a calm, soothing manner of touching and speaking
- Pay scrupulous attention to detail and accuracy

Carla Hawkins has been a dispensing optician for the past five years. "I started because I needed a job I'd be interested in," she says. "I'd just decided that I'd had enough of community college and wanted to take a break because without knowing what I wanted to do, I was wasting time." Carla found a job with Lenscrafters, an optical superstore where customers can have their glasses made for them, on-site, in less than an hour. She started at minimum wage. "The funny thing is that I mainly needed to earn money," she says. "I learned more about the work and liked it a lot. I ended up deciding to go back to school for opticianry."

Like most dispensing opticians who work in retail stores, Carla works a mix of shifts. She is in school so she doesn't work full-time, but dispensing opticians who work a full-time schedule generally put in a forty-hour week, and often evening and weekend shifts are worked into that schedule. Salaried employees are sometimes asked to work overtime, but not regularly. Dispensing opticians who are paid by the hour get paid time and a half for overtime work. Retail opticians are usually expected to work some holidays.

Dispensing opticians work in pleasant surroundings—clean, well-lit areas, whether in retail stores or in doctors' offices. They spend their time alternately sitting and standing. To aid customers in finding suitable frames, the dispensing optician generally points out several frames in different parts of the store. Customer information and medical history is generally taken sitting down. Fitting glasses and contacts involves sitting. The workspace of a dispensing optician is not too hot or too cool. It is all indoor work.

Carla works with a variety of people all day every day. Her job is to make sure the customers who walk into her store feel that they've received the best possible service and care. "People like to feel taken care of," Carla says. "For many people, buying glasses is a new experience. It's not just any purchase—people have to feel comfortable with their choice of frames because they're going to wear them. I know from experience that the customer is very grateful if they feel you've spent a lot of time with them, getting to know them a bit so that you can make suggestions. People feel vulnerable when it comes to glasses."

The work Carla does requires a lot of patience because of the different people she encounters every day, but the work is not physically difficult. Before the store opens in the morning, she makes sure that the store is neat and attractive, that new merchandise has been registered on the computer, that the customer records are up to date. There is behind-the-scenes work that allows the dispensing optician to devote her full attention to the customer when the customer is there.

Once the initial duties have been taken care of and the store is open, Carla's day revolves around the people she serves. "Some days it's all people who've never worn a pair of glasses before," she says. "This happens a lot with older people who need corrective eyewear as they age. Some days it might be all older people and little kids. Other days I get people who've all been wearing glasses for years. It's easier to help people who know what it's like to wear them, but I like to help people get over the hump of adjusting to themselves in glasses."

Carla finds out about the customers' lifestyle: what type of work they do, what type of hobbies, whether or not they play sports. "Unless they're getting two pairs of glasses or glasses and contacts, you need to do your best to make the glasses fit their lifestyle. For someone really active and involved in contact sports, you can't suggest some small delicate frame. It'll never survive." Carla works to make the customer at ease. Once she knows a bit about what they're looking for, she is ready to suggest some frame options. She goes through the store selection, suggesting frames she thinks suit the customer's needs and appearance. "It's important to find a style that's flattering to the face of the person," she says.

Some dispensing opticians spend time on the phone talking with the labs that make their glasses and contacts, but at Lenscrafters where Carla works, the lab is on site. "I know a lot of places call you up when your glasses are ready and you go in for a fitting or a lesson in how to wear contacts," she says, "and sometimes people come back for their glasses here, but mostly I deal with same-day service since we can make most prescriptions in an hour."

Like every dispensing optician, Carla is meticulous about the prescription she gets for glasses or contact lenses. Lenscrafters has an on-site optometrist, the doctor who prescribes the strength of corrective eyewear, but just as many people come with prescriptions from their own doctors. Because of the importance of the accuracy of prescriptions, Carla tests the customer's eyes before sending the prescription to the lab, and then later fits the customer with the glasses or contacts and has them do a reading test. "You have to be certain that they can see," she says. "After all, this is the reason they've come in the first place." She laughs.

## have I got what it takes to be a dispensing optician?

"Dispensing opticians have to like people," says Carla. "All day long you are interacting with people, and you can't get upset or impatient if things get frustrating. You need to be able to adjust yourself to get along with anyone." Children can be the most frustrating to deal with, but also the most rewarding, she says. "Kids come in here dragging their feet and it's not like we transform them into great fans of glasses every time," she says, "but it is very rewarding to watch their faces when they put their glasses on and you know they can really see. It's the best part of the job."

Opticians need to have steady hands and good hand-eye coordination for the tiny, tedious tightening and adjustment jobs that need to be done to glasses. They need to be patient and meticulous when it comes to measuring the cornea for contact lenses or making sure by checking their lens machine that someone's prescription is exactly right. A soothing and confident manner is beneficial for an optician. "People are very sensitive when it comes to their eyes," Carla says. "When you're asking someone to sit still for a few minutes so you can check the accuracy of their prescription through the lensometer, they feel more at ease with a quiet voice. I've been told that." Opticians touch their customers, when checking the fit of frames, when checking their eyes, when adjusting the finished glasses or inserting someone's contacts for them initially. A fine sense of touch is an attribute of a good dispensing optician.

**Employment distribution of dispensing opticians**
Data collected from respondents to 1995 OAA salary survey.

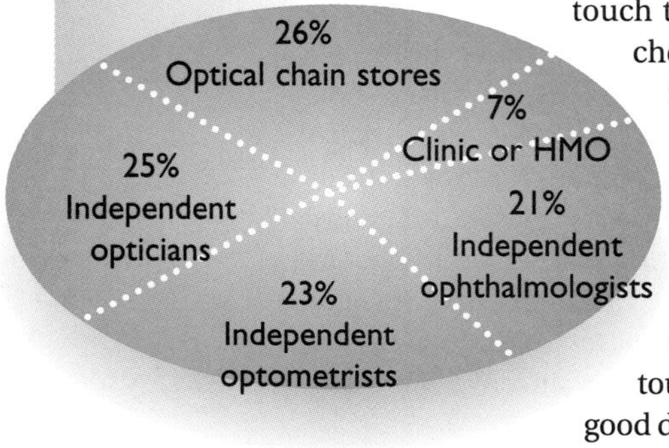

26% Optical chain stores
7% Clinic or HMO
21% Independent ophthalmologists
23% Independent optometrists
25% Independent opticians

Since dispensing opticians very often play a large role in a customer's frame selection, it is important that the optician has a good sense of color and feels comfortable advising the customer. Taking hair and skin color as well as face shape into account is what helps the optician to point out flattering and practical frames for the customer. "You wouldn't believe how many people want your opinion on the frames they choose," says Carla. "Some people have very strong ideas, but there are just as many who almost want you to choose for them. I guess they figure that since you work with eyewear, you must know best."

There are many qualities important to a dispensing optician, but perhaps the most important of all is good communication skills. To be able to obtain the information needed to help customers through the eyewear process, to make them feel at ease and in a friendly environment, to have an unthreatening and efficient manner: these are the crucial skills.

## how do I become a dispensing optician?

### education

#### High School

A high school diploma is preferred but not absolutely mandatory in the field of opticianry. While it's necessary to have completed high school to go on to a program offered by a community college, it is not impossible to work as an apprentice without a high school diploma. Students interested in entering the field will find it much easier having earned their diploma, simply because employers tend to attribute a higher level of knowledge and competency to students who've successfully completed their high school studies.

Courses such as algebra, geometry, and physics will help the aspiring optician later, as well as any course work in mechanical drawing. Communications classes are also valuable, because opticians need to be able to relate to a variety of people on all different levels.

#### Postsecondary Training

There are two ways of entering the field of opticianry. One is through a two-year associate's degree in optical dispensing from a community college. The other is by serving a supervised multiyear apprenticeship with an optician or optical association. Once an optician finishes a

## in-depth

### Transforming Light Energy into Vision

If you could look through the opening in the front of your eye (the pupil), and see to the very back surface, you would see your retina. On your retina are all the sense receptor cells that enable us to see. There are none anywhere else in the body.

On the paper-thin retina are two distinct sense receptors (nerve endings) called rods and cones. The cones are concentrated at a tiny spot on the retina called the fovea. If the focusing machinery of the eye (cornea, lens, etc.) is working just right, light rays from the outside have their sharpest focus on the fovea. Surrounding the fovea is a yellowish area called the macula lutea. Together, the fovea and macula lutea make a circle not much bigger than the head of a pin.

The cones are responsible for all color vision and fine detail vision, such as for reading. Beyond the circumference of the macula lutea, there are fewer and fewer cones, and rods become the dominant structures on the retina. It is estimated that there are roughly ten million cones on each retina, but more than ten times as many rods. The rods are responsible for vision in dim light and peripheral vision. Because the rods are not sensitive to colors, our night vision is almost completely in black and white.

The momentary interval after the cones stop working but before the rods begin to function is explained by a curious pigment in the eye called visual purple. This substance is manufactured constantly by the rods, and must be present for them to respond to dim light, but it is destroyed when exposed to bright light. Thus, after entering a darkened room, it takes a few moments for visual purple to build up in the retina. One of the principal constituents of visual purple is vitamin A, which is why this vitamin is said to increase our capacity to see in the dark.

Every nerve ending is part of a larger unit, a neuron or nerve cell, and the rods and cones are no exception. Like all nerve cells, each rod and cone sports a long nerve fiber leading away from the site of reception. In each eye, the fibers converge at a certain spot just behind the retina, forming the optic nerve. There are no rods or cones at the point where the optic nerve exits from behind the retina at the back of the eyeball. That is why everybody has a "blind spot" at that point. An image passing through that spot completely disappears.

From the retina, each eye's optic nerve goes almost directly to the middle of the brain, and the two converge a short distance behind the eyes. The optic nerve trunk then proceeds toward the rear of the head where the occipital lobes, the brain's "centers for seeing," are located. Just before reaching the occipital lobes, the optic nerve splits again into thousands of smaller nerve bundles that disappear into the visual cortex or "outer bark" of the occipital lobes. Only at this point are the bits of light energy that have stimulated our rods and cones transformed into images that our brain can "see."

program, he or she sits for licensing (that is, if located in one of the twenty-one states that require licensing to practice opticianry). If there is no licensing requirement, the new optician sits for an exam with the American Board of Opticianry.

Community colleges and trade schools concentrate on mechanical and geometric optics, opthalmic dispensing procedures, contact lens practices, business concepts, communications, and mathematics, as well as lab work in grinding and polishing lenses.

Opticians who work as apprentices learn all aspects of the business by following in the footsteps of another optician and gradually working up to taking on tasks and responsibilities individually. The process of working as an apprentice tends to take longer than the community college or technical school training; however, apprentices are paid while they learn, while students of opticianry must pay for the training they receive.

## certification or licensing

To be considered a practicing optician, twenty-one states require dispensing opticians to have a state license. The requirements for these licenses vary from state to state, but all require passing a written examination. Opticians who work in doctors' offices are not required to have state licensing.

Regardless of the requirements of the states in which they live, many opticians choose to be certified at the national level by either the American Board of Opticianry or the National Contact Lens Examiners. Certification is achieved after successful completion of examinations and is maintained by attending continuing education classes that meet with the approval of the associations. Opticians are expected to renew their certification every three years. Examinees are not required to have had formal training or experience to sit for these examinations, but some experience or education improves your chances of passing.

Many opticians choose to become members of professional organizations, such as the Opticians Association of America. These organizations offer the dispensing optician good resources regarding professional networking, continuing education, and career enhancement, as well as possible sources of employment (*see* "Sources of Additional Information").

## who will hire me?

Of the sixty thousand dispensing opticians in the United States, more than half work in the offices of optometrists and opthalmologists, and in hospitals and eye clinics. Others work in small retail optical stores, in optical departments of department stores and large discount chains, and in optical superstores that have laboratories on-site and offer to make the customer's glasses in an hour.

In urban areas, employment opportunities are particularly good because of the number of eyewear stores and departments in the area. There are also more hospitals and more doctors' offices in larger urban areas than in smaller towns or rural areas.

Community colleges and technical schools with programs in opticianry generally have good career placement centers to help their graduates find work. Apprenticeships often turn into full-time, well-paying jobs for the certified optician. The Opticians Association of America and other professional organizations also help opticians to find work in their field (*see* "Sources of Additional Information").

## where can I go from here?

When Carla began working at Lenscrafters, she was taking a break from community college. Once she realized that she loved what she was doing, she went back to school full-time and graduated with a bachelor's degree in health sciences. She is presently enrolled in graduate school for optometry. "It's a four-year degree," she says. "It's been great working at Lenscrafters. I started knowing nothing. They are good about promoting you. I've been promoted as I've been going to school. I'm a day manager now, while I'm becoming an optometrist."

Why did she decide to go into optometry? " I became very interested in the field and realized that I could go a lot further as a doctor," she says. "I guess I always knew I'd finish school, but I didn't know I'd go this far. It's been incredible." Carla's schedule is anything but easy. She works

**Advancement Possibilities**

**Managers of eyewear stores** perform the duties of a dispensing optician as well as the day-to-day duties of any manager in charge of running a retail store.

**Optometrists** examine eyes to determine the nature and degree of vision problems or eye diseases and prescribe corrective lenses or procedures.

**Opthalmologists,** or **oculists,** are medical doctors who diagnose and treat diseases and injuries of eyes.

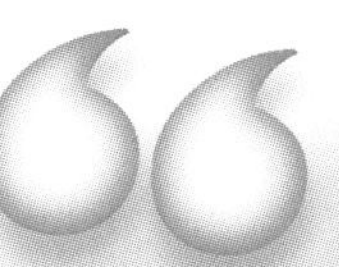

twenty-five hours a week at Lenscrafters, attends school full-time, studies, and works on campus at the eye clinic. “Working at an eye clinic is interesting,” she says, “but since it’s on campus, nothing we see is very serious. People who are working in the inner city or in poor areas of the country get a lot of exposure to serious eye problems.

Will she stay on at Lenscrafters once she has her degree? “I don’t know,” Carla says. “If there’s an opening here for an optician, I may stay for a while. I’m most interested in going into private practice. That’s when you have the most freedom and make the most money.” She laughs. “I have to get some patients first, though. Right now I don’t have any patients. It’ll probably be slow going for the first few years.”

> All day long you are interacting with people. You need to be able to adjust yourself to get along with anyone.

## what are some related jobs?

The U.S. Department of Labor classifies dispensing opticians under the headings *Miscellaneous Sales Occupations, Not Elsewhere Classified* (DOT) and *Crafts: Structural* (GOE). Also listed under these headings are department store managers, telemarketers, customer service clerks, watch and clock repair clerks, and a wide variety of people who work with their hands, such as front-end mechanics, roofers, glaziers, and carpet layers. For those who enjoy dealing with customers and performing delicate work, there are other fields of potential interest, including artificial eye making, jewelry making, ophthalmology, orthodontics, and prosthetics.

| Related Jobs |
| --- |
| Optometric technicians |
| Optometric assistants |
| Jewelers |
| Opthalmic laboratory technicians |

## what are the salary ranges?

In a 1995 survey conducted by the Opticians Association of America, all opticians surveyed reported salaries that combined to make a national average of $27,900. In states with mandatory licensure, the salary average is $29,700. States that don’t require licensing reported an average of $26,000. The average salary of managers in the field came to $30,376 per year. Apprentice opticians reported an average salary of $19,000 per year.

“As is the case in most professions, women opticians don’t make as much as men,” says Jacqueline Fairbarns, the communications director for the Opticians Association of America. “In our survey, average salary reported was $27,900. Men reported an average of $31,390 per year, while women reported $23,472. Often times, the reason given for salary discrepancy based on gender is that men have been in the business longer, but for this survey, we looked at an average age of forty for women and forty-four for men, so these women have always been part of this field.”

Out of a reported estimate of sixty thousand dispensing opticians in the United States, over half of them work in doctor’s offices. Privately employed opticians make more money than those who work in the retail business for large or small companies.

## what is the job outlook?

Job opportunities for dispensing opticians are expected to increase as fast as the average for all occupations through 2006. As the population grows older due to advances in medical technology, and as the population grows in general, there are more people around who need glasses. With 60 percent of the population requiring some type of corrective eyewear, and 98 percent of all people sixty-five and older needing corrective lenses, dispensing opticians should be busy for a long time to come. No matter what the state of the economy, dispensing opticians remain necessary. The optics field does not experience wide layoffs, although employment does fall somewhat during recessionary times.

The optics industry will also grow because of the influence of fashion. With a wider selection of colors and styles and materials of frames to choose from than ever before, people are interested in owning more than one pair of glasses. The optical superstores repeatedly offer specials encouraging people to buy two pairs of glasses, or contacts and glasses, for a price barely higher than the price of one pair.

Demand in the industry will also increase because of innovative ideas along the lines of special lens treatments and new lens and protective materials, as well as innovations relating to contact lenses, such as extended wear and disposable lenses.

To be competitive for the future of opticianry, students are encouraged to follow a course of study through a community college or tech school, and to become certified. The future holds good jobs for dedicated dispensing opticians.

## how do I learn more?

### professional organizations

Following are organizations that provide information on dispensing optician careers, accredited schools, and employers.

**American Optometric Student Association**
243 North Lindbergh Boulevard
St. Louis, MO 63141
314-991-4100

**Association of Schools and Colleges of Optometry**
6110 Executive Boulevard, Suite 690
Rockville, MD 20852
301-231-5944

**Commission of Opticianry Accreditation**
10111 Martin Luther King, Jr. Highway, Suite 100
Bowie, MD 20720-4299
301-459-8075

**National Academy of Opticianry**
10111 Martin Luther King, Jr. Highway, Suite 112
Bowie, MD 20720-4299
301-577-4828

**National Association of Optometrists and Opticians**
18903 South Miles Road
Cleveland, OH 44128
216-475-8925

**Opticians Association of America**
10341 Democracy Lane
Fairfax, VA 22030-2521
703-691-8355

## bibliography

Following is a sampling of materials relating to the professional concerns and development of dispensing opticians.


#### Books

Belikoff, Kathleen M. *Opportunities in Eye Care Careers.* Lincolnwood, IL: VGM Career Horizons, 1998.

#### Periodicals

*American Optician.* Quarterly. Features news reports and educational information for association members. Opticians Association of America, 10341 Democracy Lane, Fairfax, VA 22030-2521, 703-691-8355.

*Eyecare Business.* Monthly. Dedicated to vision care and related professionals. Cardinal Business Media, Inc., 1300 Virginia Drive, Suite 400, Fort Washington, PA 19034, 215-643-8000.

*Journal of the American Optometric Association.* Monthly. Features educational articles and news reports for optometry and related professions. American Optometric Association, 243 North Lindbergh Boulevard, St. Louis, MO 63141, 314-991-4100.

*20/20 Magazine.* Monthly. Directed at optometrists, opticians, and optical supply companies and manufacturers. Adams/Jobson Publishing Corp., 1180 Avenue of the Americas, 11th Floor, New York, NY 10036, 212-827-4700.

# drafter

**Definition**

Drafters take the ideas, rough sketches, specifications, and calculations from architects, engineers, and designers and translate them into a set of precise working instructions from which products can be made.

**Alternative job titles**

CAD operators

Drafting specialists

Layout specialists

Technical illustrators

**Salary range**

$18,000 to $31,250 to $50,750

**Educational requirements**

High school diploma; associate's degree recommended

**Certification or licensing**

None

**Outlook**

Little change or more slowly than the average

GOE
05.03.02

DOT
017*

O*NET
22514B

## High School Subjects

Art
Computer science
Mathematics
Physics
Shop (Trade/Vo-tech education)

## Personal Interests

Computers
Drawing
Fixing things
Sculpting

being on the construction site of the latest high-rise skyscraper. Sixty feet above the ground a steelworker confidently directs the crane below to hoist a twenty-ton steel beam precisely into place. The huge beam swings easily from the wire and fits perfectly. Glancing down at his blueprints, the steelworker studies the drawing to see how this beam is meant to be attached, and places the correct bolts. Another floor done.

Meanwhile, concrete is being poured, plumbing and electrical systems are installed, curtain walls and windows are installed . . . a small city of five hundred workers is putting together the various parts of the building. How do they know which piece fits where or how long the beams should be or how the windows fit into the wall? They are all relying on the drafter for his or her expertise in making the technical drawings that show them how each and every part of the building is put together. Without the drafter's drawings the workers haven't got a clue what to do next. It's up to the drafter to show them.

## what does a drafter do?

Anything that needs to be built needs to have a set of plans or detailed instructions. For example, an engineer might have an idea about how to build a new rocket, but before it can be started, a set of very detailed instructions has to be drawn explaining just how the engineer's idea can be realized. Like the old cliche, "a picture is worth a thousand words," it is often much easier to draw someone a picture of exactly what you have in mind than it is to tell them about it. *Drafters* take the ideas, rough sketches, specifications, and calculations from architects, engineers, and designers and translate them into a set of precise working instructions from which products such as rockets can be made.

The drafter prepares detailed plans and specifications that show the technical details of the product to be made. The specifications also indicate what materials are to be used, how the product is to be put together, and other information needed to carry out the job. Drafters might refer to technical manuals, tables, and other charts for more information in preparing the plans. They not only have to understand the theory and principles of these professions but also the tools and materials the builders are going to use in order to create the final product.

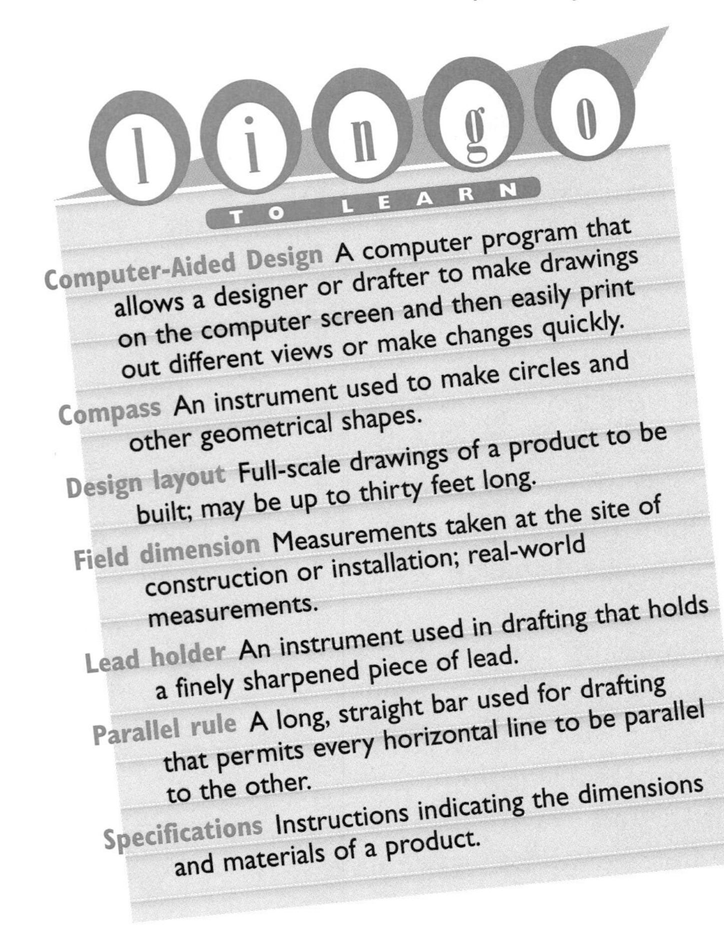


**Computer-Aided Design** A computer program that allows a designer or drafter to make drawings on the computer screen and then easily print out different views or make changes quickly.

**Compass** An instrument used to make circles and other geometrical shapes.

**Design layout** Full-scale drawings of a product to be built; may be up to thirty feet long.

**Field dimension** Measurements taken at the site of construction or installation; real-world measurements.

**Lead holder** An instrument used in drafting that holds a finely sharpened piece of lead.

**Parallel rule** A long, straight bar used for drafting that permits every horizontal line to be parallel to the other.

**Specifications** Instructions indicating the dimensions and materials of a product.

Drafters link the people who have an idea or design with the people who make the finished product. They act as an interpreter and engineer as well as builder, scientist, and architect.

Drafters are often classified according to the type of work they do or their level of responsibility. *Senior drafters* use the preliminary information and ideas from engineers, architects, or designers and create the initial drawings or design layouts. They may assign the task of drawing the complete set of drawings to *detailers*, who indicate the dimensions, materials, and assembly instructions on the drawings. *Checkers* examine the completed drawings and check for any inaccuracies in recording the dimensions and specifications.

Drafters may also specialize in a wide variety of particular fields, such as electrical, electronic, aeronautic, structural, or architectural drafting. Although a drafter's basic task—drawing—is not too different from one area of specialization to another, the types of objects the drafter draws vary considerably.

*Commercial drafters* do all-around drafting, such as plans for building sites, layouts of offices and factories, and drawings of charts, forms, and records.

*Civil drafters* make construction drawings for roads and highways, river and harbor improvements, flood control, and other civil engineering projects. *Structural drafters* draw plans for bridge trusses, plate girders, roof trusses, trestle bridges, and other structures that use structural reinforcing steel, concrete, masonry, and other materials (*see* chapter "Civil Engineering Technician").

*Electrical drafters* make schematics and wiring diagrams to be used by construction crews working on equipment and wiring in power plants, communications centers, buildings, or electrical distribution centers.

*Electronics drafters* draw schematics and wiring diagrams for television cameras and TV sets, radio transmitters and receivers, computers, radiation detectors, and other electronic equipment.

*Aeronautical drafters* prepare engineering drawings for planes, missiles, and spacecraft. *Automotive design drafters* and *automotive design layout drafters* both turn out working layouts and master drawings of components, assemblies, and systems of automobiles and other vehicles. Automotive design drafters make original designs from specifications, and automotive design layout drafters make drawings based on prior layouts or sketches.

A design team working on electrical or gas power plants and substations may be headed by a *chief design drafter*, who oversees architectural, electrical, mechanical, and structural drafters. *Estimators* and drafters draw specifications and instructions for installing voltage transformers, cables, and other electrical equipment that delivers electric power to consumers.

FYI

Drafters create many different kinds of drawings. A *plan* is a graphic representation of the design, location, and dimensions of the project, seen in a horizontal plane viewed from above. An *elevation* shows the vertical elements of a building, either exterior or interior, seen on a vertical plane. A *section* cuts through the object, whether it is a house or a piece of stereo equipment, to show someone how the inside and the outside fit together.

## what is it like to be a drafter?

Ray Miller has been working for an architectural woodworking company for the past five years. The company specializes in constructing furniture and other custom woodworking for architects. Ray's job involves taking an architect's sketch or idea and making it come to life as a real piece of furniture.

The first thing that Ray does when he gets a job is to visit the site and take the field dimensions of the place where his piece is going to be installed. "Taking accurate field dimensions is very important. It can be disastrous if we've gone through all of the work in creating this piece of furniture, and when it arrives to be installed it doesn't fit." Ray also visits the site so that he can visualize how the furniture is going to look in its setting. "Once I visit the site I can see how it all fits together. I can't do that just working from an architect's drawing."

Once he's back at the office, Ray sits down at his drafting board and makes very detailed drawings showing how he thinks the pieces of furniture must go together. For the kitchen cabinets Ray is currently working on, he has made a total of fifteen drawings. The drawings are made on his drafting board. Ray uses a parallel ruler, which is a long straight bar attached to wires that allow it to slide up and down on the drafting board. Each horizontal line Ray draws is then parallel to all of the other horizontal lines. For vertical lines, Ray usually uses a right angle triangle.

Ray also uses an assortment of other tools such as a compass, for circles and ellipses; a French curve, for complicated curves; scales, for measuring precise dimensions; and some pretty complicated-looking pencils. "Actually, we call them lead holders. They grip onto a long piece of lead which makes it easier to make a really sharp point using this lead pointer." Ray twirls the lead around in the lead pointer and pulls it out, showing a dangerously sharp point. "I need to use a really sharp point because of all the detail I'm putting into the drawings and lettering. Not only that, but if someone wants to take a measurement off of these drawings, like down in the shop during the construction, the drawings have to be precise so that their measurement of my drawing is the actual dimension that it is supposed to be."

Not only do the drawings show how long each piece of wood needs to be or what length of screw holds the pieces together, but Ray has different views of the cabinets that give a precise picture of how the cabinets fit together. "I would say about 75 percent of my time is spent making and going over the drawings. The builders in the shop have to have reliable instructions if they are going to make a piece of furniture. I look at it as though I'm making a map for someone to follow. If I give them the wrong directions, they're going to get lost. In this case the finished cabinets have to fit just right into the kitchen we're working on."

After Ray has completed all of the drawings, he must specify how much and what kind of material the builders are going to need and eventually call the lumber supply store to place the order. Ray is responsible for the entire project from beginning to end. He doesn't have a supervisor looking over his shoulder giving his approval. Ray's ability and the experience he has learned on the job allow him to be able to work on his own. "I can't say that there is a typical day with this job. Every day something new comes along that I have to deal with. I guess the hardest part is being able to juggle all of the different responsibilities at once."

Paul Jones started his career as a drafter. Now he's an architect developing his own ideas and creating new buildings. When Paul got out of college, his first job was as an architectural drafter, a common entry-level position for an architect. Paul spent his first year on the job learning how the various functions of a building work together, such as the plumbing and heating systems, for example. He also drew details of construction plans—and drew them, and drew them, and drew them. "Drafting can be long and tedious sometimes," Paul says, "and you have to be able to work on your own for long periods of time. Of course, you're at the bottom of the pecking order in the architectural firm, so everyone throws work your way that they don't want to do."

Being an architectural drafter requires learning a great deal about how buildings and other structures are put together. Much of this is taught in school, but for the real nuts and bolts of building, there is no classroom like experience. "Since becoming a drafter, I think I've spent at least twice as many hours just learning, studying—and of course drawing—all of the different systems and how they work together than I did when I was in school," Paul says.

Then, as now, Paul learned a lot from senior drafters and other architects. His office works as a team to make sure that each building is constructed properly. "The team concept is really important in an architecture firm. We don't just hand a project over to an architect and let him do the entire project on his own. There are people here who specialize in drafting and teach the interns the principles and theories of the building systems. Then there are the specifiers who make decisions on what sort of materials the building is going to need . . . accountants, lawyers, contractors, and on and on. It takes a lot of different people working together to construct a building. As an intern, you may discover that you have a talent for designing the detail work of a building and will be encouraged to become a senior drafter."

**To be a successful drafter, you should**

- Be able to think and to visualize in three dimensions
- Pay a lot of attention to detail
- Be able to communicate your ideas clearly and efficiently, as well as be able to visualize and comprehend what someone else is asking for
- Have a strong aptitude for mathematics
- Be able to work unsupervised and be a self-starter

## have I got what it takes to be a drafter?

As Ray described before, about 75 percent of a drafter's time is spent drawing. Creating different views of the same drawing involves an ability to see a drawing in three dimensions. A drafter, for example, needs to be able to visualize what the chair you might be sitting in would look like if you drew it from the floor's perspective. One of Ray's favorite pastimes is sketching, which he says also helps him out on the job. "Sometimes a builder in the shop will have a question about a job which doesn't show up on one of the drawings. A lot of times I can answer his question with a quick sketch of that detail and that's all he needs to keep going."

A drafter's job is to translate the designer, architect, or engineer's idea into an actual product. In order to make sure that he understands what he is supposed to be making is accurate, Ray has to talk a great deal with the architect he is working for. "They might have an idea of what they want, but I have to be able to talk to them to make sure *I* know what to make." This means being able to communicate your ideas clearly and efficiently, as well as being able to visualize and comprehend what someone else is asking for. When it comes time to order all the materials, Ray also spends a lot of time on the telephone with suppliers, making sure that the builders in the shop will be getting exactly what they need to make the piece of furniture.

In order to make sure that both the drawings and the product come together as the designer wanted, a lot of attention to detail is needed. A drafter has to be patient so he doesn't rush his drawings and risk making a mistake or making the drawings unreadable. Making these drawings takes quite a while. Ray spent nearly five days on the fifteen drawings of the cabinets he was working on. A drafter spends a lot of time working on his or her own.

Another part of being a drafter is making different calculations. Ray uses his math skills every day in measuring and scaling and in figuring out angles, using geometry and so on. Paul used a lot of math when he was an architectural drafter and still uses math as an architect. Precise drawings and calculations rely heavily on the drafter's ability to make calculations.

All in all, a drafter's job is varied and at the same time meticulous. You have to pay a lot of attention to details both in the drawing and drafting, as well as in the measurement and installation of the final product.

# how do I become a drafter?

Paul's goal was always to be an architect, and he spent five years in college earning his professional degree. He plans on getting a master's degree soon. "There's so much to learn in the field of architecture that you almost always are studying something."

Ray's goal was always to be a drafter. After becoming interested in woodworking and other shop classes in high school, he went to a junior college to learn how to draw the various plans that are needed to be a drafter in an architectural woodshop.

> You always have to be mentally and physically prepared for an abnormality or emergency of any sort.

# education

## High School

You don't need to be a rocket scientist in order to make a rocket. Drafters spend a good deal of time learning how to do their job in the workplace. But a high school diploma is the first step. High school courses in algebra, trigonometry, and geometry are prerequisites for mechanical drawing. Ray says that he uses something from all of these classes every day, from measuring dimensions in the field to figuring out what a particular angle is on a drawing.

Physics, art classes, and wood, metal, or electrical shop courses also are beneficial. Above all, Ray stresses that the most important aspect of his training was his commitment and interest in the field. Ray's interest in woodworking started when he was in high school and continues today in his own shop at home.

## Postsecondary Training

After high school, further training at a community or junior college is not essential to finding a job as a drafter, but it can make the job search shorter and more rewarding. Ray went to a local junior college to earn an associate's degree in mechanical drawing and drafting before he landed his current job. By earning an associate's degree, Ray was able to start being a drafter much earlier and with less on-the-job training.

Many employers look for beginners to have up to two years of college-level study in industrial or engineering technology, engineering graphics, drafting and design, or architectural drafting.

Acquiring skills in computer-aided drafting (CAD) is also extremely helpful and highly recommended, as nearly all drafting will eventually be done on the computer. Ray is currently enrolled in a CAD class so that he will be ready to go when his company starts using a computer system in a few more years. The use of the computer will greatly reduce the amount of drawing revisions Ray now has to do. For example, in Ray's present system, if the architect calls for a change, or there is a change in the dimensions, Ray has to go back to the drafting table and completely redo all of his drawings. With the computer, Ray will be able to make his drawings once and then print out various views of the object without redrawing the whole product. In some areas of specialization, such as automotive design, the computer is attached to a model-making machine. After the design is made on the computer, the machine can automatically create a complete scale model of the automobile right there in the studio! Other areas of specialization, such as electronic drafting, are already done completely on the computer (*see* chapter "Computer-Aided Design Technician").

If you want to become an architectural drafter, or a drafter in another field that requires a lot of specialized knowledge, a four- or five-year college program is required. Paul spent five years at a college to get a professional degree in architecture. This allowed him to enter into an internship at an architectural firm as an architectural drafter. Most of the architectural drafting positions are filled this way. At the end of his three- or four-year internship, Paul can then become a licensed architect. Until he passes the licensing test, all of Paul's work is the responsibility of a more senior, licensed architect. Ray, on the other hand, does not have to go through an internship or licensing of any kind. He uses his skills and expertise in order to get the job done for his client, usually an architect, who has final approval on whether Ray has done the job correctly.

## who will hire me?

Many drafters find jobs in the industrial sector. Since so many new products come to the market every year, drafters are needed to help in the design stages of the products. They draw plans of how the product will be produced on the assembly line and of the procedure for putting the product together.

About one-third of all drafters work in the engineering and architectural service sector. Another third work for areas of industry that manufacture durable goods, such as computers, televisions, appliances, and metal products. Drafters can also be found working for construction, transportation, communications, and utility companies. Drafters working for the federal government are mostly employed by the defense industry but can be found working on highways, bridges, dams, and other government projects. State and local governments hire drafters for various projects that they undertake, such as streets and sanitation projects, water management projects, or state construction projects like courthouses and jails.

If you have a specific interest, such as aeronautics or automobiles, you may want to check out companies that make those particular products.

Many of the trades, such as construction, woodworking, or engineering, have magazines or newsletters with listings for jobs and more information on what is going on in those fields (*see* "Bibliography").

Ray got his first job as an assistant to a drafter in a large architectural wood shop. He did some woodworking and helped the head drafter make details for drawings. As he gained experience, Ray was given more and more responsibility. A year after graduating with an associate's degree in drafting and mechanical drawing, Ray now works as a senior drafter, with full responsibility for creating the products the architectural woodworking company takes on.

Paul started as an architectural drafter in an architectural firm, along with two other recent graduates from college, and worked for a senior drafter for about a year before moving on in his internship. He's now a project architect but still relies on the skills and knowledge he learned as an architectural drafter.

Employment breakdown for drafters

33% Construction, transportation, communications, utilities

33% Industry

33% Engineering and architecture

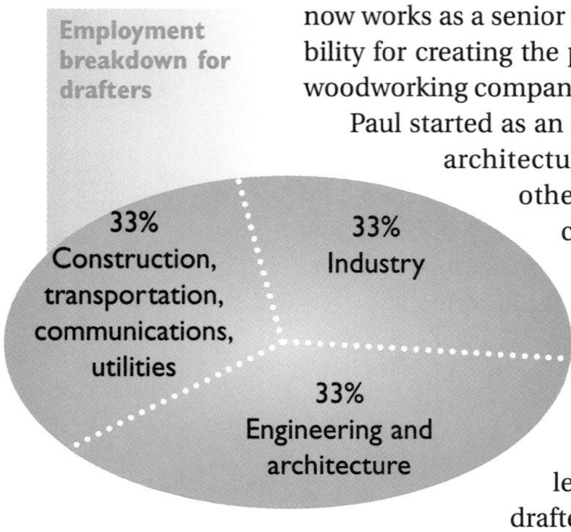

## where can I go from here?

"Being an architect is probably the ideal situation for me," Paul says. "I can be creative, and yet the precision and detail that goes into a building really make me think. I started off as an architectural drafter, so I know how all of the various components of a building fit together. But now I get to be the one who creates those scribbles for the drafters to try and figure out," he says, laughing.

Depending on the level of education, drafters usually take about three to four years to learn the skills and knowledge necessary to advance. Normally, advancement starts from beginning drafter, to detailer, to designer. Ray did not have to spend as much time being a beginning drafter because he got an associate's degree in drafting and mechanical drawing before getting a job as a drafter.

In larger organizations, a drafter's job title might tell something about what he or she does—layout drafter, checker, detailer, senior drafter. Advancement in larger firms may depend on how many drafters are employed, but drafters usually advance when they show a particular expertise A drafter who can make particularly good drawings might advance as a designer, or a drafter who knows a great deal about the possible materials that can be used in a project might advance as a specifier.

Depending on whether you are more interested in the drawing aspect of drafting or the technical aspect, with further study, a drafter could easily advance to become a graphic artist, designer, or technical illustrator, or a technical or specification writer. Graphic artists design the art or advertising layouts that are found in newspapers and magazines. A technical illustrator makes very detailed production drawings used in the manufacturing process. Technical writers, or specification writers, have become experts about the materials in their field and are responsible for selecting and describing the materials used in products.

Ray hopes to use his experience—following a piece of furniture all the way through from start to finish—to become a project manager. A project manager is the person who is responsible for completing an entire building from start to finish.

**Advancement Possibilities**

**Graphic artists,** or **designers,** design art or advertising layouts for publication in magazines, books, and newspapers.

**Technical illustrators** make very detailed production drawings used in the manufacturing process.

**Technical writers,** or **specification writers,** are experts in their specialized field responsible for describing and selecting the materials of a product.

**Project managers,** oversee all aspects of a construction or manufacturing project.

## what are some related jobs?

In the *Dictionary of Occupational Titles,* the U.S. Department of Labor classifies drafters under the headings of their various specialties, such as electrical/electronics (003), aeronautical (002), architectural (001), surveying/cartographic (018), mechanical engineering (007), civil engineering (005), mining and petroleum engineering (010), and not elsewhere classified (017).

The U.S. Department of Labor also classifies drafters under the heading *Engineering Technology: Drafting* (GOE). In this group are others who prepare drawings to convey engineering ideas and information, as well as technical illustrators, auto design checkers, and drafting specialists.

Individuals interested in drafting might explore the world of cartographic (mapmaking) drafters, photogrammetrists, specifications writers, graphic designers, and illustrators. They might also look at the work of bank note designers, set designers, stained glass artists, copyists, and other art and design workers.

| Related Jobs |
| --- |
| Computer-aided design technicians |
| Cartographers |
| Construction inspectors |
| Technical illustrators |
| Industrial engineering technicians |
| Packaging engineers |
| Surveyors |
| Automobile design checkers |
| Drafting specialists |

## what are the salary ranges?

The median annual salary of drafters who worked full-time in 1996 was $31,250, according to the 1998–99 *Occupational Outlook Handbook.* The middle 50 percent earned between $23,400 and $41,500 annually. The top 10 percent earned more than $50,750, while the bottom 10 percent earned less than $19,000. Drafters working in the federal government earn slightly less than those employed in private industry.

## what is the job outlook?

The employment outlook for drafters is expected to be slower than the average rate for all other occupations through the end of the 1990s. As new and more complicated products come to the marketplace, more complex designs will call for more drafting services. However, the use of computer-aided design (CAD) systems allows one CAD operator to do the work of several drafters, particularly that of lower-level drafters who do routine work. CAD can produce more and better variations of a design, which could stimulate additional activity in the field and create opportunities for drafters who are willing to switch to the new techniques.

Employment trends for drafters are sensitive to fluctuations in the economy. In the event of a recession, fewer buildings and manufactured products are designed, which could reduce the need for drafters in architectural, engineering, and manufacturing firms.

# how do I learn more?

## sources of additional information

Following are organizations that provide information on drafting careers, accredited schools, and employers.

**American Institute of Architects**
1735 New York Avenue, NW
Washington, DC 20006
202-626-7300
http://www.aia.org

**American Design Drafting Association**
PO Box 799
Rockville, MD 20848
301-460-6875

**Industrial Designers Society of America**
1142 Walker Road, #E
Great Falls, VA 22066-1836
703-759-0100

# bibliography

Following is a sampling of materials relating to the professional concerns and development of drafters.


### Books

Rowh, Mark. *Opportunities in Drafting Careers.* Lincolnwood, IL: VGM Career Horizons, 1993.

### Periodicals

*Computer Aided Design Report.* Monthly. Newsletter focusing on the application of computers by engineers in manufacturing design. CAD/CAM Publishing, Inc., 192 Turquoise Street, Suite 320, San Diego, CA 92109-1268, 619-488-0533.

*Design Drafting News.* Monthly. Newsletter published for association members. American Design Drafting Association, Box 799, Rockville, MD 20848-0799, 301-460-6875.

# electroneurodiagnostic technologist

**Definition**

Electroneurodiagnostic technologists and technicians obtain accurate recordings of the electrical activity in various body parts. They run tests like electroencephalograms (EEGs), electromyograms (EMGs), and electrocardiographs (ECGs). They prepare patients for tests, take medical histories, monitor equipment while tests are running, make note of any irregularities, and ensure the reliability of results.

**Alternative job titles**

Electroencephalographic (EEG) technologists and technicians

**Salary range**

$23,200 to $31,100 to $40,000

**Educational requirements**

High school diploma; associate's degree recommended

**Certification or licensing**

Highly recommended

**Outlook**

Much faster than average

GOE 10.03.01

DOT 078

O*NET 32923

## High School Subjects

- Anatomy and Physiology
- Chemistry
- Biology
- English (writing/literature)
- Mathematics
- Physics

## Personal Interests

- Figuring out how things work
- Helping people: physical health/medicine
- Science

in two weeks, Margaret pushes the electroencephalograph machine up to room 134 in the intensive care unit. "Another tragic ending," she says to herself, "and so young, too."

The young woman in 134 was brought in eighteen days ago suffering cardiac arrest. The physician and staff had run all sorts of tests on her in desperate attempts to figure out what was wrong, but she didn't respond favorably to any of them, and then she entered a coma. As her condition deteriorated, death seemed certain to follow.

Margaret is sure she has been asked to perform this last electroencephalograph (EEG) so the doctor can officially pronounce the young woman dead. Since flat EEGs are evidence of total lack of activity in the brain, they are routinely used for this purpose.

Feeling disheartened, Margaret opens the door of 134. To her great surprise, the young woman is sitting up in her bed, watching television.

"We never really made a sure diagnosis," Margaret later explains. "The doctors thought it might have been some kind of viral infection. All I know is that she is healthy and productive now and that the day she came out of it turned out to be one of the more heartwarming days of my career."

## what does an END technologist do?

Electroneurodiagnostic (END) technology is the branch of medicine that deals with using tracings (graphic recordings) of the brain's electrical waves in order to diagnose and determine the effects on the body of certain diseases and injuries, including brain tumors and accidental injuries, strokes, Alzheimer's, epilepsy, and various infectious diseases. By interpreting END test results, physicians are able to prescribe needed medicines or surgeries in the attempt to cure or relieve the suffering of a patient.

*Electroneurodiagnostic (END) technologists and technicians* are responsible for obtaining accurate tracings of the brain's electrical activity so that a doctor can use them in diagnosing patients. The term "technologist" usually refers to those END professionals who are registered with The American Board of Registration of EEG and Evoked Potential Technologists, Inc., known as ABRET, while "technician" refers to those who are not. As explained later on, registration is crucial for advancement and sometimes even regular employment.

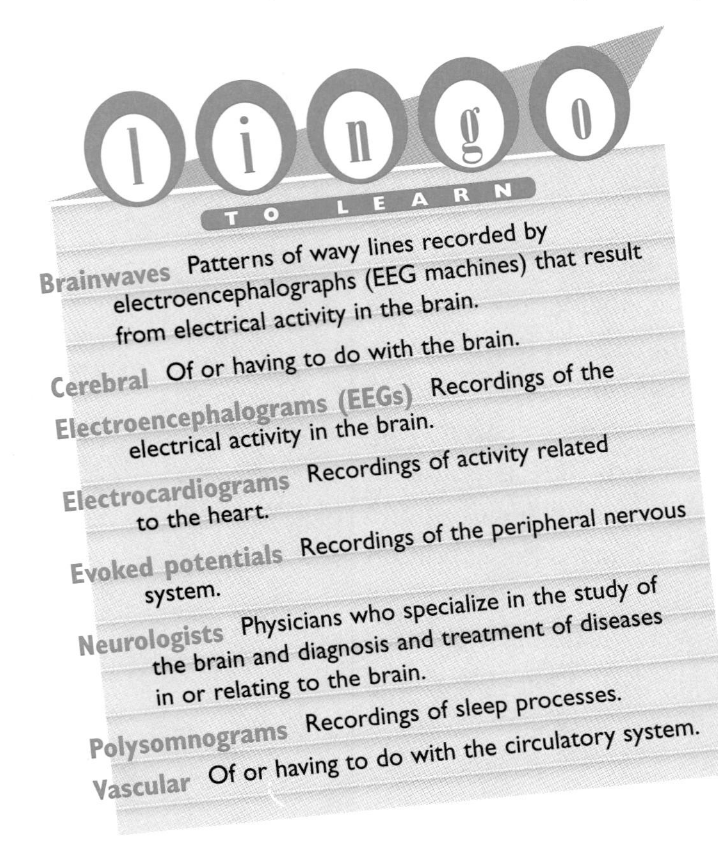


**Brainwaves** Patterns of wavy lines recorded by electroencephalographs (EEG machines) that result from electrical activity in the brain.

**Cerebral** Of or having to do with the brain.

**Electroencephalograms (EEGs)** Recordings of the electrical activity in the brain.

**Electrocardiograms** Recordings of activity related to the heart.

**Evoked potentials** Recordings of the peripheral nervous system.

**Neurologists** Physicians who specialize in the study of the brain and diagnosis and treatment of diseases in or relating to the brain.

**Polysomnograms** Recordings of sleep processes.

**Vascular** Of or having to do with the circulatory system.

END technologists conduct a variety of electroneurological tests. Electrocardiograms (ECGs) are tracings of electrical activity associated with the heart ("cardio" refers to the heart). Electrooculograms are similar but have to do with the eyes, and electromyograms (EMGs) have to do with the muscles. The most common ones, electroencephalograms (EEGs), are tracings of the brain's electrical activity. END technologists generally perform some or all of these tests.

END technologists also administer newly developed tests like the Evoked Potential (EP), which records the electrical reaction from the brain, spinal nerves, and/or sensory perceptors in response to external stimuli. They might conduct the polysomnogram (PSG), an electroneurodiagnostic procedure that combines an EEG with other physiologic measures like heart rate, eye movements, and blood oxygen levels, in order to diagnose and treat sleep disorders.

All of these tests must be administered methodically and with close attention to detail. END technologists start by preparing a patient for a test. They write down the patient's medical history, listening carefully for family or personal illness that might influence either testing procedures or tracing interpretation and diagnosis. Next, they explain each step of the procedure to the patient so that he or she can remain calm throughout the test. Making the patient comfortable in this way is very important since his or her relaxation level can directly affect test results. Technologists must practice good "bedside manner," develop compassion for the patient's concerns, and learn to answer respectfully any questions the patient might have.

Next, the more technical part of the test begins. Technologists fasten electrodes to the patient's head or to other prescribed locations and connect them to the monitor. They make educated decisions about exact placement of electrodes and combination of machine settings, which they note in the patient's file. Once the machine is activated, it begins collecting data, which is amplified and recorded on moving strips of paper or on optical disks. The resulting graph is the "tracing" of the brain's electrical activity.

During the test or "study," technologists must keep an attentive eye on both the patient and the machines, taking notes about the patient's reactions and making any necessary adjustments to the equipment. They may keep track of vital signs like heart rate, blood pressure, and breathing, to make sure the patient stays out of danger. They are trained in basic emergency care in case something does go wrong with the patient during the test. If drugs have been prescribed, they are frequently responsible for administering them.

At the conclusion of the test, some technologists simply pass results on to the prescribing physician, while others evaluate them and note whether they are normal or abnormal. Still others might participate in the diagnostic process with the doctor by writing up a report summary. A technologist's degree of involvement in the interpretative process depends on the doctor who ordered the test originally and the education and experience of the technologist.

Although technologists do not necessarily interpret tracings, they are responsible for identifying abnormal brain activity and any extraneous readings collected by the machine during the test. In order to do so properly, they consider how an individual should perform on the test based on his or her medical history and current illness.

Technologists can also detect mechanical malfunction or human error in faulty recordings. They can fix some mechanical problems themselves but may have to call on the assistance of a supervisor or equipment technician.

## what is it like to be an END technologist?

Margaret Walcoff is the director of the Neurodiagnostic Sleep Disorder Clinic of the Catawba Memorial Hospital in Hickory, North Carolina. As the laboratory manager, she is responsible for ensuring that the lab functions smoothly and successfully. Needless to say, this is a big job.

Her managerial duties include ordering supplies when they are low, assigning job duties to laboratory employees, working with hospital administration on hospital-wide policies and procedures, and training employees on new or improved equipment and lab techniques.

Margaret is also responsible for the lab's annual budget, a task which is probably the most difficult part of her job. "With all the health care reform changes, it gets harder and harder to allocate funds. There's so much new equipment to buy, so many education workshops to hold for the lab employees, and so many governmental regulations to follow."

> *You always have to be mentally and physically prepared for an abnormality or emergency of any sort.*

Margaret also maintains a full patient load of prescheduled tests, or "studies." "Our schedules have to be flexible enough, though, to accommodate emergency requests when they come in." Margaret generally performs EEGs and EPs, although she is trained in other procedures as well. Her lab is also equipped to do video EEGs, the latest in EEG technology.

A typical EEG lasts about forty-five minutes to an hour. When administering an EEG study, Margaret finds it extremely important to put a patient at ease and she makes every effort to do so. She chats chummily with the patient, goes through his or her medical history, makes careful calculations about electrode placement, hooks up the equipment, and engages the machine. While the machine is collecting data, she watches for danger signs and makes any needed adjustments. "You can never go on automatic pilot with this job," explains Margaret. "You always have to be mentally and physically prepared for an abnormality or emergency of any sort."

The lab's daily test load, as well as Margaret's own schedule, vary greatly. She might conduct four or five EEGs one day, and have none scheduled for the next. "One thing I like about my job is the diversity of my duties and daily schedule," Margaret says. Working in a county hospital, she encounters patients of all ages and backgrounds. "One minute I'm working with a ninety-nine-year-old woman and the next a two-year-old little boy—I like that."

In addition, she has a very good working relationship with the doctors; some even rely on her for initial test result evaluation. "Doctors tend to have a reputation for being hard to work with, but that has not been my experience. At least some people in every profession can be difficult, doctors don't have the monopoly on that."

Like most END technologists, Margaret works a forty-hour work week in a clean, well-lighted laboratory. In some emergency situations, she may be asked to work late, but even this is rare. Some END technologists, especially those who work a lot in emergency rooms or in surgery, may be on call during some evening, weekend, or holiday hours. END technologists who are mostly involved with diagnosing sleep disorders work many of their regular hours at night. Most technologists spend about half their hours on their feet and may be required to push testing equipment around the hospital or clinic. In addition, END technologists may do some bending and lifting in order to assist ill patients.

As if her full-time job doesn't keep her busy enough, Margaret also works as an in-house educator for a company that manufactures END equipment. "It's nice to travel a bit and see how other labs are run, what's going on in other places." As an educator, Margaret follows new equipment to the hospital or clinic that bought it in order to train their employees on its optimal use. "I like being involved in the teaching side of things. In this field, there are always a lot of new machines and procedures to keep up on."

## have I got what it takes to be an END technologist?

"You have to be a people person in this business," emphasizes Margaret, "whether it's patients of all ages and backgrounds, other hospital personnel, supply vendors, or sales representatives, you are constantly interacting with someone." The ability to communicate effectively with people should be accompanied by quick thinking skills since technologists often have to change tasks abruptly. For example, if an emergency request is made for an EEG on a thirty-year-old, a technologist might have to interrupt his or her work on a sixty-five-year-old woman to do it. The technologist must then make rapid mental adjustments to accommodate physical differences and test expectations of the new patient.

The best END technologists generally have strong manual dexterity, coupled with an aptitude for working with mechanical and electrical machines, good vision, and good abstract thinking skills. "Technology is growing at a fast pace. Without a solid foundation in the basics of what we do, that is, understand and operate technical machines, a technologist will fall behind professionally," says Margaret.

Technologists should have an eagerness to learn and the ability to learn quickly. Margaret identifies the endless task of keeping up with technology as one of the hardest parts of her job. "I want to learn everything I can but sometimes time constraints make it seem impossible," Margaret says.

Technologists generally write observations, comments, and summaries in a report composed primarily of very technical language. Writing skills are thus very important for this job. Verbal communication skills are equally essential. "Since we try to explain the procedures to all patients, we have to speak in ways that each individual patient can understand," Margaret says.

Working in a hospital or clinical laboratory can prove to be quite stressful, especially in the larger and busier ones. Just like their TV counterparts, doctors, nurses, and technologists of all specialties go in a million different directions and the effect can ruin concentration. Therefore, technologists should be good stress managers and should be able to perform well under pressure.

**To be a successful END technologist you should**

- Have strong manual dexterity skills and good vision
- Enjoy working with people of all ages and backgrounds
- Be adept at working with mechanical and electronic equipment and understanding technical data
- Communicate effectively, both verbally and in writing
- Possess good stress management skills

## how do I become an END technologist?

END technology was a major career move for Margaret, who had already earned a master's degree in education before going to work in a neurodiagnostic research lab. While she did sit in on some college neuroanatomy classes, her training has been mostly on-the-job.

## education

### High School

A high school diploma is required of all prospective END technologists. Science classes, especially biology, chemistry, physics, and anatomy, are particularly helpful, helping students to build a solid foundation in scientific knowledge, methods, and reasoning. Mathematics courses should include at least algebra and are equally important since they encourage students to develop strong analytical and abstract thinking skills. Three or more years of both science and math are recommended.

English and speech courses provide good opportunities to enhance verbal and written communication. Social science classes, like psychology and sociology, help students begin to think about the affective, or emotional, needs of patients. This understanding leads prospective END technologists to develop a good "bedside manner."

In some high schools, vocational or technical courses dealing with mechanical and electrical equipment are offered. These courses can provide students with a head start on understanding END equipment and teach them how to approach and solve a technically oriented problem.

FYI

**Specialized tests**

Sometimes, other more specialized diagnostic tests are needed to help a physician visualize the structure of the patient's brain and spinal cord. The CT scanner (for computed tomography) takes computer-generated X-rays of different "slices," or areas, of the brain. For information on careers in this area of diagnostic testing, see the chapter "Special Procedures Technician."

### Postsecondary Training

There are two types of postsecondary training for prospective END technologists. The first is on-the-job training, as Margaret had, and the other is formal classroom training at either a hospital, medical center, technical school, college, or four-year university. Margaret, who is in charge of hiring new END technologists, says that postsecondary education is important, but "the number one quality I look for in an applicant is demonstrated ability and enthusiasm to learn." Often, intellectual ability is proven through completion of a postsecondary degree; it can also be proven through work and volunteering experience. As technology continues to change, however, it is generally considered preferable to pursue formal education.

On-the-job training usually lasts anywhere between three months and one year, depending on the nature and scope of the employer. Margaret was originally hired as a research assistant in a neurodiagnostic lab and so continued to learn as she went along. "In that job, as in my current position and positions since then, I never really stopped learning so I cannot say that my training really ever came to end!"

During on-the-job training, END technologists learn first-hand how to operate the equipment and carry out specific testing, maintenance, and repair procedures. They receive instruction from doctors and senior END technologists. They may be required to do some reading at home.

Formal training in electroneurodiagnostic technology consists of both academic course work and clinical practice. The length of formal programs varies between one and several years and results in either a certificate or an associate's degree. At the university level, curricula for a bachelor's degree in END technology is being developed in some areas, but degrees are already available in many related health professions. The American Society of Electroneurodiagnostic Technologists offers a nationally comprehensive list of formal educational programs in END technology (*see* "Sources of Additional Information").

Academic instruction is usually concentrated on basic medical subjects, like human anatomy, physiology, neuroanatomy, clinical neurology, neuropsychiatry, clinical and internal medicine, and psychology. Courses specific to END technology include normal and abnormal pattern recognition, patient preparation, recording techniques, and electronics and instrumentation. The clinical practice incorporates what is taught in on-the-job training programs, but may do some of it in staged, not authentic, situations.

## certification or licensing

Certification of END technologists is regulated at the state level and so varies from state to state. Past attempts at national certification programs have not been entirely successful, so now most END technologists are encouraged to obtain official registration instead. Individuals who have completed one year of on-the-job training or have graduated from a formal training program and been in the field for at least one year and completed a required number of recordings may apply for registration with the American Board of Registration of EEG and Evoked Potential Technologist, Inc. (ABRET) *(see* "Sources of Additional Information").

After application, the individual takes a two-part examination. The first part is written and the second oral. Only students who pass the written part may take the oral. Both sections comprehensively cover subjects related to the work of an END technologist.

ABRET offers two registrations: R.EEG T., or registered electroencephalographic technologist, and R.EP T., or registered evoked potentials technologist. Though not required for employment, registration is official acknowledgement of a technologist's expertise. It also makes advancement and financial promotion easier.

## scholarships and grants

There are currently no scholarships or grants reserved exclusively for individuals seeking careers in END technology. END technologists who receive on-the-job training are usually paid regular salaries, so tuition is not an obstacle. Those who pursue formal education in hospitals and medical centers should contact them directly about costs and financial aid opportunities. Technical school or university students should seek financial aid from their school's financial office. These offices usually maintain a complete list of available awards, as well as a profile of requirements. They should be contacted directly.

If an individual is currently employed in a health- or medical-related business and wants to pursue a degree in END technology, he or she may be eligible for tuition reimbursement programs offered by his or her employer. Interested employees should contact their company's personnel or benefits office.

fyi

Electrical impulses discharged from the brain, commonly known as brainwaves, were first discovered in England in 1875 by Richard Caton.

## internships and volunteerships

It is extremely difficult to find internships or part-time jobs in the field of electroneurodiagnostic technology since most clinics and labs are staffed by fully trained, full-time professionals. Prospective END technologists have to rely on the clinical practice that is part of both formal education and on-the-job training for in-depth exposure to daily routines. In some formal education programs, however, there may be some paid internships or co-op work (apart from regular clinical practice) organized by the school or clinic as part of the curriculum. Information about these kinds of opportunities can be obtained through the program administration.

Volunteerships are readily available at local hospitals, clinics, and medical centers. Experience of this type may be helpful to prospective END technologists for several reasons. First, hospital volunteers, who witness the everyday trials of hospital work, discover first-hand if a medical profession is right for them. Second, some volunteers may be allowed to specify that they would like to work in a electroneurodiagnostic area, providing them with a sort of direct, advanced training. Third, hospital volunteerships show employers and education program administrators that an applicant is serious, interested, and enthusiastic about the health professions. Individuals should contact local hospitals and clinics for more information. ABRET and ASET may also have lists of specialized END labs across the country.

## labor unions

There are no labor unions designated specifically for electroneurodiagnostic technologists. In some areas or certain hospitals and clinics, an allied health union may operate. If an END technologist is employed in such an area, he or she may be required to join the union.

## who will hire me?

Margaret got into END technology almost by accident. Having just completed her master's degree in education, she was actually searching for a job in that field when she met a medical researcher who needed an assistant in neurodiagnostics. "I learned very quickly, and was very excited by the field," she says. After her research position, she was hired as an END technologist at Catawba Memorial Hospital and completed her on-the-job training there.

Electroneurodiagnostic technologists work in any doctor's office, clinic, group medical center, health maintenance organization, urgent care center, emergency center, psychiatric facility, lab, or hospital that has the equipment necessary to perform END tests. Most END technologists, however, hold full-time positions in hospitals.

On-the-job trainees usually have no problem finding a job after training, since their employer/trainer hires them directly. END technologists who complete formal education in this field work closely with the placement office of the medical center, technical school, or university where they study. Such offices have long-standing working relationships with area employers and are often successful at placing graduates in full-time jobs.

For all END technologists, whether trained on-the-job or through formal education, newspaper and magazine classified ads are a good place to look for job openings in the area. Prospective applicants can also contact the personnel offices of the various medical facilities directly.

ASET publishes a quarterly magazine and a monthly newsletter, both of which contain a list of employment opportunities. Membership in ASET provides the occasion to become part of a close-knit network of employed professionals, who may also be able to pass on information about job openings. Margaret, who is currently the vice president of ASET, says that membership is beneficial in several ways, not the least of which is networking for employment opportunities. ABRET may also have job lists and job hunting ideas (*see* "Sources of Additional Information").

### Advancement Possibilities

**Electroneurodiagnostic technologist supervisors** administer the more difficult or specialized lab tests and supervise other END technologists. They might also be involved in training new employees on testing techniques specific to the lab. They may share some administrative duties with the chief END technologist.

**Chief electroneurodiagnostic technologists** usually work directly under the direction of a neurologist, neurosurgeon, or other physician. They supervise or manage the END laboratory, establish lab procedures, make work and appointment schedules, keep records, and order supplies. They are also involved in teaching and training new employees from END technology education programs.

**Electroneurodiagnostic technology educators or researchers** teach in academic programs focused on END technology or in on-the-job training programs in hospitals and clinics. Researchers may work on improving, developing, and inventing electroneurodiagnostic procedures.

## where can I go from here?

As lab director, Margaret has already advanced about as far as she can go with her current employer. "The only place I could really go from here is to work in a bigger lab, with more employees and heavier schedules." As for now, she has no plans to do that but is more interested in pursuing the education side of her field. As mentioned above, she currently works part-time for a manufacturer of END equipment, traveling around to different medical sites in order to teach new users how to operate the machines correctly and how to improve their testing techniques overall. She can foresee becoming a full-time educator, either for an on-the-job or formal training program, once she retires from her present position.

After registration and several years' experience, an electroneurodiagnostic technologist is eligible for promotion to a supervisory position. Supervisors generally perform the more complex tests or the regular tests on potentially problematic patients. They might also be called upon to conduct tests in emergency situations. Depending on the size of the lab, some supervisors may be involved with training new employees and have some administrative duties like job assignment, patient and employee scheduling, and supply ordering.

Supervisory positions may be followed by promotion to chief electroneurodiagnostic technologist, a title equivalent to Margaret's position as director. The chief technologist or lab director has much more immediate contact with doctors and hospital administration than do technologists; they may consult on the diagnosis of a case, offer input into the establishment of hospital regulations and policies, and manage the lab's budget. Although they still work with patients, they are in effect lab or departmental managers, with managerial and fiscal responsibilities.

Years of experience and a solid educational foundation put certain END technologists, like Margaret, in a prime position to move into education. Some technologists may go to work for a manufacturing company, training their customers (medical END labs) in the proper use of END equipment. Others might teach in the on-the-job or formal training programs offered at medical centers and hospitals. People with advanced degrees might go to work in technical schools and universities.

## what are some related jobs?

The U.S. Department of Labor classifies electroneurodiagnostic technologists and technicians under the headings *Occupations in Medical and Dental Technology* (DOT) and *Child and Adult Care: Data Collection* (GOE). Also under these headings are people who aid doctors, nurses, and patients during child birth, people who assist dentists in teeth cleaning and examinations, and people who provide home health care to patients not needing hospitalization. Nurses and medical assistants fall under these headings, as do most aides whose work relates to medical care giving, like physical therapy aides, orthopedic aides, and emergency medical technicians. Many technician-oriented jobs are also found here, including people who administer liver dialysis treatment, others who test hearing loss, others who administer ophthalmological (eye disease) examinations, and still others who perform X-ray tests, radiation therapy, and cardiovascular evaluations. Teaching supervisors, responsible for on-the-job training of people employed in these positions, also fall under these headings.

| Related Jobs |
|---|
| Audiometrists |
| Birth attendants |
| Cardiopulmonary technologists |
| Child care attendants |
| Chiropractor attendants |
| Dental assistants |
| Dialysis technicians |
| Electrocardiograph technologists and technicians |
| Emergency medical technicians |
| First aid attendants |
| Nurses, nurses' aides |
| Optometric assistants |
| Orderlies |
| Orthopedic attendants |
| Physical therapy aides |
| Radiation therapy technologists |
| Radiologic technologists |
| Surgical technicians |

## what are the salary ranges?

The median annual base salary for full-time END technologists was $26,800 in January 1997. The middle 50 percent earned between $23,200 and $30,100. Salary steadily increases as a function of both education and years of experience. It jumps significantly if the technologist is registered by ABRET. R.EEG T.s and R.EP T.s earn between $6,000 and $10,000 more a year than nonregistered people of equivalent experience.

The top salaries for END technologists average $40,000 a year. END lab supervisors, chief technologists, teachers and training program administrators, and researchers tend to command the highest salaries.

END technologists working in hospitals generally receive full benefits packages that may include health insurance, paid vacations, and sick leave. In some locations, benefits may also include educational assistance, pension plans, and uniform allowances. Technologists working outside the hospital setting usually receive benefits as well, though they vary in content with different employers.

## what is the job outlook?

The U.S. Department of Labor estimates that the number of positions in electroneurodiagnostic technology will increase much faster than average through 2006. One area of strong growth will be in surgical units. Surgical teams find it more and more useful to have an END technologist, prepared to do an EEG, in the operating room for the duration of certain surgeries. END testing is proving to be an effective way to monitor the patient's condition and reactions to various surgical procedures.

END technology is also expanding within traditional uses. New procedures are being tested, developed, and implemented, and old ones are being improved rapidly. Evoked potentials (EP) testing, for example, has only been investigated in detail for twenty years. Technological advances in this, as well as all other testing areas, is sure to translate into more and more job opportunities for qualified individuals.

Private neurologists' offices and neurological clinics were formerly predicted to be the area of greatest increase in demand for END technologists. However, this will certainly not be the case. The initial prediction was based on a trend in the medical community of doctors buying or leasing their own END equipment. This meant that if a doctor ordered an EEG, the patient would be able to have the test performed on the doctor's premises instead of going to the hospital or a specialized lab.

Recent legislation in Congress makes this process of self-referral illegal. The concern was that since the END tests are so expensive to the patient and insurance companies and profitable to the doctors, the latter might tend to order tests when they were not really needed. Prospective END technologists should pay attention to health-related legislation at both the federal and state levels, since it can have an enormous impact on employment growth and outlook.

## how do I learn more?

### sources of additional information

Following are organizations that provide information on electroneurodiagnostic careers, accredited schools and scholarships, and possible employment.

**American Board of Certified and Registered Encephalographic Technicians and Technologists**
Neurodiagnostic Department
Mercy Hospital and Medical Center
4077 5th Avenue
San Diego, CA 92103
619-294-8111

**American Board of Registration of EEG and EP Technologists**
PO Box 916633
Longwood, FL 32791
407-788-6308

**American Society of Electroneurodiagnostic Technologists**
204 West 7th Street
Carroll, IA 51401
712-792-2978
http://www.aset.org/

**Joint Review Committee on Electroneurodiagnostic Technology**
Route 1, Box 63A
Genoa, WA 54632

# bibliography

Following is a sampling of materials relating to the professional concerns and development of electroneurodiagnostic technologists and technicians.

## Periodicals

*Brain Injury.* Bimonthly. Discusses all topics related to brain injuries. Taylor & Francis Ltd., 1900 Frost Road, Suite 101, Bristol, PA 19007-1598, 800-821-8312.

*Brain Research Bulletin.* Monthly. Concentrates on all aspects of brain research. Elsevier Science, Inc., Box 945, New York, NY 10159-0945, 212-633-3730.

*Neuroimaging Clinics of North America.* Quarterly. Focuses on brain, spine, head, and neck imaging. W. B. Saunders Co., Curtis Center, Independence Square West, Philadelphia, PA 19106-3399, 215-238-7800.

# electronics technician

**Definition**

Electronics technicians work individually or with engineers to help design, produce, improve, maintain, test, and repair a wide range of electronic equipment. Equipment varies from consumer goods like televisions, computers, and home entertainment components, to industrial, military, and medical goods, like radar and laser equipment.

**Salary range**

$15,000 to $32,000 to $40,000

**Educational requirements**

High school diploma; associate's degree recommended; bachelor's or advanced degrees for engineering and other advanced positions

**Certification or licensing**

Recommended

**Outlook**

Faster than the average

GOE 05.01.01

DOT 003

O*NET 22505A

## High School Subjects

Chemistry
English (writing and literature)
Mathematics
Physics
Shop (Trade/Vo-tech education)

## Personal Interests

Building things
Figuring out how things work
Fixing things
Reading
Science

don't realize how much electronics play a part in their lives until the day when something breaks down," says Fred Smith, a long-time certified electronics technician. "Then they call up in a panic because their TV, VCR, or CD player went out." As a part-time freelance electronics technician, Fred receives several calls a week. "One time a woman called me at 10 AM. She had to have her television set fixed by the time her favorite show came on, at 1 PM. She really couldn't understand why it might take any longer than that. When men call, it's usually so they can have their TVs fixed before a big game or movie. They rely on electronics just as much as anybody else," explains Fred.

As a full-time electronics teacher, Fred always tells his students they should be trained and ready for anything. As high tech machines and gadgets originally introduced as luxury items become more and more standard in American homes, keeping up with technology is a very challenging endeavor for electronics technicians.

Maybe that's why Jacob Straw, who owns two electronics businesses, went back to school to earn an associate's degree in electronics technology. "I was already in the business, a sort of self-taught electronics tech, but I wanted the schooling," says Jacob—or Jay as he's called by friends. "I wanted to learn the mathematics of electronics, because that's what it is. A voltage is a number, you know, and I was committed to understanding the technology of what I was doing."

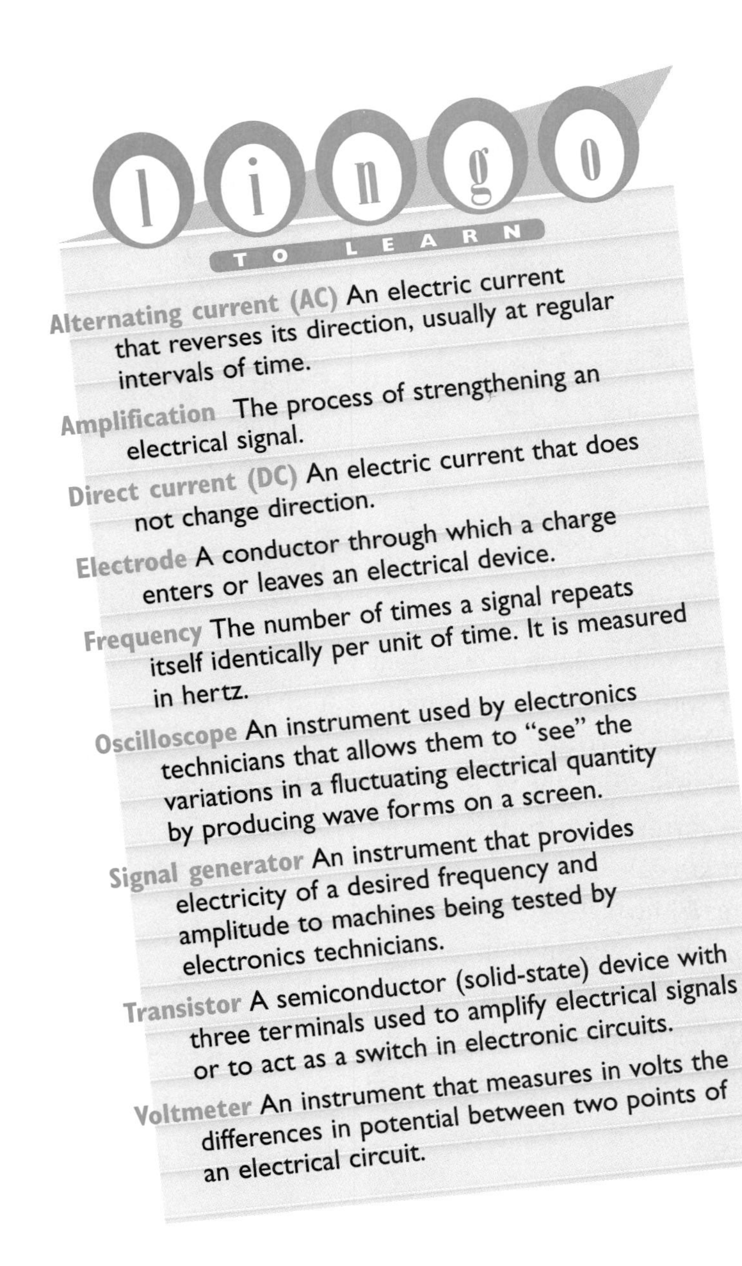


# what does an electronics technician do?

Electronic devices play a part in practically every business and even many leisure activities found around the globe. Such diverse activities as NASA space missions, sophisticated medical testing procedures, and car and airplane travel would be impossible without the use of electronic equipment. So would a band concert. *Electronics technicians* help develop, assemble, install, manufacture, maintain, and service industrial and consumer electronics, including sonar, radar, computers, radios, televisions, stereos, and calculators.

The products made by the electronics industry can be divided into four basic categories. The first one, government products, represents a high percentage of sales in the industry. Missile and space guidance systems, communications systems, medical technology, and traffic control devices are just a few of them.

The second area comprises industrial products, which include large-scale computers, radio and television broadcasting equipment, telecommunications equipment, and electronic office equipment.

The third area, consumer products, is the most commonly known and includes such things as televisions, VCRs, and radios.

The fourth category comprises the smaller pieces that all electronics are made of—components. Capacitors, switches, transistors, relays, television picture tubes, and amplifiers are among the most well known.

Electronics technicians are needed to work in various capacities on all of these kinds of products; the nature of the work done on a specific product generally falls into one of three areas—product development, manufacturing and production, and service and maintenance.

*Electronics development technicians* generally work with an engineer or as part of a research team. Engineers draw up blueprints for a new product; technicians build a prototype, or an original model, according to specifications. This job involves using hand tools and small machines to construct complex parts or components. In some cases, technicians are encouraged to offer suggestions for improvement and modifications of the design of the product.

After the prototype is finished, technicians work with engineers in testing the product for performance in various stressful conditions, analyzing how a component reacts, for example, in extreme heat and cold. Tests are run using complicated equipment and detailed, accurate records of every aspect of the test must be kept. Technicians in development may also have to construct, install, modify, or repair laboratory test equipment.

*Electronics drafters* do work closely related to that of development technicians. They are responsible for converting rough sketches as well as written and verbal communications, all provided by engineers, into easily understandable diagrams to be used in manufacturing the product. They prepare lists of all parts and components needed in the production of various products.

*Cost-estimating technicians* work in an area closely related to development. They are responsible for reviewing new product proposals in order to determine the approximate total cost of production for that product, including labor, materials, and equipment costs. Other people in the company, like sales representatives, then use these figures to decide whether or not production is economically feasible.

*Electronics manufacturing and production technicians* deal with any problems arising from the production process. They install, maintain, and repair assembly or test line machinery. In quality control, they inspect and test products at various stages in the production process. When a problem is discovered, they are involved in determining what the problem is and in suggesting solutions.

*Electronics service and maintenance technicians* diagnose and repair malfunctions in a variety of electronics products from all four categories described above. Generally, the same procedure is followed regardless of the product in question. First, the technician gathers information from the consumer or user about the nature of the problem. Using these details, the technician investigates and fixes minor problems on the spot. If the problems are too complicated, the machine must be taken back to the electronics laboratory, where test equipment like voltmeters, oscilloscopes, and signal generators attempt to locate the difficulty.

**To be a successful electronics technician, you should**

- Be neat in appearance and honest when dealing with customers
- Work patiently, methodically, and persistently
- Enjoy using tools and electronic equipment; have strong manual dexterity skills
- Be enthusiastic about keeping up with technological advances
- Demonstrate a strong aptitude for math and science

## what is it like to be an electronics technician?

You might say that Jay Straw dabbles a little in all the different areas electronics technicians could work. Jay's career as an electronics tech started out when he was a musician buying a whole lot of electronic amplifiers and expensive sound equipment. "Out of necessity, I learned how it worked and how to fix it if it broke down," says Jay. "As I got better at using the equipment, the technology that went into it got more and more complicated. That's one reason I had to go back to school."

Jay's musical ambitions led him into the business of setting up for bands and live musical shows. "I provide the sound and lighting equipment and all the people who set it up and run it." Big name bands like Guess Who, Blue Oyster Cult, and Joan Baez have employed Jay's Hi Tech Audio and Lighting company. He also sets up for orchestras and musical productions. "Sometimes I go on tour with them," Jay says, "but it's usually cheaper for shows to hire people in the area, as they go, rather than haul equipment around."

Jay runs another business, Hi Tech Electronics. "If my equipment breaks down, I can bring it into the shop, analyze the problem, and solve it." Hi Tech Electronics does some installation of PA and sound systems in schools and hospitals. "But mainly," says Jay, "we do repairs on musical instruments for shops in town." Jay spends a lot of time rewiring guitars and replacing tubes on power amplifiers.

He's also in the process of designing an original device to stop the feedback on guitars. "When guitars are plugged in, they feed a constant high or low frequency hum back through the speakers," says Jay. "I'm developing what's called a notch filter so the guitar won't feedback, and it's almost perfected." Once complete, Jay plans to market his creation.

Fred Smith's been in electronics a long time. As he looks back over his career, he can safely say he's held as many different positions as an electronics technician can have. Fred's an electronics teacher and CET (Certified Electronics Technician) test administrator in San Luis Obispo, California. He also does some freelance electronics service work on the side.

Fred's first job was at a Sears Service Center, where he initially did field work. "I would go to the office in the morning, pick up my service calls, and start out on the road." It was hard to estimate the amount of time each stop would take, since he really didn't know the extent of the problem. "Electronics technicians have a bad reputation for not showing up when they say they are going to, but the thing is, no matter how well you try to schedule things, something always puts you behind, " Fred explains.

Most all of Fred's service calls were handled in a similar manner. He would talk to the customer about the problem and the performance history of the machine. It's important to know, for example, if a television set has been repaired for the same problem anytime in the past. Then, he would inspect the machine on the premises. If some wiring was lose or a simple circuit needed replaced, he could usually finish the repair there. If the problem was more complicated, he would load the machine in his van to take back for more tests and further repair in the service center.

Fred had to keep detailed records of everything he did—which tests he ran, which parts he used, time spent on each call—and hand them in to his supervisor at the end of every day. Fred's next job was as a bench work technician. With this job, he no longer went on calls outside the lab but worked instead on machines brought back to the lab for further inspection. The electronic problems he encountered there were more difficult to solve. He would inspect the machine, run tests on it if needed, estimate the cost of repair, including labor and parts, and call the customer back with the information.

Fred had to offer honest advice about whether or not he thought the repair was worth the cost or if the client should just replace the machine. "Customers tend to be distrustful of servicing technicians, thinking we are all out to rip them off. I always tried to put them at ease by explaining exactly what the problem was and how I was going to fix it."

Fred's current job is less hands-on. Much of his day is spent teaching classes on electronics. The program he works for lasts three years and the curriculum includes basic math, science, and electronics classes as well as more complex circuit-design and electrical machinery classes. He has a fair amount of equipment in his laboratory, which provides hands-on experience to his students.

Electronics technicians usually work forty-hour weeks, and hourly-wage earners are paid extra for overtime. Depending on the company, technicians employed in manufacturing and production may have to work any of the shifts during which production runs.

Most electronics laboratories are clean, well lighted, and well equipped. A technician may have his or her own office or may occupy one desk among several in a larger room or lab. Technicians who work extensively with assembly lines may spend part of their time in damp, greasy, or noisy surroundings. Service technicians spend a great deal of time on the road.

FYI

The most rapid expansion of electronics technology occurred during World War II, when the need for long-distance communications equipment and missile-guidance systems increased greatly.

## have I got what it takes to be an electronics technician?

"With the constant advances and changes in technology, you just have to have a solid foundation in the basics of electronics and the ability and desire to learn quickly," explains Fred. Jay would agree. "Electronics is very complex stuff. Circuit design is complicated, and it's changing all the time," he says. Electronics technicians have to keep their skills up-to-date, and the ability to learn quickly and a naturally inquisitive mind are good qualities in this field. An aptitude for math and science is also important, as is the capacity to understand technical writing and drawing.

Keeping up with technological advances can be difficult. "It's the most frustrating part of my job," says Fred, "I know I am outdated, but I just don't have the time to keep up." Electronics technicians have to learn to deal with this frustration effectively; if not, it can become a great source of stress. "I'm supposed to get educational leave from my employer, but with budget cuts, I never see it. I just do as much as I can and stay satisfied with that."

Many tasks assigned to electronics technicians require patience and methodical, persistent work. Good electronics technicians work well with their hands, paying close attention to every detail of a project. Some technicians are bored by the repetitiveness of some tasks, while others enjoy the routine.

Since many technicians in industry work as part of an engineering team or directly for an engineer, they must also be good team players.

Individuals planning to advance beyond the technician's level should be willing to and capable of pursuing some form of higher education.

## how do I become an electronics technician?

Once Jay realized he needed more and better knowledge of electronics, he applied to a two-year program at the University of Montana in Missoula, Montana, where he lives. "I went to school eight hours a day for two years, working my audio and lighting business at the same time." In the summer, Jay did concerts. The rest of the year, he hit the books. "I put in a lot of late-night studying and long hours," says Jay. "My social life was nothing for two years, but it was worth it." Jay graduated with an associate's degree in electronics technology.

After working as an electronics technician in the military for many years, Fred earned his associate-level CET certification just before being discharged. "The military let me take the CET exam to make sure I was qualified for a real job when I got out." Later he went on to successfully complete several other exams and credentials.

## education

### High School

A high school diploma is necessary for anyone wishing to build a career as an electronics technician. You should focus on mathematics and science courses. Algebra and geometry are both important skill-builders, as are chemistry and physics, particularly in their coverage of electron theory and properties of electricity. Courses offered in electronics or electricity are obviously pertinent to the field as well. Participation in electronics, math, or science clubs in high school might boost your understanding of these very crucial, yet basic, subject areas.

English and speech courses provide invaluable experience in improving verbal and written communication skills. Since some technicians go on to become technical writers or teachers, and since all of them need to be able to explain technical matter clearly and concisely, communication skills are important.

Any shop classes that emphasize electronics or other technical principles in laboratory work allow you to get a head start in understanding the field. Since electronics technicians work with technical drawings all the time, mechanical drawing courses are also helpful.

### Postsecondary Training

While some current electronics technicians entered the field without prior academic training, it is increasingly difficult to do so. The demand for technicians with education and knowledge of the most sophisticated electronics products and systems is only going to increase. To stay competitive, you need to graduate from a two-year electronics technology program and then continue to gain more advanced training in the field. If you aren't able to afford a two-year technical program, don't overlook opportunities offered by the military. In addition, some major companies, particularly utilities like phone companies, still hire people straight out of high school and train them through in-house programs. Others promote people to technicians' positions from lower-level jobs, provided they attend educational workshops and classes sponsored by the company.

Most electronics technicians complete a two-year degree in electronics technology at a community college or a technical school. Though programs do vary, a typical first-year curriculum includes courses in physics for electronics, technical mathematics, electronic devices, communications, AC/DC circuit analysis, electronic amplifiers, transistors, and instruments and measurements. The second year of study may include courses in communications circuits and systems, introductory digital electronics, technical reporting, control circuits and systems, electronic design and fabrication, new electronic devices, and industrial organizations and institutions.

As a certified educator, Fred teaches many of these courses to his students. "The more advanced classes build directly on previous ones. That's just one more good reason to have a solid foundation in the basics, to make all future learning easier." Without a good grasp of basic electronics taught in the first year of technical school, the second year can be difficult. "If you keep up with your studies and follow the logical progression of things, you'll do fine," says Fred.

There are many reputable education programs in electronics technology. Interested individuals should consult the local telephone directory, electronics technicians employed in the area, or one of the professional organizations listed here for more information (*see* "Sources of Additional Information").

Some electronics technicians decide to pursue advancement in their field by becoming an engineer or branching off into research and development. These higher-level and higher-paid positions usually require the completion of at least a bachelor's degree in electronics or another engineering or technology discipline. Technicians with significant experience can sometimes finish such degrees through correspondence school, while others attend night or weekend classes.

## certification or licensing

Certification as an electronics technician is generally not mandatory, although for some specific subfields like radio and television transmission it is strongly recommended. Though voluntary, certification as a Certified Electronics Technician (CET) at any level is proof of professional dedication, determination, and know-how. "I'm convinced I got my first job at the Sears Service Center only because I had my associate-level CET," says Fred. "If it comes down to two applicants, one a CET and the other not, the employer is going to hire the first one," he says.

CET certification is awarded to those who successfully complete a written, multiple-choice examination. The associate-level CET test is designed for technicians with less than four years of experience. It is composed of the basic electronics portion of the advanced test and must be passed with a score of 75 percent or better.

After four years of education or experience, a technician can take another CET exam in order to earn a journeyman-level CET certification. This test comprises the same basic electronics portion, plus one or more of several specialty options. The specialty option tests evaluate a technician's knowledge of advanced theoretical material applicable to that specialty. Some specialty options include consumer, industrial, communications, computer, audio, medical, radar, and video.

The Certified Appliance Technician (CAT) certification is a new program considered to indicate a technician's proficiency at the journeyman level in household appliances. The exam consists of multiple-choice questions covering electrical circuits and components, refrigeration systems, laundry equipment, cooking equipment, dishwashers, and trash compactors.

For more information, including a catalog of CET exam study guides and a list of test sites and test administrators in every state, interested individuals should contact either the ISCET or the ETA, who both offer CET certification (*see* "Sources of Additional Information").

## scholarships and grants

There are no outstanding scholarships or grant awards earmarked for those wishing to pursue education in electronics technology. However, Siemens Corporation, one of the largest electronics companies in the world, offers apprenticeship programs with on-the-job training at Siemens plants and paid tuition at selected schools. For more information on the opportunity offered by Siemens, write or call the company (*see* "Sources of Additional Information"). Community college and technical school students should contact the financial aid office of their school to obtain information about financial aid options, including scholarships offered by companies directly through schools and federally subsidized and commercial educational loans.

If you're currently employed and plan to pursue an education in electronics technology, you should contact your personnel or benefits office for information about any tuition reimbursement programs available.

## internships and volunteerships

Community college and technical school students may be able to secure off-quarter or part-time internships with local employers through their college placement office. Internships for people not in school are difficult to find. You should look in the help wanted sections of local newspapers and magazines in search of opportunities. You might also contact the personnel department of industrial employers and governmental agencies for information. Some internship opportunities may be included in job lists published by ISLET, ETA, and other professional organizations (*see* "Sources of Additional Information").

**Advancement Possibilities**

**Electronics technician supervisors** are responsible for bigger or more complicated projects than technicians. They have some administrative duties, like scheduling, ordering supplies, and overseeing lab procedures. They supervise other technicians and may train new employees.

**Engineering technicians** work with teams of engineers and other technicians on developing new products or improving current ones. Their knowledge and experience are wider in scope than that of electronics technicians; their education may include exposure to the theories of electronics and engineering to complement hands-on experience.

**Production test supervisors** analyze a factory assembly line production process and determine what tests should be run on the emerging products and where. They may also be in charge of designing the equipment set-up used in production testing.

## labor unions

There are no unions organized exclusively for electronics technicians. However, those working in electronics product development and manufacturing companies are often required to join the union operating within the company. The principal unions involved are the International Union of Electrical, Radio, and Machine Workers; International Brotherhood of Electrical Workers; International Association of Machinists and Aerospace Workers; and the United Electrical, Radio, and Machine Workers of America.

Employers in other areas may have unions as well. Many local, state, and federal governmental agencies operate with unions, and technicians employed by them are required to become members.

## who will hire me?

Fred got his first electronics technician job doing bench work repair with Sears Service Center by answering an advertisement in the local newspaper. "They hired me even though I had no experience outside the military. The CET certification was key, but I also agreed to take a course on electronic consumer products at DeVry," says Fred.

Electronics technicians are employed in one of three broad areas of electronics: product development, manufacturing and production, and service and maintenance. Manufacturers of electronic equipment employ the most electronics technicians, and these manufacturers are highly clustered in eight states—California, New York, Illinois, Massachusetts, Pennsylvania, Indiana, New Jersey, and Texas.

There are a wide range of electronics companies that make equally diverse products, such as consumer, aircraft, or broadcast electronics; hospital and/or medical equipment; satellites; computers; marine navigation, audio/visual, test, quality control, or telecommunications equipment; commercial radar; office equipment; and farm machinery. In addition, companies that specialize in the development and servicing of such products hire electronics technicians as well.

Utilities companies nationwide hire a large number of electronics technicians, especially telephone, telegraph, electric light and power, and nuclear energy companies. Local, state, and federal governments, as well as all branches of the active and reserve military, are big employers of electronics technicians. Federal agencies such as the U.S. Department of Defense, the National Aeronautics and Space Administration, and the National Institutes of Health hire a particularly high number of electronics technicians.

Going into business for yourself, as Jay did, is another option. The electronics field is vast and growing, and opportunities to offer a service, like High Tech Audio and Lighting does, or to service and repair consumer products like TVs, stereos, and computers, are there for the entrepreneurial individual.

Many community college and technical school students find their first full-time position through their school's placement office. These offices tend to develop good working relationships with area employers and therefor offer you excellent interviewing opportunities. Such offices provide resume writing and interviewing seminars and advice on how and where to pursue job searches.

ISCET publishes a monthly job list, entitled "Career Opportunities in Electronics," which provides information about job openings around the country. "ISCET really does a good job of keeping you informed about what's going on. Their job list seems to be pretty comprehensive and reliable," says Fred. Other associations listed also compile such lists. Trade magazines about electronics, particularly those about computer-related technologies, generally publish lists of job opportunities in the back of each issue (*see* "Bibliography").

## where can I go from here?

As an electronics technology educator and CET test administrator, Fred has advanced about as far as he can go in his career. "After over thirty years in the business, I'm looking forward to retirement," he says. To get where he is today, Fred has worked hard and continued his education, obtaining several journeyman CET specialties, as well as credentials in related technical areas.

Starting at Sears as a field work electronics technician, Fred was promoted to bench work after several years. Then, he went into private industry as a senior electronics technician. His next promotion was to the calibration lab, where he was a technician and later senior supervisor/manager. Having done some teaching in the military, he eventually decided to get back into it.

Fred's career path illustrates a typical one taken by electronics technicians. After obtaining the journeyman CET certification, technicians may become eligible for supervisory positions. The particular duties associated with these jobs vary according to the nature of the employer. In general, however, *electronics technician supervisors* work on more complex projects than technicians. They also have some degree of administrative responsibility; they may make the employee work schedule, assign laboratory projects to various technicians, oversee the training progress of new employees, and keep the work place clean, organized, and well stocked. Electronics technician supervisors, moreover, tend to have more direct contact with project engineers or managers.

Some electronics technician supervisors may decide to further their education in order to become *engineering technicians* or engineers. Electronics engineering is concerned with the research and development of new products and the improvement and revision of old ones. It also focuses on methods of ensuring production efficiency and quality control.

Electronics technicians might also look for advancement in production testing or quality control. *Production test supervisors* make detailed analyses of production assembly lines in order to determine where production tests should be placed along the line and the nature and goal of the test.

*Quality control supervisors* determine what kinds of quality control tests should be conducted after the completion of production in order to ensure that only high-quality products go out.

With further education in business, some electronics technicians take the management route, moving further and further away from hands-on technical work each time they are promoted.

## what are some related jobs?

The U.S. Department of Labor classifies electronics technicians under the headings *Electrical/Electronics Engineering Occupations* (DOT) and *Engineering: Research* (GOE). Also listed under these headings are people who solve manufacturing, industrial, and business problems using ceramic, chemical, electrical, and mechanical engineering. There are also people who do research, both as engineers and technicians, on optical fibers, engines, lasers, and instrumentation. Related workers are people who do radio and television broadcast, planning, and cable engineering. Others work in manufacturing, either deciding when and where assembly line tests should be run or determining the scope of product-sampling and testing procedures for quality assurance testing. Others help develop and test new industrial and consumer products.

| Related Jobs |
| --- |
| Computer laboratory technicians |
| Development instrumentation technicians |
| Drafters |
| Electronics communications technicians |
| Engineering technicians |
| Engineering development technicians |
| Fiber technologists |
| Laser technicians |
| Nuclear reactor technicians |
| Optomechanical technicians |
| Production test technicians |
| Semiconductor development technicians |
| Systems testing laboratory technicians |
| Quality assurance technicians |

## what are the salary ranges?

Electronics technicians starting in entry-level jobs earn an average of $17,000 a year. Those who have completed a two-year academic program in electronics technology earn more on their first jobs, with annual salaries falling between $18,000 and $22,000. Technicians whose specialties focus on more sophisticated and complex technologies might start at slightly higher salaries, about $22,800 a year.

Successful completion of the associate-level CET certification program may translate to a higher starting salary in any subfield, while salaries offered by the federal government are slightly lower in all areas.

Electronics technicians with several years of experience make annual salaries in the median range of $32,700. Those with journeyman certification earn the highest of these salaries. Average yearly salaries for all electronics technicians range between $33,000 and $49,000.

The highest-paid electronics technicians can make anywhere from $40,000 to $65,000 annually. Salaries at this level are a function of experience, education, certification, specialization, and position.

Most electronics technicians receive a full benefits package, including paid vacation, higher compensation for overtime, health insurance, and pension plans. Some employers may offer educational reimbursement incentives, as well as paid time off for special training workshops and academic classes.

## what is the job outlook?

The U.S. Department of Labor estimates that opportunities for electronics technicians will grow faster than average through 2005. Potential for growth of the industry is linked in part to the increase of sophisticated electronic consumer goods. Personal computers, video cassette recorders, and home entertainment systems complete with large-screen TVs and surround sound speakers are extremely popular and are found in more and more American homes. The design, production, modification, maintenance, and repair of such consumer goods require crackerjack electronics technicians.

On the flip side, technology is improving at such a fast pace that the need for repair is diminishing for some products. Also, since prices drop for older products when new products come out on the market, it is often less expensive to throw away the older machine and just buy a new one; this practice reduces the need for repairs by electronics technicians. However, when the economy is down, people are more likely to have old equipment repaired rather than replaced. Given these influences, electronics technicians specialized in consumer goods should keep their training up-to-date.

All electronics technician positions, whether in industry, consumer goods, or the military, depend on a variety of factors. Foreign competition, general economic conditions, and government spending all have an impact on the field. Prospective electronics technicians should begin paying attention to certain factors that might affect the area they are thinking about working in. For example, if you are planning to work for the military or for a military contractor or subcontractor in radar technology, keep an eye on federal legislation concerning military spending cuts or increases.

The electromedical and biomedical subfields, dealing with technical hospital machines, have been identified as areas that will provide excellent job opportunities through 2005 (*see* chapter "Biomedical Equipment Technician").

The electronics industry is undeniably indispensable to our lives, and so while there will be fluctuations in growth for certain subfields, there will always be a need for qualified personnel in other fields. The key to success for an electronics technician is to stay up-to-date with technology and to be professionally versatile. Building a career on a solid academic and hands-on foundation in basic electronics enables you to remain competitive on the job market.

## how do I learn more?

### sources of additional information

Following are organizations that provide information on electronics technician careers, accredited schools and scholarships, and employers.

**American Electronics Association**
5201 Great American Parkway, Suite 520
PO Box 54990
Santa Clara, CA 95054
408-987-4200

**IEEE Consumer Electronics Society**
c/o Institute of Electrical and Electronics Engineers
345 East 47th Street
New York, NY 10017
212-705-7900

**International Society of Certified Electronic Technicians (ISCET)**
2708 West Berry Street, Suite 3
Fort Worth, TX 76109-2356
817-921-9101

**The Electronics Technician Association International (ETA-I)**
602 North Jackson Street
Greencastle, IN 46135
765-653-8262

**Electronic Industries Association**
2500 Wilson Boulevard
Arlington, VA 22201
202-457-4900

**Junior Engineering Technical Society**
1420 King Street, Suite 405
Alexandria, VA 22314
703-548-5387

**Siemens Corporation**
1301 Avenue of the Americas
New York, NY 10019
212-258-4000
http://www.siemens.com

# bibliography

Following is a sampling of materials relating to the professional concerns and development of electronics technicians.

## Books

Faissler, William L. *An Introduction to Modern Electronics*. New York: J. Wiley, 1991.

Floyd, Thomas L. *Electronics Fundamentals: Circuits, Devices, and Applications*. 3rd ed. Englewood Cliffs, NJ: Prentice Hall, 1995.

Horn, Delton T. and Abraham I. Pallas. *Basic Electricity and Electronics*. rev. ed. New York: Macmillan/McGraw-Hill, 1993.

Metzger, Daniel L. *Electronics for Your Future*. Lambertville, MI: Technology Training, 1994.

Reis, Ronald A. *Becoming an Electronics Technician: Securing Your High-Tech Future*. 2nd ed. New York: Prentice Hall, 1997.

## Periodicals

*Electronic Business*. Biweekly. Reports on current economics. Cahner's Publishing Co., 275 Washington Street, Newton, MA 02158-1630, 617-964-3030.

*Electronic Products*. Monthly. Discusses new developments in the industry. Hearst Business Publishing, UTP Division, 645 Stewart Avenue, Garden City, NY 11530, 516-227-1300.

*Electronic Servicing & Technology*. Monthly. News for consumer electronics servicing personnel. CQ Communications, Inc., 76 North Broadway, Hicksville, NY 11801, 516-681-2922.

*Electronic World News*. Biweekly. Internationally directed magazine. CMP Publications, Inc., 600 Community Drive, Manhasset, NY 11030, 516-562-5000.

*Electronics Weekly*. Weekly. General newspaper for electronics users and manufacturers. Reed Business Information, Quadrant House, The Quadrant, Sutton, Surrey SM2 5AS, England, 44-181-652-3649.

*Global Electronics*. Monthly. Covers the worldwide electronics industry. Pacific Studies Center, 222B View Street, Mountain View, CA 94041, 415-969-1545.

*Journal of Electronic Materials*. Monthly. Discusses all aspects of electronics research. Mineral, Metals & Materials Society, 420 Commonwealth Drive, Warrendale, PA 15086, 412-776-9080.

*Popular Electronics*. Monthly Contains new projects and concepts. Gernsback Publications, Inc., 500 Bi-Country Boulevard, Farmingdale, NY 11735, 516-293-3000.

# embalmer

**Definition**

Embalmers prepare the dead for burial. They wash the body and replace blood with embalming fluid as a means of preservation. When necessary, they reconstruct diseased or disfigured bodies. They dress the body and transfer it to a casket. Many embalmers also direct funerals or assist in other aspects of funeral service.

**Salary range**

$18,512 to $30,680 to $55,744

**Educational requirements**

High school diploma; some college; one- to three-year apprenticeship

**Certification or licensing**

Varies but required by most states

**Outlook**

Excellent

GOE 02.04.02

DOT 338

O*NET 39014

## High School Subjects

Anatomy and Physiology
Biology
Business
Chemistry
English (writing/literature)
Speech

## Personal Interests

Business
Helping people: emotionally
Sculpting

is grieving and struggling with the loss of their loved one," says Tom Elliott. "Those of us working in funeral service make a very difficult situation as easy as possible on families given the circumstances." Elliott is an *embalmer* and *funeral director* at Elliott Mortuary in Hutchinson, Kansas.

"Since most embalmers are also funeral directors, our job doesn't end in the preparation room. We're involved in all aspects of the service."

"A lot of people think that all embalmers do is prepare bodies. Others think that we only wear suits and drive limousines all day. But our job is much more involved than that, especially in a mid-sized firm like this one. Today, for instance, I started by sweeping the driveway of the funeral home and vacuuming the front hall. Then, because it rained, I helped rewash our vehicles before the afternoon service began. Later on I'll do a bit of bookkeeping and visit with another family to consult about tomorrow's funeral. Sure, I do embalming too, but that's not all I do."

## what does an embalmer do?

*Embalmers* prepare the human body for funeral services and for burial. First the body is washed with an antibacterial soap and dried. Next a tube is inserted into an artery and blood is evacuated from the body while embalming fluid is pumped into the body. Fluids and wastes are removed from body organs. All incisions are closed using sutures. Embalming fluid, a scientifically-formulated mixture, slows down decomposition, prevents the spread of disease, and maintains body firmness for cosmetic purposes.

Embalmers also repair body parts that have been disfigured due to illness or injury. This is called restoration. Depending on the nature of the repair, a number of materials may be involved in restoration. These include clay, cotton, mortuary putty, plastics, plaster, wire, and wax.

After restoration, the deceased person's hair is shampooed and styled. Cosmetics are applied in order to give the person as natural an appearance as possible. The deceased are dressed appropriately, their bodies are positioned, and they are placed in caskets.

During the embalming process, records are kept of chemical usages. Safety measures are scrupulously followed in order to comply with OSHA (Occupational Safety and Health Administration) regulations. To protect themselves from disease and prevent the spread of infection, embalmers wear protective clothing, which may include surgical-type masks, gloves, face shields, goggles, and shoe covers. Following the process, the preparation room is cleaned, equipment and clothing is sanitized, and all waste is disposed of according to government standards.

Since many embalmers also work as *funeral directors,* their other job duties may include greeting family members and friends of the deceased, arranging and conducting funeral services, writing obituary notices, serving as or organizing pallbearers, arranging with cemeteries to open and close graves, preparing grave sites for outdoor services, transporting the deceased in hearses, driving family members to and from funerals, transporting floral arrangements, handling various forms of paperwork, doing administrative tasks, and looking out for basic building maintenance at the funeral home.

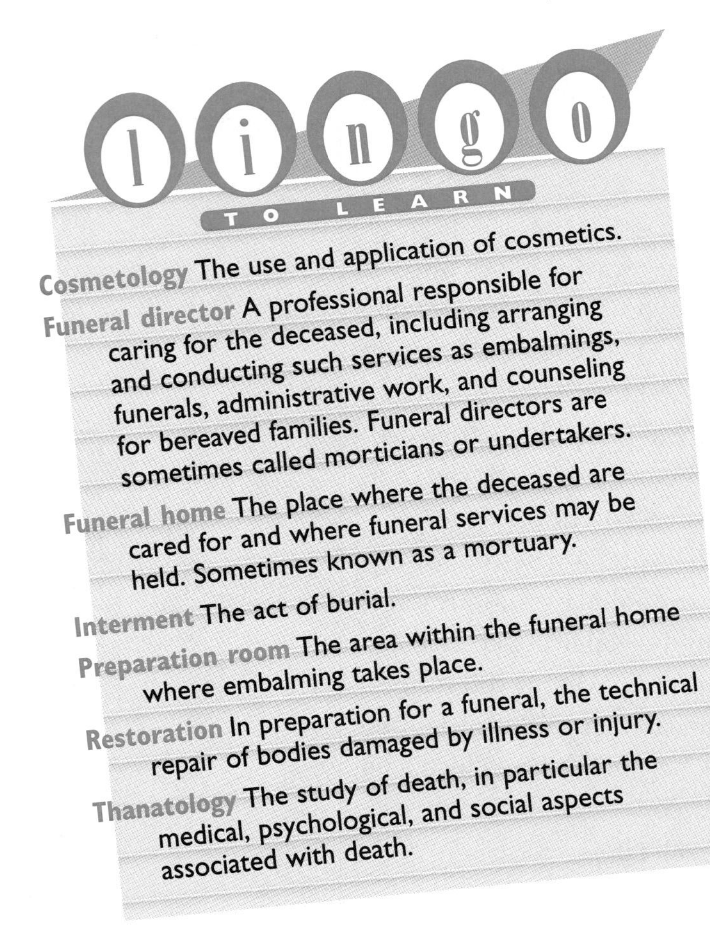


## what is it like to be an embalmer?

Millions of people die each year in the United States, at every hour of the day and night. As a result, embalmers and funeral directors are on call twenty-four hours a day, seven days a week. They often work long, irregular hours, including weekends and evenings. Today, on a Sunday in spring when most people are enjoying a day off and relaxing with their families, Tom Elliott is at work, speaking with families who request his firm's embalming and funeral services. There are already three funerals scheduled for Monday at his funeral home, four scheduled for Tuesday, and two for Wednesday.

"Very few embalmers," Elliott explains, "spend eight hours a day in the preparation room. Those who do usually work for very large firms where each job is extremely specific and you're hired to do only that one task. But here, as in most firms, you get a little bit of everything. That's good, of course, because no one gets bored and you get to follow the family and help them through the entire process."

"There's no typical day at a funeral home," Elliott adds. "My day might begin with picking up a body and bringing it here for embalming. Or I might assist the coroner during an autopsy and then complete the embalming process and clean the preparation room. Later on, I might discuss music for a funeral with a family or explain to them how to claim veterans' and social security benefits. Sometimes I'm arranging floral sprays in the chapel or transporting bouquets to gravesides prior to services there."

"Sometimes I'm doing government paperwork or basic bookkeeping. Or I might be on the telephone with casket manufacturers, checking on availability. Or I could be outside shoveling snow or raking leaves from the driveway."

He leans back in his chair and sighs. "Of course, a big part of my job is basic listening. Listening is tremendously important. So many times there's not much you can say really. The spoken word can do so little. Mostly families simply need to tell their stories." He adds, "At times of stress, families aren't likely to remember everything you say so follow-up is an important part of our service. A week or two after the funeral we contact the family and help them with insurance, legal odds and ends, government benefits and such. We also have a family services coordinator on staff who helps with grief counseling and offers a support group to families who request that service."

Elliott explains that the actual embalming process is highly regulated. "All employees go through hazardous materials training, including basic handling and safe disposal techniques."

"Years ago, embalmers didn't even wear gloves, and exposure to chemicals was a real problem for them. Now we have a ventilation system, emergency showers, and an eyewash station, all in order to keep ourselves safe and comply with governmental regulations."

Working in the embalming room requires a mixture of art, aptitude, and practice. "You can learn the basics of embalming from a textbook or in a college class, but applying what you've learned to a human body is something else again," Elliott says. "You start with a basic knowledge of anatomy, and then you need to be skillful with your hands. Embalmers aren't in the same league as artists, but over time, you learn to be good at what you do."

He continues, "But every body is different. How we deal with the body of an eighty-year-old man who's been ill and living in a nursing home for many years is quite different from how we deal with the body of a thirty-seven-year-old man who burned to death in a fire. Sometimes there's only so much restoration that you can do, and that's hard to explain to a family."

One of the things Elliott appreciates about his profession is that through his work as an embalmer, he has been able to become a licensed *eye enucleator*. When directed to do so, he recovers the corneas from deceased people, packs them in refrigerated containers, and ships them to hospitals where they are used for cornea transplants. Other embalmers may become involved in other types of tissue or organ recovery.

"Very few embalmers spend eight hours a day in the preparation room."

## have I got what it takes to be an embalmer?

Good embalmers are good listeners. They pay attention when families describe how their loved ones looked while alive and make every effort to honor that memory throughout the entire embalming and restoration process. They are genuinely respectful when handling the dead, treating each deceased person as if he or she were a cherished parent or sister or brother.

But such attention to the needs of others can take its toll. "You are what you feel," says Tom. "You have to be truly compassionate. It can't just be a hat that you put on when you arrive at work. Still, it's important that you be able to separate yourself a bit from what's going on. You need to accept that what you're doing is a way to assist families and that by doing a good job, you help them through a difficult time."

Embalming is also a physically demanding job, requiring strength and endurance. Exposure to hazardous chemicals and disease is another occupational concern. The AIDS virus, for instance, continues to live for an unknown amount of time after death within an infected individual's body. Even after death, many body fluids, including tears, saliva, urine, and blood, can carry disease.

Embalmers need to be well trained, use common sense, and maintain basic sanitary standards. However, embalming and its associated field, funeral directing, can be very rewarding. It's a good job for someone who is outgoing and enjoys working with people, especially someone who is sensitive to the unique needs of others during times of grief. Embalmers and funeral directors perform a function that is very basic to society. Their work and the ceremonies surrounding death help make the loss of loved ones more bearable for the living.

Embalming is also a good job for someone who likes challenging work and who is easily bored by repetitive types of labor. No two embalming jobs are ever the same. Each body brought into the preparation room presents a unique challenge.

Job security is also a big factor to consider. Funeral homes exist in every area of the country and qualified embalmers are always in demand. The armed forces and the federal government also have staff embalmers.

## how do I become an embalmer?

Tom Elliott grew up in the funeral business. His grandfather started the firm in 1935, and his father, uncle, and brother continued the tradition. Tom didn't set out to study embalming or to become a funeral director, however. Nevertheless, during his junior year of college while studying for a degree in history, he decided that embalming was the career for him after all. He finished his bachelor's degree and then entered a two-year mortuary science program. "I'm lucky," Tom says. "Some people enter mortuary school and they've never even seen the inside of a funeral home, but I already had a lot of background. Still, I learned so much at school and even more during my apprenticeship."

**To be a successful embalmer, you should**

- Be a good listener
- Enjoy working with people
- Be tactful in difficult circumstances
- Have a genuine respect for the dead
- Have compassion for the feelings of others
- Be respectful of different religious beliefs and cultural practices
- Have good manual dexterity
- Be physically fit and able to lift heavy objects

## education

### High School

Receiving a high school diploma is the first step toward a career in embalming. Students should take as many laboratory science classes as possible in high school, including chemistry, biology, and physics. English, public speaking, math, and business courses are also helpful, as are computer science and typing classes. In addition, students should be sure to fulfill the entrance requirements of the college, university, or mortuary science program they plan to attend.

### Postsecondary Training

Currently there are more than forty schools offering mortuary science degrees and enrolling approximately two thousand students a year. Each program is accredited by the American Board of Funeral Service Education. Some embalming programs are associated with mortuary colleges, and some are associated with community colleges or universities. Some programs offer certificate-level credentials, while others grant associate's or bachelor's degrees.

The amount of postsecondary education necessary to become an embalmer varies from state to state. States require either one or two years of college plus one or two years at a mortuary college; or an associate's degree in mortuary science; or a high school diploma plus one year of mortuary college.

General course work includes classes in anatomy, biology, chemistry, business, mathematics, computer science, and English.

Students also learn about pathology, microbiology, public health, embalming theory, restorative art, cosmetology, mortuary chemistry, funeral service counseling, grief psychology, mortuary law, funeral service merchandising, and mortuary management.

State board examinations follow completion of postsecondary course work. After successfully passing the examination, all embalmers must serve an apprenticeship under the guidance of a licensed embalmer. Apprenticeships are determined by state regulation and last from one to three years.

## in-depth

### History of Embalming

Every civilization must deal with the death of its individual members. Some societies burn their dead, while others bury them in caves, place them high in trees or in rock niches, or sink them beneath the ocean.

Egyptians were the first to develop embalming techniques. Around 4,000 BC, embalmer-priests held respected positions within their society. They soaked bodies in chemical solutions, rubbed them with oils and spices, and wrapped them in linen shrouds before burial. Following the decline of Egyptian culture, other societies continued to anoint their dead with essences of cypress, frankincense, rosewater, or musk, but few practiced full-scale embalming.

In the United States, embalming was not practiced widely until during the Civil War. Then it became necessary to transport deceased soldiers over long distances in order that they might be buried on their home ground. Standardized chemical solutions and the discovery of formaldehyde in 1867, combined with the large number of battle casualties resulting from the Civil War, greatly accelerated the acceptance of embalming practices in the United States.

Throughout the early part of this century, embalming was often done in the homes of the deceased, using portable embalming tables. Now the procedure is conducted in specially-designed preparation rooms, immaculately clean facilities that are equipped with many types of special equipment and are used only for this purpose.

## certification or licensing

Licensing for embalmers varies according to the state of residence. Most states require graduation from an accredited mortuary science program, one to three years' apprenticeship, and successful completion of the state or national examination. Contact your state's Board of Funeral Directors and Embalmers or Bureau of Professional Licensure for more information.

## scholarships and grants

Most colleges and universities offer general scholarships. Institutions teaching mortuary science also set aside monies to be used exclusively for students entering this field. In addition, individual state and national funeral directors associations provide scholarships for future embalmers. Information on scholarships available to students in funeral service or mortuary science is available through the American Board of Funeral Service Education (*see* "Sources of Additional Information").

## internships and volunteerships

"The greatest experience for any student is to work at a funeral home," Tom says. "You can learn only so much in the classroom. As with any trade, there's no substitute for on-the-job training. Even though I came from a funeral business background, it was good to see how a different firm operated. Any experience someone can gain before committing full-time to the profession is invaluable. Take a part-time job cutting the grass at your local mortuary. Or wash the limousines. If you don't have some idea about what the job entails before you enter it, you're not likely to have a long tenure as an embalmer."

## who will hire me?

Tom has been an embalmer for over a decade. Before he joined his family's firm, he served an apprenticeship with an established funeral home near Kansas City Community College where he studied mortuary science.

The average funeral home has been in its community for over forty years. Often funeral homes are multigenerational businesses, passed down from parent to child, but you needn't be part of a family business in order to start a

Advancement Possibilities

**Funeral directors** arrange and direct funeral services.

**Funeral home owners** are usually licensed funeral directors and embalmers who perform those functions in addition to those of running a business.

**Mortician investigators** inspect mortuaries for sanitary conditions, equipment, and procedures, and competence of personnel. They examine the registration credentials of trainees and licenses of funeral directors to ensure compliance with state licensing requirements. They investigate complaints and recommend license revocation or other action.

successful career as an embalmer. Only five of the sixty students Elliott attended mortuary science school with had any sort of funeral business background. Nationwide only about one-third of all funeral service students have any connection to family businesses. With or without family connections, building a career as an embalmer starts with successful completion of a mortuary science program, passage of state exams, and serving of an apprenticeship at an established funeral home under the guidance of a practicing embalmer.

Women are entering the profession in increasing numbers. In the recent past women had been discouraged from entering funeral service in general and from becoming embalmers in particular. However, 50 percent of some mortuary science classes now are female.

Most schools provide placement services for their graduates. Sometimes students find full-time work at the same firms where they served their apprenticeships. *The Director Magazine*, a publication of the National Funeral Directors Association, offers classified ads for prospective employees and employers (*see* "Bibliography").

## where can I go from here?

Without a license, embalmers cannot independently practice their trade. Every state except Colorado requires a license. Entry-level technicians (apprentices gaining work experience prior to official licensing) remain under the supervision of certified embalmers until they complete their apprenticeships and successfully pass required exams. Without a license, technicians will remain on the lowest rung of funeral service.

Most embalmers eventually become licensed funeral directors, too. Funeral directors oversee and participate in all aspects of funerals. Embalmers must have some amount of college and/or mortuary science education. Funeral directors are regulated by individual state boards and must pass an examination to qualify for licensing. The goal of many embalmers is to become an owner/operator of a funeral home. This opportunity is most available to those people who have credentials as both an embalmer and a funeral director, have several years of experience in the funeral business, and have developed a reputation for friendliness, professional courtesy, and respect.

## what are some related jobs?

Some careers that are related to embalmers are funeral director, grief counselor, prearrangement counselor, and morgue attendant.

The U.S. Department of Labor classifies embalmers under the headings *Embalmers and Related Occupations* (DOT) and *Laboratory Technology: Life Sciences* (GOE). Also under these headings are people who work in laboratories and help scientists, medical doctors, researchers, and engineers in their work.

| Related Jobs |
|---|
| Agriculture cytotechnologists |
| Ballistics experts |
| Biological aides |
| Biomedical equipment technicians |
| Chemistry technologists |
| Criminalists |
| Forensic paleontological helpers |
| Laboratory technicians |
| Medical technologists |
| Pharmacists |
| Photo optics technicians |
| Public health microbiologists |
| Radioisotope production operators |
| Scientific helpers |
| Tissue technologists |
| Ultrasound technologists |
| Veterinarians |

## what are the salary ranges?

Experience, employer, and location determine the salary range for embalmers. In general, salaries are higher in metropolitan areas and lower in rural areas. Licensed embalmers start with salaries averaging around $25,000. The 1998-99 *Occupational Outlook Handbook* reports that embalmers who also work as funeral directors earned salaries that ranged from $18,512 to $55,744 in 1996. Embalmers who work as funeral directors and who also own their own funeral homes can earn considerably higher salaries.

## what is the job outlook?

Employment prospects for embalmers are expected to be excellent through the year 2006; the number of mortuary school graduates will not match the number of openings available in the field.

There are more than twenty-two thousand funeral homes in the United States and all employ embalmers. Funeral homes are located in every metropolitan area and in many small towns. As the average age of the general population rises and as diseases such as AIDS take their toll, demand for services provided by embalmers grows. The only exception is in areas where cremation is used extensively and where state laws do not demand embalming prior to cremation. Nevertheless, because of the nature of the funeral business, qualified embalmers will always find work, regardless of location.

## how do I learn more?

### sources of additional information

Following are organizations that provide information on embalming careers, accredited schools, and employers.

**American Board of Funeral Service Education**
PO Box 1305
Brunswick, ME 04011
207-798-5801

**Associated Funeral Directors International**
11691 Oval Drive
Largo, FL 33774
813-593-0709

**Federated Funeral Directors of America**
1622 South MacArthur Boulevard
Springfield, IL 62704-3623
217-525-1712

**National Funeral Directors and Morticians Association**
1800 East Linwood Boulevard
Kansas City, MO 64109
816-921-1800

**National Funeral Directors Association**
11121 West Oklahoma Avenue
Milwaukee, WI 53227-4033
414-541-2500

**National Selected Morticians**
5 Revere Drive, Suite 340
Northbrook, IL 60062-8009
847-559-9569

### bibliography

Following is a sampling of materials relating to the professional concerns and development of embalmers.

#### Books

Jones, E. Ray. *Standard Funeral Manual.* Cincinnati, OH: Standard Publishing, 1991.

Mayer, Robert G. *Embalming: History, Theory, and Practice.* 2nd ed. Norwalk, CT: Appleton & Lange, 1996.

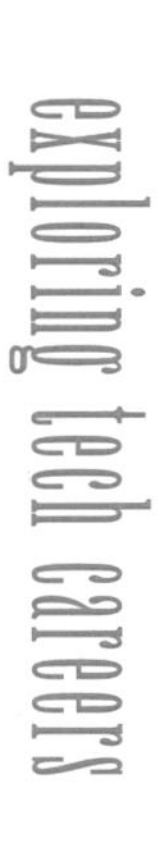

## Periodicals

*Director.* Monthly. Conveys topics related to funeral services, including education, public relations, and management. NFDA Publications, Inc., 11121 West Oklahoma Avenue, Box 27641, Milwaukee, WI 53227, 414-541-2500.

*Funeral Service Insider.* Weekly. Newsletter of interest to funeral home personnel. Atcom, Inc., 1541 Morris Avenue, Bronx, NY 10457-8702.

*Mortuary Management.* Eleven times per year. For all funeral industry personnel. Abbott & Hast Publications, 761 Lighthouse Avenue, Suite A, Monterey, CA 93940-1033, 408-657-9403.

FYI

### Causes of Death

About 39 percent of the total yearly deaths in the United States are caused by cardiovascular diseases and about 29 percent are caused by cancer. Accidents account for about 7 percent of deaths. Nearly half of all fatal accidents involve motor vehicles. Other leading causes of death include chronic lung diseases, pneumonia, diabetes mellitus, chronic liver diseases, and kidney diseases.
Suicide accounts for about 2.1 percent of all deaths, and homicide about 1.6 percent.

# emergency medical technician

**Definition**

Emergency medical technicians give immediate first aid treatment to sick or injured persons both at the scene and en route to the hospital or other medical facility. They also make sure that the emergency vehicle is stocked with the necessary supplies and is in good operating condition.

**Salary range**

$21,200 to $24,650 to $27,400

**Educational requirements**

High school diploma; standard training course of 100 to 120 hours

**Certification or licensing**

Required

**Outlook**

Faster than the average

GOE 10.03.02

DOT 079

O*NET 32508

## High School Subjects

Anatomy and Physiology
Biology
Computer science
English (writing/literature)
Health
Physical education

## Personal Interests

Helping people: physical health/medicine
Helping people: protection

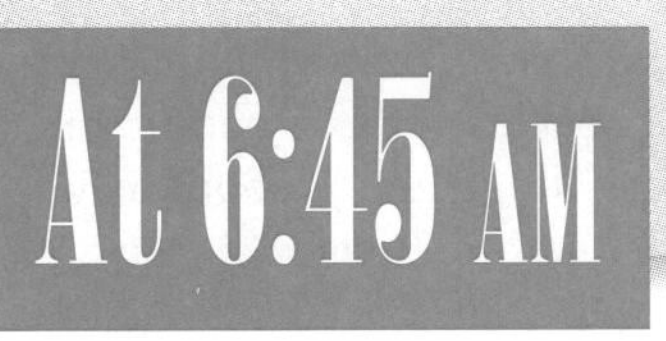


Kelly Richey clocks in, says hello to the EMTs going off duty, and goes to the Coke machine for her first Diet Coke of the day. It is half gone when the first call comes in, and fifteen minutes later she is inside a crumpled black Camaro, placing an oxygen mask on a critically injured teenager.

"Are you still in school?" she asks him as she works over him. "What grade? Senior?" She is trying to determine his level of consciousness, and she has to speak in a near-shout because the fire department is cutting the roof off the car while she works. As soon as the roof comes off, she and her co-worker must work quickly to get the patient out of the car and on the way to the hospital. He has injuries to the chest and both legs and is losing blood quickly. If he doesn't reach surgery within an hour, his chances for survival are greatly diminished.

In the ambulance, while her partner drives, Kelly continues to talk to the young man, taking his vital signs every four to five minutes. She calls in to the hospital emergency room to notify them that they have a serious trauma patient coming in. "We have a seventeen-year-old patient, conscious and alert," she begins. "Involved in a single vehicle accident. Patient has chest wounds and injuries to both legs. Vital signs are stable. We have an ETA of five minutes."

At the hospital, she and her partner unload the patient and, delivering him to the emergency staff, give a brief report to the doctor who will be on his case.

Kelly gets a blank "run report," and her second Diet Coke, and starts to document the first run of her day.

**IV or intravenous** Administered by an injection into the vein

**Cardiac arrest** The complete stoppage of the heartbeat.

**Endotracheal intubation** The insertion of a tube into the trachea, or windpipe, to provide a passage for the air, in case of obstruction.

**Backboard** A long, flat, hard surface used to immobilize the spine in the case of neck or spinal injury.

**Defibrillator** An apparatus consisting of alternating currents of electricity, with electrodes to apply the currents to heart muscles in order to shock the muscles into operation. Requires the operator to interpret the heart rhythms and apply the shock at the proper time.

**Amkus Spreader** A hand-held rescue device used to free trapped victims by pulling crumpled metal apart.

**Amkus Cutter** A hand-held rescue device, similar to scissors, used to free trapped victims by cutting through metal.

**Amkus Rams** A hand-held rescue device used to free trapped victims by pushing or pulling obstructions, such as dashboard and seats, away from the victim.

## what does an emergency medical technician do?

If you are sick or hurt, you usually go to a doctor; if you are very sick or hurt you may go to the emergency room of a local hospital. But what if you are alone and unable to drive, or you are too badly injured to travel without receiving medical treatment first? It often happens that an accident or injury victim needs on-the-spot help and safe, rapid transportation to the hospital. *Emergency medical technicians* are the ones who fill this need.

Emergency medical technicians, or EMTs, respond to emergency situations to give immediate attention to people who need it. Whether employed by a hospital, police department, fire department, or private ambulance company, the EMT crew functions as a traveling arm of the emergency room. While on duty, an EMT could be called out for car accidents, heart attacks, work-related injuries, or drug overdoses. He or she might help deliver a baby, treat the victim of a gunshot wound, or revive a child who has nearly drowned. In short, EMTs may find themselves in almost any circumstance that could be called a medical crisis.

Usually working in teams of two, they receive their instructions from the *emergency medical dispatcher,* who has taken the initial call for help, and drive to the scene in an ambulance. The dispatcher will remain in contact with the EMT crew through a two-way radio link. This allows the EMTs to relay important information about the emergency scene and the victims and to receive any further instructions, either from the dispatcher or from a medical staff member, if need be. Since they are usually the first trained medical help on the scene, it is very important that they be able to evaluate the situation and make good, logical judgements about what should be done in what order, as well as what should not be done at all. By observing the victim's injuries or symptoms, looking for medic alert tags, and asking the necessary questions, the EMTs determine what action to take and begin first aid treatment. Some more complicated procedures may require the EMT to be in radio contact with hospital staff who can give step-by-step directions.

The types of treatments an individual is able to give depend mostly on the level of training and certification he or she has completed. The first and most common designation is *EMT-Basic* or *EMT-Ambulance.* A basic EMT can perform CPR, control bleeding, treat shock victims, apply bandages, splint fractures, and perform automatic defibrillation, which requires no interpretation of EKGs. They are also trained to deal with emotionally disturbed patients and heart attack, poisoning, and burn victims. The *EMT-Intermediate,* which is the second level of training, is also prepared to start an IV, if needed, or use a manual defibrillator to apply electrical shocks to the heart in the case of a cardiac arrest. A growing number of EMTs are choosing to train for the highest level of certification—the *EMT-Paramedic.* With this certification, the individual is permitted to perform more intensive treatment procedures. Often working in close radio contact with a doctor, he or she may give drugs intravenously or orally, interpret EKGs, perform endotracheal intubation, and use more complex life-support equipment.

In the case where a victim or victims are trapped, EMTs first give any medical treatment, and then remove the victim, using special equipment such as the Amkus Power Unit. They may need to work closely with the police or the fire department in the rescue attempt.

If patients must be taken from the emergency scene to the hospital, the EMTs may place them on a backboard or stretcher, then carry and lift them into the ambulance. One EMT drives to the hospital, while the other monitors the passenger's vital signs and provides any further care. One of them must notify the hospital emergency room, either directly or through the dispatcher, of how many people are coming in and the type of injuries they have. They also may record the blood pressure, pulse, present condition, and any medical history they know, to assist the hospital.

Once at the hospital, the EMTs help the staff bring in the patient or patients, and assist with any necessary first steps of in-hospital treatment. They then provide their observations to the hospital staff, as well as information about the treatment they have given at the scene and on the way to the hospital.

Finally, each run must also be documented by the acting EMTs for the records of the provider. Once the run is over, the EMTs are responsible for restocking the ambulance, having equipment sterilized, replacing dirty linens, and making sure that everything is in order for the next run. For the EMT crews to function efficiently and quickly, they must make certain that they have all the equipment and supplies they need, and that the ambulance itself is clean, properly maintenanced, and filled with gas.

"I like knowing I'm helping people. I love to know I may save someone's life."

## what is it like to be an emergency medical technician?

Kelly Richey is a calm, soft-spoken twenty-nine-year-old woman who may save a life as a routine part of a day's work. For the past three years, she has worked for the hospital in an average-sized Indiana city as an EMT. Although her father had been an EMT for years, Kelly had no intention of working in Emergency Medical Services when she first took the training course. "I just took the course for knowledge," she says. "I didn't really plan to do it as a job." However, during the training course clinicals, when she got the opportunity to actually go on runs, she changed her mind. "I just kind of got the fever," she says, laughing. "I decided it was something I wanted to do. I decided it was something I *could* do, and I thought if I can do this and I can save a life, I should maybe make this my profession."

Kelly's unit gives its EMTs the option of working two twenty-four-hour shifts or one eight-hour and two sixteen-hour shifts each week. Kelly has done both, but prefers the shorter hours, which she currently works. This number of hours is fairly common for employees of ambulance firms and hospitals. EMTs who work for fire and police departments may be scheduled for as many as fifty-six hours per week.

EMT work can be physically demanding. EMTs often work outside, in any type of weather, and most of their time on a run is spent standing, kneeling, bending, and lifting. It is also stressful and often emotionally draining. "I'm pretty tired at the end of a shift," Kelly says. The sort of shift an EMT has depends almost entirely on how many and what type of calls come in. Some shifts are incredibly busy—up to fifteen runs, according to Kelly, although not all of those are emergency runs. Her ambulance is frequently dispatched to carry nursing home patients to and from the hospital for treatment, as well as to transport stable patients from the local hospital to the larger medical complex an hour away for scheduled surgeries. Other shifts are considerably slower, with maybe only seven runs in the entire sixteen hours.

At Kelly's hospital, the average length of a run is an hour and a half from dispatch till return to ready status. The location of the emergency and its seriousness greatly affect that time frame, however. There is a perception among many people that most calls involve life-or-death situations. That is not really the case, in Kelly's experience.

"On a monthly basis, only about 10 to 15 percent of my runs are actually life-threatening," she says. Other calls run the gamut from serious injury to cut fingers. "You wouldn't believe the stuff I've been called out for," Kelly says. "I've gotten calls for things you could put a Band-Aid on and go right about your business."

No matter what the call, the EMTs must respond. Each team of two has a certain area of the city that it is assigned to, but any team will take any call, if the designated team for that area is already out. Kelly's hospital employs approximately seventy EMTs and paramedics, including several who work on a part-time schedule, and it staffs up to five ambulances during the busiest parts of the day.

All the EMTs in Kelly's unit are required to take a driver training program when they are hired, so they are all qualified to drive. Kelly says that she usually drives on every other run, alternating turns with her partner. The team schedule is set up in such a way that the same partners always work together. That way, they get used to working as a team and communicating with each other. Also, they are able to become accustomed to each others' techniques and work habits and are able to function more efficiently.

Not every emergency requires that a patient be taken to the hospital. If it isn't necessary, or if the patient refuses to go, the EMTs give treatment on-site and return to await the next call. If a patient does need further attention, he or she is transported to the emergency room and unloaded by Kelly and her partner. The attending doctor is briefed on the case and the treatment. Whether a patient is transported or not, each run must be documented on a state-issued form called a run report.

Between calls, Kelly and her co-workers wait in the crewhouse. Complete with a kitchen, living room with TV, and two sleeping rooms, it has a comfortable, homey atmosphere. The ambulances are kept in an adjoining garage, and during each shift, if time allows, the crew is responsible for washing its vehicle and making sure it is fully stocked and ready to go. "Some people like to restock after every run," Kelly says. She and her partner prefer to take stock at the end of their shift and make sure everything is replaced. "That way you know you're leaving it in good shape for the shift coming on."

**To be a successful emergency medical technician, you should**

- Have a desire to serve people
- Be emotionally stable and clearheaded
- Have good manual dexterity and agility
- Have strong written and oral communication skills
- Be able to lift and carry up to 125 pounds
- Have good eyesight and color vision

## have I got what it takes to be an emergency medical technician?

EMTs regularly encounter situations that many people would find upsetting. Because they are faced with unpleasant scenes, crises, and even death, they need a certain emotional capability for coping. They must have stable personalities and be able to keep their heads in circumstances of extreme stress.

The stress level is the hardest thing about the job for Kelly. She also warns that the potential EMT must be able to deal with death. "There have been times when the family members have hung out in the E.D. [Emergency Department] after we brought patients in and that really gets to me," she says. "I feel like they're looking at me and saying `Why didn't you do more?' when I know I've done everything I could do." At Kelly's hospital, the staff has initiated a debriefing process to help EMTs work through a bad experience on a run. It's important that the EMT be able to cope with such bad experiences without suffering lasting negative results. The stress and the emotional strain can take its toll; there is a high turnover rate in the Emergency Medical Services field.

On the other hand, the job of EMT can be extremely fulfilling and rewarding. Kelly says that she tends to feel more confident and secure in all situations, just knowing she is trained to deal with many medical emergencies. But that's not the best part of it. "I like knowing I'm helping people," Kelly says. "I love to know I may save someone's life." It's important for EMTs to have a willingness to help people, even when the patients they deal with are difficult and abusive. "Oh, I've been hit, kicked, spit on," laughs Kelly. "Drug overdoses and people with head injuries fight you a lot of the time." There is also always the possibility that a patient is infected with the AIDS virus, or another contagious disease. Although rubber gloves are commonplace, and the emergency staff at Kelly's hospital are now beginning to wear protective masks to cover their mouths, fear of disease is still something to consider. Regardless of the circumstances, the patient must be treated, so a genuine desire to serve is important to success in this field.

Kelly's job is never routine. She enjoys the fact that she doesn't know from one day to the next, or even from one hour to the next, what she will be doing. "It's important to be flexible," she says. She also says it's important to be able to work closely with others. The partners have to rely very heavily on each other and communicate easily and well to be a good emergency team. She describes the successful EMT as a person who can handle stress well, is confident in his or her abilities, communicates well, has physical strength and stamina, and really wants to help others.

## how do I become an emergency medical technician?

All states offer EMT training programs consisting of 100 to 120 hours of training, usually followed by ten hours of internship in a hospital emergency room. To be admitted into a training program, it is necessary to be eighteen years old, be a high school graduate, and hold a valid driver's license. Exact requirements do vary slightly in different states and in different courses.

## education

### High School

A high school diploma, or its equivalent, is required for admission into the EMT training program. While most high school studies will not yield experience with emergency medical care, health classes may offer a good introduction to some of the concepts and terms used by EMTs. It may also be possible to take courses in first aid or CPR through the local Red Cross or other organizations. This sort of training can be a valuable background, giving the student advance preparation for the actual EMT program. Some science classes, such as biology, can also be helpful, in that students can become familiar with the human body and its various systems.

Driver's education is recommended as well for anyone who is interested in a career as an EMT. The ability to drive safely and sensibly in all different types of road conditions, and a firm knowledge of traffic laws is essential to the driver of an ambulance. English is a desirable subject for the potential EMT, since it is important to have good communication skills, both written and verbal, along with the capacity to read and interpret well. Finally, depending on what area of the country the potential EMT might work in, it might be very helpful to have a background in a foreign language, such as Spanish, to assist in dealing with patients who speak little or no English.

### Postsecondary Training

For the high school graduate with a strong interest in the Emergency Medical Services, the next step is formal training. The standard training course was designed by the Department of Transportation and is often offered by police, fire, and health departments. It may also be offered by hospitals or as a nondegree course in colleges, particularly community colleges.

The program teaches EMTs-to-be how to deal with many common emergencies. The student will learn how to deal with bleeding, cardiac arrest, childbirth, broken bones, and choking. He or she will also become familiar with the specialized equipment used in many emergency situations, like backboards, stretchers, fracture kits, splints, and oxygen systems.

If you live in an area that offers several different courses, it might be a good idea to research all the options, since certain courses may emphasize different aspects of the job. After completing the basic course, there are training sessions available to teach more specialized skills, such as removing trapped victims, driving an ambulance, or dispatching.

EMTs who have graduated from the basic program may later decide to work toward reaching a higher level of training in the EMS field. For example, the EMT-Intermediate course provides thirty-five to fifty-five hours of further instruction to allow the EMT to give more extensive treatment, and the EMT-Paramedic course offers an additional 750 to 2,000 hours of education and experience.

## certification or licensing

After the training program has been successfully completed, the graduate has the opportunity to work toward becoming certified or registered with the National Registry of Emergency Medical Technicians (NREMT). (*See* "Sources of Additional Information.") All states have some sort of certification requirement of their own, but many of them accept registration in NREMT, in place of their own certification. Applicants should check the specific regulations and requirements for their state.

Whether registering with the NREMT or being certified through a state program, the applicant for an EMT-Basic title will be required to take and pass a written examination, as well as a practical demonstration of skills. The written segment will usually be a multiple-choice exam of roughly 150 questions. After passing the exam, EMTs usually work on a basic support vehicle for six months before certification is awarded.

While it is not always required in every state that EMTs become certified or registered, certification, at least, is a common requirement for employment in most states. It certainly is recommended, as it will open up many more job possibilities. The higher the level of training an EMT has reached, the more valuable he or she will become as an employee.

EMTs who are registered at the Basic level may choose to work on fulfilling the requirements for an EMT-Intermediate certification. After the mandatory thirty-five to fifty-five hours of classroom training, as well as further clinical and field experience, another examination must be taken and passed. The EMT who wants to earn Paramedic status must already be registered as at least an EMT-Basic. He or she must complete a training program that lasts approximately nine months, as well as hospital and field internships, pass a written and practical examination, and work as a Paramedic for six months before receiving actual certification.

All EMTs must renew their registration every one to two years, depending upon the state's requirement. In order to do so, they must be working at that time as an EMT, and meet the continuing education requirement, which is usually twenty to twenty-five hours of lecture and practical skills training.

## labor unions

Some EMTs may have the opportunity to join a union when they become employed. The unionization of this industry has been fairly recent and is not yet widespread. Therefore, membership in a union will not likely be a requirement for employment, and may not even be offered as an option. However, the EMT unions are growing rapidly, particularly in the public sector.

One of the principal unions of EMTs and Paramedics is the International Association of EMTs and Paramedics, which is a division of the National Association of Government Employees, AFL-CIO. The other is the National Emergency Employee Organization Network.

EMTs with membership in a union pay weekly or monthly dues, and receive in return a package of services designed to improve working conditions, which include collective bargaining for pay and benefits, governmental lobbying, and legal representation.

## who will hire me?

Kelly was lucky when she started to look for a job as an EMT. "I just went to the hospital and filled out their application," she says. "They called me in for an interview and I got the job. I didn't even have to apply anywhere else." Not everyone will be lucky enough to get a job on the first try, but currently the statistics are in the favor of EMTs. The demand exceeds the number of persons trained to do the work.

Roughly two-fifths of EMTs work in private ambulance services. An estimated third are in municipal fire, police or rescue departments, and a quarter more work in hospitals or medical centers. Also, there are many who volunteer, particularly in more rural areas, where there often are no paid EMTs at all.

Because new graduates will be in heavy competition for full-time employment, it may be easier to break into the field on a part-time or volunteer basis . By beginning as a volunteer, or part-timer, the new EMT can gain hours of valuable experience, which can be useful in landing

FYI

**EMTs on TV**

Not every job receives the media attention that the professions in the emergency medical services field do. The American television viewer can currently watch ten to twelve full hours each week of either fictional shows based on hospital emergency departments or actual emergency personnel in action. The only occupation that seems to fascinate the public more is law enforcement, with police shows totalling more than twenty-five hours of air time weekly.

a paid, full-time position later. The competition is also stiffer for beginning EMTs in the public sector, such as police and fire departments. Beginners may have more success in finding a position in a private ambulance company. There are also some opportunities for work that lie somewhat off the more beaten path. For example, many industrial plants have EMTs in their safety departments, and security companies sometimes prefer to hire EMTs for their staff. Most amusement parks and other public attractions employ EMTs in their first aid stations, and in many cities there are private companies who hire EMTs to provide medical coverage for rock concerts, fairs, sporting events, and other gatherings.

One good source of employment leads for an EMT graduate is the school or agency that provided his or her training. Job openings may sometimes be listed in the newspaper classifieds under "Emergency Medical Technician," "EMT," "Emergency Medical Services," "Ambulance Technician," "Rescue Squad," or "Health Care." It may be worthwhile for students nearing the end of their training course to subscribe to a local paper. Also, many professional journals, and national and state EMS newsletters list openings. Finally, the National Association of Emergency Medical Technicians (NAEMT) prints each new member's resume in their monthly newsletter, upon request, so that prospective employers may see it. (*See* "How Do I Learn More?")

It is also a good idea for the graduate to apply directly to any local ambulance services, hospitals, fire departments, and police departments. The best approach is usually to send a current resume, complete with references, and a letter of inquiry. The letter should consist of a brief description of the applicant's situation and interests, and a request for an application. Most agencies have specific applications and employment procedures, so the resume and cover letter alone is not necessarily adequate. It is important to remember that most employers will accept applications to keep on file even if there is no specific job open at the time.

## where can I go from here?

Kelly decided that she wanted to pursue more advanced levels of training shortly after she started working as an EMT. At present, she has already passed the paramedic training course and is getting ready to test for her certification. Eventually, she says, she might like to get into EMS education and train EMTs to do the job she does now.

Moving into the field of education and training is only one of several possible career options. For an EMT who is interested in advancement, usually the first move is to become certified as a paramedic. Once at that level, there are further opportunities in the area of administration. Moving into an administrative position usually means leaving fieldwork and patient care for a more routine office job. An EMT-Paramedic can pursue such positions as supervisor, operations manager, administrative director, or executive director of emergency services. Or, like Kelly, he or she may be interested in a career in education and training. Also, several new areas of specialization in EMS have recently received more emphasis. Quality control, safety and risk management, communications, and flight operations are some examples of these up-and-coming administrative areas.

Some EMTs move out of health care entirely and into sales and marketing of emergency medical equipment. Often, their experience and familiarity with the field make them effective and valuable salespersons. Finally, some EMTs decide to go back to school and become registered nurses, physicians, or other types of health workers. Kelly has seen EMTs take many different paths. "One girl who was a paramedic when I started at the hospital is a doctor now," she says, "so you can see how far you can go with it."

## what are some related jobs?

The job of EMT is similar to many other positions in the medical field in that there are very specific procedures and treatments that EMTs are trained and authorized to perform. The U.S. Department of Labor classifies Emergency Medical Technicians under the headings *Occupations in Medicine and Health, Not Elsewhere Classified* (DOT) and *Patient Care* (GOE). Also under this heading are people who work under the supervision of a doctor or registered nurse to assist in medical treatment in a variety of areas. The areas of specialization that these medical assistants, or aids, could work in include orthopedics, psychiatry, optometry, dentistry, occupational therapy, physical therapy, podiatry, and surgery.

| Related Jobs |
|---|
| First-aid attendants |
| Ambulance attendants |
| Birth attendants |
| Chiropractor's assistants |
| Dental assistants |
| Nurse's aides |
| Licensed practical nurses |
| Occupational therapy aides |
| Orderlies |
| Orthopedic assistants |
| Physical therapy aides |
| Podiatric assistants |
| Psychiatric aides |
| Surgical technicians |
| Optometric assistants |

**Advancement Possibilities**

**Training directors** plan and oversee continuing education for rescue personnel, design and implement quality assurance programs, and develop and direct specialized training courses or sessions for rescue personnel.

**Emergency medical services coordinators** direct medical emergency service programs, coordinate emergency staff, maintain records, develop and participate in training programs for rescue personnel, cooperate with community organizations to promote knowledge of and provide training in first aid, and work with emergency services in other areas to coordinate activities and area plans.

**Physician assistants** work under the direction and responsibility of physicians to provide health care to patients: they examine patients; perform or order diagnostic tests; give necessary injections, immunizations, suturing, and wound care; develop patient management plans; and counsel patients in following prescribed medical

## what are the salary ranges?

The salaries for emergency medical technicians have shown an increase in recent years of approximately 4 percent annually. Since the demand for qualified EMTs exceeds the supply at present, salaries will most likely continue to increase.

According to a 1997 salary survey published by the *Journal of Emergency Medical Services*, overall salary for an EMT-Basic is approximately $24,670 a year. An EMT just starting out in the field can expect to make $21,200 on the average and may eventually earn a salary in the mid-twenties while still at Basic status. An average starting salary for an EMT-Intermediate is $23,250, while the average starting salary for an

EMT-Paramedic is $25,200. The average top salaries for EMT-Intermediates and EMT-Paramedics are $30,265 and $34,240, respectively.

The most significant factor in determining salary is whether the EMT is employed in the private or the public sector. Private ambulance companies and hospitals traditionally offer the lowest pay, while fire departments pay better. Geographical location is another significant influence. The average salary for an EMT-Basic varies from $22,050 in the Southeast to $30,800 in the Northwest. And the percentage of employers offering life and medical insurance has declined over the last few years. Now, about 82 percent of employers offer life insurance, and around 87 percent offer major medical insurance. Other benefits often offered are uniform allowance, retirement or pension plans, and paid seminars and conferences.

## what is the job outlook?

Employment opportunities for EMTs are expected to grow faster than average for all occupations through the year 2006. One of the reasons for the overall growth is simply that the population is growing, thus producing the need for more medical personnel. Another factor is that the proportion of elderly people, who are the biggest users of emergency medical services, is growing in many communities. Finally, many jobs will become available because, as noted earlier, the EMT profession does have a high turnover rate.

The job opportunities for the individual EMT will depend partly upon the community in which he or she wishes to work. In the larger, more metropolitan areas, where the majority of paid EMT positions are found, the opportunities will probably be best. In many smaller communities, financial difficulties are causing many not-for-profit hospitals and municipal police, fire, and rescue squads to cut back on staff. Because of this, there are likely to be fewer job possibilities in the public sector. However, since many of the organizations suffering cutbacks opt to contract with a private ambulance company for service to their community, opportunities with these private companies may increase.

The trend toward private ambulance companies, which have historically paid less, is an important one, since it is likely to influence where the jobs will be found, as well as what the average pay is. One reason for the growth of the private ambulance industry is a health care reform concept called managed care. As health care costs increase, Americans are more and more often leaving their private health care plans for "pool" systems, which provide health care for large groups. The reason this affects the EMT profession is that medical transportation is one of the major services typically contracted for by these pool systems, or managed care providers. As managed care gains popularity, there is a greater need for the private ambulance contractors.

Because of America's growing concern with health care costs, the person considering a career as an EMT should be aware of the fact that health care reforms may affect all medical professions to some extent. Also, as mentioned before, the increase and growth of private ambulance services will almost definitely change the face of the emergency services field. In looking at a future as an emergency medical technician, both of these factors are worth keeping in mind.

## how do I learn more?

### sources of additional information

Following are organizations that provide information on EMT careers, accredited schools, and possible sources of employment.

**American Medical Association**
Division of Allied Health Education and Accreditation
515 North State Street
Chicago, IL 60610
312-464-4696

**International Association of EMTs and Paramedics**
159 Burgin Parkway
Quincy, MA 02169
617-376-0285

**National Association of Emergency Medical Technicians**
408 Monroe Street
Clinton, MS 39056-4210
601-924-7744

**National Registry of Emergency Medical Technicians**
Box 29233
6610 Busch Boulevard
Columbus, OH 43229
614-888-4484

# bibliography

Following is a sampling of materials relating to the professional concerns and development of emergency medical technicians.

## Periodicals

*JEMS*.Carries classified advertising; publishes salary surveys and a yearly careers supplement that may include a directory of employers. Mosby-Yearbook, Inc., 11830 Westline Industrial Drive, St. Louis, MO 63146-3318, 314-453-4351.

*Rescue*. Carries classified advertising and job listings. JEMS Publishing, Inc., PO Box 2789, Carlsbad, CA 92018, 619-431-9797.

# energy conservation technician

**Definition**

Energy conservation technicians measure the energy used to heat, cool, and operate buildings, including homes, businesses, and industrial facilities. They conduct audits to determine energy efficiency and recommend corrective steps. They install corrective measures, or repair and modify heating, ventilating, cooling, and electrical systems and equipment.

**Alternative job titles**

Building maintenance technicians

Environmental technicians

Weatherization technicians

**Salary range**

$16,000 to $22,000 to $39,000

**Educational requirements**

High school diploma; three to twenty-four months specialized training

**Certification or licensing**

Recommended

**Outlook**

About as fast as the average

GOE
05.07.01

DOT
959

O*NET
24505E

## High School Subjects

Computer science
English (writing/literature)
Mathematics
Physics
Shop (Trade/Vo-tech education)
Speech

## Personal Interests

Building things
The Environment
Fixing things
Science

**John Maggs** is almost finished with his first home energy audit of the day. He checked the customer's furnace with a draft gauge and tested carbon monoxide levels. He completed a blower-door test to detect where air is leaking in from outdoors, and took indoor and outdoor pressures which he will later enter into a computer. He has checked the temperatures at which the hot water heater is set. Now he is discussing the results of his audit with the homeowner and recommending repairs—some weather-stripping, some new storm windows, and some adjustments to the furnace. Before he leaves, he will give her some brochures describing energy-saving measures she can do herself. Once the repairs are made, she will see her energy bills, including gas and electricity, drop anywhere from 13 to 30 percent.

# what does an energy conservation technician do?

*Energy conservation technicians* work in a wide variety of settings. They may inspect homes, businesses, and industrial buildings to identify conditions that cause energy waste; recommend ways to reduce the waste; and help install reduction measures. They may be employed in power plants, research laboratories, construction firms, industrial facilities, government agencies, or companies that sell and service equipment.

There are four general areas of energy activity: research and development, production, use, and conservation. *Research and development technicians* usually work in laboratories testing mechanical, electrical, chemical, pneumatic, hydraulic, thermal, or optical scientific principles. Such laboratories are found in institutions, private industry, government, and the military.

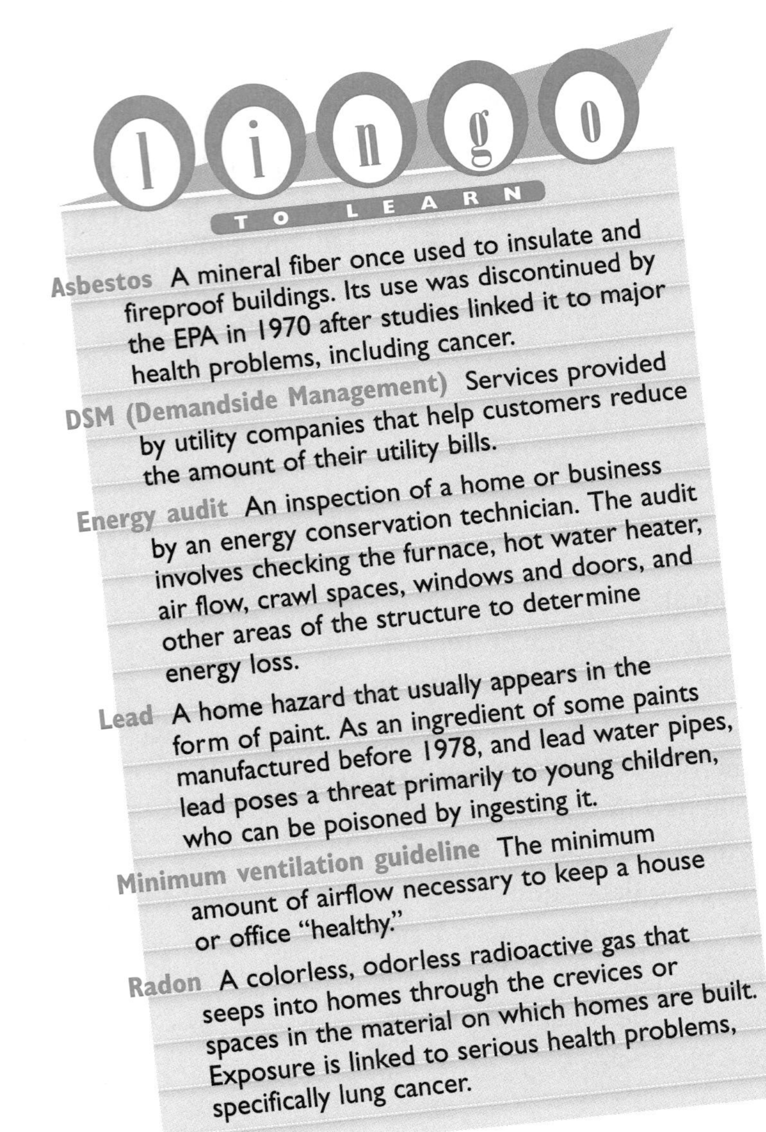


Technicians in laboratories assist engineers, physicists, or chemists with experiments. They help record data and analyze it using computers, and they may also be responsible for maintaining and repairing equipment.

Companies that are involved with energy production include manufacturers, installers, and users of solar energy equipment; power plants; and process plants that use high-temperature heat, steam, or hot water. In these settings, energy conservation technicians work with engineers and managers to develop, install, operate, maintain, modify, and repair systems and devices that convert fuels or other resources into useful energy.

In the field of energy use, energy technicians might work in an industrial facility to improve efficiency in engineering and production-line equipment. They also maintain equipment and buildings for hospitals, schools, and multifamily housing.

In the area of energy conservation, manufacturing companies, consulting engineers, energy-audit firms, and energy-audit departments of public utility companies use energy technicians to determine building specifications, modify equipment and structures, audit energy use and efficiency of machines and systems, then recommend modifications to save energy. These types of energy technicians work for municipal governments; hotels; architects; private builders; and heating, ventilating, and air-conditioning equipment manufacturers.

More and more utility companies are providing services, called demandside management (DSM) programs, that help customers reduce the amount of their electric bill. DSM programs use energy conservation technicians to visit customers' homes and interview them about household energy use, such as type of heating system, number of persons home during the day, furnace temperature setting, and prior heating costs.

Whether a technician works for a utility, a nonprofit agency, or a private company, the energy audit process is similar. In addition to the client interview, technicians draw a sketch of the house, measure its perimeter, windows, and doors, and record dimensions on the sketch. They inspect attics, crawl spaces, and basements and note any loose-fitting windows, uninsulated pipes, and deficient insulation. Technicians read hot-water tank labels to find the heat-loss rating and determine the need for a tank insulation blanket. They examine furnace air filters and heat exchangers to detect dirt in filters and soot buildup in exchangers that might affect furnace operation.

Energy conservation technicians then discuss problems with the home owner and recommend actions to reduce energy waste, such as weatherstripping, reducing hot water temperature setting, adding insulation, and installing storm windows. They may also suggest new lighting systems, such as compact fluorescent lamps, fixtures that reflect more light onto work surfaces, electronic ballasts, or sensors that detect daylight and occupancy, and control lighting accordingly.

Technicians give customers literature describing energy conservation improvements and sources of loans. They record the results of their audit on a computer and use software programs to calculate heat-loss estimates. A printout of the data is usually mailed to the customer. Some companies conduct one-visit residential programs, in which the technician conducts an energy audit, installs energy-saving measures, such as low-flow showerheads, insulation wrapped around a water heater, caulking, and weatherstripping, and then inspects the installation. Some companies have technicians that only conduct audits, while others employ one team of technicians to do audits and another team to do installations.

## what is it like to be an energy conservation technician?

John Maggs is an energy auditor for Step, Inc., a large nonprofit agency that does weatherization for low-income housing. He starts out the week by planning and scheduling audits, which he conducts Monday through Wednesday. "In one day, my round trip is up to 100 to 125 miles around the ten counties our agency covers. I do a complete audit on each home, including a furnace test and a blower door test, and I audit two or three homes a day. I return to the office and start writing up estimates on the computer." John spends Thursday and Friday writing up and finalizing the week's activities, so he can hand in reports. "I always leave a half day for any questions or problems that might come up," he says, "and there's always something."

The U.S. Departments of Energy, and Health and Human Services fund Weatherization Assistance Programs in many states. Training courses are offered for individuals employed by local governments and nonprofit agencies, like Step, Inc., that administer home energy conservation assistance to low-income clientele.

"In one day, my round trip is up to 100 to 125 miles around the ten counties our agency covers."

"I do a preassessment on a house and furnace, and identify areas where we can save energy. I am also qualified to do asbestos and lead tests. I give the client information, including pamphlets, brochures, and advice on how to conserve energy. I communicate the problems to other technicians who do the repairs and then I inspect their work."

John uses an electronic furnace analyzer, which gives carbon monoxide levels, and a draft gauge, which shows how the furnace is drafting. He also uses a blower door, which depressurizes the house and tells where the leaks are. A smoke stick helps him locate chaseways and leaky spots. Indoor and outdoor pressures are entered into a computer, which gives a leakage rate in cubic feet per minute. "Since a home needs a certain amount of air moving through it to make it safe, we also set up a minimum ventilation guideline," John explains. "We take into account the number of smokers, pets, space heaters, and fireplaces and adjust the MVG accordingly." After weatherization measures are installed, he takes a post-weatherization CFM measurement. "That's how we justify our expenses and see how well we've done our audits."

John mentions some equipment used in residential audits. Energy conservation technicians may use other technologically complex machinery that may contain electric motors, heaters, lamps, electronic controls, mechanical drives and linkages, thermal systems, lubricants, optical systems, microwave systems, pneumatic and hydraulic drives, pneumatic controls, and in some instances, even radioactive samples and counters.

Rhode Islanders Saving Energy has seven technicians who serve about one thousand home owners. They use state-of-the-art equipment, such as infrared scans, polaroid equipment, and Bacharach testing equipment. All technicians use computers and portable printers on site.

Great Lakes Control Energy Corporation in Elk Grove Village, Illinois, specializes in building automation and control systems, energy management systems, lighting control systems, and energy audits for commercial properties.

The company employs energy technicians who travel to sites coast to coast. Some are engineers who do surveys and audits, and some are installers.

Bonnie Esposito of the Center for Energy and Environment in Minneapolis, Minnesota, says, "Technicians range along a whole continuum of skills. Half of our technical staff do research and testing, and also sometimes do audits. Some are engineers who don't do audits, but some do audits on commercial properties. I also have technicians who are full-time auditors. We use different technicians for each project, depending on the service we need to provide for that particular job."

Although many energy technicians have normal daytime work schedules, they may also have to work in shifts, especially in energy-production or energy-use plants. Many technicians have to travel to customer locations and work indoors and out. Sites are often dirty and noisy, and technicians must pay attention to safety measures, including protective clothing and equipment.

## have I got what it takes to be an energy conservation technician?

Energy conservation technicians must have an interest in machines—how they are made, how the parts fit and work together, and how they operate. But what's the most important skill for a technician? "Communication," says John Maggs. "You have to be able to talk to people of all kinds. I can talk to anybody of any race or creed. Our clients are mainly low-income, but since I've become more specialized in asbestos and lead testing, I also deal with higher-income people. It doesn't matter. We've been very well trained, are knowledgeable, and can speak confidently about weatherization."

John Clune, of Eastern Utilities Association Citizens Conservation, has also done residential audits on multifamily units. He says, "You move around a lot and meet a lot of people in residential work. You need good personal skills. Some homeowners are going to be drunk or nasty. But some will treat you really well. You may see three or four customers per day, all unique. In addition to mechanical skills, you must have good verbal and communication skills to relate to clients."

David Fenton, manager of program marketing for Rhode Islanders Saving Energy, agrees, "Communication skills are definitely important in dealing with the public—equally as important as knowledge of the industry."

Problem solving is one of the most challenging aspects of the work energy conservation technicians do. They often have to respond to urgent calls and crisis situations involving unexpected breakdowns of equipment. They have to interrupt other duties to fix machinery that is not working properly.

Maggs sometimes finds it cumbersome to comply with the state guidelines that regulate government agencies. "Politics often get in the way. Sometimes you have shut down a job or just say it's beyond our scope, instead of attempting to do something to make it better."

To be a successful energy conservation technician, you should

- Have excellent communication skills
- Be people-oriented
- Have mechanical aptitude
- Pay attention to detail

## how do I become an energy conservation technician?

Energy conservation technicians must understand the physical sciences and have mathematical aptitude. They must be able to understand operation and maintenance manuals, schematic drawings, blueprints, and formulas. They must be able to communicate with clients, and also make themselves clear in scientific and engineering terms.

Energy technicians have a wide range of educational backgrounds, from high school to a four-year engineering degree. At least two years of formal technical education beyond high school is recommended.

# in-depth

**Weatherization Programs Try to Establish Standards**

Currently there is no national standard for weatherization training. Leaders from the weatherization/energy retrofit community, however, are working to develop professional-skills-development standards to improve the quality of work provided by energy-efficiency professionals. Some of the areas they are addressing include energy efficiency, health and safety, affordability, comfort, durability, environmental concerns, competency-based graduated skills designation, on-site verification of competency, and assessment of field performance by certified practitioners with periodic recertification.

Though some weatherization officials would like to see national or international standards, it is more likely that the process will proceed on a regional or statewide level. New York State has already begun to work on a standardization process, beginning with the New York State Weatherization Program under the direction of Director Rick Gerardi. The program will gradually expand to include neighboring states, utility employees, and private contractors. Work is also underway in Washington State with possible regional cooperation in the future, and Iowa may be a center for a possible Midwest organization.

## education

### High School

It is recommended that high school students take at least one year of algebra and geometry, physics, chemistry, and ecology, including laboratory work. Computer science and mechanical or architectural drafting are practical and necessary for those who want to pursue an engineering degree.

Students should look for opportunities to participate in special projects on solar equipment, energy audits, or energy-efficient equipment.

John Maggs says, "Public speaking, communication skills, and machine shop are probably important in high school. But it's hard to say. I dropped out of high school and later joined a youth program. I was lucky because I happened to get placed in a summer job program with weatherization when I was seventeen. I've been in weatherization now for eleven years. I did some other carpentry work prior to coming to the youth corps, including laying tile and carpet. I took a year to do some roofing and construction work, but then came back to the weatherization program." Although the weatherization program was good for John, he recommends students finish high school and get some technical training.

### Postsecondary Training

Bill Van der Meer, director of Pennsylvania's Weatherization Training Center, says, "Energy technicians often come up through the ranks without college education. There is a wide variety of positions, from crew-technician level through scientist positions to selling energy-efficiency products."

"We train people under a federal grant to work for nonprofit organizations that provide weatherization for low-income housing. We teach a variety of skills, including carpentry, installation of insulation, air sealing to reduce infiltration, and installation of storm windows. We teach them to use diagnostic equipment and tools, such as blower doors. They must understand the house as a system and know potential interactions that go on."

Training is both theoretical and practical hands-on instruction. After students fulfill competency requirements and pass examinations and laboratory exercises, they are awarded certificates. Courses may include weatherization tactics, diagnostics, residential heating systems, combustion analysis and retrofit, home-energy auditing, blower-door tests, duct-leakage tests, and installation of energy conservation materials.

Some regions have youth corps, like the one Maggs joined. The South Bronx in New York City has a Youth Energy Corps, a weatherization agency that does air sealing, replaces windows, and blows high-density cellulose insulation. The workers are all high school dropouts who alternate one week of work with one week of school for one year. Youth Energy Corps teaches them weatherization skills and helps them earn a General Equivalency Diploma (GED), get a job, or both.

Maggs speaks highly of state-funded weatherization training programs. "Weatherization training centers are funded very well. Trainers go to all the conferences and are on top of new technologies." Though the training is good, he suggests students get the education and start their own businesses, since funding for nonprofit agencies can be unstable.

Although many employers do not require energy technicians to have a degree, further education definitely expands career options. A degree from a technical institute, community college, or other specialized two-year program, can lead to a higher-paying position with a private firm. A technical degree includes courses in physics, chemistry, mathematics, the fundamentals of energy technology, energy-production systems, fundamentals of electricity and electromechanical devices, heating systems, air conditioning, and microcomputer operations.

Additional courses might include mechanical and fluid systems; electrical power and illumination systems; electronic devices; blueprint reading; energy conservation; codes and regulations; technical communications; instrumentation and controls; and energy economics and audits.

Many technical colleges, community colleges, and technical institutes provide two-year programs specifically in energy conservation and energy-use technology, or energy-management technology. There are related programs in solar power, electric power, building maintenance, equipment maintenance, and general engineering technology. Graduates with associate's degrees in energy conservation and use, instrumentation, electronics, or electromechanical technology will normally enter the workforce at a higher level, with more responsibility and higher pay. Jobs in energy research and development almost always require an associate's degree.

Dick Walsh of Great Lakes Control Energy Corporation says, "We don't require that anyone be a graduate of a trade school or any kind of an apprenticeship program. But our employees have a great deal of experience. We have a journeyman pipe fitter who has an in-depth understanding of HVAC and mechanical equipment, so it's easy for him to do audits. Electrical installers have learned enough on the job about mechanical equipment to do audits as well, and we also have auditors with four-year engineering degrees who have studied mechanical or electrical engineering."

"Some of our installers are field laborers who pull wire and bend conduit, but we also have an ongoing training program on the crew, so if workers are capable of learning and progressing, they can."

David Fenton prefers that technicians have a background in building or home construction, any type of heating including home heating and solar heat, plumbing, or insulation.

Bonnie Esposito says, "Field experience is more important than technical training. I can teach employees calculations, computer skills, and the latest technologies with electronic ballasts, but to have real field experience to deal with owners, contractors, vendors, to know buildings, to know businesses—that's invaluable."

Bonnie believes the future trend will be to hire more graduates with four-year degrees. "There will always be lower-level technical jobs," she says, "but with a four-year degree you have a lot more options."

## certification or licensing

Certification is not required in every state, but many states approve certain training programs and award certificates to students who complete classes and tests in residential, commercial, or multifamily audits and installations. A certificate from the National Institute for Certification in Engineering Technologies or a degree or certificate of graduation from an accredited technical program, are highly regarded credentials, though not required by most employers (*see* "Sources of Additional Information"). Some employers require technicians to undergo security clearances and psychological tests.

## internships and volunteerships

Part-time or summer work is available to technicians in large hospitals, major office buildings, hotel chains, universities, and large manufacturing plants.

Summer youth programs are offered in some regions and communities. John Maggs worked on a crew for about three months and then got placed in a weatherization program. There are many opportunities for on-the-job training in energy conservation.

### Advancement Possibilities

**Heating plant superintendents** supervise workers who produce and distribute steam heat and hot water in commercial or industrial facilities, and maintain mechanical equipment.

**Home energy consultants** supervise and coordinate activities of technicians who conduct audits for customers of public utilities.

**Customer service representatives,** or **field representatives,** for utilities investigate customer complaints concerning gas leakage or low pressure.

## labor unions

Jobs offered to high school graduates in electrical power production are represented by employee unions such as the International Brotherhood of Electrical Workers or the International Association of Machinists and Aerospace Workers. Workers often start as apprentices with a planned program of work and study for three to four years.

Union membership is not usually required for government-agency workers, research and development technicians, and conservation technicians.

## who will hire me?

Institutions, private industry, government, and the military employ research and development technicians. *Energy-production technicians* are employed by solar-energy equipment manufacturers, installers, and users; power plants; and process plants that use high-temperature heat, steam, or hot water.

*Energy-use technicians* work in hospitals, schools, and apartments. Energy conservation technicians are hired by manufacturing companies, consulting engineers, energy-audit firms, and energy-audit departments of public utility companies. Technicians are also hired by municipal governments, hotels, architects, private builders, and heating, ventilating, and air-conditioning equipment manufacturers.

Technicians who were trained or employed in the military can find opportunities in building or equipment maintenance. Former U.S. Navy technicians are particularly sought after in the field of energy production.

Power companies need technicians to perform energy audits. Technicians in these firms assist energy-use auditors and energy-application analysts while they make audits or analyses in businesses or homes of their customers.

Industrial facilities hire energy technicians to manage machinery and equipment, including its planning, selection, installation, modification, and maintenance.

There are also sales and customer-service opportunities with manufacturers of energy-efficient or energy conservation equipment.

John Clune says, "Nonprofit organizations have different requirements than profit companies, and there is a wide spectrum of opportunities available for technicians, trained and untrained. But more and more companies are hiring people with engineering degrees. I got involved in the mid-1970s after the energy crisis. I am not an engineer, but because of my twenty years of experience, I now have about the same knowledge as a licensed professional engineer."

## where can I go from here?

The opportunities for advancement in government-funded, nonprofit agencies are few. John Maggs is considering starting his own business. "Getting the necessary training to go into the private sector will be one of the booming things in the next five to ten years," he predicts.

Technicians with experience and money to invest may start their own businesses, selling energy-saving products or providing audits, weatherizations, or energy-efficient renovations.

Technicians in private companies can move into supervisory and management positions. Some become trainers or add training responsibilities to their job duties. Hotels, restaurants, and stores hire experienced technicians to manage energy consumption and then provide training in energy-saving procedures.

Experienced technicians can also move into sales or customer service as representatives for producers of power, energy, special control systems, or equipment designed to improve energy efficiency.

## what are some related jobs?

Energy technicians have the skills and experience to work in construction, remodeling of existing buildings, or retrofitting. The U.S. Department of Labor classifies energy conservation technicians under the headings *Occupations in Production and Distribution of Utilities, Not Elsewhere Classified* (DOT) and *Engineering, Quality Control: Structural* (GOE). Also listed under these headings are people who inspect buildings, railroads, and bridges, and the electronic and mechanical equipment that runs them.

| Related Jobs |
|---|
| Building equipment inspectors |
| Gas meter checkers |
| Fire equipment inspectors |
| Locomotive inspectors |
| Oil pipe inspectors |
| Shop estimators |
| Bridge inspectors |

## what are the salary ranges?

As with other technical jobs, salaries vary with formal training and experience. High school graduates with little or no experience who begin as trainees will earn $15,400 to $19,800 a year in the private sector. Those who have several years of experience can earn $35,000 to $40,000 a year. Technicians with engineering degrees who travel across the country can earn up to $50,000. Technicians with military experience and three to six years of technical experience earn from $19,500 to $33,000 annually. Graduates of postsecondary technician programs earn the same.

Energy technicians with five to ten years of experience in research, engineering design, or in machinery engineering and maintenance can earn up to $39,000 annually.

Government and nonprofit jobs pay less than those in the private sector. Beginning technicians in weatherization programs earn a little more than $5 an hour, or about $12,000 a year. Experienced technicians in nonprofit agencies earn $12 an hour or about $25,000 a year.

Most companies offer technicians paid vacations, group insurance benefits, and employee retirement plans. In addition, many firms offer support for continuing education.

## what is the job outlook?

The economy and prices of oil and other energy sources will affect the availability of jobs for energy technicians. Since energy use constitutes a major expense for industry, commerce, government, institutions, and private citizens, the demand for technicians trained in energy conservation is likely to remain strong. In addition, the United States is still dependent on foreign fuel supplies of petroleum, so energy conservation will continue to be an issue for all energy consumers.

Utility companies are moving toward deregulation, opening new avenues for energy-service companies. Utilities have begun to work with manufacturers and government agencies to establish energy-efficiency standards, for example to produce energy-efficient appliances. The Consortium for Energy Efficiency in 1992, a collaborative effort involving a group of electric and gas utilities, government energy agencies and environmental groups, is developing programs for commercial air-conditioning equipment, clothes washers, lighting, geothermal heat pumps, and industrial motor systems. These efforts will create job opportunities for technicians.

Utility DSM programs, which have traditionally concentrated on the residential sector, are now focusing more attention on commercial and industrial facilities. They are expanding to work with contractors, builders, retailers, distributors, and manufacturers with the goal of realizing larger energy savings, lower costs, and more permanent energy-efficient changes.

Another factor affecting the field of energy conservation is the growing awareness of environmental and global warming issues. The financial costs of purchasing natural resources will always be a major concern, but now we have the added reality of the physical costs of depleting those resources.

The next decade will bring substantial change in energy careers. "The field is really changing," says Bonnie Esposito. "I have laid off over thirty people in the last year. The company is expanding, but we're hiring only engineers now. We are doing a lot more research and smaller projects that are much more highly specialized than in the past. We used to go out with field data forms, and now my auditors all take laptops out in the field."

# how do I learn more?

## sources of additional information

Following are organizations that provide information on energy conservation technician careers, accredited schools, and employers.

**American Society of Heating, Refrigerating, & Air Conditioning Engineers, Inc.**
1791 Tullie Circle, NE
Atlanta, GA 30329-2305
404-636-8400

**Energy Efficiency and Renewable Energy Clearinghouse**
PO Box 3048
Merrifield, VA 22116
800-363-3732

**National Institute for Certification in Engineering Technologies**
1420 King Street
Alexandria, VA 22314-2794
703-684-2835

## bibliography

Following is a sampling of materials relating to the professional concerns and development of energy conservation technicians.

### Books

O'Callaghan, Paul W. *Energy Management.* New York: McGraw-Hill Book Co., 1993.

### Periodicals

*Energy Conservation Digest.* Biweekly. Covers all aspects of the energy conservation industry. Editorial Resources, Inc., PO Box 21133, Washington, DC 20009, 202-783-2929.

*Energy Conservation News.* Monthly. Focuses on energy conservation in industry. Business Communications Co., 25 Van Zant Street, Suite13, Norwalk, CT 06855, 203-853-4266.

**An Energy Conservation Timeline**

Coal and petroleum were abundant and relatively inexpensive early in the twentieth century. Petroleum eventually displaced coal because it was inexpensive and easily produced heat, steam, electricity, and gasoline. The United States became dependent on foreign oil for half its energy supply.

The field of energy conservation began during the energy crisis of 1973-74 when many foreign oil-producing nations stopped shipments of oil to the United States and other Western nations. Fuel costs suddenly soared. Although oil prices are more moderate today, they are still very volatile. There is always the threat of oil-price increases or a cutoff of supplies.

Bill Van der Meer, director of Pennsylvania's Weatherization Training Center, says "Energy conservation is not as hot an issue now as it was in the late 1970s. If fuel prices increased, there would be renewed interest in energy efficiency, but the focus has changed to environmental issues, such as burning fossil fuels and its effect on the environment."

"We are running out of fossil fuels," comments Dick Walsh of Great Lakes Control Energy Corp. "People have to conserve whether they like it or not. Natural gas is more expensive. Electricity is more expensive and will continue to get more expensive. It's incumbent upon facility owners and managers to conserve any way they can with building automation control systems and more efficient mechanical equipment. It's not a luxury, it's a necessity."

*Energy Today.* Monthly. Trends Publishing, Inc., 1079 National Press Building, Washington, DC 20045, 202-393-0031.

*Home Energy.* Bimonthly. Discusses all aspects of residential energy conservation. Energy Auditor & Retrofitter, Inc., 2124 Kittredge Street, Suite 95, Berkeley, CA 94704, 510-524-5405.

*Industrial Energy Conservation.* Monthly. Assesses worldwide industrial energy conservation. National Technical Information Service U.S., U.S. Dept. of Energy, 5825 Port Royal Road, Springfield, VA 22161, 703-487-4630.

*Journal of Energy and Development.* Semiannual. Deals with all aspects of energy, including conservation, management, finance, etc. International Research Center for Energy and Economic Development, 909 14th Street, Suite 201, Boulder, CO 80302, 303-492-7667.

# fire protection engineering technician

**Definition**

Fire protection engineering technicians design and draft plans for the installation of fire protection systems. They also estimate the cost and oversee the installation of their fire protection plans.

**Alternative job titles**

Fire alarm superintendents
Fire inspectors
Fire insurance adjusters
Fire insurance inspectors
Fire safety technicians
Plant protection inspectors
Sprinkler layout technicians

**Salary range**

$17,700 to $18,550 to $54,800

**Educational requirements**

High school diploma

**Certification or licensing**

Required by certain states

**Outlook**

Faster than the average

GOE
05.03.02

DOT
019

O*NET
22132B

## High School Subjects

Chemistry
Computer Science
Mathematics

## Personal Interests

Computers
Drawing
Figuring out how things work

a fire protection engineering technician in Tucson, Arizona, has a problem. A business has contacted him to develop a fire sprinkler system for its office's computer room. The computer room houses the company's large mainframe computers where many irreplaceable business files are stored. The company is concerned that the sprinklers might accidentally set off, flooding and spraying the room with water. While such accidents are rare, they can happen and would ruin not just the desks and lamps, but also the expensive mainframe computers, too.

Jim's dilemma is that he has to design a sprinkler system that eliminates the risk of an accidental activation, yet will perform if indeed there is a fire. As he bends over his drafting table, the solution comes to him. He'll place heat sensors on the sprinkler heads. With these sensors, the sprinklers will activate only when the temperature in the room is so hot that only a fire could generate such heat. And as a fail-safe, he'll install a valve in the water pipes, a buffer that will only feed the water to the sprinkler head once the heat sensors have gone off.

## what does a fire protection engineering technician do?

Each year, fire causes billions of dollars of damage to property and claims many thousands of lives. Helping to prevent fires and to reduce the destruction they cause is the chief concern of the *fire protection engineering technician.*

Fire protection engineering technicians help design fire alarm and sprinkler systems to protect commercial, residential, and industrial buildings from the hazards of fire. They are employed by local fire departments, fire insurance companies, government agencies, and businesses that design, fabricate, install, or sell fire protection devices and equipment.

The primary role of the fire protection engineering technician is to draft plans for the installation of fire protection systems for buildings and other structures. They must create plans that adhere to strict fire protection and building codes, will perform as designed in the event of a fire, and provide cost-effective solutions for the property owner.

Fire protection engineering technicians often take their cues from engineers and architects who develop specifications and plans and design standards. Once the design decisions have been determined, the technician creates working plans and layout drawings. The technician analyzes the blueprints and specifications to select the appropriate type and size of fire protection system.

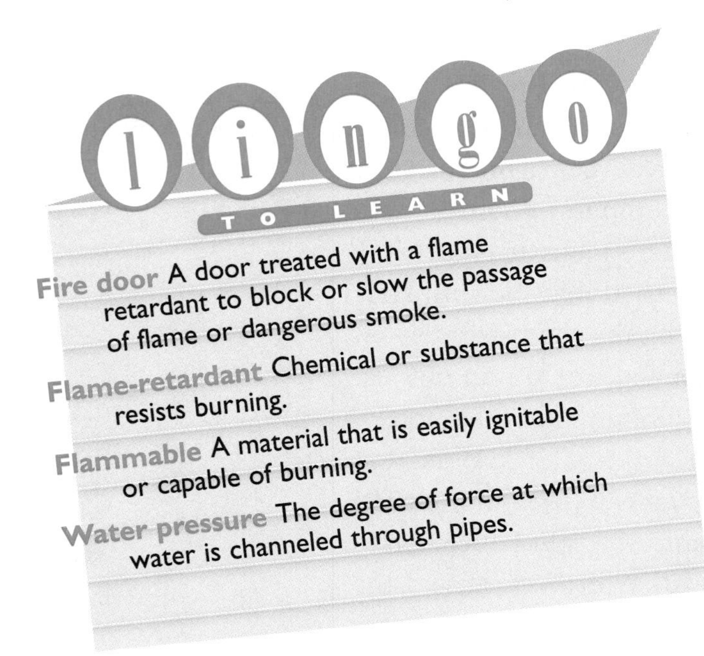


Fire protection engineering technicians also assist in the installation of fire protection systems to ensure that the working plans are being followed. It is not unusual for technicians to specialize in one kind of fire protection system, such as foam, water, dry chemicals, or gas.

Fire protection engineering technicians work in the larger field of fire prevention. Within the field, there are many specialists, scientists, and others who work together to minimize the risk of fire.

*Fire protection engineers* have completed a higher level of education than fire protection engineering technicians. They evaluate information to determine the extent of fire and explosion potential. Fire protection engineers may also conduct studies on fabric flammability, fire safety in high-rise buildings, or evaluate the damage caused to property by fire.

Insurance companies employ a number of fire protection professionals. *Fire insurance inspectors* work for insurance companies. They inspect buildings to determine the risk of fire and make recommendations for improving fire safety. *Fire insurance adjusters* investigate claims for loss due to fire, compute rates for adjustment, and settle the claims.

*Fire inspectors* check firefighting equipment and report any potential fire hazards. They recommend changes in equipment, practice, materials, or methods to reduce fire hazards.

*Plant protection inspectors* inspect industrial properties to assess the potential for fire and report their findings and make recommendations to eliminate any fire hazards.

*Fire alarm superintendents* inspect fire suppression systems and fire alarms in public, private, and government buildings.

*Sprinkler layout technicians* plan how sprinklers and piping should be installed in buildings. They work with blueprints and building plans designed by engineers and architects and must make sure their working plans conform to the standards and specifications indicated.

# what is it like to be a fire protection engineering technician?

To be a successful fire protection engineering technician, you should

- Have patience
- Pay attention to detail
- Have computer skills
- Be able to understand complex equations

Jim Tierney has worked as a fire protection engineering technician for the past fifteen years. During the last two of those years, he has been employed by Cutler Fire Protection, a company that designs, fabricates, and installs fire sprinkler systems in Tucson, Arizona.

Jim typically works from 8:00 AM to 4:00 PM, although he can work as many as fifty-five hours a week when the company has several different projects due at one time. "During the summer there's usually more construction going on," says Jim. "That means our business picks up a little, too."

For the most part, Jim spends his day indoors working at his drafting table or on a computer. Sometimes he goes to a site to inspect the installation of the fire sprinkler systems he designs, but he estimates that the majority of his day, about 80 percent, is spent in the office.

Jim's job begins when an architect sends him the blueprints for a house or office building or industrial warehouse. Often, the blueprints are for a structure that is scheduled for construction, but sometimes the structure is already built and in the process of renovation or upgrading. It's Jim's role to design a fire sprinkler system that will protect the building in the event of a fire.

With his knowledge of blueprints and construction techniques, Jim determines how much physical space he'll have in which to design and install a functioning fire sprinkler system. He must also determine which type of fire suppressant the sprinkler system will release in the event of a fire. While most sprinkler systems spray water to extinguish fires, water isn't always the most effective fire suppressant. If the structure being built were an oil storage tank, for example, Jim would design a system that sprayed foam or other chemical retardants. Or, as was the case with the design he created for the computer room, he must consider variables that might cause the sprinkler system to accidentally activate, causing hundreds of thousands of dollars in damage to the structure and its contents.

Once he's determined the type of sprinkler system that would best meet the needs of the blueprints, Jim must then calculate how much water pressure will be required to make the sprinklers work effectively. Jim uses a computer program to help calculate the thickness and amount of pipe that will be required in order to carry enough water from the city's water system, through the structure and to the sprinkler heads. He must find the most cost-effective way of installing the system, yet ensure that the system will be successful in putting out fires.

Not only must Jim understand how these variables will affect the final design of his sprinkler system, but he must also be aware of the state's fire regulations. These regulations dictate the number of sprinkler heads that must be present in a building, the range of feet or yards the sprinkler heads need to spray water, and the thickness of the pipe that carries the water from its source to the sprinkler heads.

After carefully considering all these factors, Jim sits down at his drafting table and draws the system on an overlay that will then be attached to the blueprint. The overlay serves as a guide for the contractor doing the building construction and installation of the sprinkler system.

Once he has completed drafting the sprinkler system, Jim sends the design to the state fire regulatory commission. The commission reviews Jim's calculations and design to ensure that they conform to the state's fire code. Once it has passed this inspection, the design is then returned to Jim, who purchases the sprinkler heads and have the pipe cut to the dimensions specified in his design. Finally, the materials and the draft are sent to the contractor who will install the sprinkler system.

## have I got what it takes to be a fire protection engineering technician?

Fire protection engineering technicians provide a service that can save lives and protect property in the event of a fire. Because their work is so important, fire protection engineering technicians must be competent in the skills that the career requires.

If you want to be a successful fire protection engineering technician, you should have basic drafting skills and be able to understand and grasp spatial relationships. Communication skills are important as well—if you can't convey your ideas or plans clearly or accurately, your fire protection plans aren't going to go far. You must be detail oriented, mechanically inclined, and computer savvy.

The hardest part of the job, says Jim, is learning the different and complex set of regulations that govern the type of materials that can be used in a fire sprinkler system. "It was very difficult learning all the codes when I was first starting out," he says.

Another challenge is being able to work with complex equations and calculations. The size and width of the pipe, the amount of pipe required, and the spray area of the sprinkler head are all variables that Jim must consider when designing a fire sprinkler system. While many of the calculations are performed using computer programs specifically designed for the field, much of the work must still be done by hand.

Jim likes that fact that every fire sprinkler system he designs is unique. That means there's always plenty of variety in the job and new challenges that help him keep his skills fresh.

## how do I become a fire protection engineering technician?

Jim's first job out of high school was a training position with a company that manufactured fire sprinklers. There, he further developed his mechanical drafting skills and found that he really enjoyed the challenges of the work.

## education

### High School

A high school diploma is mandatory for anyone considering a career as a fire protection engineering technician. High school courses can provide the basic skills that fire protection engineering technicians use each day on the job, such as science and mathematics.

High school courses in physics, mathematics, algebra, calculus, and geometry are important to learn the basic principles involved in making calculations and measurements. Courses in computer science and mechanical drafting are also recommended, since drafting is such an essential skill for fire protection engineering technicians, and most all of them work with computers. Additionally, classes in computer-aided design (CAD) will come in handy, since the use of CAD systems is growing in popularity.

Don't think, however, that you can forget about English and focus on math and science. Written and verbal communication skills are crucial for fire protection engineering technicians.

### Advancement Possibilities

**Fire protection engineers,** or **fire loss prevention engineers,** advise and assist private and public organizations and military services for purposes of safeguarding life and property against fire, explosion, and related hazards.

**Fire prevention research engineers** conduct research to determine causes and methods of preventing fires and prepare educational materials concerning fire prevention for insurance companies.

**Fire insurance claim examiners** analyze fire insurance claims to determine the extent of an insurance carrier's liability and settle claims with claimants in accordance with policy provisions.

Summer or part-time jobs with fire alarm installation companies or sprinkler system service companies will expose you to the daily rigors of fire protection engineering technicians. Insurance companies may also have job openings in their fire insurance divisions.

### Postsecondary Training

The most common way of becoming a fire protection engineering technician is to receive on-the-job training. As one's knowledge and skills develop, it is then possible to find better-paying and more interesting work with companies that design and install fire extinguishing systems.

According to Jim, when he reviews applicants for positions with his company, he is most interested in seeing how well developed the applicant's drafting skills are. "I pay attention to how clean the lines they draw are, and whether they use pen instead of a pencil. Working in pen is very important—it shows confidence in your abilities."

Some colleges and training schools offer two-year programs in fire protection engineering technology. In addition to learning skills in such courses as mechanical drafting and computer science, students may also find work-study opportunities that will make it easier to find employment after graduation.

The curriculum in an associate's degree program in fire protection engineering technology includes classes in industrial fire hazards, chemistry, codes and standards, water supply analysis, and fire protection design, just to name a few. Associate's degrees in fire science or mechanical engineering technology are also applicable in fire protection engineering technology.

If you aspire to become a fire protection engineer, you'll need a bachelor's degree from a four-year college. Check with your school counselor for additional information or contact the Society of Fire Protection Engineers or the National Fire Protection Association (*see* "Sources of Additional Information").

**FYI**

What do fire sprinklers and pianos have in common? In 1874, American Henry S. Parmalee invented sprinklers to protect his piano factory. While pianos remained safe, human beings sometimes did not. Sprinklers were used primarily to protect buildings, such as factories and warehouses, until the 1940s and 1950s, when some large fires took many lives. In Boston in 1942, a fire in the Coconut Grove Nightclub killed 492 individuals. In 1946, a fire in the LaSalle Hotel in Chicago left 61 people dead, and a fire in the Winecoff Hotel in Atlanta took the lives of 119 individuals. Since then, fire sprinkler systems have become as important for protecting lives as for safeguarding buildings and material wares.

## certification or licensing

Licensing requirements for fire protection engineering technicians vary from state to state. Certification is conducted by the National Institute for Certification in Engineering Technologies (NICET), a division of the National Society of Professional Engineers (NSPE). Fire protection engineering technicians must pass written examinations and accumulate five years of work experience or related education to become Certified Engineering Technicians (CET) in Fire Protection Engineering Technology. Subfields in automatic sprinkler system layout, special hazards systems, and fire alarm systems are also available for those who wish to specialize. A CET in automatic sprinkler system layout, for example, is qualified to independently prepare working plans and layouts from an engineer or architect's designs. While the CET certification is not always required, it can give you a competitive edge in the job market.

## scholarships or grants

Scholarship and grant opportunities are often available from colleges with fire science or fire protection engineering programs. Contact your advisor for further details. You may also find scholarships through professional organizations or large manufacturers of fire equipment.

## internships and volunteerships

Internships or cooperative work-study programs where students study part time and work part time may be available through college programs. Large insurance companies or manufacturers may also sponsor internships. It would be a good idea to check with your departmental advisor or the school's placement office.

If you are interested in the hands-on fire-fighting end of fire protection, it may be possible to join a group of volunteer fire fighters. This may be a good way to work your way into the field, gain some experience and firsthand knowledge of the cause and prevention of fires, and determine which specialization may be right for you.

## who will hire me?

The most common employer of fire protection engineering technicians are private companies that design and install fire extinguishing systems. These companies work as subcontractors, members of the construction team that build or renovate commercial, industrial, or residential property. The size of these companies varies—it may be a "one-man shop" or a large business that employs fifty or more fire protection engineering technicians. These companies may also specialize in fire sprinkler systems, fire alarm systems, restaurant systems, and more.

Fire protection engineering technicians may also work for private consulting firms or insurance agencies. At these companies, the role of the fire protection engineering technician is to assist in the evaluation of a structure's fire hazards and suggest steps that should be taken to correct them.

## where can I go from here?

Five years from now, Jim would still like to be working as a fire protection engineering technician. "I like the work," he says. He also likes the challenge of designing fire extinguishing systems for different types of structures. Some years ago, he designed the fire sprinkler system for the Biosphere, an experimental science station in southern Arizona where a group of researchers lived sealed inside a glass-and-steel-domed structure.

While Jim is happy with his current employer and the variety of projects he works on, many fire protection engineering technicians decide to explore other advancement opportunities. Some find employment with state or local fire regulatory agencies and inspect fire alarms and fire sprinkler systems in buildings.

With further education, some technicians choose to become fire protection engineers. Fire protection engineers have typically completed a four-year course of study in fire science or industrial engineering. Fire protection engineers are more deeply involved in identifying fire hazards and often conduct experiments to determine a material's flammability. They also strive to improve or develop new methods of fire suppression.

## what are some related jobs?

According to Jim, fire protection engineering technicians have skills that would allow them to work in several different careers.

With their knowledge of and experience with drafting and blueprints, fire protection engineering technicians could find employment with an architectural firm. Because of their knowledge of fire codes and regulations, many fire protection engineering technicians choose to work in the permit and regulatory offices in federal, state, and local governments. Finally, some fire protection engineering technicians choose to go to work for companies that manufacture fire sprinklers and suppression systems as members of their sales teams.

The U.S. Department of Labor classifies fire protection engineering technicians under the headings *Occupations in Architecture, Engineering, and Surveying, Not Elsewhere Classified* (DOT) and *Engineering Technology: Drafting.* Also listed under these headings are technicians who work with biomedical equipment, lasers, and calibration equipment, as well as a wide array of drafters and detailers.

| Related Jobs |
| --- |
| Aeronautical drafters |
| Aircraft layout workers |
| Architectural drafters |
| Automotive design drafters |
| Detailers |
| Heating and ventilating drafters |
| Mechanical equipment engineering assistants |
| Mosaicists |

Photogrammetrists
Plumbing drafters
Specification writers
Technical illustrators
Tool drawing checkers

## what are the salary ranges?

Salaries for fire protection engineering technicians are comparable to those of other engineering technicians. According to the 1998–99 *Occupational Outlook Handbook,* earnings of engineering technicians at the junior level ranged from $17,700 to $22,800 a year in 1995. Those with more experience had median earnings of $32,700, and engineering technicians in senior level positions made about $54,800.

According to the National Fire Sprinkler Association, Incorporated (NFSA), fire protection engineering technicians with CET certification in the fire sprinkler field earn $30,000 to $50,000 a year.

Fire protection engineering technicians generally receive full benefits packages from their employers.

## what is the job outlook?

The outlook for a career as a fire protection engineering technician should be good in the years ahead. Since the 1960s, the field of fire prevention and fire safety has been growing. Many businesses and industries have discovered that it is more cost efficient to invest in fire prevention systems, such as sprinklers and alarms, than it is to replace a building or property destroyed by fire.

To some extent, the amount of work available for fire protection engineering technicians will depend upon the construction industry. When construction is booming, the demand for the services that a fire protection engineering technician provides will also increase. But because older buildings also need fire suppression systems or have systems that need upgrading, fire protection engineering technicians should be able to find work throughout the year.

# how do I learn more?

## sources of additional information

The following organizations provide information on careers in fire protection engineering technology as well as scholarship and certification opportunities.

**National Fire Protection Association**
1 Batterymarch Park
Quincy, MA 02269
617-770-3000
http://www.nfpa.org

**National Fire Sprinkler Association**
Robin Hill Corporate Park, Route 22
PO Box 1000
Patterson, NY 12563
914-878-4200
http://www.nfsa.org

**National Institute for Certification in Engineering Technologies**
1420 King Street
Alexandria, VA 22314
888-IS-NICET
http://www.nicet.org

**Society of Fire Protection Engineers**
7315 Wisconsin Avenue, Suite 1225W
Bethesda, MD 20814
301-718-2910
http://www.wpi.edu/Academics/Depts/Fire/SFPE/sfpe.html

## bibliography

Following is a sampling of materials relating to the professional concerns and development of fire protection engineering technicians.

### Books

Cassidy, Kevin A. *Fire Safety and Loss Prevention.* Boston: Butterworth-Heinemann, 1992.

## Periodicals

*Fire & Flammability Bulletin*. Monthly. Focuses on the international fire scene. Rubber and Plastics Research Association of Great Britain, Interscience Communications, Ltd., 24 Quentin Road, Blackheath, London SE13 5DF, England.

*Fire News*. Bimonthly. Covers news regarding fire protection and prevention. National Fire Protection Association, One Batterymarch Park, Quincy, MA 02269, 617-770-3500.

*Fire Protection Contractor*. Monthly. Includes fire protection systems and services. H. B. Brumbeloe & Associates, Inc., 12972 Earhart Avenue, Suite 302, Auburn, CA 95602-9538, 916-823-0706.

*Fire Safety Journal*. Bimonthly. Elsevier Science, Regional Sales Office, Box 945, New York, NY 10159-0945, 212-633-3730.

*Firewatch*. Quarterly. Discusses fire protection engineering systems. National Association of Fire Equipment Distributors, One East Wacker Drive, Chicago, IL 60601, 312-923-8500.

*Journal of Fire Protection Engineering*. Quarterly. Society of Fire Protection Engineers, 1 Liberty Square, Boston, MA 02109-4825, 617-482-0686.

*Journal of Fire Sciences*. Bimonthly. Technical topics about combustion, flammability, retardance, etc. Technomic Publishing Co., Inc., 851 New Holland Avenue, PO Box 3535, Lancaster, PA 17604, 717-291-5609.

*Sprinkler Age*. Monthly. Covers general topics relating to sprinklers and people in the field. American Fire Sprinkler Association, 12959 Jupiter Road, Suite 142, Dallas, TX 75238, 214-349-5965.

*U.S. Fire Sprinkler Reporter*. Monthly. Covers laws, new products, building codes, etc. Wakeman/Walworth, 300 North Washington Street, Alexandria, VA 22314, 703-549-8606.

# fluid power mechanic

**Definition**

Fluid power mechanics manufacture, assemble, install, test, maintain, and service fluid power systems, which utilize the pressure of a liquid or gas to transmit or control power.

**Alternative job titles**

Fluid power technicians

Hydraulic technicians

Pneumatic technicians

**Salary range**

$18,500 to $33,250 to $48,000

**Educational requirements**

Some postsecondary training

**Certification or licensing**

Voluntary

**Outlook**

Much faster than the average

GOE 05.05.07

DOT 600

O*NET 95021

## High School Subjects

Computer science
Mathematics
Physics
Shop

## Personal Interests

Figuring out how things work
Fixing things

"Looks like a hydraulic fluid leak," Stan Durnal, a fluid power mechanic, says as he examines a huge orange and green combine in the repair shop of Claas. The combine, although new, had been left outside an equipment dealership through a harsh Iowa winter and was sent back to Claas's North American headquarters for repair. The greasy liquid that should be inside the hydraulic cylinder covers Stan's hands. "We'll fix it up, and it'll be good as new," he says, wiping his hands with a soft cloth.

The warehouse/repair room is filled with farm equipment of all shapes and sizes, most of which uses fluid power to move, lift, or pull. Even the forklift used to stock the warehouse shelves uses fluid power.

# what does a fluid power mechanic do?

From agriculture and construction to industry and recreation, fluid power is used to make life easier and industry more efficient. Anywhere there is moving equipment, fluid power is most likely part of that equipment and helps to push, pull, rotate, regulate, or drive.

There are two types of fluid power machines. Hydraulic machines use water, oil, or another liquid in a closed system to transmit the energy to do needed work. Pneumatic machines use air or another gas instead of a liquid. *Fluid power mechanics* may work with either or both of these types of machines. They operate, maintain, repair, and test fluid power equipment or components.

Fluid power mechanics who are responsible for testing fluid power systems typically work with a team or group that includes other mechanics, engineers, or scientists. The mechanic is responsible for setting up fluid power equipment as well as testing that equipment under operating conditions. This is accomplished by connecting the equipment to testing units that measure fluid pressure, flow rates, and power loss from friction or wear. Once the equipment is established and working, the mechanic gathers data for development and quality control and may recommend changes to the system.

Many times, fluid power mechanics are involved in the maintenance and repair of fluid power systems. They are the official troubleshooters, and when a machine breaks down or needs adjusting, the mechanic is called to detect and fix the problem. Knowledge of hydraulics, pneumatics, electric motors, control systems, and mechanical devices is necessary for the fluid power mechanic to properly diagnose and repair problems.

Fluid power mechanics often work in plants or factories where fluid power systems are used in manufacturing processes. For example, pneumatically controlled machines that might bolt together products on an automated assembly line may be in need of repair or service, a job for the fluid power mechanic.

Research and development teams also need the help of fluid power mechanics to find better ways to develop and use fluid power systems. Mechanics may help set up and test equipment under operating conditions.

Fluid power mechanics can also work in sales and services for companies that make and sell fluid power equipment to industrial plants. With their knowledge and background, mechanics are ideal sales and service representatives and can travel to plants, providing customers with specialized information and assistance with their equipment needs.

Fluid power mechanics analyze blueprints, drawings, and specifications, as well as communicate with the rest of the team both verbally and with written technical reports.

Fluid power mechanics use many types of equipment and machinery, including milling, shaping, grinding, and drilling machines to make precision parts for fluid power equipment. They also use sensitive measuring instruments to measure parts and hand and power tools to assemble the components in the fluid power system.

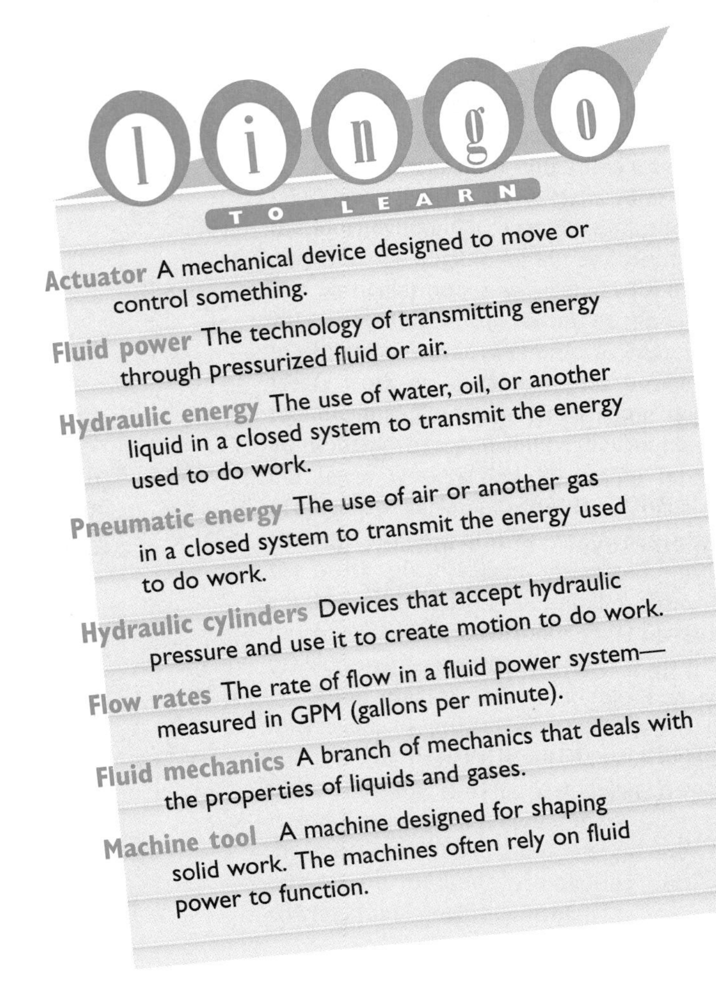

## what is it like to be a fluid power mechanic?

Stan is a fluid power mechanic and service manager at Claas's North American headquarters in Columbus, Indiana. Claas is the world's largest manufacturer of harvesting equipment, and, as is common throughout the agriculture industry, fluid power is used in almost every aspect of Claas's equipment.

Stan's primary duties as a fluid power mechanic involve the maintenance and repair of fluid power farm machinery. Components and machines from all over North America are sent to Stan's repair shop to be fixed and sent back to the owner or retailer or are sold as used farm equipment at the headquarters facility.

Stan must be familiar with many types of fluid power systems—mostly hydraulic—from the largest combine to the smallest handheld equipment.

"Fluid power is the thing that makes our machinery move," Stan says. "Our machines operate on a stop/go basis, which means if everything's okay then the machine is moving, and if one thing is wrong, it's not. When the machine's not moving, the operator is obviously not happy."

Keeping that operator (and customer) happy is Stan's job, so efficiently analyzing and fixing fluid system problems are essential. With many years of experience, Stan is familiar with the basic problems that arise in hydraulic farm equipment systems. "Bad fluid, a faulty seal, or a leak are mostly the causes of breakdown. Because the hydraulic cylinder moves so many times during operation, it is bound to be where problems occur," Stan says. "We basically do whatever we have to do to get the machine running. The good news is that fluid power systems are very fixable; finding the problem is usually the hard part."

Stan also coordinates on-site training at Claas. In the training room and on the repair shop floor, Stan must communicate with other employees, teaching them how to repair and analyze the machines.

Although Stan's work involves mostly repair, technicians and mechanics in other industries and facilities may work with manufacturing, testing, or developing fluid power equipment.

Stan's repair shop is very large and includes huge garage doors to accommodate the immense size of the agriculture equipment. The different work areas are often dotted with grease and other fluids that have leaked out during hydraulic repair.

> You can't be afraid to get your hands dirty.

Many fluid power mechanics work in industrial settings. Some mechanics work as *fluid power field mechanics*. Field mechanics travel to different customer locations to repair equipment for a manufacturer, and others may work in a laboratory or test facility to develop new ways to use fluid power.

Fluid power mechanics work a typical forty-hour workweek with many opportunities for overtime, which generally includes extra compensation. Some overtime may be required because of machine breakdown or an urgent situation and may involve weekend or holiday work.

Working on large machinery and components involves physical strength and may also require working in areas where safety regulations must be followed, such as the floor of a manufacturing facility or the inside of a large assembly machine.

## have I got what it takes to be a fluid power mechanic?

Fluid power mechanics must be good at understanding and analyzing mechanical systems. Because of the troubleshooting nature of the job, the ability to meet a challenge is also extremely important.

Roger Rigney, a former instructor of fluid power technology classes at General Motors, explains that mechanics must be good problem solvers. "When you are called to a machine, all the operator generally knows is that the machine quit. It is up to the mechanic to find out why, and sometimes there are not very many clues available."

The work of a fluid power technician is interesting, challenging, and not at all repetitious. "I really like the challenge of the job," Stan says. "You always have to be flexible and look at each repair individually." The ability to work well with others is important as well, since fluid power mechanics often work on teams.

## in-depth

Machinery that operates on fluid power has been used for thousands of years. Have you seen rotating paddle wheels with water flowing through? In Roman times, they were used to produce power necessary for milling.

Nowadays, however, fluid power systems are much more sophisticated. They are also more pervasive. You may not be aware of this, but fluid power plays a major role in our daily lives. Automatic door closers, bicycle pumps, brakes on your car, and spray guns all use fluid power systems to operate. Hydraulic-powered machines make plastic containers for milk, shampoo, and motor oil.

Fluid power systems are faster and less costly to operate than manual power, which means the cost savings can be passed on to the consumer.

"You can't be afraid to get your hands dirty," adds Stan, explaining that, for some people, the potential grimy nature of fluid power work might not be appealing. Also, the physical strength required to work on some of the larger machines and systems might be considered a drawback.

"I enjoy fixing things and finding out what's wrong with them," Stan says. "I think a natural curiosity and talent for understanding complicated systems are most important to this job."

## how do I become a fluid power mechanic?

Many older fluid power mechanics like Stan were hired by companies on the basis of some technical expertise and were then trained in fluid power technology. Currently, however, postsecondary training and certification are extremely beneficial if you want to become a fluid power mechanic.

**FYI**

Many great movies have fluid power controlled "actors." Every toe of King Kong, for example, was activated by a separate hydraulic cylinder, and the ferocious shark in *Jaws* was completely pneumatic.

## education

### High School

High school subjects that provide a good background for fluid power mechanics include computer science, mathematics, physics, shop, English, and drafting. With the movement toward integrating fluid power and electronics, an electronics class would also be helpful. Joining organizations where you can gain skills working on fluid power equipment or any type of machinery would be beneficial. Working on automobiles or farm machinery will give you experience working with your hands and learning how to solve mechanical problems.

### Postsecondary Training

Although a college diploma is not mandatory in order to become a fluid power mechanic, most employers prefer to hire those who have had at least two years of postsecondary training. Postsecondary training may be pursued at vocational or community colleges or may be provided by employers at their own training facilities or at outside educational facilities. Classes may include fluid power fundamentals and components, calculations (the application of math concepts), hydraulic components, accessories and circuits, electricity/electromechanical control, basic computers, systems analysis, and labs. There are only about thirty technical training programs that focus specifically on fluid power technology, but many provide related courses in mechanical or electronics technology, for example. These programs may lead to diplomas, certificates, or associate's degrees, depending on the school and the program.

Stan teaches in a company-sponsored training program that focuses on hands-on activities as well as classroom work. Taking up an entire wall of Stan's training room is a complete hydraulic system from a combine. This model can simulate real problems and allows trainees to actually see the system in progress. "We really try to present realistic problems so that the mechanics can learn in a nonthreatening classroom situation what may work when they're on the floor," Stan says.

To be a successful fluid power mechanic, you should

- Have excellent mechanical aptitude
- Have an inquisitive and analytical mind
- Be able to work well and communicate with others
- Enjoy challenges and troubleshooting problems

## certification or licensing

Certification for fluid power mechanics is not mandatory but is available through the Fluid Power Certification Board, a division of the Fluid Power Society (*see* "Sources of Additional Information").

Certification consists of an optional educational review session followed by both a written and practical examination. The Fluid Power Society offers certification for fluid power mechanics and also fluid power engineers, specialists, and technicians. Technologies included in the certification programs include hydraulics, pneumatics, electronic control, and vacuum. The Fluid Power Certification Board will soon offer certification programs for hydraulic specialists and pneumatic specialists. Certifications are valid for five years and then the mechanic must be recertified based on career achievement over those past five years.

Advantages to certification include providing the mechanic with a credential that could open career doors, offering the employer another way to measure the mechanic's value to the company, and a recognition of the mechanic's skills, efforts, and professional expertise.

## scholarships and grants

Information regarding scholarship opportunities may be available from professional associations. One source is the Fluid Power Education Foundation, a division of the National Fluid Power Association (*see* "Sources of Additional Information").

Training facilities or colleges may also offer scholarships and financial awards. Contact the financial aid offices or your advisor for additional details.

## internships and volunteerships

Internship opportunities may be available at large manufacturing plants or through colleges. Consult your advisor or career placement office for information and possible leads.

A few employers may offer on-the-job training programs, although most prefer to hire candidates with some previous experience or training. It may be to your benefit to contact large companies you are interested in working for and inquire about internship opportunities.

## labor unions

Many mechanics working in industry and manufacturing may belong to labor unions. For example, the United Auto Workers of America would be the union for any technician or mechanic working in the automotive field. Union membership and dues vary from industry to industry. There is no specific union for fluid power mechanics.

**Advancement Possibilities**

**Fluid power instructors** teach the basics of fluid power, including hydraulic and pneumatic systems.

**Field service representatives** advise, repair, and work with the equipment of a manufacturer at the customer's location.

**Sales/Marketing personnel** market and sell fluid power systems and components.

**Application engineers** design fluid power systems to accomplish required tasks.

**Fluid power consultants** work with different companies to analyze, design, or improve fluid power systems.

## who will hire me?

Fluid power is used in practically every branch of industry. Machine tools, cars, farm machinery, airplanes, boats, and kitchen appliances all use fluid power. It is one of the most versatile means of transmitting power and creating movement. This versatility also means that fluid power mechanics work in every facet of American industry, including aerospace, agriculture, construction, distribution, government, machine tools, manufacturing, marine, mining, packaging, plastics, and more.

Fluid power has always been used in farm equipment to provide the horsepower and speed needed for crop cultivation and harvesting. Energy is another area that utilizes fluid power in offshore oil drilling and other energy applications.

Mechanics in automation and industry help keep the integrated manufacturing systems working smoothly. In the area of construction, machines that dig, crush, level, and move earth and rock rely on fluid power equipment. In fact, the construction industry is one of the largest users of fluid power.

Stan started working at Claas because of a basic interest in agriculture that later developed into more specified training in fluid power. Most mechanics today will obtain their jobs through schools and placement offices.

Organizations such as the Fluid Power Society and the Fluid Power Educational Foundation have lists of their corporate members that can be used for job location (*see* "Sources of Additional Information").

## where can I go from here?

Stan has already realized a few of his advancement goals since he is now the service manager at Claas and also plays a part in training other mechanics. Another area that he has an interest in is field service. Very similar to the job he currently holds, field service representatives travel to different areas of the country to solve problems with Claas equipment that is not shipped to the headquarters facility. Fluid power mechanics may also wish to move into sales or marketing positions. Although sales jobs are more administrative than mechanical, the representative will be able to use his or her knowledge of fluid power systems to help customers make appropriate purchasing decisions.

Fluid power mechanics with additional education such as a bachelor's degree can advance into positions as application engineers or fluid power consultants. Application engineers apply engineering principles to design fluid power systems. Fluid power consultants are called upon by various companies to assist in analyzing, designing, or improving fluid power systems.

## what are some related jobs?

The U.S. Department of Labor classifies fluid power mechanics under the headings *Machinists and Related Occupations* (DOT) and *Craft Technology: Machining* (GOE). Also under these headings are people who support the aircraft-aerospace manufacturing industry, the automotive service industry, foundries, and people who work on electrical equipment, engines, and turbines.

| Related Jobs |
|---|
| Machinists and maintenance machinists |
| Tool grinders |
| Automotive machinists |
| Sample body builders |
| Saw makers |
| Steam and gas turbine assemblers |
| Metal patternmakers |
| Tool and die makers |
| Fixture makers |
| Tool and machine set-up operators |

## what are the salary ranges?

Because fluid power mechanics work in a variety of industries, it is difficult to pinpoint specifics regarding salaries.

Fluid power mechanics who work on mobile heavy equipment earned a median weekly income of $613 in 1996. The lowest 10 percent made less than $383 a week, and the top 10 percent earned more than $981 a week.

Those who work in the aircraft industry earned a median annual salary of $35,000 in 1996. The lowest 10 percent made less than $23,000, and the top 10 percent earned more than $48,000 a year. According to the Airline Employment Assistance Corps., machine tool operators in the aircraft manufacturing industry earned $10 to $16 an hour, assemblers and installers made $8 to $16 an hour, and electrical installers and technicians earned $8 to $25 an hour.

Information from the Blue Mountain Community College in Oregon indicates that fluid power mechanics in industrial maintenance technology start at an average hourly wage of $11. More experienced mechanics can earn $15.53 an hour. Spokane Community College in Washington indicates that entry-level wages for fluid power mechanics in the Puget Sound region of Washington start at $1,800 a month. Mechanics with at least three years of experience can earn between $2,400 and $2,700 a month.

## what is the job outlook?

Because of its use in so many different industries and businesses, the need for fluid power mechanics is growing rapidly and at a rate that currently exceeds the supply of trained mechanics.

"Last year, we had twenty-six students graduating from our program," Duane Olson says,"and there were over 180 employment opportunities." Duane indicates that this may be an excellent time to become a fluid power mechanic.

Fluid power mechanics possess specialized skills that can be applied to many industries, so the job outlook should be positive for fluid power mechanics especially those employed in the construction and aircraft industries.

## how do I learn more?

### sources of additional information

Following are organizations that provide information on fluid power mechanic careers, accredited schools and scholarships, and possible employers.

**Fluid Power Society**
2433 North Mayfair Road, Suite 111
Milwaukee, WI 53226
414-257-0910
http://www.ifps.org

**Fluid Power Education Foundation**
3333 North Mayfair Road, Suite 311
Milwaukee, WI 53222
414-778-3364
http://www.fpef.org

**Fluid Power Distributors Association**
PO Box 1420
Cherry Hill, NJ 08034
609-424-8998
http://www.fpda.org

**Institute for Fluitronics Education**
PO Box 106
Elm Grove, WI 53122
414-782-0410

**National Fluid Power Association**
3333 North Mayfair Road
Milwaukee, WI 53222
414-778-3344

# bibliography

Following is a sampling of materials relating to the professional concerns and development of fluid power mechanics.

## Periodicals

*Fluid Power Service Center.* Quarterly. Directed at hydraulic maintenance and repair facilities. Penton Publishing, 1100 Superior Avenue, Cleveland, OH 44114-2543, 216-696-7000.

*Hydraulics & Pneumatics.* Monthly. Provides coverage of the use of fluid power components and systems in industrial machinery and mobile, marine, farm, construction, and aerospace equipment. Penton Publishing, 1100 Superior Avenue, Cleveland, OH 44114-2543, 216-696-7000.

*National Fluid Power Association Reporter.* Bimonthly. Newsletter featuring information on hydraulic and pneumatic product standards. National Fluid Power Association, 3333 North Mayfair Road, Suite 311, Milwaukee, WI 53222, 414-778-3347.

# food technologist

**Definition**

Food technologists use knowledge of the physical, microbiological, and chemical makeup of food to develop new or improved food products; devise new or improved food production, packaging, storage, and distribution technology; and test foods for compliance with industry and government specifications and regulations.

**Alternative job title**

Food scientists

**Salary range**

$28,000 to $40,500 to $60,000+

**Educational requirements**

Bachelor's degree

**Certification or licensing**

Voluntary

**Outlook**

About as fast as the average

GOE 02.02.04

DOT 041

O*NET 24505B

## High School Subjects

Chemistry
Biology
English (writing/literature)
Family and Consumer Science
Foreign language
Mathematics
Physics

## Personal Interests

Cooking
Figuring out how things work
Plants/gardening
Science

**What happens** to food after farmers grow it and before it hits the shelves at your neighborhood store? Who develops new brands of soup or cake mix, or things like freeze-dried coffee, powdered soup, and breakfast cereal? Who discovers new ways to use things like soybeans or tofu?

And who helps make sure things like fruit and vegetables don't spoil before they get to the store? Who determines the shelf life of a can of soup? Who helps develop the chemicals that control things like mold or rancidity? Who's responsible for all those labels on food? Finally, who develops the packaging that keeps food fresh, preserves it for as long as possible, helps it hold up during shipment, and makes it attractive to consumers? Most of us take the food we eat for granted, rarely thinking about the complexities of its production. But the fact is, behind almost every food we buy is a history of research, development, testing, refining, and packaging. This is the world of the food technologist.

## what does a food technologist do?

*Food technologists* help to ensure the safety and nutrition of processed foods. They develop new products, production methods, preserving techniques, and food distribution methods. They also create new packaging to improve the shelf life and preserve the nutritional value or other qualities of food. In addition, they develop food standards, nutrition labeling, safety and sanitary regulations, and waste management and water supply specifications.

Americans' constant appetite for new foods is part of what drives the profession of food technologist. But another important influence is advancing scientific knowledge. Food technologists help to apply advances in science and engineering to all phases of food development and production, improving the quality of the food we eat. This includes everything from developing ways to retain more nutrients in product processing to formulating low-fat and low-salt products, for example.

As the Institute of Food Technologists (IFT), a major trade group, points out, food technologists also come up with ways to feed an exploding population. For example, they explore ways to use inexpensive food sources like soybeans, cereal grains, and fish meal to feed hungry people in developing countries.

A food technologist generally specializes in one area, such as basic research, quality control, product development, processing, packaging, labeling, technical sales, market research, production, technical management, or standards, laws, and safety. Other areas include technical writing, teaching, or consulting. A food technologist also may focus on a specific category of food, such as cereal grains, meat and poultry, fats and oils, seafood, animal foods, beverages, dairy products, flavors, sugars and starches, stabilizers, preservatives, colors, and nutritional additives. In addition, there are many different types of companies to work for—food processing, agricultural science, chemical and biochemical, canning/preserving, bakery and confectionery, dairy, pasta, and so forth.

Increasingly, this is an industry of scientists—even in positions formerly held by people with degrees in business or marketing. "More and more of the industry is being run by food scientists, people in research and development, rather than business people," says IFT's Dean Duxbury. "I used to work for a big company years ago and the salespeople weren't trained in science. If there was a question or problem, they'd bring an R&D (research and development) person in. That's changed now. At American Maize, for example, all of the sales reps are degreed scientists."

Duxbury also observes that fast R&D and marketing is the rule today. "In the past, individual processing companies had top secret R&D; they did it all themselves," he says. "Now, with cutbacks and other economic concerns, processors and suppliers are working together to develop the formulas; it's a joint effort. It's much quicker and cheaper to do it this way, and much more efficient. When I was in the field, development time for a new product might be five years. Now it might be five months."

Conditions of work vary depending on the specific position and industry. You might work out in the plant overseeing food production, and it might be bustling and noisy. On the other

**Biotechnology** As applied to food production, the science of plant breeding, gene splicing, microbial fermentation, plant cell tissue culturing and other activities to produce enhanced raw products for processing.

**Emulsify** To turn into an emulsion, or liquid.

**FDA** Acronym for the Food and Drug Administration, a government agency that regulates food safety, nutrition, and labeling for all foods except meat and poultry.

**Pasteurize** To sterilize a food or beverage (cheese, milk, wine) by exposing it to high temperatures that destroy microbes or germs.

**Rancidity** A condition that some foods reach when the fat reacts with oxygen and breaks down into component parts like hydroxy acids, causing the "stale" smell and taste of spoiled food.

**Preservative** A chemical substance or preparation used to slow down the decomposition of food.

**Stabilizer** A substance added to food during processing to prevent or arrest unwanted chemical changes in the food.

**Test kitchen** A facility for testing new food formulations.

hand, you may work in a quality control or research lab that is clean, quiet, and temperature-controlled.

> You can be a good technician, but if you're missing the art of it, you're not going to produce good food.

## what is it like to be a food technologist?

Food technologist Bob Benson of Vanee Foods in Berkeley, Illinois, agrees that the length of time to develop a new product has dropped dramatically. "Maybe some of the big companies can still take two or three years," he says. "But here, it's more like three to four months."

Unlike many U.S. food producers, Vanee Foods is a relatively small, family-owned company of about 120 workers. Here, #10-sized (large) cans of sauces, gravies, beef stew, chili, corn beef hash, and other foods are produced, along with paste-like concentrated soup bases. Vanee is a customizer of food, which means it works with the customer to develop specific formulas to meet the customer's needs. Although it sells some food directly to end-users, Vanee primarily sells to other food producers, such as Kraft and A&W. "Almost all of what we produce is private-label," Bob says.

As Vanee's director of research and development, Bob has his fingers in a lot of pies. "It's a small company, so we all wear many hats," he notes. "But my primary job here is to come up with, develop, and modify new food ideas." Bob spends most of his time in Vanee's research and development lab, but if there's a problem on the production line, he'll go out to the plant floor to troubleshoot a machine or otherwise help out.

"I started out here as a one-man department, but now I have two assistants doing R&D (research and development) with me," Bob says. Other food technicians and technologists who work at Vanee include about ten to fifteen quality assurance (QA) and quality control (QC) people.

Vanee's R&D lab is a simple, efficient one with a scale, a stove, and various instruments for checking things like salt, pH, and other factors. Here, says Bob, "we may come up with forty new-product formulations a year, although only ten may come to fruition." On a typical day, "I'll look over various formula modifications, maybe get the marketing people in on it, and decide what changes should be made.

"People sometimes lose sight of it, but food is basically nothing but chemicals," Bob says. "In development, we're working with an understanding of what's going on at the chemical level. So when you're combining foods, for example, you must understand the chemical reactions that can take place—what will happen when these two foods are put together. There's also an art to it; you have to be a bit of a chef: do the flavors interact well together? Does it taste and smell good? You can be a good technician, but if you're missing the art of it, you're not going to produce good food.

"Say we're making a new cream sauce," he says. "Just like a cook, you'll get your flour, starch, and other ingredients together, all of them with the right properties. Sometimes you'll have an ingredient modified by the supplier, such as a preservative to retard spoilage. Then you test to see which combinations provide the cream sauce you want. When the proteins and the sugars in the starches and flours combine, for example, you don't want the sauce to turn brown; it's got to stay a light color. Here's where chemistry comes in: you have to understand how the products interact in order to avoid that brown color.

"A lot of this is trial and error," he acknowledges. "In time, you might know something won't work from past experience. But a lot of it is simply trying something, checking it, and then modifying it and trying again."

Throughout both the R&D and production phases, Vanee Foods tests for quality, safety, shelf life, and numerous other factors. "Probably the single biggest test is to make sure it looks the same as it did before," Bob says. "There's a great need for consistency. That's because ultimately the consumer is not going to be doing the sophisticated tests we do; he or she just wants it to taste and look the same as before."

Vanee Foods' products primarily are thermoprocessed, and part of Bob's job is to make sure the food is properly sterilized for staying on shelves for extended periods of time. The testing process includes detecting the temperature of the thermal lining inside the can. The USDA and FDA require food producers to run such tests, and Bob has developed something of a side specialty in it.

"Also, when a new product is finished, part of my responsibility is to help develop the labeling, which the quality assurance manager then submits to the USDA and FDA," Bob says. He also works closely with customers and may be involved with meetings or presentations at customer sites.

## have I got what it takes to be a food technologist?

"First, a food technologist is usually someone with an interest in science—chemistry, biology, the life sciences," Duxbury says. "You really need that in this field; it's really all science. Second, the person usually has an interest in the preparation and the formulation of food—they like cooking food and especially coming up with their own recipes."

Bob agrees with this assessment. "First and foremost, you have to have a scientific approach to things—an inquisitive mind, the desire to know how things work," he says. "And almost always, people who get into this field have had some interest in food, at least at some point in their life. For example, the quality control manager here grew up on a farm; I'm from a small town, and growing up we had a garden where we raised fresh vegetables. I still have a vegetable garden today. I don't think of myself as a chef but I do like to cook, grow fresh foods, and so forth. There are exceptions to this—some people at the very technical end may not have that interest—but most of us do."

Others have an interest in the government end of things, says Duxbury, such as regulation, inspection, and so forth. Duxbury himself got into the field while in the army, when for a time he considered being a meat inspector.

The IFT says that good traits for a food technologist are similar to those for any type of scientific and research environment: curious, analytical, detail-oriented, tactful, self-confident. Food technologists also need to be able to communicate ideas to others, verbally or in writing. Good business and marketing sense can help those who seek growth in the field; after all the R&D, companies need to sell the food, too.

To be a successful food technologist, you should

- Like food and cooking
- Be innovative and inquisitive
- Enjoy trying to recreate precise chemical reactions
- Excel in science, especially chemistry
- Be methodical, detail-oriented, observant, and patient

## how do I become a food technologist?

"I'm not sure exactly how people decide to join the industry, but they find it," Duxbury says. "One of the major reasons why is opportunity—people will switch into it from other degree programs or fields because of all the opportunity. After all, food is the world's biggest industry. And it's a responsible position."

Bob is a good example of someone who happened onto the field. "I started out as a chemistry major in college," he says. "Sophomore year, I was having some problems in one chemistry course, putting in forty hours a week in the lab. I went to my adviser and said, 'I like chemistry, but not the dull, dry aspects of it.' He said, 'Have you ever considered applying chemistry or microbiology to food? Imagine applying science to ketchup. . .' I went to interview the head of the food science department and switched over. I'd never heard of food science before that; I've always been grateful that adviser steered me in this direction."

### education

The educational requirements for being a food technologist are fairly stringent. Generally, you need at least a bachelor of science (BS) degree, usually in food science, although some food technologists have some other type of science degree. The quality control manager at Vanee Foods, for example, has a chemistry degree. In addition, according to IFT, about half of all food scientists also have an advanced degree.

Those who are interested in the field but aren't sure they want to earn a four-year degree can enter as a food technician, which generally requires only a high school diploma. "The usual breakdown is that the technologist has supervisory and decision-making responsibilities, and the technician doesn't," Duxbury says.

#### High School

Those considering a career in food technology should take a wide variety of science courses while in high school, especially chemistry. "Chemistry is extremely important," Bob says.

## in-depth

### What Do Food Companies Think of Food Labels?

Before the Nutrition Labeling and Education Act, food producers were disclosing the nutritional value of their products in a wide variety of formats, if at all. They could put them in any order. They could leave things out. Or they could just not say how much protein, fat, carbohydrates, or other components you were eating. Now there are very strict, very uniform guidelines for nutrition labeling. (You've probably noticed that all the labels look exactly the same.) How do food companies feel about having to conform to this law? The National Food Processors Association, which has five hundred member companies and among other things helps to represent the food processing industry in Washington, DC, worked with Congress in the development of the Act. "The food industry was very supportive of it," says Tim Willard, a spokesperson for NFPA. "We made comments and suggestions on it."

Will the act provide more opportunities for food technicians and technologists? "Absolutely," he says. "It was a big project to get existing foods labeled. And now, for every new food coming on the market, there's a need for full nutrition analysis. People with that kind of technical knowledge will certainly be in demand."

"It provides the broad base for understanding everything that goes on with food.

"As a food technologist, you've also got to be able to communicate well, so English classes are very important," he adds. "You must be able to write reports and communicate technical information to your peers. Also, you'll need to be able to say it in a nontechnical way" for communicating with marketing people or customers who aren't scientists themselves.

The Institute of Food Technologists' recommended high school courses include English, biology, physics, chemistry, social science, math and—partly because of the expanding opportunities for food technologists in the global market—a foreign language.

### Postsecondary Training

As mentioned above, food technologists earn a four-year degree in food science or a related area such as chemistry, microbiology, or even engineering. "There's actually a food engineering degree now that allows you to focus your studies in that area," Bob says.

According to IFT, undergraduates should take physics, biology, chemistry, nutrition, microbiology, mathematics and statistics, food chemistry and analysis, food microbiology, and food processing operations and engineering. A list of fifty colleges offering training in food technology is available from IFT (*see* "Sources of Additional Information").

With a bachelor's degree in science, technologists may work in research and development, quality control and assurance, technical sales, marketing, or other positions.

Those who go on to earn an advanced degree increase their opportunities and may become managers, supervisors, company executives, researchers, and professors. Common advanced degrees in the field include MS, Ph.D., and MBA degrees.

In a recent IFT survey, most respondents named food science/technology (41 percent) as the field of their highest degree, followed by chemistry (11 percent) and business/marketing (9 percent).

Bob attends workshops to keep up with new developments in product development and production technology, and he is a member of IFT and the Institute for Thermoprocessing Specialists. He also reads trade magazines like *Food Engineering, Food Technology,* and *Food Processing,* to keep up with ongoing change in the field (*see* "Bibliography"). "There's always something new to learn," he says. "New regulations, new technologies. Just when you think you've got the rules down, they change."

## scholarships and grants

Contact individual colleges and universities for school-specific loans and scholarships. Nationally, a good source for undergraduate and graduate- level scholarships is IFT's Scholarship/ Fellowship Program, available to students who attend a university with a food science/technology department or program approved by IFT. Currently, there are more than fifty such IFT-approved schools, including several in Canada and Mexico. The 1998-99 program offers twenty-five freshman scholarships, twenty-four sophomore scholarships, sixty-four junior/senior scholarships, and thirty-four graduate fellowships. Some of the scholarships are sponsored by big food companies, such as Coca-Cola, Frito-Lay, Nabisco, and Gerber. A brochure about the Scholarship/Fellowship Program is available from IFT (*see* "Sources of Additional Information").

## who will hire me?

Bob's first job was with a now-defunct company called Continental Coffee, where he worked in research and development and had three promotions during his tenure. He first learned of Vanee Foods, his present employer, when he came to the Vanee plant to do a test for Continental Coffee.

Food technologists are part of the food processing industry, which is one of the largest employers in the country. The IFT says about 7 5 percent of all U.S. food scientists work for private companies. The rest work for federal government agencies like the FDA, USDA, Department of Health and Human Services, Defense Department, State Department, and Commerce Department; state and local government agencies; and international agencies like the World Health Organization of the United Nations. Those who work for the government may become involved with such things as developing and testing standards and regulations for producing, labeling, and distributing food; inspection; or research.

As noted earlier, food technologists typically specialize in one area. For example, they may focus on research or applied research; quality control or production; processing or engineering; sales and marketing; or management. They may concentrate on one kind of food or beverage, like meat and poultry, or sugars and starches, or additives; or develop a specialty like nutritional labeling or sensory testing. They may be employed by suppliers or processors. Teaching, consulting, and writing also are possibilities.

"There are a lot of 'floaters,' or independent consultants," says Duxbury. "They're usually former researchers, hired because of their expertise in a specific area. They may be hired by the hour, day, week, project, or other basis. They can help food companies meet their goal of faster R&D."

In private industry, employers can include:

**Processing plants**. This is where foods and beverages are produced—converted from raw agricultural and other materials into the finished consumer goods we buy at the store. There are different kinds of processing plants for different foods, such as those for cereal; chips and other snack foods; meat, fish, or game; pop and other beverages; fruits and vegetables; and so on.

**Food ingredients plants.** Have you checked out the ingredient listing of the typical store-bought food lately? Ingredients plants are where the various things added to foods (or that you can buy to put on or in them) are made—vitamins, preservatives, salt and pepper, spices, flavorings, and more.

**Food manufacturing plants.** This is where new foods are created out of unexpected raw materials. Cheese made from soy products, tofu pizza, nondairy creamers, and other foods are manufactured in these types of plants.

FYI

**Basic versus Applied Food Research**

Food technologists in basic research study food's structure and composition by observing changes that take place during storage or processing. What they learn enables them to develop new sources of proteins, determine the effects of processing on microorganisms, or isolate factors that affect the flavor, appearance, or texture of foods. Technologists in applied R&D have the task of creating new food products and developing new food methods. They also continue to work with existing foods to make them more nutritious or flavorful, or to improve their color or texture.

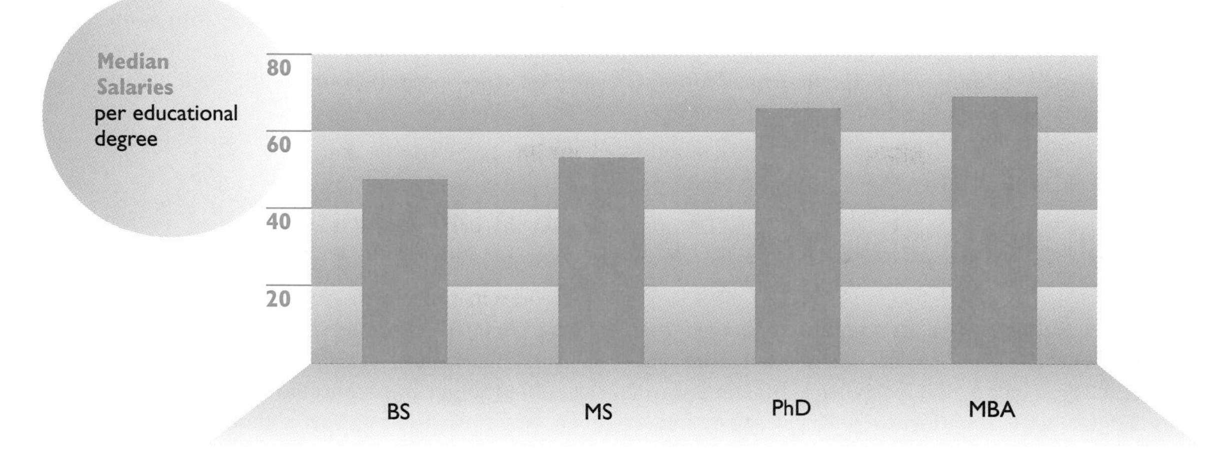


After many years of being "the industry that people stumble into," the food science industry now is actively recruiting young people to join their ranks. Although white males still dominate the field, the percentage of women is increasing. In a recent IFT employment and salary survey, about 68 percent of the respondents aged twenty to twenty-nine were women, compared with 13 percent in the group aged fifty to fifty-nine. Positions in the industry are well suited or can be adapted for people with disabilities.

## where can I go from here?

Steps up for the food technologist may include heading up a department, such as quality control or production; leading R&D teams; getting into product management; moving into regulatory work; or becoming involved with the sales end of the business. Increasingly, food sales and management positions are held down by those with education and training in food science.

Food technology is an area in which it definitely pays to receive a master's or doctorate degree (see salary section), and about half of the people in the industry do so. Food chemistry and technology are quite sophisticated, and there is a need for advanced knowledge to help propel new research and development, manufacturing techniques, and distribution methods. Corporate management/administration can be high paying, in the $70,000 to $80,000 range and up.

There are good opportunities for those who have developed expertise in a specialized area. Specialists include people like sensory evaluation experts, who measure consumers' response to the look, smell, and taste of food. Nutrition labeling is another specialty that currently is in high demand. Another possibility is to become a food marketing specialist, who might help a U.S. food company understand and penetrate international markets. Consulting is another option, and gives the food technologist the opportunity to apply his or her knowledge to problem solving for a wide range of companies.

## what are some related jobs?

The U.S. Department of Labor classifies food scientists under the heading *Occupations in Biological Sciences* (DOT) and *Life Sciences: Food Research* (GOE). Occupations in these categories typically involve applying knowledge of science to everyday human and animal needs. Also under this heading are toxicologists, who study the effect of toxic substances on humans, animals, and plants; environmental epidemiologists, who are involved with studying cases of disease in industrial settings and the effect of industrial chemicals on human health; and public health microbiologists, who check for the presence of harmful bacteria in water, food, and the general environment. Biologists,

| Related Jobs |
| --- |
| Biochemists |
| Biophysicists |
| Environmental epidemiologists |
| Microbiologists |
| Mycologists |
| Pharmacologists |
| Physiologists |
| Plant breeders |
| Plant pathologists |
| Public health microbiologist |
| Toxicologists |
| Zoologists |

biochemists, entomologists, geneticists, plant breeders, pharmacologists, and other occupations also are under this heading. Related areas include agronomy, horticulture, animal science, food engineering, food law, and packaging engineering.

## what are the salary ranges?

A 1995 survey by IFT of its members showed that those with a BS degree earn an average of $29,340 a year. Those with a master's earn $38,000 a year, and those with twenty or more years of experience earn an average of $48,000 a year. Top pay for food technologists is $175,000 or more, with administrators earning as much as $180,000 to $200,000. Women tended to earn somewhat lower salaries than men, even at the entry level, although the survey showed that salary differences between men and women are decreasing every year. IFT found the highest salaries were paid in the Pacific and South Atlantic states.

## what is the job outlook?

The food nutrition labeling laws now in place (see sidebar) opened up a number of opportunities for food technicians and technologists, as companies scrambled to get into compliance with the law. Even now nutrition analysis will be needed for each new food that comes onto the market.

Also keep an eye on HACCP, pronounced "hay'-cep." This is an acronym for hazard analysis critical control point. "It's a prevention technology—a means of preventing the contamination

### Advancement Possibilities

**Consultants** are hired on a per-project or other short-term basis to provide expertise in a specific area, such as product development, quality control, and so on.

**Directors of research and development** plan, oversee, and direct activities for creating and testing new products or production methods or improving existing ones.

**Food marketing specialists** apply knowledge of sales and marketing to help food companies gain new consumers for their products, such as overseas markets.

**Nutrition labeling specialists** plan and direct tests to document food nutrition content and direct labeling activities to comply with the U.S. nutrition labeling law.

**Product managers** are specialists in a specific food brand or family of brands.

**Quality assurance or quality control managers** ensure company compliance with standards for safety, cleanliness, consistency, and other factors as established by state, federal, and other law; company policy; and/or customer requirements.

**Sensory evaluation experts** measure consumer response to food qualities, such as appearance, texture, taste, flavor, and smell.

of food by identifying at the microbiological level the critical point at which the food begins to break down," says Tim Willard, spokesperson for the National Association of Food Processors (NAFP). "The industry is actively exploring this area right now, and anyone studying food technology and science in the near future will hear about it. Certainly, it's going to be important through the end of the century."

Everyone eats; there's always going to be a demand for food. But the fortunes of specific food-related companies rise and fall depending on changing consumer tastes, competition, and other factors. In recent years, sources of growth have included a rising population and global market to sell to, improved production technology, and the demand for healthier foods (low-fat, low-salt, or cholesterol-free, for example). These factors have provided significant opportunities for food technologists.

There can be a lot of competition among food companies for market share, though. (Think about the number of breakfast cereals on the market today.) Also, generic brands, house brands, and specialty foods have cut into the profits of some standard-brand companies. Still, food technologists can be valued weapons in food manufacturers' arsenal, speeding R&D time and helping to improve production and marketing. Product management and sales-related areas like marketing and advertising are cited by some experts as good growth areas. International opportunities with U.S. food producers opening branches abroad also may be a good bet.

# how do I learn more?

## sources of additional information

Following are organizations that provide information on food technologist careers, accredited schools and scholarships, and possible employers.

**Institute of Food Technologists**
221 North LaSalle Street, Suite 300
Chicago, IL 60601
312-782-8424

**National Food Processors Association**
1401 New York Avenue, NW, Suite 400
Washington, DC 20005-2154
202-639-5900

## bibliography

Following is a sampling of materials relating to the professional concerns and development of food technologists.

### Books


Chalmers, Irena. *The Food Professional's Guide.* A look into the people, products, and services of the food industry. New York, NY: Wiley, 1990.

Jones, Julie Miller. *Food Safety.* An overview of food safety, including regulation and implementation. St. Paul, MN: Eagan Press, 1992.

Sims-Bell, Barbara. *Foodwork: Jobs in the Food Industry and How to Get Them.* Santa Barbara, CA: Advocacy Press, 1994.


### Periodicals


*Beverage World.* Monthly. Provides coverage of the beverage industry, with emphasis on marketing, production, and packaging. Strategic Business Communications, 226 West 26th Street, New York, NY 10011, 212-633-3730.

*Chilton's Food Engineering.* Monthly. Relates new developments in food and beverage processing, packaging, and operating. Chilton Publishing, One Chilton Way, Radnor, PA 19089, 610-964-4447.

*Food and Chemical Toxicology.* Monthly. Addresses issues on the research and regulation of food, cosmetics, agriculture, etc. Elsevier Science, Regional Sales Office, Box 945, New York, NY 10159-0945, 212-633-3730.

*Food Processing.* Biweekly. Discusses news, trends, and analysis of the food industry. Putman Publishing Co., 301 East Erie Street, Chicago, IL 60611, 312-644-2020.

*Food Technology.* Monthly. Includes technical information dealing with problems in food manufacturing. Institute of Food Technologists, 221 North LaSalle Street, Chicago, IL 60601, 312-782-8424.

*Journal of Food Composition & Analysis.* Quarterly. Collates data on the composition, use, storage and distribution of food. Academic Press, Inc., 525 B Street, Suite 1900, San Diego, CA 92101, 619-230-1840.

*Journal of Food Processing and Preservation.* Bimonthly. Explores the chemistry of food in relation to processing and preservation. Food & Nutrition Press, Inc., 2 Corporate Drive, Box 374, Trumbull, CT 06611, 203-261-8587.

# forestry technician

**Definition**

Forestry technicians help plan, supervise, and conduct operations necessary to maintain and protect the country's forests. They also manage forests and wildlife areas, control fires, insects, and disease and limit the depletion of forest resources.

**Alternative job titles**

Biological aides

Information and education technicians

Survey assistants

Technical research assistants

Wildlife technicians

**Salary range**

$15,000 to $19,000 to $28,500

**Educational requirements**

Some postsecondary training

**Certification or licensing**

May be required for chemical handling and surveying

**Outlook**

Little change or more slowly than the average

**GOE**
03.02.02

**DOT**
452

**O*NET**
24302A

## High School Subjects

Biology
Computer science
English (writing/literature)
Geology
Mathematics

## Personal Interests

Camping/Hiking
The Environment
Exercise/Personal fitness
Science

**It's been** a long day, and at 9 PM forestry technician Jim Ets Hokin is ready to hit the hay. That's when his radio crackles to life. It's the dispatcher from the National Forest Service, and she is calling to report a possible forest fire in the Coronado National Forest in southern Arizona where Jim is stationed. Pulling on his boots, Jim hops into his truck and drives up one of the old rutted forest service roads until he comes to the coordinates where the dispatcher said the fire had been spotted.

Sure enough, as he approaches the coordinates Jim can see the distinctive glow in the distance of flames scorching the brush and timbers. It's a small fire, only two-tenths of an acre in size, nothing like the monsters that burned the summer before in the nearby Rincon Mountains, consuming more than twenty-eight thousand acres of forest. Still, the wind is gusting on this night, and Jim knows that even a small fire can quickly turn dangerous. He radios to one of his fire crews for assistance, and using chain saws and a tool called a pulaski (a combination hoe and ax), the two clear a fire lane, removing most of the brush and timbers that fuel the fire. Then they wait for the fire to burn itself out.

In the morning, a fire engine will come to the site and spray the area with water, ending this fire's threat to the forest. But that won't mean Jim can rest easy. It's only the beginning of spring. In a few months, the forest will be dry from lack of rain. That's when it becomes a real powder keg. A lightning strike, a careless campfire, and Jim and his crew will be back in action again.

## what does a forestry technician do?

Hundreds of years ago, much of America was covered with dense, majestic forests. These natural woodlands gave shelter and sustenance to the peoples and wildlife indigenous to the continent. With the arrival of European settlers, the forests were cleared for farming, and timber was cut to build houses and boats, furniture and tools. With the clearing of the forests, coupled with natural forces such as fire and disease, the forests began to be depleted faster than they could grow.

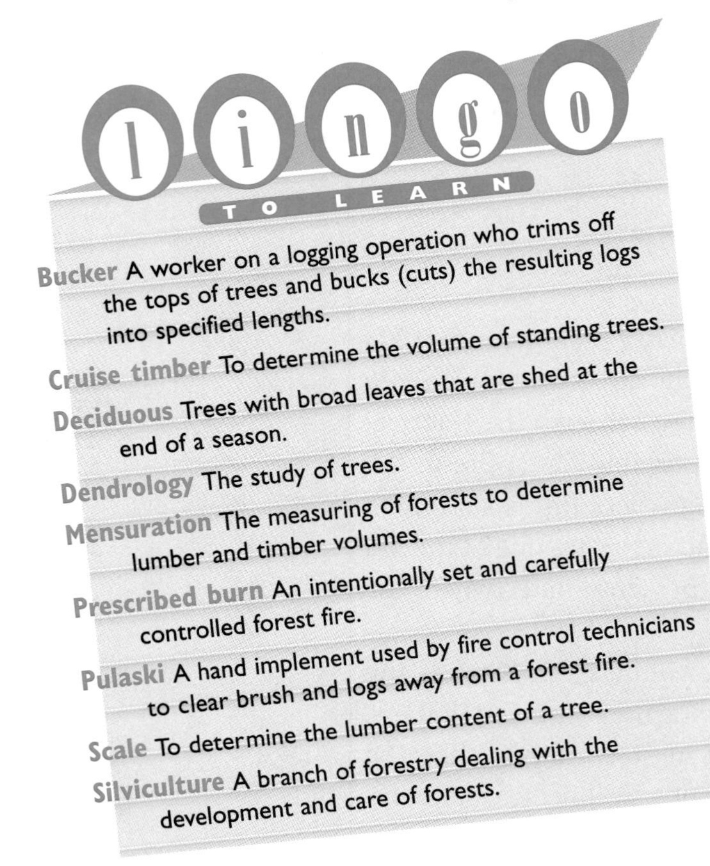


To diminish the loss of the forests and help in their restoration, the United States government enacted a series of conservation measures in the early 1900s. Among those measures was to train and establish a corps of forestry workers to protect and replenish the nation's fragile forestry reserves.

*Forestry technicians* work as members of a forest management team, which includes professional foresters and forestry scientists. Some also work for private companies in the forest industry.

The day-to-day duties of a forestry technician are based on a range of activities for managing and harvesting the nation's forest resources. They plant trees to replenish forest reserves that have become depleted through harvesting, fire, or disease. They care for maturing trees by spraying them with pesticides (if disease infested), by thinning to ensure the best growth of the forest, and by protecting them from fire and other dangers. They inventory (or scale) trees to determine the amount of lumber to be gained through harvesting. They then harvest forests and assist in the marketing of the lumber. And finally, they recondition the harvested forest and plant new trees, beginning the cycle anew.

Forestry technicians work under the supervision of professional foresters and managers who are responsible for developing an overall plan to manage a forest area. Each forest area is managed with a particular objective, and this affects the duties of the forestry technician. For example, a forest area might be tended for producing large sawlogs, in which case the planting, thinning, and care of the forest is different than it would be if the management plan called for producing timber as a habitat for wildlife.

Because the management plan and the care cycle of each forest area differs, forestry technicians often must perform various duties that require different skills. And while many forestry technicians are employed by the Federal Government, some also work for private companies such as lumber yards and sawmills.

Forestry technicians work under a variety of titles. *Biological aides* work in insect and disease control. They run experiments, analyze data, and map damage to forests caused by parasites. *Buyers* for sawmills purchase high-grade lumber for milling and furniture manufacture. *Information and education technicians* prepare media releases and serve as public relations specialists in nature centers. *Lumber inspectors* grade and determine the quantity of hardwood and softwood lumber at mills. *Pulp buyers* work at paper mills and purchase pulp wood for use in paper products. *Survey assistants* determine and mark boundary lines. They may also assist in the clearing of forests and in the construction of roads for logging operations. They consult aerial photographs and prepare maps of surveys, and assist in land appraisal and acquisition for federal, state and private employers. *Technical research assistants* gather and analyze field data to help researchers understand problems that relate to timber, watershed, and wildlife management. *Tree-nursery management assistants* assist in the operation of tree nurseries. They run seed tests and assist in planting, and in the supervision of personnel. *Wildlife technicians* conduct field work for the nation's various fish and game commissions. They track and tag animals for census making, and help to implement animal control policies.

**To be a successful forestry technician, you should**

- Enjoy working outdoors
- Be able to think clearly and act calmly in stressful and dangerous situations
- Be able to understand and execute scientific procedures
- Be able to identify plants and trees and the diseases that attack them
- Be able to read and interpret maps and aerial photographs

# what is it like to be a forestry technician?

Jim Ets Hokin has been a forestry technician since 1976. "I've always liked the outdoors," he says. So when he returned from a stint in the army, he looked into careers with the National Forest Service, the government agency that oversees the country's 154 national forests.

Each of the national forests is divided into districts, the number of which depends upon the size of the forest. Jim works in the Palisades Administrative District of the Coronado National Forest, located near Tucson, Arizona. He lives in the forest, in quarters furnished by the federal government. Typically, he works from 9 AM to 6 PM, but those hours can change when there are emergencies to tend to, such as a fire or a lost hiker.

For Jim, the most exciting part of his job as a forestry technician is when he and his crew have to fight a forest fire. "It can be dangerous work," he says. "But if you remember your training and pay attention to all the factors that you've been trained to pay attention to, you can stay out of trouble."

When mistakes happen, people can die. In 1994, for example, fourteen forest fire fighters died fighting a blaze in Colorado. Because the work can be so dangerous, forestry technicians who work in fire-fighting functions have to be alert to conditions that can cause forest fires, and understand how fires spread.

"There are three basic factors you have to consider (when fighting a forest fire)," Jim says. "You have to understand the weather, how the wind can make a fire spread, and the natural topography of the area (where the fire is burning)."

The last factor includes the types of fuel feeding the fire. "A fire in a ponderosa pine forest will behave differently than a desert fire," Jim says.

While the work isn't too physically demanding, Jim says, there are times when the job takes its toll on the human body. "When there's a desert fire it might be 105 degrees outside, and on the fire line it might be more like 180 degrees. You almost have to be a little nuts to do this type of work," he says with a laugh.

After a fire has been successfully extinguished, Jim returns to the district office. There he must complete a report detailing how the fire was treated and assess the damage that occurred to the forest. These reports help Jim's supervisors evaluate fire-fighting procedures and determine whether changes in those procedures should be made.

Sometimes, Jim sets fires as part of a carefully developed forest management plan to ensure the forest's long-term health. These fires, or "prescribed burns," are set to thin out sections of the forest, thereby eliminating a fuel source from which an accidental fire could escalate into a large-scale forest fire. And since fires return vital nutrients to the soil, some prescribed burns are set to enrich the soil so that it will continue to sustain the forest.

Summers are the busiest season for Jim because of the number of fires that occur, and also because more people visit the forest during the spring and summer months. During the winter, the work load drops off significantly, although Jim keeps busy by doing trail maintenance, by holding training sessions for fire crew members, and by catching up on paperwork and reports that will ensure the fire crew has the necessary material supplies and personnel for the coming fire season.

Jim likes the diversity that being a forestry technician offers. He also says that by working for the federal government, he receives an excellent benefits package. And while each district office is run by a district forester, Jim says he works independently much of the time. This means he has to show initiative to see that the district's forest management plan is followed, and that he fulfills his job's obligations.

## have I got what it takes to be a forestry technician?

Being a forestry technician can be dangerous work. When Jim and his fire crew head out to battle a blaze, each carries along a portable fire shelter, a specially designed pup tent built to withstand extreme heat. Each crew member carries a fire shelter in case he should suddenly find himself trapped by an approaching fire. "It's your last resort," Jim says. "Luckily, I've never had to use one."

Because of the dangerous nature of the firefighting duties, a forestry technician has to be able to think clearly in stressful situations. A rushed decision at a crucial moment could endanger the lives of the forestry technician and his or her crew.

Not all forestry technicians fight fires like Jim. Depending upon the forest management plan that's been enacted for the forest in which they work, forestry technicians may perform more research and data-gathering duties. While the chance for bodily harm is less, these forestry technicians have to have a firm grasp of scientific procedures. They must also be skilled in recognizing insects and other diseases that might harm the forest.

FYI

The first professional forester in the United States was Gifford Pinchot (1865-1946). A personal friend of President Theodore Roosevelt, he was one of the first people to advocate planned conservation of natural resources.

Forestry technicians who work in national forests work under the supervision of a district manager. They must be able to follow direction and adhere to a carefully designed forest management plan that dictates how their forest will be managed. In some cases, a forestry technician might supervise a staff, too. During fire season, Jim supervises a crew of twenty fire fighters who work on a seasonal basis.

Ultimately, forestry technicians have to care for one of the planet's most important natural resources. Trees absorb carbon dioxide and release oxygen, playing an important role in providing the planet with the air we breathe. They provide shelter for birds and small animals, beneficial insects and some reptiles. Trees are also used to provide wood and lumber so that we may have shelter ourselves, and tools and other implements. Forestry technicians are committed to ensuring that our country's forest areas serve all these functions.

## how do I become a forestry technician?

For Jim, the decision to become a forestry technician was an easy one. He loved working in the outdoors, and when he returned to the United States after serving in the army, he looked for a job that would let him follow his passion. He worked for a summer in a national park in Colorado, and decided that he'd like to pursue a career as a forestry technician.

## education

### High School

To be a forestry technician, most employers require candidates to have completed a two-year postsecondary training program. These programs require applicants to have a high school diploma.

To prepare yourself for a career as a forestry technician, take high school courses in mathematics, computer science, chemistry, botany, zoology, ecology, and the social sciences. Good writing and communication skills are also needed to advance in the field.

### Postsecondary Training

After high school, you'll need to attend and graduate from a forest technology program. Most of these programs offer an associate's degree in forest technology. The Society of American Foresters currently recognizes twenty-four educational programs (twenty-one in the United States; three in Canada) that offer a two-year associate's degree in forest technology or some equivalent (*see* "Sources of Additional Information").

While in a forest technology program, students receive instruction in mathematics, communications, botany, engineering, surveying, dendrology (tree identification), forest soils, computer applications, elementary business principles, timber harvesting, map drafting, outdoor recreation and environmental control, wildlife ecology, insect and disease control, forest measurement, and aerial photographic interpretation.

## certification or licensing

Forestry technicians who work with chemicals may need to be certified or licensed in some states. Technicians who do surveying work usually need a license to survey property for public records.

## internships and volunteerships

In addition to course work, nearly all forestry technician programs require students to get first-hand experience in the field. Students may work in nearby forests for the summer or during part of the school semester. They might also work at a sawmill or lumber mill.

High school students can get a sense of what it's like to work in a forest by participating in the Youth Conservation Corps, a summer employment and training program for students aged fifteen to eighteen. Students spend approximately six weeks in a national park, forest, or wildlife refuge where they perform such duties as trail maintenance and building, tree pruning, campsite clearing and maintenance, and watershed maintenance. Students receive room and board and a small stipend. For more information, contact the following:

**United States Department of Agriculture**
Forest Service
Youth Conservation Corps
PO Box 96090
Washington, DC 20090
703-235-8861

"It might be 105 degrees outside, and on the fire line it might be more like 180 degrees."

## who will hire me?

About half of all forestry technicians work for government agencies. These include the Forest Service and the Natural Resources Conservation Service of the U.S. Department of Agriculture. They also work for the Bureau of Land Management, the Bureau of Indian Affairs, the National Park Service and the Fish and Wildlife Service of the U.S. Department of the Interior. States also have forest areas where forestry technicians are employed (*see* "Sources of Additional Information").

Other forestry technicians work in the private sector. They are employed by logging, lumber, plywood, and paper companies. Others do reforestation work for mining, oil, and utility companies.

The Society of American Foresters can direct you to resource guides to jobs in forestry. They can also provide you with employment statistics and information on forestry training programs (*see* "Sources of Additional Information").

## where can I go from here?

"I have to say I'm happy where I am," Jim says. "I really don't want to move again." Jim's plan is to remain in the Coronado National Forest working as a forestry technician for another fifteen years. Then, he'll be fifty-five and able to retire with full benefits.

A forestry technician's advancement often means having to move from one forest district or even national forest to another. As federal employees, forestry technicians are placed in ranks or grades. Workers advance to a higher grade after attaining a certain number of years' experience in their current grade. Advancement often means that the forestry technician assumes more administrative duties, which can keep him or her from working in the field as much as the lower grades do.

One advancement option is for a forestry technician to become a forester. This requires additional education—foresters complete a four-year course of study at an accredited university. Foresters are most often those responsible for developing the forest management plans that the forestry technicians help to implement.

## what are some related jobs?

Forestry technicians have knowledge and skills that can make them attractive employees in the logging and sawmill industry. Some positions in the logging industry call for individuals skilled in assessing the quality and volume of lumber in a forest. Working for a logging operation is physically demanding, and calls for an understanding of specialized machinery and hand tools that forestry technicians may already be acquainted with.

The U.S. Department of Labor classifies forestry technicians under the headings *Forest Conservation Occupations* (DOT) and *General Supervision, Plants and Animals: Forestry and Logging* (GOE). Other related jobs include farmers, logging supervisors, fire rangers, tree planters, smoke jumpers, and fire wardens.

| Related Jobs |
| --- |
| Logging supervisors |
| Fallers |
| Cruisers |
| Buckers |
| Tree cutters |
| Log markers |
| Firefighters |

## what are the salary ranges?

Forestry technicians who work for the federal government receive a salary determined by the current grade they hold. Typically, entry-level forestry technicians who work for the federal government earn slightly less than those working in the private sector, but their salaries soon become comparable after a few years on the job.

Forestry technicians who have graduated from a two-year training program usually start their first job earning $17,000 to $18,000 a year. After several years' experience, forestry technicians can expect to earn around $21,000 a year. Those who have advanced to supervisory positions and have specialized earn about $30,000 or more a year.

In addition to their annual salary, forestry technicians who work for the government also receive numerous benefits. These include paid vacations, medical insurance and sick days, and retirement plan benefits. Some also receive financial assistance for further education and training so that they may keep abreast of changes occurring in the field.

## what is the job outlook?

Employment in forestry-related occupations is expected to grow more slowly than the average for other careers through the year 2006. This means competition for jobs as a forestry technician will be high. There are only 154 national forests in the United States, and while each is divided into districts, all of which employ forestry technicians, the number of available openings at any time is limited.

### Advancement Possibilities

**Foresters** manage and develop forest lands and resources for economic and recreational purposes.

**Fire wardens** administer fire prevention programs and enforce governmental fire regulations throughout specific forest and logging areas.

**Silviculturists** establish and care for forest stands.

**Smoke jumper supervisors** supervise and coordinate activities of airborne firefighting crews engaged in extinguishing forest fires.

Because jobs as a forestry technician can be difficult to come by, many interested in the career work on a seasonal basis in order to gain experience in the field, making them more attractive candidates when openings do occur. Typically, the demand for seasonal workers occurs during the late spring and summer. This means that these individuals usually have jobs during the winter months as well in order to support themselves year round.

On the bright side, more people are enjoying our national forests than ever before. As recreational use increases, so will a need for forestry technicians skilled in managing forest resources. Further, as demand for wood products continues to grow, so too will the need for a steady supply of timber. A forestry technician's forest management skills will play an important role in ensuring that this demand is met. Their knowledge of planting and forest care will create opportunities for forestry technicians in the years to come.

**FYI** The first national park in the world, and the largest national park in the United States, is Yellowstone National Park, situated mainly in northwest Wyoming. It contains more than 3,450 square miles and was designated a national park by Theodore Roosevelt in 1872.

# how do I learn more?

## sources of additional information

Following are organizations that provide information on forestry technician careers, accredited schools, and employers.

**The Society of American Foresters**
5400 Grosvenor Lane
Bethesda, MD 20814-2161
301-897-8720

**United States Department of Agriculture**
Forest Service
PO Box 96090
Washington, DC 20090
202-205-1661

**United States Department of Agriculture**
Natural Resources and Environment
14th and Independence Avenue, SW
Washington, DC 20250
202-720-7173

**United States Department of the Interior**
Bureau of Indian Affairs
Trust Responsibilities
Main Interior Building
Washington, DC 20240
202-208-5831

**United States Department of the Interior**
Bureau of Land Management
Land and Renewable Resources
Main Interior Building
Washington, DC 20240
202-208-4896

**United States Department of the Interior**
Fish and Wildlife Service
Main Interior Building
Washington, DC 20240
202-208-4717

**United States Department of the Interior**
National Parks Service
PO Box 37127
Washington, DC 20013
202-208-4621

## bibliography

Following is a sampling of materials relating to the professional concerns and development of forestry technicians.

### Books


Leuschner, William A. *Forest Regulation, Harvest Scheduling, and Planning Techniques.* New York: Wiley, 1990.

Leuschner, William A. *Introduction to Forest Resource Management.* Malabar, FL: Krieger Pub. Co., 1992.

Muir, John. *A Thousand-Mile Walk to the Gulf.* The travels of noted conservationist and naturalist John Muir. New York: Sierra Club Books, 1992.

Ramakrishna, Kilaparti and George M. Woodwell (eds.). *World Forests for the Future: Their Use and Conservation.* New Haven: Yale University Press, 1993.

Sharma, Narendra P., ed. *Managing the World's Forests: Looking for Balance Between Conservation and Development.* Dubuque, IA: Kendall/Hunt Pub. Co., 1993.

### Periodicals

*American Forests.* Bimonthly. General information about forests, trees, conservation, etc. American Forests, Box 2000, NW, Washington, DC 20005, 202-667-3300.

*Consultant.* Quarterly. Directed toward forestry consultants and owners. Association of Consulting Foresters of America, Inc., c/o Mason, Bruce & Girard, Inc., 1005 Yuba Street, Redding, CA 96001, 916-246-2455.

*Environmental History.* Quarterly. Focuses on the North American forest industry. Forest History Society, 701Vickers Avenue, Durham, NC 27701-3147, 919-682-9319.

*Forest Industries Newsletter.* Bimonthly. Safety and health issues. National Safety Council, 1121 Spring Lake Drive, Suite 558, Itasca, IL 60143-3201, 630-775-2281.

*Forest Science.* Quarterly. Explores the latest in forestry research. Society of American Foresters, 5400 Grosvenor Lane, Bethesda, MD 20814, 301-897-8720.

*Journal of Forestry.* Monthly. Discusses numerous aspects of forest use, including protection and management. Society of American Foresters, 5400 Grosvenor Lane, Bethesda, MD 20814, 301-897-8720.

*Journal of Sustainable Forestry.* Quarterly. Includes varied aspects of forest issues. Haworth Press, Inc., 10 Alice Street, Binghamton, NY 13904, 607-722-5857.

# geological technician

**Definition**

Geological technicians assist geologists and engineers by gathering, plotting, and storing samples and technical data.

**Alternative job titles**

Environmental technicians
Geological aides
Hydrologic technicians
Petroleum technicians
Seismic technicians

**Salary range**

$20,000 to $30,000 to $50,000

**Educational requirements**

High school diploma; bachelor's degree highly recommended

**Certification or licensing**

None

**Outlook**

About as fast as the average

GOE 02.04.01

DOT 024

O*NET 24111A*

## High School Subjects

Earth science
Geology
Mathematics

## Personal Interests

The Environment
Science

**Laura** Burnett Blass looks through the morning's scout cards as if she's checking the winning lottery numbers. A few days earlier, an oil crew began drilling for oil in an area north of Houston. They picked that site to drill because research gathered by Laura indicated there might be pockets of crude oil located there. The crew may drill as many as 13,000 feet down into the earth searching for oil, which takes a considerable amount of time and money. Laura is anxious to see what the scout card, a report of the crew's findings, will show, since her work directly affected the crew's choice of drill site. As she scans the preliminary data, her eyes light up. The crew struck oil, and Laura can feel proud for a job well done.

# what does a geological technician do?

*Geological technicians* work in the field of the geosciences. Geosciences are the sciences dealing with the earth, including geology, geophysics, and geochemistry. Usually, they work under the supervision of a geologist or geoscientist who is trained to study the earth's physical makeup and history. Often, geoscientists specialize in one area of study. Among these specialties are environmental geology, the study of how pollution, waste, and hazardous material affect the Earth and its features; geophysics, the study of the Earth's interior and magnetic, electric, and gravitational fields; hydrology, the investigation of the movement and quality of surface water; petroleum geology, the exploration and production of crude oil and natural gas; and seismology, the study of earthquakes and the forces that cause them.

Petroleum technicians most commonly work with petroleum geologists to determine where deposits of oil and natural gas may lay buried beneath the Earth's surface. Since petroleum and natural gas play such a significant role in our daily lives, it's important that we find new deposits of these fossil fuels. There are small companies and large multinational corporations that search all over the world for deposits of oil and gas. They send teams of geologists and drill operators to scour farms, deserts, even beneath the ocean floor for new deposits of oil.

Using data gathered from workers in the field, the geological technician drafts maps displaying where drilling operations are taking place and creates reports that geologists use to determine where an oil deposit might be located.

When drafting maps, petroleum technicians pinpoint the exact location where a drilling crew has dug a well. The map may also indicate the depth of the well and whether oil was found or, if not, that it was a "dry well." Once this information is included on a map, geologists are better able to determine the size of the oil deposit that lies buried beneath the surface of the Earth.

When creating reports, petroleum technicians must analyze several types of raw data. Often, crews detonate carefully planned explosions that send shock and sound waves deep into the Earth. These waves are recorded by microphones, and by studying the patterns of the waves, it is possible to determine the composition of the rock beneath the surface. Geological technicians take these patterns and remove any background noise—even an airplane flying overhead can create a sound wave that will be picked up by the receiving microphones—and then write a report that summarizes what the sound patterns indicate.

Some geological technicians work in the field of environmental engineering and are often known as *environmental technicians*. They assist geologists in studying how man-made structures—such as roads and commercial, residential, and industrial development and landfills—can affect the geological landscape. The information they gather is used in environmental impact statements that are then used by developers, government officials, and private land owners to minimize damage to the environment.

Environmental technicians also study how pollution and other materials affect the Earth by gathering and testing samples of soil, water, or other substances. For instance, environmental technicians monitor wells and sample groundwater, checking for contamination. If the water is contaminated, environmental technicians then assist with site remediation or cleanup.

*Hydrologic technicians* are concerned with weather patterns, precipitation, and water supply. They study the properties, distribution, and circulation of waters on or beneath the

**Dry well** A well that has produced no oil or natural gas deposits.

**Fossil fuel** A nonrenewable energy source, such as oil or gas, derived from organic remains of past life.

**Groundwater** The water within the earth.

**Natural gas** A combustible mixture of methane and hydrocarbons used chiefly as a fuel.

**Hydrology** The science of the properties, distribution, and circulation of water in the atmosphere and on and below the earth's surface.

**Petroleum** An oily black substance that is found beneath the earth's surface and can be processed for use as gasoline, grease and plastic. Also called crude oil.

**Scout card** A preliminary report from an oil drilling crew.

**Seismic** Of or relating to an earth vibration caused by an earthquake, a planned explosion or an impact.

**Seismograph** An instrument designed to measure the severity and intensity of earthquakes. It records the seismic waves generated by earthquakes.

earth's surface as well as in the atmosphere. Hydrologic technicians also assist hydrologists in the study of precipitation, including the form and intensity and the rate at which it penetrates the soil. They also research how precipitation moves through the Earth and returns to the ocean and atmosphere. Hydrologic technicians collect, analyze, and interpret hydrologic data to assess the availability and quality of water, including lakes, streams, groundwater supply, and more.

*Seismic technicians* help seismologists gather data so they can better understand and predict earthquakes and other vibrations of the earth. Seismic technicians use seismographs and other geophysical instruments to collect and analyze data.

**To be a successful geological technician, you should**

- Be able to pay close attention to detail
- Have good time management skills
- Have organizational skills
- Be able to communicate well with other people

## what is it like to be a geological technician?

Laura Burnett Blass has been a geological technician for the past fifteen years. For most of those years she has worked for gas and oil companies in central Texas. Today, she works at the Geotechnology Research Institute, a research facility that helps petroleum companies develop better and more efficient methods of finding and retrieving fossil fuels.

While some geological technicians frequently go out into the field where drilling operations are taking place, Laura's days are spent indoors at the institute's offices in Woodlawns, Texas. Typically she works an eight-hour day, five days a week, but when several projects are under way she must often put in overtime.

Each day she has certain specific duties she must perform. She first gathers the results of the previous day's seismic tests and enters the data into a database on her computer. Depending upon how active the seismic crews were the day before, this task can take anywhere from ten minutes to four hours.

She also examines the drilling crews' scout cards. The scout cards are preliminary reports that reveal several types of information. They tell how deep into the Earth the crew drilled, what type of rock was found at which depths, and whether oil was or was not found. Laura takes this information and places it on a map, which is called "spotting a well."

Often, she must draw her own maps on a drafting table in her office, but with advances in computer software technology, much of her work can now be performed on her computer.

Once she has compiled and written reports based on the various data she has gathered from the drilling and research crews, Laura gives the reports to a petroleum geologist. The geologist then analyzes her findings and may also ask Laura for maps showing all of the wells currently being drilled in a specific region. With the information Laura has compiled, the geologist then recommends where future drilling operations should take place.

Mark Meeks works as an environmental technician for Earth Systems Consultants in San Luis Obispo, California. He focuses on groundwater contamination and has two main duties: he monitors wells and collects groundwater samples, and he helps with site remediation when necessary. When a site is suspected to be contaminating the ground, such as leaking gas tanks, Mark's company is sent out to dig wells in various spots surrounding the area. The wells are monitored regularly and the groundwater tested to determine how far the contamination has spread, known as the plume, and the severity of the pollution.

On a typical day, Mark may visit up to fifteen wells. Sometimes he works with a partner, depending on the size of the project and the number of wells. When he arrives at the well site, Mark opens the well cover, drops down an instrument designed to measure the distance down to the groundwater, which is usually four to twenty-five feet beneath the surface, then bails out the old water, which can entail bailing out buckets of stagnant water by hand. Mark then lets the wells recharge up to 80 percent, fills up three to four jars, then delivers the jars to a laboratory to test for contaminants, such as crude oil, benzene, or lead.

Sites are visited at different times—some are checked weekly, while others are visited semiannually. They are also located in many different areas. "If you do a gas station site," says Mark, "you have to cone off, and you have to watch for people trying to run you over. And then when I do the tank farm, which is semiannual, you have to put all your equipment in a backpack and hike out to the wells because they're inaccessible to traffic."

Mark also assists with site remediation, or cleanup, once the extent of the contamination is determined. "Once we know how big the plume is," Mark explains, "and how deep and how badly it's contaminated, then we put in vapor extraction systems or pump and treat systems." Vapor extraction systems clean the soil by sucking the fumes from the water and running them through a system that burns off the contamination. This process usually takes a year or two. With a pump and treat system, the water is run through carbon filters to clean it.

## have I got what it takes to be a geological technician?

Geological technicians working in the petroleum industry are an important link in a company's search to find hidden pockets of fossil fuels. They take the data that field crews gather from seismic tests and other measures and process that information into a report that helps the company narrow its search. Because drilling for oil can be a costly and time-consuming venture, geological technicians must be competent in their ability to correctly interpret data.

Laura admits that working with such a weighty responsibility can sometimes be intimidating. Yet this responsibility can also be one of the most satisfying aspects of the job. "You actually get to see what happens to your work," Laura says. "They dig a well based on information you've provided. I know that my efforts make a contribution. And it's a good feeling knowing that they depend on you and trust your judgment."

To be a successful geological technician requires an ability to pay attention to detail since mistakes can cost time and money. "You also need to have very good time management and organizational skills," says Laura. "I wasn't a very organized person before I started working as a geological technician. But this job made me into one." Geological technicians working in environmental geology should be mathematically inclined, says Mark, since you're constantly dealing with calculations and volumes and distances. Mark also emphasizes the importance of being physically fit—there is a lot of lifting and outdoor activity involved in the job of an environmental technician.

## how do I become a geological technician?

Laura began her career as a geological technician in 1981 when an opening became available at the oil company she was working for in Woodlawns, Texas. Because she had no formal training in geology, she had to acquire her skills on the job.

Over the years, she's learned all about map reading and map making, data evaluation, and performing complex calculations—all the skills that a geological technician must use each day on the job. Because of recent technical innovations, she's now learning how to make 3-D representations of oil fields and oil wells on a computer.

"It's fascinating," she says. "There's always new technology coming out, which means you're always learning something new."

## education

### High School

A high school diploma is mandatory for anyone considering a career as a geological technician. High school courses in geology, geography, and math—including algebra, trigonometry, and statistics—are all recommended. The science courses will provide you with an understanding of the Earth and the forces that affect it, while

**Advancement Possibilities**

**Geologists** study the composition, structure, and history of the Earth's crust

**Soil engineers** study and analyze surface and subsurface soils to determine characteristics for construction, development, or land planning.

**Environmental geologists** study the effects of human activity, the atmosphere, the geosphere, and the hydrosphere on the Earth to provide solutions to problems such as pollution, erosion, and waste management.

**Petroleum geologists** are involved with exploring for oil and natural gas resources.

the math courses are important to learn how to perform complex calculations and statistical analysis. In addition, drafting courses can teach the skills used in map making, and since many companies now design maps using computer software, courses in computer science are also recommended. Because the geological technician serves in many respects as a liaison between the field crews and the geologists, courses that develop communication skills are also beneficial.

Part-time or summer employment may be available with large oil and gas companies, environmental engineering firms, or other companies that employ those in the geological field. Summer camps with an educational focus may also expose you to various facets of geology and the environment.

You might also consider joining a club or organization concerned with rock collecting or the environment. Local amateur geological groups may sponsor events or activities such as local hikes or trips to sites of geological interest.

### Postsecondary Training

While a college degree is not required for anyone considering a career as a geological technician, it is highly recommended, and many companies will not hire technicians who lack a degree. Mark does not possess a college degree, but he advises, "If you want to become a full-blown geological technician, you need to have a college degree in geology or soils engineering or something."

Some two-year colleges offer associate's degrees in geological technician programs, but most offer basic geology programs with credits that are transferable to four-year schools. Bachelor's degree programs in geology and the geosciences are readily available at most colleges and universities.

## certification

There is no certification specifically for geological technicians. Some environmental technicians, however, may need to acquire certification for handling and sampling hazardous materials such as crude oil. Certification requires attending a forty-hour class, with an eight-hour refresher course every year thereafter. Mark has found that most employers will finance the training.

## scholarships and grants

For those planning to attend college, there are plenty of scholarship opportunities. You may wish to contact your financial aid office or advisor for information on scholarships. There may be general school scholarships or department awards.

Professional organizations such as the Geological Society of America or the American Geological Institute offer scholarships and lists of scholarship opportunities (*see* "Sources of Additional Information").

## internships and volunteerships

Some large corporations offer internship programs, and professional associations are excellent resources for information on internship opportunities. Contacting companies or organizations directly and inquiring about internships would be a good idea.

Students in two-year or four-year programs may find it easy to locate summer or part-time internship opportunities through the school placement program or departmental advisor. Completing an internship assures employers that you have hands-on experience and will make you a more desirable candidate.

Volunteer opportunities are plentiful, especially if you are interested in environmental issues. Many nonprofit groups sponsor activities and field trips, and volunteers are frequently needed to help geologists or geoscientists working in the field. The U.S. Geological Survey also offers volunteer opportunities. Volunteers must be eighteen years or age or older (*see* "Sources of Additional Information").

## who will hire me?

When Laura landed her first job as a geological technician, she was working as a secretary at a small, independent oil company in central Texas. "They needed someone to be a geological technician," she says, "and I was willing to give it a try."

Today, such stories are rare, she says. "It's really very tough right now [to find a geological technician position]," Laura says. But that doesn't mean jobs as a geological technician aren't out there, and there are some resources available to help find those positions.

A good place to begin looking for a position as a geological technician is with the major oil and gas companies. Because of their size—most are multinational corporations—they are more likely to have several geological technicians on staff. Chevron, Arco, Shell, Texaco, and Exxon are some of the largest gas and oil companies in the United States, and each may have overseas departments where geological technicians work as well.

The federal government or state government agencies employ geologists and geoscientists and geological technicians. The U.S. Geological Survey, a division of the Department of the Interior, the Department of Energy, the Forest Service, the Department of Agriculture, the U.S. Army Corps of Engineers, and state geological surveys are just some of the departments that may have job opportunities for geological technicians. Because the competition for federal and state jobs is fierce, most agencies require that applicants possess a bachelor's degree.

Many environmental engineering companies and environmental consulting firms also hire geological technicians. With these firms, geological technicians assist in creating environmental impact studies and monitoring environmental resources. These companies can be found under the Environmental Engineering or Engineering headings in most cities' Yellow Pages.

There are several professional associations that may be able to provide further information about employment opportunities. Most of these professional organizations, as well as large corporations and governmental agencies, have web sites that provide information on job openings (*see* "Sources of Additional Information").

## where can I go from here?

"I'd really like to train or teach others to be geological technicians," Laura says.

Supervisory positions are available to those geological technicians who have several years of on-the-job experience. Typically, these geological technicians train new staff in proper procedures and methods, as well as check their work for accuracy before sending it to the geologist.

A geological technician's advancement depends partly on the size of the company he or she works for. At smaller companies and consulting firms, geological technicians may be called upon to perform some tasks that geologists might do, as well as perform some clerical duties. At larger companies, they might specialize in performing one set task—such as well spotting—or they might supervise a staff of several geological technicians.

With additional years of schooling, geological technicians can go on to become engineers or scientists, such as geologists, environmental geologists, soil engineers, or petroleum geologists. A master's degree and/or doctorate are generally needed for these positions. Geologists are concerned with the physical nature of the Earth, while soil engineers study surface and subsurface soils. The work of soil engineers is especially helpful for land planning, construction, and land development. Environmental geologists focus on solving problems such as pollution, waste management, flooding, erosion, and urbanization. They accomplish this by studying and analyzing the interaction between human activities and the atmosphere, geosphere, and hydrosphere. Petroleum geologists work for companies in the petroleum and gas industry and are involved with the exploration for natural gas and oil resources. They also assist with the production of oil.

## what are some related jobs?

Geological technicians have many specialized skills that are best suited for their own field. With their knowledge of maps and their drafting abilities, however, some geological technicians might find employment with companies that value those skills, such as public utility companies and architectural firms. Some private consulting firms and developers might also have a need for the skills that a geological technician has acquired.

The U.S. Department of Labor classifies geological technicians under the headings *Occupations in Geology* (DOT) and *Laboratory Technology: Physical Sciences* (GOE). Also listed under these headings are other geology professionals, such as geologists, geophysicists, and paleontologists, and laboratory workers such as laboratory testing technicians, diagnostic medical sonographers, coal mining technicians, quality control technicians, pharmacists, and chemical technicians.

| **Related Jobs** |
|---|
| Diagnostic medical sonographers |
| Coal mining technicians |
| Quality control technicians |
| Pharmacists |
| Laboratory testing technicians |
| Chemical technicians |
| Geologists |
| Geophysicists |
| Paleontologists |

## what are the salary ranges?

Geological technicians working in the petroleum and gas industry generally make the highest wages, according to the 1998–99 *Occupational Outlook Handbook*. Science technicians earned an average annual income of $27,000 in 1996, with the lowest 10 percent making less than $15,500 and the highest 10 percent earning more than $49,500.

Also according to the *Occupational Outlook Handbook*, the average annual income for physical science technicians working for the federal government in early 1997 was $35,890. Hydrologic technicians made $33,230.

Those with bachelor's degrees in geology and geological sciences earned an entry-level average salary of $30,900 in 1997, according to the National Association of Colleges and Employers.

According to Mark, salaries tend to vary by region, company size, and work experience. "On average," Mark believes, "geological technicians can make $18 to $20 an hour." Entry-level technicians, Mark says, earn an average of $14 to $16 an hour and most possess bachelor's degrees.

## what is the job outlook?

Mark feels the job outlook for environmental technicians is positive. "They're coming out with new laws every year," Mark reasons, "so there's a need for environmental technicians." The 1998–99 *Occupational Outlook Handbook* confirms Mark's beliefs. Technicians will be in demand to help monitor compliance with environmental regulations, clean up polluted and contaminated sites, and collect and analyze samples for contaminants.

The *Occupational Outlook Handbook* indicates that job growth for science technicians should increase about as fast as the average through the year 2006. Technicians who have experience and training on equipment used in laboratories and production facilities should find themselves in the highest demand.

The petroleum industry is prone to cyclical fluctuations and is affected by the country's economic climate. Although the 1980s and early 1990s saw a decrease in productivity in the petroleum industry, the outlook is now considerably better. As the demand for oil and gas and energy resources grows, so do job opportunities for geological technicians. Stronger economies throughout the world, coupled with increasing populations, create the need for new energy sources. The petroleum industry, therefore, should enjoy growth through the year 2006.

## how do I learn more?

### sources of additional information

Following are organizations that provide information on geological technician careers, accredited schools, and employers.

**American Association of Petroleum Geologists**
PO Box 979
Tulsa, OK 74101
918-584-2555
http://www.aapg.org

**American Geological Institute**
4220 King Street
Alexandria, VA 22302-1507
703-379-2480
http://www.agiweb.org

**American Institute of Professional Geologists**
7828 Vance Drive, Suite 103
Arvada, CO 80003
303-431-0831
http://www.nbng.unr.edu/aipg

**Association of Engineering Geologists**
323 Boston Post Road, Suite 2D
Sudbury, MA 01776
978-443-4639
http://aegweb.org/

**Geological Society of America**
PO Box 9140
Boulder, CO 80301
303-447-2020
http://www.geosociety.org

**U.S. Geological Survey**
601 National Center
Reston, VA 20192
http://www.usgs.gov

# bibliography

Following is a sampling of materials relating to the professional concerns and development of geological technicians.

## Books

Montgomery, Carla W. *Fundamentals of Geology.* 2nd ed. Dubuque, Iowa: Wm. C. Brown Publishers, 1993.

O'Reilly, Gerald. *Planning for Field Safety.* Alexandria, VA: American Geological Institute, 1992.

## Periodicals

*Abstracts with Programs.* Seven times per year. Published papers from the national meetings of the Geological Society of America. Geological Society of America, 3300 Penrose Place, PO Box 9140, Boulder, CO 80301, 303-447-2020.

*Earth Surface Processes and Landforms.* Eight times per year. John Wiley & Sons, Inc., 605 Third Avenue, New York, NY 10158, 212-850-6645.

*Economic Geology & Bulletin of Society of Economic Geologists.* Eight times per year. Examines geological deposits focusing on their economic importance. Economic Geology Pub. Co., 5808 South Rapp Street, Suite 209, Littleton, CO 80120-1942, 915-533-1965.

*Environmental and Engineering Geoscience.* Quarterly. Technical articles. Association of Engineering Geologists, 323 Boston Post Road, Suite 2D, Box 132, Sudbury, MA 01776, 508-443-4639.

*Environmental Geology.* Quarterly. Geological management. Springer-Verlag, 333 Meadowlands Parkway, Secaucus, NJ 07094, 212-460-1500.

*Exploration & Mining Geology.* Quarterly. Discusses worldwide Earth science studies. Elsevier Science, Inc. Regional Sales Office, PO Box 945, New York, NY 10159-0945, 212-633-3730.

*Geological Magazine.* Bimonthly. Well-known magazine which discusses all aspects of geological science. Cambridge University Press, 40 West 20th Street, New York, NY 10011, 212-924-3900.

*Geology.* Monthly. Contains international information regarding current developments. Geological Society of America, 3300 Penrose Place, PO Box 9140, Boulder, CO 80301, 303-447-2020.

*GSA Today.* Monthly. Geological Society of America, 3300 Penrose Place, PO Box 9140, Boulder, CO 80301, 303-447-2020.

*Journal of Geoscience Education.* Five times per year. Directed toward improving the teaching of Earth sciences. National Association of Geoscience Teachers Inc., PO Box 5443, Bellingham, WA 98227-5443, 360-650-3587.

*Shale Shaker.* Bimonthly. Focuses on the petroleum industry. Oklahoma City Geological Society, 227 West Park Avenue, Oklahoma City, OK 73102, 405-236-8086.

# hazardous waste management technician

**Definition**

Hazardous waste management technicians help identify, remove, and dispose of hazardous wastes. Technicians may collect water, soil, or other samples for laboratory analysis and monitor waste control systems.

**Alternative job titles**

Environmental technicians

Environmental technologists

**Salary range**

$16,000 to $25,000 to $45,000

**Educational requirements**

Some postsecondary training

**Certification or licensing**

Voluntary

**Outlook**

Faster than the average

GOE 11.10.03

DOT 168

O*NET 24505E

## High School Subjects

Biology
Chemistry
Earth sciences
English
Mathematics

## Personal Interests

The Environment
Science

**The instrument** is narrow and tubelike, three feet long, with a screw-on cap at the top and holes at the bottom. Carl carefully slides it into the monitoring well and draws exactly one-third of a liter of water from the ground, following strict procedures defined by the Environmental Protection Agency, the local county government, and his employer. The data from the test must be absolutely accurate, so precise that it can stand up in court if necessary. Carl is a detective today, looking for any sign of water pollutants such as benzene, a compound that causes cancer in humans.

"We work with very tiny numbers, very tiny compounds," he says. "For example, for benzene, one part per billion in the water is considered too much."

This well and others nearby are sampled every day to ensure that the thousands of barrels of fuel stored at the facility are not leaking. Carl will take his samples to a laboratory operated by his employer, the water will be tested, and if unacceptable levels of hazardous substances are present, the owners of the fuel storage facility will need to clean up the problem and prevent further leaks.

## what does a hazardous waste management technician do?

A hazardous waste site can be anything from a neighborhood gas station to an old garbage dump to a huge industrial complex that once produced military equipment. Broadly defined, a hazardous waste is any substance that threatens human health or the environment. The list includes common products, such as the solvents in paint and the used oil from automobiles, and substances not found in the average home, such as items contaminated by radiation at a nuclear power plant. *Hazardous waste management technicians* help protect human beings and the environment from these substances by cleaning up sites that have been polluted, monitoring for signs of pollution, responding to chemical spills and other emergencies, compiling data, and transporting and storing wastes.

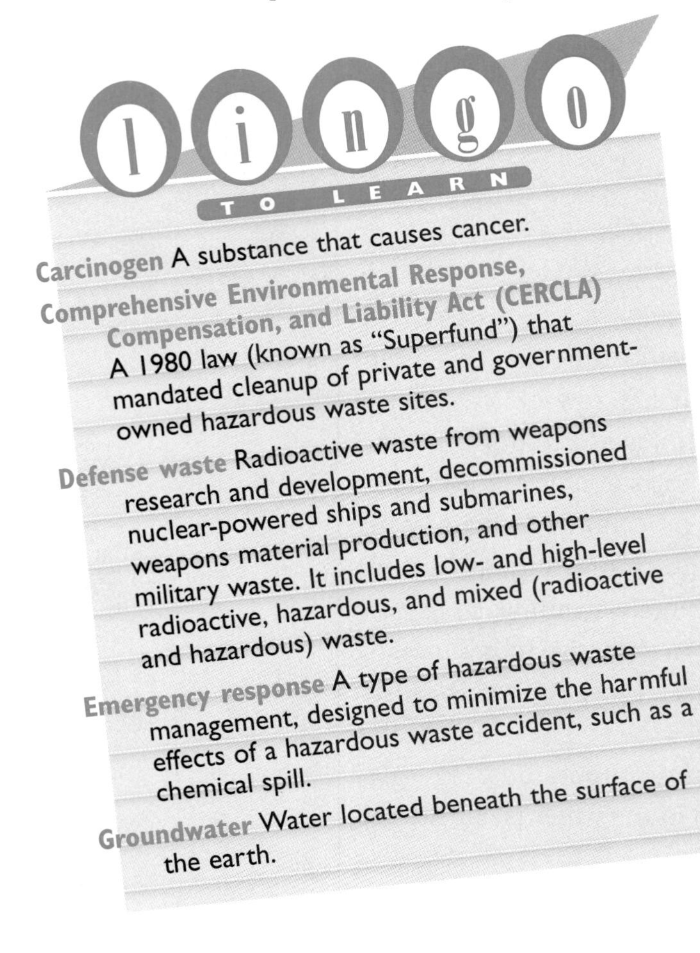

**lingo TO LEARN**

**Carcinogen** A substance that causes cancer.

**Comprehensive Environmental Response, Compensation, and Liability Act (CERCLA)** A 1980 law (known as "Superfund") that mandated cleanup of private and government-owned hazardous waste sites.

**Defense waste** Radioactive waste from weapons research and development, decommissioned nuclear-powered ships and submarines, weapons material production, and other military waste. It includes low- and high-level radioactive, hazardous, and mixed (radioactive and hazardous) waste.

**Emergency response** A type of hazardous waste management, designed to minimize the harmful effects of a hazardous waste accident, such as a chemical spill.

**Groundwater** Water located beneath the surface of the earth.

Hazardous wastes are generated by industries that produce food, textiles, metals, petroleum, plastics, and many other goods and services. Private industry is a major polluter, but the government's military bases and military production sites generate an enormous amount of hazardous materials. Cities and towns also generate a significant amount.

The harmful effects of these wastes were not a cause of widespread concern until a few decades ago, when the nation began to realize that pollution was taking its toll on human health and the environment. An oil spill off the California coastline, a river in Ohio so polluted that it caught on fire, and the shocking discovery of toxic chemicals seeping under a residential neighborhood in Love Canal, New York, were just a few of the catastrophes that triggered a public outcry and a series of new laws overseen by the U.S. Environmental Protection Agency.

The passage of the Comprehensive Environmental Response, Compensation, and Liability Act (CERCLA) in 1980 was a landmark moment. The legislation became known as "Superfund" because it allocated billions of dollars toward cleaning up hazardous wastes in general and inoperative or abandoned dumpsites in particular. The 1986 Superfund Amendment and Reauthorization Act (SARA) established standards and procedures that serve as the basis for the hazardous waste management industry.

Hazardous waste management technicians help ensure compliance with those laws and clean up sites that were polluted in the past. Since many of the worst industrial messes have been decontaminated, technicians now most often are involved in lower priority cleanups, monitoring waste, and preventing accidents.

Technicians assess hazardous waste sites to determine what substances are present and how widespread the pollution is. Some technicians specialize in investigating a site's history to learn what substances were generated or dumped there and who was in charge of the operation. Most technicians, however, focus more on identifying and monitoring chemicals in the soil, air, or water by taking field samples and performing laboratory tests. They also drill wells for monitoring water in the ground, and they help prevent the pollution from spreading. Engineers and other environmental professionals use data collected by technicians to develop plans for cleaning up the site and preventing further contamination.

Sometimes technicians help with bioremediation, placing nutrients in the soil to promote natural decomposition of hazardous wastes. For example, an oil spill would dissipate more quickly if nutrients were added to the soil; microbes would consume the nutrients, multiply, and break down the spilled oil at the same time. Technicians also help remove wastes from the site, transport them, treat them, and store them if necessary.

Some technicians do nothing but respond to accidents or emergencies; they are usually some of the first people on the scene. It's their job to stop the spread of the hazardous substance as quickly as possible. They identify the substance, decide what type of protective gear workers will need to use at the site, and organize the chain of authority called the "incident command system."

> I like being part of the cleanup efforts —after all, it's my drinking water too.

## what is it like to be a hazardous waste management technician?

Carl Patterson has worked for four years as a technician for CH2M Hill, one of the biggest environmental consulting firms in the United States. The company does analysis, cleanup, and follow-up projects for industrial companies, cities, and the government. Carl works at the company's office in Deerfield Beach, Florida, where he performs highly accurate measuring, monitoring, and sampling to ensure that clients are complying with environmental laws.

Carl specializes in identifying, correcting, preventing, and monitoring contamination in surface water (such as lakes, streams, and canals) and groundwater (layers of water beneath the soil, which supply wells and springs). When fuel, chemicals, or other substances are spilled, they can soak through the soil and pollute the groundwater, especially in Florida's marshy environment, where water is often only a few feet beneath the surface of the ground. Carl performs groundwater readings and other field tests in accordance with models developed by engineers. He typically completes about a hundred field tests in one or two months, then spends the rest of the year compiling data. He often performs the same tasks three times to verify the accuracy of the results.

Since the engineers and project managers don't visit the sites as often as technicians, they rely on the technicians to report on what they observe at the scene. "People want to hear what I have to say," Carl says. "They may not always act on every idea, but they listen. They want to know, was the site lush or dead-looking? Is this slope higher than we thought? Is there runoff down to the creek? There's an almost unbelievable number of possibilities for contamination, and understanding what the site is like can help them determine what's wrong."

Carl does not work only in Florida, but in a variety of sites. "It's hard to express how broad it is, how there's projects almost everywhere," he says. "Sites are indoors and outdoors and range from the obvious, like an oil refinery, to the less obvious, like a carpet factory. Metal manufacturing can create dust from cutting, bending, or other processes; when you sweep or wash the dust out of the plant, it has the potential to reach the groundwater and contaminate it. Cork isn't toxic, but the chemicals that hold it together can be. When someone says, 'We're going to Hawaii to do testing at a pineapple factory,' or whatever, my first thought is always apprehensive: 'What's there?' It's my responsibility to find out and take the necessary precautions. You have to be aware of your company's health and safety policy, and your own, but also that of the company site you're visiting.

"There's lots of travel to this job, and sometimes real travel, all over the world, but sometimes you're traveling only to see the armpit of a country, the nasty side. And the work day is long. Sometimes you're at it twelve to sixteen hours a day. You're under budget and time restrictions; you may have to do two weeks' worth of work in one week. On the road, the company might work me ten days straight. I'll take a break and then work another ten days straight. Other jobs are even longer. In Alaska they can only test in spring; they get as many people together as they can and work for weeks straight.

"Of course," Carl adds, "one day I might go out to get surface water samples from a lake. I'll go out in a speed boat. It's warm and sunny. It's beautiful out there. And at the end of the day, you think, 'I didn't really work. I spent the day at the lake.'"

## have I got what it takes to be a hazardous waste management technician?

Hazardous waste management technicians must be alert, accurate, and able to follow orders. Some positions require technical and scientific aptitude. You should be able to cooperate and communicate with others, because you'll usually be working as part of a team. Reasoning skills, analytical thinking, and the ability to respond to unexpected conditions are important. A respect for safety is vital.

There are standards and guidelines for every part of the hazardous waste management technician's job, and you must follow them. If a standard says you must obtain a half-liter sample, you have to get exactly that. When you don't know or don't follow the regulations, you can cost your employer or client a great deal of money in fines or you could endanger someone's life.

At the same time, good technicians are adaptable and quick-thinking. "Sometimes, once you get out to the site, you find that the ground's not dirt like everyone expected, it's mud," Carl says. "There were plans for specific tests, but now it's pouring rain. Or there are power lines that the managers or engineers didn't anticipate."

Good technicians also are conscientious. "When you get to a site, it's not just, 'let's get the drill rig and get a hole in the ground,'" he says. "It's, 'let's find the best place and get the best sample possible.' Or at a monitoring well, 'let's get a decent sample, and not take water that's pooled near the mouth of the well.'"

Testing water, air, and soil in the great outdoors is not the only work a technician does. Some sites are indoors, and many are unpleasant. "Some indoor sites are so noisy that you can't communicate. By the end of the day, you're frazzled," Carl says. "Some really smell bad. Some of these places are producing some really nasty effluents. I'm sometimes stunned at what people do with their wastes.

"Sometimes you're all bundled up in special suits, a paper hat, a respirator," he continues. "The respirator's just an air-purifying mask with filters, but I'm claustrophobic, and it bothers me. It makes it difficult to communicate. The ones I can't deal with are the zip-on suits. You're sealed in a suit from head to toe, like an astronaut. They're for Level A work, for substances with a very high dermal threat, those that will burn your skin." Luckily, he adds, CH2M Hill doesn't do Level A testing.

"We also do some testing in and around Miami, and some of the sites are in bad areas. I look at it as another aspect of health and safety. You and your partner have to watch out for each other," Carl notes.

Still, he enjoys many aspects of his job. "Each project is almost a new world unto itself," he says. "And I like being part of the cleanup efforts. After all, it's my drinking water, too."

**To be a successful hazardous waste management technician, you should**

- Be alert, accurate in your work, and able to follow orders
- Follow safety procedures to avoid harming yourself or others
- Have technical and scientific aptitude
- Be cooperative and able to work as part of a team
- Communicate well with others
- Have good reasoning and analytical skills
- Be flexible and able to respond to unexpected conditions at a site

## how do I become a hazardous waste management technician?

Carl did a stint in the Air Force before winding up with CH2M Hill, where he received extensive on-the-job training. Although it is possible to enter the field with only a high school diploma, a two-year degree or more advanced education from a college or technical institute is becoming important for some hazardous waste management technicians.

"For some field or monitoring work, two-year degrees may be needed," says Steve Wilson of the Air and Waste Management Association. "Some of these jobs involve sophisticated work, like chemical analyses or working under protocols. Other technicians are essentially moving waste, like forklift drivers, warehouse workers, or drivers. They will get some particular training or instruction from the company, but they don't need a degree for this kind of work."

All technicians receive extensive health and safety training on the job, because they constantly come into contact with materials that are corrosive, toxic, and flammable. The training includes instruction in "human hygienic procedures," such as when to put on protective clothing and when to use a breathing apparatus. Technicians must also learn numerous regulations imposed by individual companies and various levels of government.

## education

### High School

To prepare for a career in hazardous waste management, you should study sciences and mathematics. "Grounding in the basic disciplines is going to be crucial," Wilson says. "You may not be able to take all of them, but biology, chemistry, physics, the natural sciences, algebra, trigonometry, and geometry are very important. It won't be a black mark if you take an additional year or two to get this education." Courses in English, speech, writing, and computers are also recommended.

### Postsecondary Training

Many community colleges, technical colleges, and vocational institutes offer two-year degrees in hazardous waste management or environmental resource management. Before you enroll in this type of school, make sure it's accredited and visit the placement office to find out what organizations have hired the school's graduates.

Courses at these schools typically include core classes, such as mathematics and communications. Students also learn how to keep records; how to classify, manage, and dispose of hazardous waste; what the regulatory requirements are for companies; and how to measure certain parameters of hazardous materials, such as toxicity. Additional training is provided for students who expect to work in a nuclear power plant or other site that will require specialized knowledge.

Hazardous waste schools and associations also sponsor workshops where employees can learn the most recent developments in technology and stay informed about legal aspects of the trade. Employers usually provide the required safety training and refresher courses for workers who have been in the profession for some time.

## certification or licensing

The Institute for Hazardous Materials Management gives a test for certifying people as hazardous materials managers. The National Environmental Health Association certifies hazardous waste specialists. In addition, a variety of accrediting bodies award professional engineering technician credentials (*see* "Sources of Additional Information").

## internships and volunteerships

Two national nonprofit groups help place people in internships in the environmental industry: the Student Conservation Association (SCA) and the Environmental Careers Organization (*see* "Sources of Additional Information").

The SCA focuses on internships in resource management, placing people in projects in federal, state, and private parks and natural lands. Each year, the SCA places about 1,500 high school students and high school graduates in summer internships and other jobs.

The ECO is a major environmental placement service for college students, recent graduates, and first-time job seekers. "About 85 percent of what we do is placement," says Lee DeAngelis, ECO Great Lakes Regional Director. "We're like the Kelly Girl or Manpower of environmental services." The ECO publishes *Resources for Planning Your Career* and *The New Complete Guide to Environmental Careers.*

Many federal and local government agencies also have internship programs, including the EPA, National Park Service, and Bureau of Land Management. In the private sector, an internship in a nonprofit organization is probably the easiest to obtain. These groups include the National Wildlife Federation and the Natural Resources Defense Council. Opportunities exist at both the local and national levels.

## labor unions

There are no national labor unions for hazardous waste management technicians, but in many cases technicians can join a company union. For example, all automotive companies have hazardous waste management technicians who join company unions. Such unions typically provide their own training. Technicians tend to leave the unions when they become supervisors or managers.

## who will hire me?

Companies that don't comply with environmental laws can be fined thousands or even millions of dollars. They often hire teams of environmental consultants, including hazardous waste management technicians, to ensure that they meet government regulations. About three out of four hazardous waste management technicians work for private companies. The government also hires private environmental consultants to clean up its numerous contaminated sites, at a cost of billions of dollars every year. Additional jobs are available with nonprofit organizations, waste disposal companies, waste disposal consulting engineering firms, and environmental consulting firms.

In this field you might be hired by a mining company to address the problem of arsenic and other hazardous minerals that leach from mines into nearby surface water and groundwater. Food and beverage companies use hazardous materials for cleaning and refrigeration. University research facilities, hospitals, laboratories, health care facilities, and pharmaceutical companies also hire technicians in this field. In fact, most industries produce hazardous waste.

It's common for large companies to have their own staff of environmental technicians. A small company might have one or two professionals on staff or hire independent consultants. The exception to this is the government, which generates a great deal of waste but tends to hire private consulting companies to clean it up.

There are companies that do nothing but dispose of other companies' wastes. Probably the largest is Waste Management, an international corporation. Smaller firms may specialize in treating one type of waste, such as waste from metal finishing.

Jobs are also available with federal agencies. The U.S. Forest Service, the National Park Service, the Bureau of Land Management, and the Fish and Wildlife Service employ numerous technicians. The EPA employs some technicians but uses more scientists and other professionals with advanced education. Of course, the government's Superfund projects have created many jobs in the field. Opportunities are also arising in Eastern Europe, the Middle East, the former Soviet Union, and other countries where hazardous waste management is becoming a priority.

## where can I go from here?

Technicians in this field are typically promoted to specialized or supervisory positions as they become more experienced on the job. You might become a specialist who studies hazardous waste management projects for the government or private industry. That position involves providing information on the treatment and containment of hazardous waste. Specialists who work for government agencies often help develop regulations for the industry.

You might become an incident commander, the person in charge of and responsible for the success or failure of projects at hazardous waste sites. Incident commanders delegate tasks to other workers and interact with state and federal regulatory authorities.

Some environmental professionals work in community relations or public affairs, informing the public about what companies are doing with their wastes. The government also employs media specialists to help spread information about regulations and cleanup efforts.

Most hazardous waste management technicians don't return to college for more advanced degrees, but if you did, you would probably earn more money and be given greater responsibilities. Carl notes, "If I had a two-year associate's degree in one of the related scientific areas, my title would be different. If I had one in geology, for example, I'd be a geo-tech." (*See* chapter "Geological Technician.") Carl has no plans to pursue a degree in hazardous waste management. In fact, he's taking night classes that he hopes will allow him to change careers and become an artist, perhaps a filmmaker.

| Related Jobs |
|---|
| Solid waste management technicians |
| Waste water treatment plant operators |
| Land and water conservation technicians |
| Landfill operators |
| Physical geographers |
| Pollution control technicians |
| Radioactive waste disposal dispatchers |
| Waste disposal attendants |

## what are some related jobs?

"Environmental health and medicine is a growing area," Steve Wilson says. "It involves things like health risk analysis, including epidermalogical (skin) studies to check for health problems due to exposure to hazardous substances." The field includes such occupations as environmental nursing and public health veterinary medicine, environmental sanitation, and industrial hygiene. *Environmental health field and lab technicians* help identify human or animal exposure to contaminants in workplaces or other locations and the ways such exposure affects health.

Another growing area is pollution prevention. "Companies are starting to look at pollution and waste from the standpoint of generation—how and why it starts, how it can be prevented or reduced, recycling possibilities, and so forth," Steve says. *Pollution prevention technicians* work on projects to reduce or eliminate air, water, or soil pollution before it has a chance to become waste.

The U.S. Department of Labor classifies hazardous waste management technicians under the heading *Regulations Enforcement, Health, and Safety.* Also under this heading are people who work in industrial health and safety. Other related occupations include *municipal water treatment plant technicians,* who test and monitor water quality at treatment plants to be sure it's safe to drink, and *radioactive disposal attendants,* who collect radioactive waste and contaminated equipment, load it on a truck using a forklift, and transport to a storage facility or landfill. These workers may help make the concrete forms used to contain waste that will be buried. They also may clean contaminated equipment for reuse, using sand blasters and scrubbers, detergents, and other tools and supplies.

## what are the salary ranges?

"Salaries vary a lot, depending on the region, the company, the specific duties of the technician, and his or her education and experience," says Mike Waxman, associate professor in the College of Engineering at the University of Wisconsin–Madison. "A lot also depends on the type of company. If you've worked for OHM and have had experience in handling big (emergency response) incidents, for example, you're going to be very valuable to another company and paid very well. You also have to be very good at your job and very good with people."

### FYI

**What Are the Risks?**

Are there significant health risks for people who work with hazardous waste?

"When you put it that way—significant health risks—no," Carl says. "Because I have the training, equipment, and unbelievably redundant systems to protect me. My company has a comprehensive quality control plan that covers general decontamination procedures and so forth. Then, there are procedures for specific sites, such as a gas station site or a metals manufacturing plant.

"And in the field, the company gives me the right to shut down the project if I ever feel there's a threat to my safety or others'," he adds. "A lot of time and money go into planning a project; I better have a good reason for shutting it down. But if we go out in the field and see a problem that would cause some risk to health, we go home, even if it's a multimillion-dollar project."

"In this job, there's safety and accuracy. Both are important. But if you have to put one first, the company says put safety first.

"Sometimes you don't always know what's out there. Some chemicals have no smell, no taste. You can be breathing something [harmful] and you don't even know it. In my opinion PCBs are some of the worst. They're big molecules, but they like to grab onto things. You have to be careful. When you're working with water, something's always going to spill; with PCBs, if you're talking with someone, not paying attention, and the water splatters and gets on your lip, you could inhale the PCBs. We decommission everything [equipment, clothing, and vehicles] so we don't bring hazardous substances home with us."

## in-depth

### Not In My Backyard

Almost everyone agrees that environmental cleanups are necessary. Unfortunately, however, there are problems that have slowed progress: the very high costs of many of the cleanup efforts, court actions involving companies that try to circumvent the regulations and disagreements over what constitutes priority cleanups and "hazardous waste."

There's also a problem popularly known as NIMBY, for "Not In My Backyard," one that Steve Wilson knows very well. He explains, "By law, states are supposed to provide a place to dispose of waste. The disposal sites must be within their borders. Before the act, states like New York were sending their waste to other states, like North Carolina. Now North Carolina won't take it anymore, in compliance with the federal legislation."

However, a commission for the state of New York researched potential disposal sites throughout the state and came up empty handed. Each of the five sites chosen by the commission was nixed by the public. "Local residents blocked them," Steve says. "They said, 'No way here.'"

The commission focuses on low-level radioactive waste, which is governed by the Nuclear Regulatory Commission, rather than the EPA. "Most of the waste cleanup is from nuclear power plants," Steve says. "Things like solidified fuel, uniforms, booties. These plants are big; there's lots of stuff under the dome. There's a need to cut it up and put it somewhere." The rest of the waste is in smaller amounts, from operations such as clinics, hospitals, and university research facilities that use isotopes in research.

"Right now, the utility companies are keeping their own wastes, warehousing it, waiting for a site to be cleared for disposal," Steve says. As for the research facilities, he adds, "They have a problem. They don't have places to store it. They're either building temporary buildings for it, or it's basically sitting in barrels out in the rain.

"That's the irony," he adds. "This material can't 'blow up'; it doesn't get warm, it just radiates; there's not enough to make a bomb. But the public would rather have it sit around in barrels in warehouses or temporary storage facilities than to bury it in a landfill. Yet there's a much higher chance that it will be mismanaged, that someone will back a truck into it or something, if it's left at the generators."

"Right now, the commission's hands are tied. New York State isn't the only one that failed to come up with landfill sites, though. Of the forty-eight contiguous states, none has come up with a site. Not in the seven or eight years since the act passed." What happens next? "I'm not sure," Steve admits. "Send the legislation back to Congress to rewrite the law? Find central sites for multiple states?

"We're not being allowed to do anything," he says. "The public is very scared of it. But hazardous waste is with us. It exists now, and we have to put it somewhere."

---

According to the *Encyclopedia of Careers and Vocational Guidance,* a hazardous waste management technician with only a high school diploma earns about $16,000 annually. With some postsecondary education, you could expect to earn $18,000 to $25,000 annually, depending on the geographic area and the amount of hazardous waste management being done there. With education and years of experience, you could make up to $45,000 annually. To earn a higher salary than that, you would need experience and advanced education in chemistry. Technicians who can perform complex field tests usually done only by professional chemists are paid the highest salaries. Most technicians also receive health insurance, paid time for vacations, and reimbursement for the cost of traveling to job sites.

According to the U.S. Bureau of Labor Statistics, the median annual salary for all science technicians in 1996 was about $27,000; and the middle 50 percent were paid $19,800 to $37,100. In 1997 science technicians with federal jobs started at $15,500 to $19,500, depending on their education and experience. The average salary for physical science technicians with federal jobs was $35,890 in 1997.

## what is the job outlook?

Hazardous waste management was one of the most rapidly growing environmental careers of the 1990s and is still a field with great opportunities. Although its growth has slowed somewhat, it has continued expanding faster than the average occupation.

Some contaminated sites have been cleaned up, but many are still being decontaminated. In addition, new hazardous materials and sites in need of attention are still being identified. The question of what to do with nuclear waste has not been resolved; the nuclear industry will probably generate jobs in the field for several decades. Numerous other industries that produce hazardous waste will also have to continue hiring technicians to ensure that they meet government regulations.

Opportunities in hazardous waste management, particularly the Superfund projects, are affected by politics and the public's willingness to see tax dollars allocated for cleaning up pollution. Although technicians can expect a fair degree of job security, their jobs may change, and the requirements for entering the field may change accordingly. The trend will probably be toward generating less waste instead of finding ways to dispose of it. Technicians will also be affected by the invention of new technology.

"For example, the industry is looking at some things that would eliminate technicians going out as a sampler," Carl says, "like fiber optic detectors that tell you directly what constituents are in the groundwater, so you don't need to do it in the lab. It saves money—there's no shipping, no transporting, no lab. There's some of these now: you stick a probe down there, and it tells you what is in the water.

"But I think technicians will always be able to move into another area," he adds. "And some testing will always require a human presence. For example, there are mobile rigs to dig fifty to one hundred feet into the ground to take a sample. Someone needs to be there to be sure the rig is set up properly, with a tight seal, and be around when it's time to pull up the sample."

## how do I learn more?

### sources of additional information

Many of the science and engineering societies are good sources of information about the environmental industry, but their main focus is not on workers at the technician level. Currently, there is no trade organization designed for hazardous waste management technicians.

**Air and Waste Management Association**
One Gateway Center, Third floor
Pittsburgh, PA 15222
412-232-3444
http://www.awma.org

**Environmental Careers Organization**
National Office
286 Congress Street, 3rd floor
Boston, MA 02210
617-426-4375

**National Solid Waste**
**Management Association**
4301 Connecticut Avenue, NW, Suite 300
Washington, DC 20008
202-244-4700

**American Institute of Chemical Engineers**
345 East 47th Street
New York, NY 10017
212-705-7338

**Association of Solid Waste Management**
444 North Capitol Street, NW
Washington, DC 20001-1512
202-624-5828

**Institute of Hazardous Materials**
5010 Nicholson Lane
Rockville, MD 20852-3108
301-984-8969

**Student Conservation Association**
PO Box 550
689 River Road
Charlestown, NH 03603
603-543-1700
http://www.sca-inc.org

# bibliography

Following is a sampling of materials relating to the professional concerns and development of hazardous waste management technicians.


## Books

Basta, Nicholas. *The Environmental Career Guide: Job Opportunities with the Earth in Mind.* New York: Wiley, 1991.

Sharp, Bill (principal author). The Environmental Careers Organization. *The New Complete Guide to Environmental Careers.* 2nd ed. Washington, DC: Island Press, 1993.

Wang, Robert L. and Rhoda G. M. Wang, eds. *Effective and Safe Waste Management: Interfacing Sciences and Engineering with Monitoring and Risk Analysis.* Discusses safety measures and management of hazardous waste. Boca Raton, FL: Lewis Publishers, 1993.

Simmons, Milagros S., ed. *Hazardous Waste Measurements.* Explains techniques of quality assurance and control for hazardous waste. Chelsea, MI: Lewis Publishers, 1991.

## Periodicals

*Environmental and Engineering Science.* Quarterly. Discusses regulation and management of hazardous waste from the economic and ecological viewpoints. Mary Ann Liebert, Inc., 2 Madison Avenue, Larchmont, NY 10538, 914-834-3100.

*Environmental Careers Bulletin.* Monthly. Opportunities from a wide selection of environmental occupations. Environmental Careers Bulletin, 11693 San Vicente Boulevard, Suite 327, Los Angeles, CA 90049, 310-399-3533.

*Hazardous Substances & Public Health.* Bimonthly. Connects hazardous waste issues and human health. U.S. Department of Health & Human Services, Agency for Toxic Substances and Disease Registry, 1600 Clifton Road NE, E-33, Atlanta, GA 30333, 404-639-0540.

*Hazardous Waste Business.* Biweekly. General news concerning hazardous waste. McGraw-Hill, 1221 Avenue of the Americas, New York, NY 10020, 212-521-6410.

*Hazardous Waste Consultant.* Bimonthly. Technological and regulatory aspects. Elsevier Science, Inc., Box 945, New York, NY 10159-0945, 212-633-3730.

*Hazardous Waste News.* Weekly. Discusses all aspects of hazardous waste disposal. Business Publishers, Inc., 951 Pershing Drive, Silver Spring, MD 20910-4464, 301-587-6300.

*Haztech News.* Biweekly. Reviews the technological aspects for treating and cleaning hazardous sites. HazTECH Publications, Inc., 14120 Huckleberry Lane, Silver Spring, MD 20906, 301-871-3289.

*Waste Management.* Quarterly. International in scope, this magazine presents information covering all aspects of waste and its disposal. Elsevier Science, Inc., Regional Sales Office, Box 945, New York, NY 10159-0945, 212-633-3730.

# heating, air-conditioning, and refrigeration technician

**Definition**

Heating, air-conditioning, and refrigeration technicians build, test, install, maintain, and repair heating, air-conditioning, and refrigeration systems and equipment.

**Alternative job titles**

Environmental control service technicians

Furnace installers

Heating, air-conditioning, and refrigeration mechanics

Heating, ventilation, air-conditioning, and refrigeration technicians

**Salary range**

$15,000 to $25,500 to $50,000

**Educational requirements**

High school diploma; two-year technical school highly recommended

**Certification or licensing**

Required for handling refrigerants

**Outlook**

About as fast as the average

GOE
05.05.09

DOT
637

O*NET
85902A

## High School Subjects

Chemistry
Computer science
Mathematics
Physics
Shop (Trade/Vo-tech education)

## Personal Interests

Building things
Figuring out how things work
Fixing things

Tom Gretka dials the combination to the shop padlock. It's 6:00 in the morning and just beginning to get light. The heavy metal doors rattle as he slides them up. Inside the shop, Tom walks past the large wooden work tables, past electric saws and power tools used to cut and join sheet metal, and past neat compartments of bolts, pipe fittings, and electrical wiring stacked fifteen feet to the ceiling. Beside a tall shelf, Tom pulls a piece of paper from a file and reads his work order for the day: "Project: Check duct runs in attic. Install 20-foot underground return air duct. Also 6-foot duct as per plan." Tom knows he has a big job ahead of him. He loads the tools he'll need into a truck and begins driving, happy to get an early start on the day.

# what does a heating, air-conditioning, and refrigeration technician do?

*Heating, air-conditioning, and refrigeration technicians* build, install, and maintain equipment necessary for comfortable, safe indoor environments. They provide heating and air-conditioning in such structures as shops, hospitals, theaters, factories, restaurants, office buildings, and private homes. They may work to provide temperature-sensitive products such as computers, foods, medicines, and precision instruments with refrigerated climates. They may also provide cooling or heating systems in modes of transportation such as trucks, planes, and trains.

Heating, air-conditioning, and refrigeration technicians work for a variety of companies and businesses. They may be employed by heating and cooling contractors, manufacturers, dealers and distributors, utilities, or engineering consultants.

When assembling and installing heating, air-conditioning, and refrigeration systems and equipment, technicians work from blueprints. Blueprints indicate how to assemble and install components. While working from the blueprints, technicians use algebra and geometry to calculate the sizes and contours of duct work as they assemble it.

Heating, air-conditioning, and refrigeration technicians work with a variety of hardware, tools, and components. For example, in joining pipes and duct work for an air-conditioning system, technicians may use soldering, welding, or brazing equipment, as well as sleeves, couplings, and elbow joints. Technicians handle and assemble motors, thermometers, burners, compressors, pumps, and fans. They must join these parts together when building climate-control units, and then connect this equipment to the duct work, refrigerant lines, and power source.

As a final step in assembly and installation, technicians run tests on equipment to ensure proper functioning. If the equipment is malfunctioning, technicians must investigate in order to diagnose the problem and determine a solution. At this time, they will adjust thermostats, reseal piping, and replace parts as needed.

Technicians who respond to service calls perform tests, diagnose problems, and make repairs. They calibrate controls, add fluids, change parts, clean components, and test the system for proper operation. For example, in performing a routine service call on a furnace, technicians will adjust blowers and burners, replace filters, clean ducts, and check thermometers and other controls.

Rather than work directly with the building, installation, and maintenance of equipment, some technicians work in equipment sales. These technicians are usually employed by manufacturers or dealers and distributors and are hired to explain the equipment and its operation to prospective customers. Other technicians work for manufacturers in engineering laboratories, performing many of the same assembling and testing duties as technicians who work for contractors. They assist engineers with research, equipment design, and testing.

Technicians may also work for consulting firms, such as engineering firms or building contractors who hire technicians to help estimate costs, determine heating and air-conditioning load requirements, and prepare specifications for climate-control projects.

Heating, air-conditioning, and refrigeration technicians may also specialize on only one type of cooling, heating, or refrigeration equipment. *Air-conditioning and refrigeration technicians*

## lingo TO LEARN

**Blueprint** A photographic print in blue and white used especially for mechanical drawings and architectural plans.

**Chlorofluorocarbon (CFC)** A chemical compound used for cooling in air-conditioning and refrigeration systems. The release of CFC contributes to stratospheric ozone layer depletion, and scientists are working to develop safer substitutes.

**Duct** A pipe or tube through which air or gasses are carried in heating, air-conditioning, or refrigeration systems.

**HVAC&R** A common abbreviation for heating, ventilation, air-conditioning, and refrigeration.

**Triangulation** A trigonometric operation for finding a position or location by means of bearings from two fixed points a known distance apart.

**Voltmeter** An apparatus for measuring the quantity of electricity passed through a conductor by the amount of electrolysis produced.

install and service air-conditioning systems and a variety of refrigeration equipment, including walk-in coolers and frozen-food units in supermarkets.

*Heating-equipment technicians* or *furnace installers* install oil, gas, electric, solid-fuel, and multi-fuel heating systems. They also perform routine maintenance and repair work.

Technicians must also be aware of the potential hazards of dealing with chlorofluorocarbon (CFC) and hydrochlorofluorocarbon (HCFC) refrigerants, which are commonly used in refrigeration and air-conditioning systems. The release of HCFCs and CFCs into the atmosphere contributes to the depletion of the ozone layer, and therefore technicians must be careful to conserve, recycle, and recover these refrigerants. By the year 2000, in fact, manufacturers can no longer produce CFC refrigerants, and regulations will prohibit the discharge of HCFCs and CFCs.

**To be a successful heating, air-conditioning, and refrigeration technician, you should**

- Have excellent communication skills
- Enjoy building and repairing things
- Have manual dexterity and be detail oriented
- Be confident using hand and power tools and be mechanically inclined
- Not be afraid of heights

## what is it like to be a heating, air-conditioning, and refrigeration technician?

Tom Gretka has worked in the heating, air-conditioning, and refrigeration field for twenty years. For the last four years he has worked at the Carlson Company, a heating and cooling subcontractor in Tucson, Arizona. "We contract work from all kinds of places," says Tom, "restaurants, businesses, and private homes. We do installation on new structures and on renovations. We also do service calls." At Carlson, Tom works a forty-hour week, from 7:00 AM to 3:30 PM or 6:00 AM to 2:30 PM Monday through Friday, depending on the season. "Of course, if it gets really busy, or if there's a deadline creeping up, I may work more, but forty hours is typical," he says.

The first thing Tom does when he arrives at Carlson in the morning is to look at his work order. The work order tells Tom what he will be doing for the day. Some days, Tom stays in the shop building duct work that will be installed later. "I work from a shop drawing or a blueprint," he says, "then I use a device called a ductolator that tells just about everything you need to know to size the ducts. The square footage of the structure determines how big the ducts should be. The ductolator helps with the figuring." When working with sheet metal to build the ducts, Tom operates brakes, shears, and other shop machines to bend and cut the material. He refers carefully to the blueprint and does quick geometric computations. Then he begins to solder. "Whether we fully assemble the ducts in the shop or put them all together at the site depends on how big they'll be. Knocked down (not assembled) the ducts might fit in one truck. Put together, they could take four."

"Basically, if you work for a contractor, you're going to be doing shop work, field work, or service work," says Tom. If his work order requires field or service work, Tom notes the name and address of the customer, which is printed clearly at the top of the order, so that he knows where he will be working. Then he reads the project description to determine the duties he will be performing and the equipment and tools he will need. "It might tell me I'm going to be removing a cooler and replacing it with a new one, or removing a gas or electric furnace, or adding air-conditioning to a house. There's a big variety, really."

"Once I'm at the job site," explains Tom, "I do whatever's necessary to fulfill the order. I install the duct work, run refrigerant tubing, position the outdoor condensing unit, do all necessary electrical work. Really, it just depends on the work order." After Tom has removed and replaced equipment per his work order, he runs tests on the system to make sure it functions properly. Once the tests are complete, Tom says, "we clean everything up and make it look nice. That's important. You can't just leave a bunch of exposed tubing. It might work fine, but no customer's going to be happy with the job if you leave things looking terrible." At the end of the workday, Tom drives back to Carlson, neatly replaces the tools in the shop, and washes up.

## have I got what it takes to be a heating, air-conditioning, and refrigeration technician?

"There's not much I don't like about my job," says Tom, "but anyone interested in doing it has to realize there'll be some dirty, hot work. That comes with the territory." When heating, air-conditioning, and refrigeration technicians install equipment at construction sites, they will often be working outside in the elements, including summer heat, winter cold, rain, and snow. In addition, of course, the construction sites will not have a climate-control system running, since that's what the technicians are there to put in! "It's true," says Tom, "we don't always benefit from the results of our labor. By the time the air-conditioning system is up and running, our job is over. It's time for us to leave." Technicians face the same situation in doing repairs, where they may work in cold basements, hot attics, or stuffy refrigerated trucks to repair malfunctioning systems.

Besides being able to work in uncomfortable temperatures, you should have good manual dexterity if you are interested in becoming a heating, air-conditioning, and refrigeration technician. Working with motors, timers, and other components often means performing small, meticulous calibrations and adjustments. Precise, steady manipulative skills are an asset on this job.

FYI

**History**
The ancient Greeks and Romans constructed buildings with central heating and ventilation systems, but this knowledge was lost during the Middle Ages (about AD 500-1500 ) in Europe.

You should also have confidence in using a variety of hand and power tools and be mechanically adept. This is particularly important if you wish to work in the installation and service facets of the field. Being in good physical condition is also important since this type of work can involve lifting and moving heavy equipment.

Heating, air-conditioning, and refrigeration technicians must also have people skills, since they often interact with customers and fellow employees. Communication skills and the ability to be courteous and tactful are crucial.

One reason Tom likes his job is that it offers him a variety of experiences. He works at different locations and performs different tasks every day. His favorite part of the job is remodeling. "I love going into an old house and putting in a new system. It's become like a hobby for me, really. Every single one is different, and it's great to get them up in working order again with a good system." If Tom were working in a lab or in sales, he speculates that his job might have less variety, which might appeal to some. "There are people who like to work in the same place, who don't want to have to haul equipment around and always wonder what their work setting will be later in the week." He suggests that these people would be happier working for manufacturers or dealers in the laboratory or sales office, or for institutions that employ their own heating, air-conditioning, and refrigeration staffs.

## how do I become a heating, air-conditioning, and refrigeration technician?

"I started at the bottom and worked my way up," says Tom. After high school, Tom took a job at an air-conditioning warehouse where he learned the names of the equipment while he took inventory, shelved and organized stock, and filled orders. He got promoted and became a helper to a technician who gave him such duties as carrying equipment, cleaning up, and insulating refrigerant lines. "My training was on-the-job. I was lucky enough to work with someone who told me, 'If you pay attention to me you can learn everything you need to know about this job.' My advice is to stick with whoever is willing to train you, watch what they're doing, and listen."

## education

### High School

While in high school, you should take at least two years of advanced mathematics, including at least one year each of algebra and geometry if you are interested in becoming a heating, air-conditioning, and refrigeration technician. "When building and installing ducts, especially, we do an awful lot of figuring," says Tom. "We work with variables and use geometric concepts such as triangulation."

You should take metal and machine shop to become acquainted with construction, blueprints and shop drawings, sheet metal, and power and hand tools. Shop courses in electricity and electronics will provide a strong introduction into understanding circuitry and wiring and will teach you to read electrical diagrams, necessary when building, installing, and repairing climate-control systems. Students who study these subjects in high school will have an advantage when looking for on-the-job training or when studying at a technical college.

Courses in computer applications and programming, including computer-aided design (CAD), are also important for high school students interested in working in this field because much climate-control equipment is now being installed with microcomputers. Physics is an important subject for heating, air-conditioning, and refrigeration technicians, who need a solid background in a variety of physical principals to understand the purposes and functions of the equipment and the power sources they use.

Finally, English composition classes will provide you with the skills you will need to be able to follow detailed written directions and to write clear reports for employers and customers outlining how equipment has been installed, what tests have been run, and what maintenance procedures have been followed.

### Postsecondary Training

Although postsecondary training is not mandatory to become a heating, air-conditioning, and refrigeration technician, employers prefer to hire technicians who have training from a technical school, junior college, or apprenticeship program.

Many technical schools and community colleges offer one- or two-year programs in which students can receive a degree or certificate as a heating, air-conditioning, and refrigeration technician. In these programs, students study physics and mathematics and how they apply to the climate-control field. The programs also cover equipment design, construction, installation, and electronics, providing hands-on experience as well as classroom demonstration and lecture. Most programs require additional courses such as English and computers. The Armed Forces also offer six-month to two-year training programs in heating, air-conditioning, and refrigeration.

Apprenticeship programs are another option for gaining education and training. These programs are often run by local professional organizations and unions, including the Air-Conditioning Contractors of America, the Mechanical Contractors Association of America, the National Association of Plumbing-Heating-Cooling Contractors, the Sheet Metal Workers' International Association, and others. These programs typically last four to five years and combine classroom instruction with on-the-job training. Students must receive a high school diploma or equivalent before applying to any of these programs.

## certification or licensing

Heating, air-conditioning, and refrigeration technicians who will be purchasing or handling refrigerants such as CFCs must pass an EPA-approved certification test. The written exam tests the applicant's knowledge about how refrigeration systems work, federal regulations and legislation regarding refrigerants, and the refrigerant products themselves. Certification is broken down into three areas, depending on the specialization area of the technician. Type I certification is for those who service small appliances. Type II certification is for technicians who deal with high-pressure refrigerants, and Type III certification is for those working with low-pressure refrigerants. The tests are administered by EPA-approved organizations, such as trade schools and unions.

Technicians enrolled in postsecondary certification programs will probably take this exam with the rest of their class at the end of the training period. Technicians who are not in collegiate, Armed Forces, or apprenticeship programs must find out how to obtain certification on their own. Tom's employer told him when and where to take the exam. He attended a seminar on the handling of chemical refrigerants and then took a test. Tom had to pay $150 to receive his certification, and after he passed the exam, he received a certification card. At Carlson, all employees who handle or purchase refrigerants must have certification cards on file in the office.

Two new voluntary certification programs now exist as well. ACE, or Air Conditioning Excellence, is a national certification program sponsored by the Air Conditioning Contractors of America (ACCA). Certification requires that the technician pass a core examination and a heating and/or air-conditioning test. The other certification program is the North American Technician Excellence (NATE) Program, which certifies technicians within various specialties, including air-to-air heat pumps, air conditioning, air distribution, gas furnaces, and oil furnaces. Certified technicians must get recertified every five years.

Technicians working in the industrial refrigeration industry may choose to become certified through the Refrigerating Engineers and Technicians Association's (RETA) program.

## scholarships and grants

Schools that offer programs in heating, air-conditioning, and refrigeration technology may provide scholarships as well. Contact your financial aid office or advisor for a complete list and additional details.

Professional associations frequently offer lists of scholarships, and some grant scholarship awards, such as the National Association of Plumbing-Heating-Cooling Contractors (*see* "Sources of Additional Information"). Large manufacturers may also offer scholarship opportunities.

## internships and volunteerships

With few exceptions, high school students lack the necessary training and work experience to be hired for summer jobs or internships in the field of heating, air-conditioning, and refrigeration. There are, however, still many ways for students interested in this field to begin developing some of the skills necessary for the job. Hobbies such as working on cars, bicycles, and small electrical appliances will help students gain confidence with tools, develop manual dexterity, practice assembling and repairing components, and learn about electronics.

High school students may also arrange to visit heating, air-conditioning, and refrigeration manufacturers, contractors, and dealers, as well as utility companies to get a direct look at what the field entails. When visiting, students may talk to technicians, helpers, engineers, and contractors to discuss job possibilities in the field. High school students may also want to contact professional organizations and unions for information (*see* "Sources of Additional Information").

Internship opportunities may be available for students enrolled in technical schools or colleges. In fact, internships may be part of the curriculum, especially during the latter stages of the school program or during the summers.

## labor unions

Many technicians employed in the field of heating, air-conditioning, and refrigeration belong to a union. Two major unions for this field are the Sheet Metal Workers' International Association and the United Association of Journeymen and Apprentices of the Plumbing and Pipefitting Industry of the United States and Canada. Some employers require technicians to become members of a trade union. At Carlson, as with many other small shops or contractors, employees are not required to be union members. In addition to unions, technicians may join any of several professional associations which offer seminars, meetings, and publications for members.

## who will hire me?

After about a year of on-the-job training as a helper, Tom became a technician, with his own truck and his own jobs. He got his job at Carlson by word-of-mouth. "Once you have your foot in the door, you hear about what's going on and who's hiring." Tom was interested in working for Carlson because he knew that Carlson frequently worked on house renovations, an aspect of the trade that Tom especially enjoys. David Carlson, Tom's boss and co owner, knew Tom and was familiar with his work skills and background. When Tom applied for the job, David was glad to hire him.

Like Tom, more than half of all heating, air-conditioning, and refrigeration technicians are employed by cooling and heating contractors who are hired to install and maintain climate-control equipment in buildings and homes. "If you're willing and able to do shop work, field work, and service work for a contractor, you have a better chance of finding work." Technicians who are interested in specializing, however, should look for work with a contractor who has that specialty. For example, a technician who prefers service should apply for work with a service-oriented contractor.

### Advancement Possibilities

**Inspectors** ensure that equipment or systems are manufactured or installed according to approvals, specifications, or codes and that installations and equipment adhere to standards of companies, associations, and/or government agencies.

**Service managers** supervise service workers in the field, publish service information, and provide technical training to employees and customers.

**Project managers** oversee a specific project or projects, including engineering design, materials control, on-site problem solving, and coordination with other companies at the job site.

**Sales managers** work with distributors and dealers, promoting and selling manufacturer's products.

While finding employment with contractors and dealers by word-of-mouth is routine, technicians can also find opportunities with manufacturers, utilities, contractors, and dealers in the classified section of the newspaper or on the Internet.

Technicians who attend technical or trade schools can often find employment through the school. Students may do an internship at a company between the first and second years of training and then be hired by that company after graduation. Even if the company does not have job openings, the people should be able to provide the student with references and possibly with information on a company that does have job openings. The school's job placement office is also an excellent way for students to find work. Sometimes these offices have job opening listings or bulletin boards that are even available to the general public. Prospective technicians who are not in a training program should contact local technical or junior colleges to see if such listings exist.

Local trade organizations are another way for prospective technicians to find employment. These organizations may publish newsletters that include job listings or may sponsor conferences at which employers meet with and interview job candidates. Tom's boss, David Carlson, recommends membership in the Refrigeration Service Engineers Society (RSES). RSES hosts local, national, and international meetings, and David notes that these meetings are a good way to break into the heating, air-conditioning, and refrigeration trade (*see* "Sources of Additional Information").

## where can I go from here?

Over the last twenty years, Tom's advancement in the field of heating, air-conditioning, and refrigeration is evident through his steady increase in responsibilities and pay. At this point, Tom has become a master craftsman in his field and has the most responsibilities he can have as a technician. He goes out on his own jobs, creates his own work strategy for completing a task, and discusses the job directly with the customer. He is responsible for overseeing the work of helpers and less-experienced technicians and for writing up the work reports when a job is completed. Tom enjoys his job and plans to stick with it, learning all that he can about renovations of old houses. "Even after twenty years, I'm always learning," he says. "That keeps it fun."

If Tom did want to change jobs and alter his duties, there are many employment possibilities open to a skilled heating, air-conditioning, and refrigeration technician. Some technicians go into business for themselves, starting their own contracting or consulting firms. Others choose to switch to a different aspect of the trade, such as moving from service into sales or from installation into manufacturing. Technicians who start out in sales may move up to sales management. *Sales managers* promote and sell products, acting as liaisons between distributors and dealers.

Technicians at large companies or factories may get increases in responsibility and salary by being promoted to *project manager* or *service manager*. Service managers oversee other technicians who install and service equipment. Project managers are responsible for specific projects and oversee the design aspect, materials control, troubleshooting, and overall coordination of projects.

Some technicians choose to leave the physical aspect of the job and move into administrative positions with contractors, dealers, or manufacturers. Some become *inspectors* and check on installations of equipment to make sure they meet specifications and codes. In addition, many technicians move into the related field of plumbing. Finally, it is possible for technicians to return to college to earn a four-year degree in electrical, mechanical, or civil engineering, which will open an even greater number of employment possibilities to them.

## what are some related jobs?

Skilled heating, air-conditioning, and refrigeration technicians with a solid knowledge of t he applications of physics and mathematics in their field, an aptitude for using tools, and an understanding of electronics and blueprints can work in a variety of mechanical and service-oriented fields. The U.S. Department of Labor classifies heating, air-conditioning, and refrigeration technicians under the headings *Utilities Service Mechanics and Repairers* (DOT) and *Craft Technology: Mechanical Work* (GOE). Also listed under this heading are people who work as industrial machinery mechanics, farm equipment mechanics, automobile service technicians, locksmiths, aircraft technicians, and diesel technicians.

| Related Jobs |
|---|
| Aircraft mechanics |
| Automobile service technicians |
| Locksmiths |
| Industrial machinery mechanics |
| Farm equipment mechanics |
| Diesel technicians |
| General maintenance mechanics |
| Computer and office machine service technicians |
| Agricultural equipment technicians |

## what are the salary ranges?

According to the 1998–99 *Occupational Outlook Handbook,* median weekly earnings of heating, air-conditioning, and refrigeration technicians were $536 in 1996. The top 10 percent made more than $887 a week, while the lowest 10 percent earned less than $287 a week.

A 1997 salary survey by the Air Conditioning and Refrigeration Institute reports that entry-level wages for heating, air-conditioning, and refrigeration technicians were $15,000 a year, with graduates of technical schools earning $16,500 to $22,500 a year. The average annual income ranged from $25,500 to $34,000, and the highest paid technicians made more than $50,000.

Wages may vary by experience and employer size and location. Large companies and contractors generally provide benefits packages that include sick leave, medical benefits, pension plans, and paid vacations.

## what is the job outlook?

According to the 1998–99 *Occupational Outlook Handbook,* the outlook for employment in this field is expected to be very good, particularly for skilled technicians. As technology advances, the use of microcomputer-controlled systems that regulate all building systems, including heating, air conditioning, lighting, and security, will increase, and so will the need for skilled technicians to install and service these systems.

Rising energy costs and the desire to conserve energy are encouraging consumers to replace or upgrade their climate-control systems with more energy-efficient equipment, again creating work for technicians. In addition, the 1990 Clean Air Act, which bans the discharge of CFC refrigerants by the year 2000, ensures that technicians will be needed to retrofit or replace existing equipment to meet new standards.

Maintenance and repair technicians' jobs are not as affected by the economy as some other jobs. While in bad economic times a consumer may postpone building a new house or installing a new air-conditioning system, existing units in homes, hospitals, restaurants, technical industries, and public buildings will still require skilled technicians to service their climate-control systems. Technicians who specialize in installation of systems have no cause for concern. With population growth and a good economy, there is often new construction of residential and commercial buildings, which means installation of climate-control systems is necessary.

Refrigeration needs are also growing due to the growth of businesses such as supermarkets. According to the 1998–99 *Occupational Outlook Handbook,* nearly 50 percent of products for sale in convenience stores require some type of refrigeration.

Skilled technicians are also fortunate to have the opportunity for mobility. Climate-control services are needed worldwide and in a variety of settings. If technicians have difficulty finding work in one region or setting, they will almost certainly be needed in another.

# how do I learn more?

## sources of additional information

Following are organizations that provide information on heating, air-conditioning, and refrigeration technician careers, schools, and employers.

**Air Conditioning Contractors of America (ACCA)**
1712 New Hampshire Avenue, NW
Washington, DC 20009
202-483-9370
http://www.acca.org

**Air Conditioning and Refrigeration Institute**
4301 North Fairfax Drive, Suite 425
Arlington, VA 22203
703-524-8800
http://www.ari.org

**American Society of Heating, Refrigerating, and Air Conditioning Engineers**
1791 Tullie Circle, NE
Atlanta, GA 30329
404-636-8400
http://www.ashrae.org

**Mechanical Contractors Association of America**
1385 Piccard Drive
Rockville, MD 20850
301-869-5800
http://www.mcaa.org

**National Association of Plumbing-Heating-Cooling Contractors**
PO Box 6808
180 South Washington Street
Falls Church, VA 22040
800-533-7684
http://www.naphcc.org

**North American Technician Excellence**
8201 Greensboro Drive, Suite 300
McLean, VA 22102
http://www.natex.org

**Refrigerating Engineers and Technicians Association**
401 North Michigan Avenue
Chicago, IL 60611
312-527-6763
http://www.reta.com

## bibliography

Following is a sampling of materials relating to the professional concerns and development of heating, air conditioning, and refrigeration technicians.

### Books

Budzik, Richard. *Opportunities in Heating, Ventilating, Air Conditioning, and Refrigeration Careers.* Lincolnwood, IL: VGM Career Horizons, 1995.

### Periodicals

*ACCA News.* Ten times per year. Newsletter providing members with association and industry news. Air Conditioning Contractors of America, 1712 New Hampshire Avenue NW, Washington, DC 20009, 202-483-9370.

*Air Conditioning, Heating and Refrigeration News.* Weekly. Widely read by workers in all areas of the industry. Business News Publishing Co., 755 West Big Beaver Road, Suite 1000, Troy, MI 48084, 810-362-3700.

*Contracting Business.* Monthly. Covers the fields of air conditioning, refrigeration, heating, and related areas. Penton Publishing, 1100 Superior Avenue, Cleveland, OH 44114-2543, 216-696-7000.

*Contractor.* Semimonthly. Directed at contractors who install air conditioning, heating, and plumbing systems. Cahner's Publishing, 1350 East Touhy Avenue, Box 5080, Des Plaines, IL 60018, 847-390-2110.

*Heating/Piping/Air Conditioning.* Monthly. Devoted to all areas of heating, air conditioning, and refrigeration. Penton Publishing, 1100 Superior Avenue, Cleveland, OH 44114-2543, 216-696-7000.

*Refrigeration Service & Contracting.* Monthly. Journal sponsored by the Refrigeration Service Engineers Society. Business News Publishing Co., 755 West Big Beaver, Suite 1000, Troy, MI 48084, 810-362-3700.

# histologic technician

**Definition**

Histologic technicians prepare tissue specimens for microscopic examination. They process specimens to prevent deterioration and cut them using special laboratory equipment. They stain specimens with special dyes and mount the tissues on slides. Histologic technicians work closely with pathologists and other medical personnel to detect disease and illness.

**Salary range**

$21,400 to 27,000 to $30,400

**Educational requirements**

Some postsecondary training

**Certification or licensing**

Required by certain states

**Outlook**

About as fast as the average

**GOE**
02.04.02

**DOT**
078

**O*NET**
32102U

## High School Subjects

Biology
Computer science
Chemistry
Health
Mathematics

## Personal Interests

Helping people:
physical
health/medicine
Painting
Science

lies unconscious on the operating room table under a blaze of lights, surrounded by doctors and nurses. The chief surgeon peers at the patient's exposed stomach, looking for evidence of disease. She thinks cancer has caused the patient's symptoms, but she needs confirmation. She has taken a biopsy of the tissue, and the operation cannot be completed until the results come back from the laboratory.

Meanwhile, in the clinical pathology department of the large hospital, Jim Pond works swiftly with his cryostat, a special medical instrument that cuts the tissue sample into slices only three to six microns thick. He has stabilized the sample by freezing it. Now he arranges the paper-thin slices of tissue on microscope slides and quickly, carefully adds a chemical stain. Like magic, the stain turns some components of the specimen a different color. What was invisible suddenly becomes apparent. Jim passes the specimen to a pathologist, who examines the tissue under a microscope and verifies the surgeon's suspicions. The results are rushed to the operating room, and the surgery proceeds.

As a histologic technician, Jim sometimes works under pressure, but he handles it well. "I thrive on that," he says. "Sometimes we must work quickly, and sometimes we have more time, but we're always careful and pay close attention to detail. Patients depend on our work."

## what does a histologic technician do?

When a physician or researcher takes a tissue sample from a human, animal, or plant and sends it away for analysis, a team of laboratory workers prepares the specimen and studies it under a microscope. Cancer, leprosy, bacterial infections, and many other disorders can be detected in this way. The professional who performs basic laboratory procedures to prepare tissues for microscopic scrutiny is a *histologic technician*. Workers in this field use delicate instruments, which are often computerized. They also perform quality control tests and keep accurate records of their work.

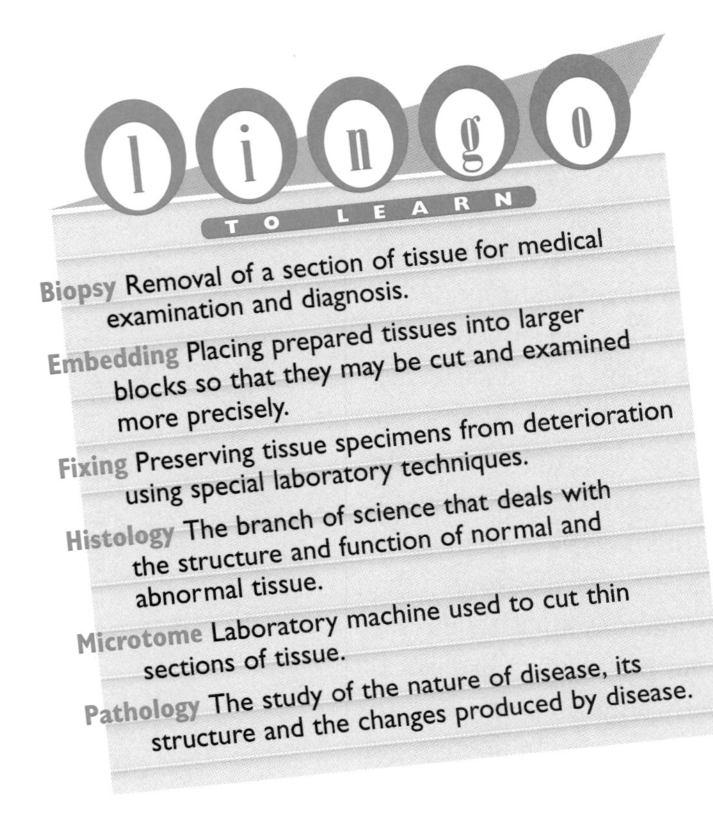


**Biopsy** Removal of a section of tissue for medical examination and diagnosis.

**Embedding** Placing prepared tissues into larger blocks so that they may be cut and examined more precisely.

**Fixing** Preserving tissue specimens from deterioration using special laboratory techniques.

**Histology** The branch of science that deals with the structure and function of normal and abnormal tissue.

**Microtome** Laboratory machine used to cut thin sections of tissue.

**Pathology** The study of the nature of disease, its structure and the changes produced by disease.

After a tissue sample is taken, the first step in preparing it for study, known as "fixation," is usually performed by a pathologist or scientist. The specimen is examined, described, trimmed to the right size, and placed in special fluids to preserve it.

When the fixed specimen arrives at the histology lab, the technician removes the water from it. The water is replaced with melted wax, which moves into the tissue and provides support for the delicate cellular structure as it cools and hardens. Then the technician places small pieces of wax-soaked tissue in larger blocks of wax, a step called "embedding." Without the wax, the tissue would collapse during the next step of the process.

The technician sections the specimen by mounting it on a microtome, a scientific instrument with a very sharp blade. The microtome cuts thin slices of tissue, often only one cell thick, a procedure that requires precision, patience, and a steady hand. The technician cuts many sections of tissue, usually one after another so they form a ribbon, which is placed in warm water until it flattens out. Then the prepared sections are laid on microscope slides.

Next, the technician stains each tissue specimen by adding chemicals and places a coverslip over the sample to protect it. Different stains highlight different tissue structures or abnormalities in the cells, which helps pathologists, researchers, and other scientific investigators diagnose and study diseases.

A second, quicker technique is used to prepare samples and make diagnoses while the patient is still in the operating room. In these cases tissue specimens are frozen instead of being embedded in wax. It's important for a technician to work swiftly and accurately and to cooperate well with the rest of the team during this procedure, because the surgery cannot be completed until the test results are delivered.

Histologic technicians are part of a team of workers. They may be supervised by *histotechnologists,* who have more education; they perform complicated procedures, such as the staining of antigenic sites inside tissues. Some histotechnologists specialize in electron microscopy, which involves the precision cutting of tissues on such a small scale that the work is done under a microscope. *Cytotechnologists* are highly skilled and educated professionals who determine the presence of disease by studying slides prepared by histologic technicians; they sometimes supervise or educate technicians. *Pathologists* are medical doctors who interpret and diagnose the effect of disease on tissues, often by examining slides prepared by histologic technicians. The technician may also work closely with other *physicians* and *researchers*.

A laboratory team may include workers in other areas of specialization, such as *medical and clinical laboratory technologists,* who perform a variety of complex tests. For instance, they examine blood and other bodily substances under microscopes; make cultures to detect the presence of bacteria, fungi, and other microorganisms; type and cross-match blood samples for transfusions; and determine a patient's cholesterol level. *Clinical chemistry technologists* prepare specimens and analyze the chemicals and hormones in bodily fluids. *Microbiology technologists* examine and identify micro-organisms such as bacteria. *Blood bank technologists* collect blood, determine its type, and prepare it for transfusions. *Immunology technologists* study the way the human immune system protects itself from disease by responding to viruses and other foreign bodies.

## what is it like to be a histologic technician?

My job as a histologic technician is never dull," says Jim Pond. "Just when I think I've seen it all, something new comes along—something medical or technical—and I get a chance to experience that. It may be working with large mechanical parts when I fix a problem with a machine, or it may be doing something new with tissue from a heart biopsy. Sometimes the specimens histologic technicians work with are so tiny that there's almost no room for error. When I'm able to solve a problem, big or small, that gives me pride in what I do."

Jim typically begins his day by receiving specimens and keeping track of the patient's identification number. "Then I process the tissue specimens and cut thin sections, using the microtome. Next, depending on the physician's instructions, I stain the tissue so that a diagnosis can be made," Jim says. He pauses and nods his head. "The true histologic technician is able to do many things. Sure, we work with a lot of machines—robotic stainers, tissue processors, cover slippers—but in this lab, if every machine crashed tomorrow, we could do it all anyway. It would take longer, but we could do it."

A large part of Jim's day is spent working with people, often resident physicians in training, who are just learning about histopathology. "This hospital is part of a teaching institution," he explains, "so we get asked to do more stains and a larger variety as well, simply because the residents are learning which techniques work best."

**To be a successful histologic technician, you should**

- Have good color vision
- Be attentive to detail
- Be able to concentrate well
- Be patient
- Be able to work under pressure and to work quickly when necessary
- Be honest and willing to admit mistakes

Orders for histology procedures on common tissue specimens might include sampling gall bladders, tonsils, cardial sacs, and some of the more typical cancers. Rarer cancers might need an electron microscope workup, while unusual diseases like leprosy might need special stains.

"You have to know what you're doing in this job and be tactful, too," says Jim. "Over time, you gain a lot of experience and information as a histologic technician. Sometimes a resident will ask me to do a certain silver stain and I'll think that it's because he's searching for signs of a fungus, since we usually do silver stains for fungus. Then the resident might mention that he's concerned about the possibility of Alzheimer's disease, and so I might suggest that we try another stain, too.

"There are lots of ways to approach taking tissue samples, and each one provides a different kind of information. For instance, histologic technicians can take multiple samples at a certain level within the tissue, or we can reorient the tissue completely, or we can not reorient it at all. Then there's all those different types of stains. One stain helps us look for infectious agents in the lungs, for example; another highlights connective tissue troubles; another tracks down inflammation in the blood vessels. The doctors make their diagnoses based on the information you and your techniques provide for them. Histologic technicians, through tact and craft, help physicians find out what's causing problems and confirm suspicions so that accurate diagnoses can be made."

Another histologic technician, Carol Bischof, directs the program at Fergus Falls Community College in Fergus Falls, Minnesota. Part of her day is spent working with students, either in the classroom or as they gain clinical experience in area hospitals. "I love it when I can get into the laboratory with students," she says. "Some of the stains histologic technicians use are so beautiful. Tri Chrome stain, for instance, is very pretty. Its colors could be used to design a quilt."

Carol explains that there is a significant amount of technique involved in staining tissue. "Your artistic abilities come out when you do this," she says. "Some stains need to be added gradually or removed slowly in order to bring out the correct details. Some stains differentiate between connective tissue and muscle tissue; others differentiate between brain and spinal tissue. It's a complex and beautiful procedure."

This is a fascinating and satisfying career, but it does have a few drawbacks. When histologic technicians work for large laboratories, they may spend a great deal of time standing or sitting in one position and performing one type of operation, but most are able to rotate the type of work they perform. Increasingly, they also must deal with governmental regulations designed to insure quality control in laboratory work. A part of each day is spent complying with these rules. Some histologic technicians, especially those working for large hospitals, may work rotating shifts, including weekends and holidays.

## have I got what it takes to be a histologic technician?

Lee G. Luna, a pioneer in histotechnology, said that histologic technicians need to be three things: scientist, mechanic, and artist. They need an interest in anatomy, biology, chemistry, and physics; an ability to repair equipment malfunctions; and an artistic eye for pattern, detail, and color. It's also important to have manual dexterity, patience, and the ability to work quickly and with precision.

As in most professions, histologic technicians tend to do a better job if they're willing to learn new techniques and concepts. Jim Pond remarks, "It helps to know when to ask questions, how to listen for answers, and how to store up information. Good pathologists will encourage histologic technicians to ask questions, and they will help them gain a bigger knowledge base so that the laboratory team can function even more efficiently."

As part of a team, technicians must also develop a sense of professional integrity. "It's essential, too, that you be professionally honest," says Eileen Nelson, former director of the University of North Dakota's histologic technician program at Grand Forks. "Sometimes the work is quite repetitive, and if you make a mistake, you don't cover it up. When you cut a slice wrong or misplace a piece of tissue, you have to say so. People's lives may depend on how well you do your job."

A lot of the work histologic technicians do uses standard procedures and is almost like following a recipe. Sometimes histologic technicians who work in larger laboratories may repeat the same part of a procedure all day long. Workers at a huge clinic might do cutting only, for example, or embedding only. This can be boring. Histologic technicians generally report that they are more satisfied with their jobs when they can spend time alternately cutting, staining, and charting tissues.

Histologic technicians work in laboratories that are well ventilated, and most of the tissue processors that they use are enclosed. This minimizes problems with odors and chemical fumes. Sometimes histologic technicians work with hazardous chemicals, but protective clothing is worn, as well as badges that will indicate overexposure and allow workers to seek immediate medical treatment should accidents occur. Histologic technicians also face the possibility of coming into contact with disease through tissue samples, but several of the steps involved in preparing specimens generally kill any living organisms.

FYI

In 1664, Robert Hooke, an English scientist, used his penknife to slice pieces of cork. He placed these thin sections under the microscope. A few years later, the Dutch naturalist Anton van Leeuwenhoek used his shaving razor to carve thin sections from flowers, a writing quill, and a cow's optic nerve. Both men wanted to observe the microscopic structure of objects. Because of their investigations, the science of histology was born.

# how do I become a histologic technician?

Carol Bischof's path to histotechnology began with college in Mississippi and Ohio and led to a job in Minnesota. "While working on my master's degree in zoology, I did research for the Environmental Protection Agency," she says. "I studied how substances in paints and plastics affected rats and watched as histologic technicians processed and embedded the rodent tissues that we used. Then a friend called and said that the lab where she worked needed someone to do histology. I was familiar with laboratory techniques and so I got the job. Later on, I went back to school and got my certificate as a histologic technician."

## education

### High School

Biology, chemistry, and other science courses are necessary for students wishing to enter histotechnology programs after graduation. Mathematics and computer science courses are also important.

### Postsecondary Training

There are three ways to become a histologic technician. It's possible to enter the profession with a high school diploma and on-the-job training only, but a college degree or formal training through an institution such as a hospital is becoming more generally recommended.

On-the-job training is still an acceptable entry into the field, according to Sumiko Sumida, a histologic technician and histotechnologist. Sumiko is technical director of the histology laboratory at the University of Washington Hospital Pathology Department in Seattle. She says, "Many laboratories will train people. It's an exciting field, because it has a lot of opportunities for people at the time they're just getting out of high school and maybe don't have the money to go to college. Most of the hospitals do require some college background, but opportunities do exist for on-the-job training," especially in rural areas.

The second way to enter the field involves completion of a one- to two-year certificate program at an accredited institution, usually a hospital. The certificate program includes classroom studies along with clinical and laboratory experience.

The third method involves earning an associate's degree from an accredited college or university and includes supervised, hands-on experience in clinical settings. General course work for histologic technicians includes classes in organic and inorganic chemistry, biochemistry, general biology, mathematics, medical terminology, and medical ethics. Additional classes include histology, quality control, instrumentation, microscopy, records and administration procedures, and studies in fixation, processing, and staining. College programs may also require core curriculum classes such as English, social science, and speech.

**Advancement Possibilities**

**Histotechnologists** hold a bachelor's degree and have obtained accreditation or have at least a year of experience. They perform more complicated procedures than histologic technicians, such as enzyme histochemistry, electron microscopy, and immunofluorescence. Histotechnologists can teach, become laboratory supervisors, or become directors in schools for histologic technology.

**Cytotechnologists** hold a bachelor's degree and are accredited or have at least five years of experience. They analyze stained tissue samples for subtle clues that indicate disease. A cytotechnologist with more education and experience may advance to become a specialist and will typically supervise or educate other employees.

**Categorical technologists** hold a bachelor's degree and have one to two years of experience. They focus on one field instead of rotating through various departments of the laboratory. They may be certified in microbiology, chemistry, blood banking, immunology, or hematology.

## certification or licensing

Certification is not required for entry-level histologic technicians. Some states do require that technicians be licensed. Histologic technicians become certified by passing a national examination offered by either the Board of Registry of the American Society of Clinical Pathologists (the main certifying organization) or by the National Certification Agency for Medical Laboratory Personnel. Applicants can qualify for the Board of Registry exam in three ways. They can complete an accredited program in histotechnology, earn an associate's degree from an accredited college or university and combine it with one year of experience, or have a high school diploma and two years of experience. If you don't live near a school that offers training for histologic technicians, you might want to enroll in a new, nontraditional, teleconferencing program that prepares students to take the certification exam. The program is operated by the Allied Health Department at Indiana University.

## scholarships and grants

Most colleges and universities offer general scholarships. The National Society for Histotechnology is the main source of information for scholarships specifically for students of histotechnology (*see* "Sources of Additional Information"). Several scholarships and awards are available each year through this organization. Occasionally, employers will provide educational funding for general laboratory workers who wish to enter the field of histotechnology.

# who will hire me?

Jim Pond worked at three hospitals before deciding to study histotechnology. First, he had a job as a clerk. Then he moved to laboratory work in phlebotomy. His third job involved basic hematology and chemistry work. At each hospital, pathologists would note the quality of his work and encourage him to get additional schooling so that he could advance into a position of more responsibility within the laboratory. "Finally, I listened to what they were recommending," Jim says. "The hospital where I was had a school for histologic technicians, and when the hospital said that they'd even pay me while I learned more about the field, I decided to give it a try."

A histologic technician has the opportunity to work in many fields of medicine and science. Most are employed by hospitals or by industrial laboratories that specialize in chemical, petrochemical, pharmaceutical, cosmetic, or household products. Some work for medical clinics, universities, or government organizations. Many biomedical companies hire histologic technicians to aid in research projects. Immunopathology, forensic medicine, veterinary medicine, marine biology, and botany are just a few of the options.

Sumiko Sumida says that people who have training in both histology and cytotechnology are in demand, especially in rural areas, which tend to have greater difficulty in attracting qualified technicians. That's partly due to the fact that salaries are often lower there. In addition, many histologic technicians are women, and it's sometimes hard for them to move their families to rural areas.

If you'd like to work part-time in the field while you go to school, consider a job with a regional laboratory for large health systems. Histology is one of many fields that have been consolidated in recent years as health groups have pooled their resources to save money. For example, the average hospital used to employ a team of three to six histologic technicians, and smaller facilities often had only one. Now, many of these institutions have combined their resources. The result is laboratories with teams of perhaps thirty to fifty histologic technicians. Because these facilities operate seven days a week, twenty-four hours a day, there's a good chance you could be assigned to shifts that would not interfere with your classes at college.

Employers sometimes schedule recruiting visits to schools that offer histologic technician programs. You can also find leads on employment in *Advance for Medical Laboratory Professional,* which publishes a list of job openings and carries advertisements for employment. The *Journal of Histotechnology* also carries job advertisements, as do other magazines for medical personnel (*see* "Bibliography").

# where can I go from here?

Some histologic technicians become supervisors. Others specialize in certain areas of histotechnology, such as orthopedic implants or diseases of the lungs.

"If you have an interest in science, you can grow with the field and use it as a jumping-off point for the future. It's a field you can use as a stepping stone to go in any direction. It's really up to the individual to take the plunge, take some risks," says Sumiko Sumida.

She suggests that a young person without much money could work as a histologic technician for a few years while attending college part-time or completing a training program at a hospital, then use that experience and training to advance to a high-paying position as a pathologist's assistant. She strongly recommends taking a certification examination after you've worked in the field for two years.

Technicians who have more education and experience are more apt to be promoted. With some employers, no matter how good your job skills might be, your chances of advancement are significantly smaller if you don't have formal training, including some college study. In the future an associate's degree will likely become the standard requirement for entering the field and being promoted. Returning to school, earning a bachelor's degree, and becoming a histotechnologist will boost your career even further and will probably increase your salary.

"If you go on to college, there are a lot of opportunities," Sumiko says. "You can go into management or become a technical representative" for a business. You could combine business classes with your knowledge of medicine and go into administration; your background in the laboratory would give you an unusual edge over other job candidates, since most administrators have never worked in a laboratory. You could also earn an advanced degree and teach sciences.

| Related Jobs |
| --- |
| Cytotechnologists |
| Dental hygienists |
| Pheresis specialists |
| Biological aides |
| Microbiology technologists |
| Laboratory technicians |
| Medical technologists |
| Phlebotomy technicians |
| Clinical chemistry technologists |

## what are some related jobs?

The U.S. Department of Labor classifies histologic technicians under the heading *Laboratory Technology: Life Sciences* (GOE). Also under this heading are medical technologists, polygraph examiners, farm crop production technicians, clinical chemists, and medical laboratory technicians.

## what are the salary ranges?

Geographic location, experience, level of education, employer, and work performed determine the salary ranges for histologic technicians. According to a 1996 wage survey conducted by the American Society of Clinical Pathologists, pay rates were highest in the Northeast and far West and lowest in the central United States.

Beginning histologic technicians had median annual salaries of $21,400 (up about 7 percent since 1994); average salaries of $27,000; and a top rate of $30,400. The pay was highest in hospitals, with median beginning salaries of $21,400 and average salaries of $27,400. Hospitals with fewer than a hundred beds paid significantly less, and those with two hundred to three hundred beds paid higher salaries than did larger facilities. Private clinics and reference laboratories paid median beginning salaries of $19,000 and average salaries of $22,000. Private doctors' practice groups paid median beginning salaries of $18,600 and average salaries of $24,200.

Histotechnologists had median beginning salaries of $26,208; average staff-level salaries of $32,032; and a top rate of $37,024. Supervisors had median beginning salaries of $30,000 and average salaries of $35,800. Staff-level cytotechnologists had median beginning salaries of $29,400 and average salaries of $36,000. Supervisory cytotechnologists had median beginning salaries of $34,800 and average salaries of $43,800. According to Sumiko Sumida, salaries for pathologists' assistants often start at about $60,000.

A median salary represents a figure at the midpoint of all earnings in any given category. Half of the medical laboratories in the survey paid below the median, and half paid above the median.

## what is the job outlook?

A few years ago there was a general shortage of personnel to work in medical laboratories, but that need has tapered off since 1994. Prospects still look good for histologic technicians, however. The ratio of job openings to full-time histologic technicians (known as the vacancy rate) almost doubled between 1988 and 1996, according to the 1996 wage survey conducted by the American Society of Clinical Pathologists. The vacancy rate for histologic technicians was 11.7 percent in 1996, up from 8.7 percent in 1994. In 1988 the vacancy rate was only 6.2 percent.

Between 1994 and 1996 the vacancy rate for histologic technicians increased slightly, despite the fact that the median beginning salary more than tripled during that time. The vacancy rate increased significantly for cytotechnologist supervisors, increased slightly for histologic supervisors, and declined substantially (from 17.4 percent in 1994 to 7.3 percent in 1996) for histologic technicians.

In the past, job prospects were particularly good at large hospitals, but many of those institutions have downsized since 1994. Nearly 40 percent of all facilities surveyed had reduced their staffs, and about 30 percent had reduced the size of their laboratory staffs. Of those, nearly one out of three were hospitals with at least three hundred beds.

If you begin your career as a histologic technician, you're apt to advance to more responsible positions through on-the-job training. About 90 percent of managers in the survey said their personnel were being trained in other technical areas within the laboratory, and 24 percent said employees were being cross-trained outside the laboratory.

"Right now, because of the consolidations that health care has been seeing in the last year or so, it's been pretty stable," Sumiko Sumida comments, but she adds, "There will probably be a huge outflux of people within the next five years," because a large number of professionals in the field are nearing retirement age.

# how do I learn more?

## sources of additional information

Following are organizations that provide information on histologic technician careers, accredited schools, and employers.

**American Medical Association**
PO Box 7046
Dover, DE 19903
800-621-8335
http://www.ama-assn.org

**American Society of Clinical Pathologists**
Board of Registry
2100 West Harrison
Chicago, IL 60612
312-738-1336
http://www.ascp.org
bor@ascp.org

**National Accrediting Agency for Clinical Laboratory Sciences**
8410 West Bryn Mawr Avenue, Suite 670
Chicago, IL 60631
773-714-8880
http://www.mcs.net/~naacls
naacls@mcs.net

**National Society for Histotechnology**
4201 Northview Drive, Suite 502
Bowie, MD 20716-2604
301-262-6221
http://www.nsh.org
histo@nsh.org

## bibliography

Following is a sampling of materials relating to the professional concerns and development of histologic technicians.

### Books

Carson, Freida L. *Histotechnology: A Self Instructional Text.* 2nd ed. Includes techniques. Chicago: ASCP Press, 1997.

Cormack, David H. *Essential Histology.* An introductory book to the field. Philadelphia: J. B. Lippincott Co., 1993.

Reid, Philip D. and Rafael F. Pont-Lezica, eds. *Tissue Printing: Tools for the Study of Anatomy, Histochemistry, and Gene Expression.* San Diego: Academic Press, 1992.

### Periodicals

*Advance for Medical Laboratory Professionals.* Biweekly. Features general articles of interest to medical laboratory personnel. Merion Publicaitons, 650 Park Avenue, PO Box 61556, King of Prussia, PA 19406-0956, 610-265-7812.

*Journal of Histotechnology.* Quarterly. Features articles dealing with education and medical news, as well as case studies and reports on techniques. National Society for Histotechnology, 4201 Northview Drive, Suite 502, Bowie, MD 20716-1073, 301-262-6221.

# home appliance and power tool technician

**Definition**

Home appliance and power tool technicians install and repair home appliances, such as ovens, washers, dryers, refrigerators, and vacuum cleaners, as well as power tools, such as saws and drills.

**Alternative job titles**

Appliance service representatives

Service technicians

**Salary range**

$21,840 to $23,920 to $41,870

**Educational requirements**

High school diploma; two-year vocational degree or on-the-job training

**Certification or licensing**

Voluntary

**Outlook**

Little change or more slowly than the average

GOE 05.10.03

DOT 637*

O*NET 85999A

## High School Subjects

Chemistry
English (writing/literature)
Mathematics
Physics
Shop (Trade/Vo-tech education)

## Personal Interests

Building things
Figuring out how things work
Fixing things

**Letting** his truck warm up, Dave Allender goes back inside the shop to review his work tickets for the day. His 9:00 AM appointment is for a GE refrigerator that isn't cooling, his 10:00 AM is a Maytag washer that won't drain, and his 11:00 AM is a stove built in 1952 that the customer wants repaired at any cost. Looking through the remainder of his appointments, Dave moves into the parts room to collect what he will need for the day's work.

Carrying the parts across the parking lot to load on his truck, he glances at his watch. It's 8:35. That refrigerator sounds like it may need a new compressor, and that will take some time. He decides to go a little early and get a head start on the day. That way, he'll be sure to make his second appointment on time.

# what does a home appliance and power tool technician do?

From toasting bread, to washing clothes, to watching the latest release on video, almost everyone relies to some extent on household appliances. When something goes wrong with an appliance we depend on for frequent use, it can significantly disrupt our day. We then call on a *home appliance and power tool technician* to put things back on track.

Often called *service technicians,* these technicians perform maintenance work on a number of different home appliances and power tools. Some service large appliances, such as washers, dryers, refrigerators, and ranges, making service calls to the customer's home.

Others repair smaller appliances and tools, such as irons, electric drills, toasters, food processors, and coffee makers. Repair work on such small, portable appliances and tools is usually done in a repair shop, as opposed to at the customer's home. Some techs install and service window air-conditioning units or central air systems, and some specialize in gas appliances, such as stoves and furnaces.

Depending upon the individual, and his or her place of employment, the repair tech's work may include all types of appliances and tools, or may be limited to a particular name brand or type of appliance.

When starting work on an appliance or tool, the repair tech must first determine the nature of the problem. He or she does this by first checking visually for visibly frayed electrical cords or broken connections. If possible, he or she runs the machine to listen for odd noises or too much vibration, and to look for fluid leaks or loose parts. In some cases, the appliance must be taken apart so that its internal parts can be examined.

In dealing with electrical systems, the repair tech may need to consult wiring diagrams, or schematics, in order to understand the circuitry and to check for shorts or faulty wiring. He or she may also need to use specific testing devices, such as ammeters, voltmeters, or wattmeters. Some repairs may require the use of a service manual or troubleshooting guide.

After the repair tech has discovered the nature of the problem, he or she must correct it. This may involve replacing or repairing defective parts, such as belts, switches, motors, circuit boards, heating elements, or gears. He or she may lubricate and clean the machinery, if necessary, or align and tighten it for maximum performance. Repair techs often use common hand tools, such as screwdrivers, pliers, and wrenches, but may also use special tools designed for work on particular appliances.

Those repair techs who service gas appliances may have to replace pipes, valves, or thermostats. They may also need to measure, cut, and thread pipes, and connect them to gas feeder lines, particularly when installing appliances. Additionally, installation could require that the repair tech saw holes in walls or floors, or hang supports from beams in the house in order to place the pipes.

Finally, *gas-appliance installers and repair techs* are often responsible for checking for gas leaks, explaining proper usage to customers, and answering emergency calls regarding leaks or other problems.

Although repair and maintenance of appliances and power tools is the primary duty, the repair tech does have certain other responsibilities. He or she may answer questions and field complaints from customers. When making service calls to customer homes, he or she may also demonstrate or explain how to care for and use appliances or tools. Repair techs may also

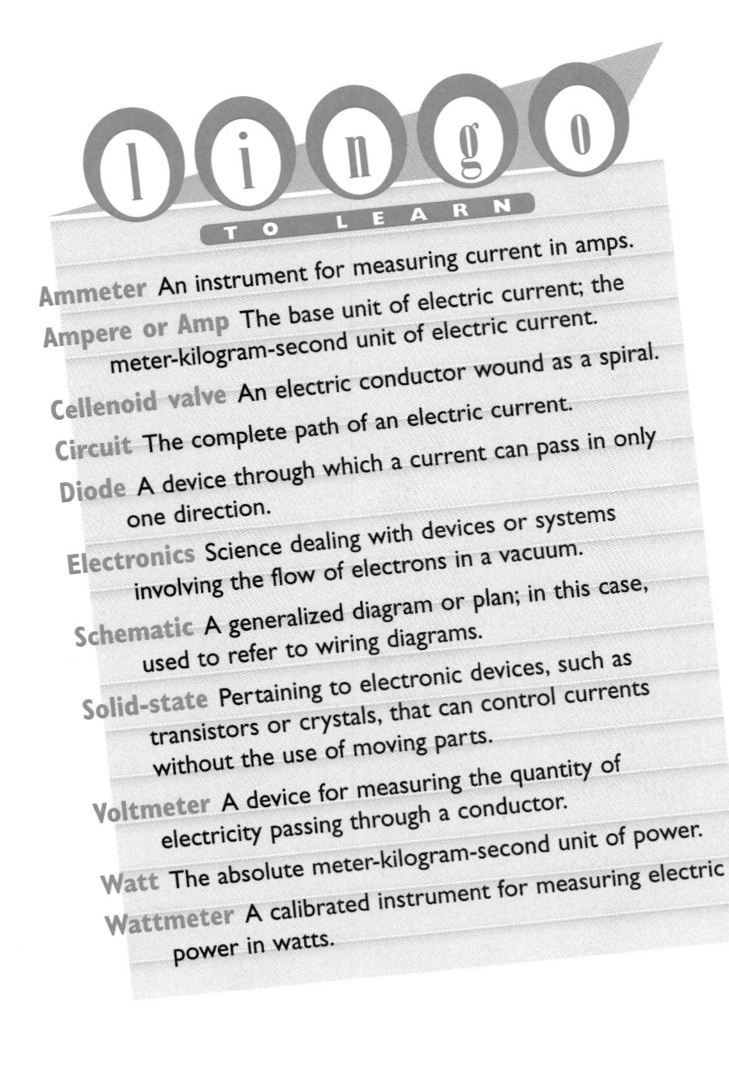

lingo TO LEARN

**Ammeter** An instrument for measuring current in amps.

**Ampere or Amp** The base unit of electric current; the meter-kilogram-second unit of electric current.

**Cellenoid valve** An electric conductor wound as a spiral.

**Circuit** The complete path of an electric current.

**Diode** A device through which a current can pass in only one direction.

**Electronics** Science dealing with devices or systems involving the flow of electrons in a vacuum.

**Schematic** A generalized diagram or plan; in this case, used to refer to wiring diagrams.

**Solid-state** Pertaining to electronic devices, such as transistors or crystals, that can control currents without the use of moving parts.

**Voltmeter** A device for measuring the quantity of electricity passing through a conductor.

**Watt** The absolute meter-kilogram-second unit of power.

**Wattmeter** A calibrated instrument for measuring electric power in watts.

have to order parts from a catalog in order to complete some jobs. They must usually record the amount of time spent on a repair, the parts they have used, and whether the appliance is under a warranty. Sometimes, they include a brief description of the work they have performed. Finally, repair techs may have to estimate the cost of the repairs, prepare bills, and collect payment for their work.

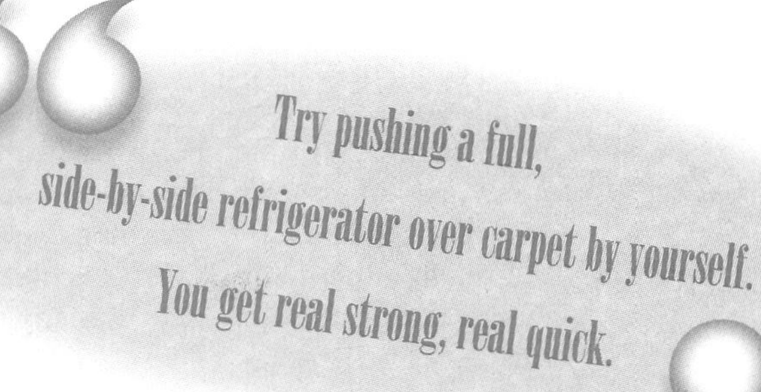


## what is it like to be a home appliance and power tool technician?

Dave Allender is a technician of major household appliances. He is certified and experienced in refrigeration and air-conditioning units and, in his eleven years of experience, has worked on nearly every brand of every kind of appliance. "I work on any name you can think of," he says. "I work on anything in the market." He does not, however, repair power tools. He says that appliance repair has kept him so busy that he hasn't needed to look for any additional work.

Dave usually works five days and one half day each week, for a total of forty-four hours. Occasionally, he puts in extra time, if it is necessary to complete a job. In warm weather, he tends to work more due to the demands for repair on air-conditioning and refrigeration units. "In the summers, I probably put in fifty to fifty-five hours a week to keep up with the refrigeration calls," he says. Because he is employed by an appliance dealership as an independent contractor, he does not receive extra compensation for overtime. As an independent contractor, he gets a flat percentage of the labor charges on each call, which adds up to a higher gross income, he says. However, he does not receive any of the benefits often associated with employment, such as health insurance, paid vacations, and sick leave. "This isn't for everyone," Dave says. "I started out in a hourly job. I could only do this after I got some experience."

**Develop a Specialty**
Most appliance repair techs agree that problems with self-cleaning ranges are the most complicated ones to diagnose. Also, because most of the new household appliances contain computerized or electronic components, technicians who can repair these will do well.

For Dave, the work day starts at the office at 8:30 AM. "Usually, first thing in the morning, we get five or six people calling in with questions, so I spend about fifteen minutes on the phone." Then he looks at each work order for the day to assess the symptoms, and makes sure he has all the necessary parts on his truck. "My first appointment is at 9:00 AM, so I'm out of the office by 8:45 AM," he says. He usually spends the rest of the day out of the office, going from house to house to make service calls.

The nature of his day depends upon the nature of the individual service calls. Some problems take more time to repair than others. "Generally, I'm in and out of a customer's home within fifteen minutes," Dave says. "That's about the average length." According to him, changing the compressor on a refrigerator/freezer is the most lengthy process, taking around forty-five minutes. Often, the problem is something as simple as replacing a belt, or unclogging a hose. Other times, the work is more physically demanding. "You should try pushing a full side-by-side refrigerator over carpet, by yourself," Dave says, laughing. "You get real strong, real quick." Some jobs also require more thought and skill than others, especially those that require the repair tech to read wiring diagrams to complete the job.

After each repair is complete, Dave writes up a brief description of the work he has done and the parts he has used, and collects payment from the customer. Often, he also answers customers' questions. "They have lots of questions," he says. "Customers absolutely do not read their manuals—that's the number one rule." He says he tries to explain to the customer how to get better performance from their appliance.

Appointments are usually scheduled one hour apart, and on an average day, Dave makes six to nine calls. After his last call is complete, he heads back to the shop. "Every day, I come in, turn in my tickets, and pick up my tickets for tomorrow," he says. He also gives the parts manager his orders for any parts that need to be ordered.

In addition to his daily routine, Dave says he spends about one day every three months at a manufacturers' seminar. Periodically, manufacturers invite the shop's servicers to a course on their new appliances. By attending these classes and reading the manuals, Dave is able to stay abreast of the latest products of many different appliance lines.

## have I got what it takes to be a home appliance and power tool technician?

The technician's most important responsibility, according to Dave, is to correct the problem properly the first time. "Any time you get a recall on the same appliance for the same problem," he says, "the customer doesn't get charged, and you don't get paid. Besides that, the customer stops trusting in your ability." He says the repair tech must be very organized in his or her diagnosing and repairing, so as not to miss something that may result in another service call, or worse. "You may be held liable if the machine catches on fire within thirty days after you work on it," he says. "I've known two people who got sued—and lost." The repair tech must work responsibly.

A technician must also be able to deal courteously with all different sorts of people. "You've got to be a people person, mostly," Dave says. "You've got to be able to talk to them so that they have a certain amount of respect and trust for you." Occasionally, the repair tech must work in less than ideal conditions, and still be able to maintain tact. "I hate to say it, but the worst part is when you walk into a really filthy house, and then have to work on a refrigerator that's been closed up, not running, for two weeks," Dave says. "You just have to go right ahead and do the work, and be polite about it."

Dave says that mechanical aptitude, or a genuine understanding of the way things work, is vital to successful repair work. "If you can't picture in your mind exactly how the stuff works, you can't operate on it," he says. The ability to read wiring diagrams, or "schematics," is equally important. "If you don't know how to read schematics, you may as well walk away," he says.

Finally, it is necessary that the repair tech be very self-directed because he or she will, in many cases, be working alone, with no supervision. "You're on your own the whole day," Dave says, "which I really enjoy." He says he likes having to rely only on himself to get the job done.

**To be a successful home appliance and power tool technician, you should**

- Have mechanical aptitude and work well with your hands
- Enjoy and be good at communicating with people
- Be able to work efficiently in varying settings
- Be willing to constantly learn about new techniques and products
- Be able to work effectively with little or no supervision

## how do I become a home appliance and power tool technician?

While still in high school, Dave started to tailor his education to get a good background for repair work. He took classes in building trades, industrial and electrical shop, and drafting. "I already knew what my base field would be when I was sixteen years old," he says. Immediately after high school, he enrolled in a technical college, and received an associate's degree in heating, cooling, refrigeration, and electrical repair.

## education

### High School

Most employers require that their repair techs at least have a high school diploma or its equivalent. For the student who plans to take further schooling at a vocational-technical school, graduation from high school is also a prerequisite.

Students who are interested in becoming home appliance and power tool repair techs can start preparing while still in high school. Shop classes are extremely valuable at providing a good background for this job. A knowledge of electronics is becoming increasingly important, so electrical shop will be particularly valuable. Working with electrical and electronic systems, or building electronic equipment from kits, should give the prospective repair tech a basic understanding of how the various systems function, as well as providing practice in reading diagrams.

Mathematics and physics are good choices to build a knowledge of mechanical principles. Chemistry classes are helpful for the student who plans to work on gas appliances. Finally, any classes that build communication skills are a good addition to the curriculum. Understanding the nature of the problem, and explaining both the problem and its solution to the customer are crucial elements of the repair tech's job. This requires the ability to listen and communicate well.

The high school student who thinks that he or she may be interested in a career as an appliance servicer might want to talk to employees of local repair shops and appliance dealerships. These employees may be able to give the student a clearer picture of the job. Another possibility is a summer or part-time job which allows the student to observe or assist in repair work. Such a job would provide a firsthand knowledge of the industry, plus valuable experience and a potential job lead.

### Postsecondary Training

Although it may be possible, in some cases, to find a job as a trainee, with no further formal education, these opportunities are becoming increasingly rare. Formal training, especially courses in electricity and electronics, are more and more necessary in the repair of today's technologically advanced appliances.

Most often, high school graduates who have an interest in appliance repair choose to take a training program at a vocational or technical school, or a community college. Such programs provide both classroom training and laboratory experience in the service and repair of appliances. The program length varies from school to school, with many lasting one to two years. Another means of obtaining an education in appliance repair is through correspondence courses, which often include both book and video training.

Finally, although it is less common, some students are able to get their training on the job. In this case, they may be hired as helpers or apprentices to observe and assist experienced repair techs, either in the shop, or on-site, at customers' homes. Some appliance and power tool manufacturers and department store chains have formal training programs, where trainees work with demonstration appliances and other training equipment to acquire experience and skill.

Regardless of the beginning repair tech's educational background, he or she will almost certainly require some further training by his or her employer, in order to be familiar with the specific work to be done. Also, as products change, so do the repair techniques, and the repair tech needs to stay up-to-date. Reading manuals and attending periodic courses given by manufacturers is a continuing responsibility of all technicians and servicers.

FYI

**No time like the present**
Currently, there is a severe shortage of trained appliance repair technicians. The field will always be lucrative, as there are major appliances in every home and apartment.

## certification or licensing

Currently, only four states—California, Florida, Washington, and New York—require some form of licensure or certification for home appliance and power tool repair techs. The student who plans to work in one of these states should check the specific requirements, since they do vary, state by state. There is no national standard of testing for certification in this industry.

Due to environmental concerns over the commonly used refrigerant, CFC (chlorofluorocarbon), since 1994 it is a federal requirement that all repair techs who work with refrigerant be certified for that type of work. In order to work legally, repair techs who plan to work with appliances that use refrigerant, such as air-conditioners, refrigerators, and freezers, must take and pass a federally mandated class and certification test. The class and test last about one day and are administered on a local level by the Air Conditioning Contractors of America (ACCA), which is a federally approved program.

Although the test is optional, for those not working with CFC, it may be well worth the repair tech's time, since the ability to work on refrigeration units will most likely make him or her a more valuable employee.

## who will hire me?

Dave found his first job through a classified ad. He says that even though he had completed the technical schooling, he lacked the hands-on experience that makes a repair tech valuable. "I went in to this guy," he says, "and I told him I was willing to take a job at starting pay, but in turn, he'd have to train me. That's how I got the job."

According to a 1992 survey, about 70 percent of the appliance repair techs in the work force are employed by retail trade establishments, such as department stores, household appliance stores, and dealers that sell and service appliances and power tools. About 14 percent are self-employed, and the remainder work for gas and electric utility companies, wholesalers, and electrical repair shops. Although there is a need for repair techs in almost every sort of community, the highest concentration of jobs will most likely be found in larger cities, where the population is high.

Technicians who have a formal education from a community college or vocational school will have the best prospects for finding jobs. Since few companies can afford to hire unskilled, untrained repair techs, increasingly employers are looking for workers who have learned the basic skills and just need to be trained on the specifics. For the prospective repair tech who has taken a technical training course, the school's placement office may be a good place to begin a job search. If the school he or she has attended doesn't offer a placement service, often teachers or administrators will have job leads or ideas about where to look.

In order to gain experience and break into the field, the prospective repair tech who would rather get his or her training on the job might want to approach employers directly, offering to work for a substantially reduced wage. Some larger dealerships and department stores have in-house training programs and may be willing to train a promising applicant.

The job searcher might also want to watch the classified ads in all local newspapers, since repair tech jobs are sometimes listed. He or she might also send resumes to the personnel departments of larger retailers or manufacturers, or to the owners or managers of smaller repair shops or dealerships. Gas or electric utility companies may also be possibilities.

**How are technicians paid?** The graph reflects the percentage of appliance technicians receiving each payment type.

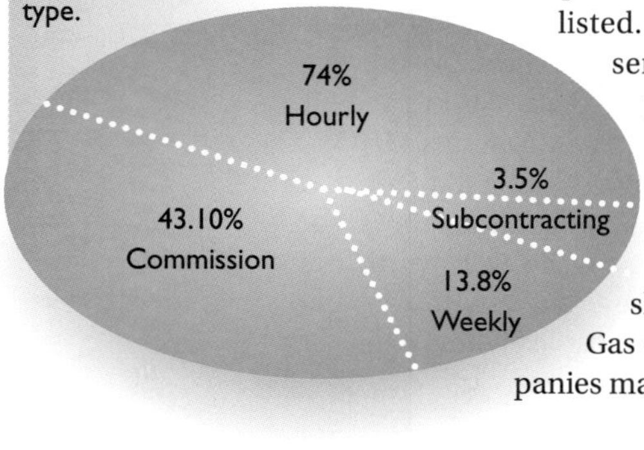

74% Hourly
3.5% Subcontracting
13.8% Weekly
43.10% Commission

Finally, there are a number of publications which are targeted specifically toward the appliance repair business. These publications often carry classified job listings, and may be a source of leads (*see* "Bibliography").

## where can I go from here?

Dave says he's happy with his employment situation for the present. However, the owner of the shop he works for is preparing to retire, and may offer to sell him a part of the business. Dave is not sure that he wants to become involved in the "hassles" of running a business, but says he'll decide when the time comes. "I don't know that I want to go any further than where I am," he says. "Essentially, I'm already self-employed, anyway."

An important factor in the advancement potential for the individual appliance repair tech is his or her place of employment. Those who are working in a small, privately owned repair business will probably not have many chances for rapid promotion. In such smaller businesses, there are usually not many levels of managerial or supervisory positions; often the owner or manager performs all administrative duties.

The repair tech who works for a large retailer, a factory service center, or a utility company will have more possibilities for promotion. Within these organizations, he or she could become a supervisor, assistant service manager, or service manager. The repair tech who demonstrates superior technical skills and the ability to deal well with customers and co-workers might eventually be promoted to a position such as regional service manager or parts manager.

Another career path for the experienced appliance repair tech involves teaching other repair techs at a factory service training school, technical school, or vocational school. This position would require a strong familiarity with a variety of products and their maintenance, as well as an aptitude for public speaking.

The repair tech who works for an appliance manufacturer might eventually move into a position in which he or she helps write service manuals. Those employed by manufacturers might also have the opportunity to become salespersons or representatives to appliance dealerships.

Finally, some repair techs may eventually decide to open their own appliance repair businesses. With experience, a knowledge of business management, and the necessary funds to invest in equipment, the repair tech can become owner and manager of his or her own shop.

## what are some related jobs?

The repair of home appliances and power tools is closely linked to their fabrication, testing, and installation. All of these functions require a basic understanding of the way the machines are put together, and the way they operate.

The U.S. Department of Labor classifies home appliance and power tool technicians under the headings *Occupations in Assembly and Repair of Electrical Appliances and Fixtures*, *Utilities Service Mechanics and Repairers*, and *Occupations in Assembly, Installation, and Repair of Large Household Appliances and Similar Commercial and Industrial Equipment* (DOT). It also classifies them under *Crafts: Electrical-Electronic* (GOE). Also listed under these headings are workers who supervise and perform the assembly, testing, and installation of both large and small home appliances, and who install and service industrial gas equipment, turbine pumps, environmental-control systems, and office equipment (*see* chapter "Heating, Air-conditioning, and Refrigeration Technician").

| Related Jobs |
| --- |
| Major appliance assembly supervisors |
| Small appliance assembly supervisors |
| Household appliance assemblers |
| Fluorescent lighting model makers |
| Lighting fixtures assemblers |
| Heating-element winders |
| Office-machine servicers |
| Statistical-machine servicers |
| Industrial-gas servicers |
| Cooler service supervisors |
| Pump erectors |
| Solar-energy-system installers |
| Environmental-control-system installers |
| Refrigerator testers |
| Control-panel testers |
| Gas chargers |

## what are the salary ranges?

The salaries for home appliance and power tool technicians have remained fairly level for the past few years, with no substantial increase or decrease. According to a 1996 survey, the overall average wage for an appliance repair tech is $11.50 an hour. For the repair tech who works full time, this translates into roughly $23,920 yearly. The middle 50 percent earn between $10.50 and $13.25 an hour. The most experienced repairers have median earnings of $20.13 an hour, or $41,870 a year.

There is a difference in average pay between repair techs who service heating, air conditioning, and refrigeration units, and those who do not. The average starting salary for appliance repair techs who do not work in these areas is significantly lower than the salary for those who do.

In addition to skill level and type of equipment serviced, earnings vary significantly with the size of the repair tech's company. Larger companies, in general, pay higher wages than do smaller companies, and frequently offer more comprehensive benefit programs as well.

Since the majority of repair techs are paid on an hourly basis, as opposed to being salaried, they are usually compensated for overtime work. In addition to their regular pay, some technicians may also receive commissions.

## what is the job outlook?

In 1997, there were around 60,000 appliance repair techs throughout the country, employed by appliance manufacturers and dealers, utility companies, and repair shops. Over the next ten years, this number is expected to decrease slightly.

The main reason for this decrease is that the appliances and tools that are currently being manufactured have been technologically improved to such a degree that they require much less repair work. The increasing use of electronic parts, such as solid-state circuitry, microprocessors, and sensing devices, has lead to a greater dependability in almost all appliances.

An additional factor is the increase of so-called "throwaway" appliances. Over the years, manufacturers have developed methods to produce various appliances much more cheaply than they originally could. This has reduced the cost of such appliances to such an extent that, in some cases, buying a new appliance is not much more expensive than repairing an existing one. Therefore, although Americans are buying increasing numbers of both home appliances and power tools, there are fewer repairs being made now than have been necessary in the past.

Even so, there will continue to be some demand for well-trained repair techs, particularly those with a solid background in electronics. Most openings will arise as a result of workers switching occupations or retiring. Workers in this field are less affected by the overall economy than are those in many other professions. The reason for this is that the need for appliance repair remains relatively steady, even during economic downturns.

# how do I learn more?

## sources of additional information

Following are organizations that provide information on home appliance and power tool technician careers, accredited schools and scholarships, and possible employers.

**Professional Service Association**
71 Columbia Street
Cohoes, NY 12047
518-237-7777

## bibliography

Following is a sampling of materials relating to the professional concerns and development of home appliance and power tool technicians.


### Books

Langley, Billy C. *Heating, Ventilating, Air Conditioning, and Refrigeration.* 4th ed. Englewood Cliffs, NJ: Prentice Hall, 1990.

Langley, Billy C. *Major Appliances: Operation, Maintenance, Troubleshooting, and Repair.* Englewood Cliffs, NJ: Regents/Prentice-Hall, 1993.

Rowh, Mark. *Opportunities in Electronics Careers.* Discusses the commercial, industrial, and home equipment repair fields. Lincolnwood, IL: VGM Career Horizons, 1992.

### Periodicals

*Appliance Service News.* Monthly. Informs owners and technical personnel about industry news. Gamit Enterprises, Inc., 110 West, St. Charles Road, PO Box 789, Lombard, IL 60148, 630-932-9550.

*Appliance Tech Talk!* Bimonthly. Precision Trax, Inc., 9940 West 59th Place, Suite 3, Arvada, CO 80004, 303-431-4976.

*HomeWorld Business.* Biweekly. Provides information relevant to the housewares industry. ICD Publications, 1393 Veterans Highway, Suite 214N, Hauppauge, NY 11788, 516-979-7878.

*Retail Observer.* Monthly. Directed toward owners and sales personnel of appliance and other related stores. Retail Observer, 1442 Sierra Creek Way, San Jose, CA 95132, 408-272-8974.


### Advancement Possibilities

**Electrical-appliance-servicer supervisors** supervise and coordinate activities of workers engaged in servicing, repairing, and installing electrical household appliances as well as order tools, supplies, and replacement parts.

**Industrial gas servicers** test, repair, and adjust all types of industrial gas equipment, including control mechanisms and other safety controls. They may advise customers on proper installation of equipment and may install and remove large industrial meters, regulators, and related equipment.

**Air-conditioner-installer supervisors** supervise activities of workers engaged in fabricating, installing, and repairing air-conditioning systems in residential and commercial buildings. They inspect and measure to determine airflow requirements, plan and draw layouts of duct work, get price quotes on materials and supplies, purchase equipment, and schedule installation times.

# home health care aide

**Definition**

Home health care aides give hands-on, one-on-one care to the elderly, ill, and disabled so they can continue to live in their own homes for as long as possible. They assist with personal hygiene, such as bathing, grooming, and dressing, transferring clients from bed, and light housekeeping and cooking.

**Alternative job titles**

Homemaker-home health aides

Home care aides

**Salary range**

$10,900 to $16,640 to $20,800

**Educational requirements**

High school diploma

**Certification or licensing**

Required by some states

**Outlook**

Much faster than average

GOE
10.03.03

DOT
354

O*NET
32505*

## High School Subjects

Anatomy and physiology
Family and consumer science
Health
Psychology

## Personal Interests

Babysitting/Child care
Cooking
Helping people: personal service
Helping people: physical health/medicine
Volunteering

Madison let herself into the downstairs apartment and glanced at the hall clock. She was running late—not unusual for her.

"That you Becky?" an anxious voice calls from the living room.

"You bet Mr. Wells," Becky answers. She came into the room and found her client holding a newspaper about three inches from his eyes, straining to see. Papers lay strewn about the floor. "Doing some reading this afternoon?"

"I'm looking for something I read about down in Missoula. They're looking for old historic photographs of people in the area. I got a bunch of 'em."

"I've no doubt you do," says Becky. "I'm going to draw your bath now."

"Well make it good and hot. And put some Epsom salts in it, will you?"

"It can't be too hot, Mr. Wells. It'll burn your skin," says Becky. At 105 years old, Max Wells's skin was like tissue paper. But he was as spry as a seventy-year-old and loved his bath. Unlike most of Becky's clients, Max Wells could easily lift himself in and out of a tub, so she filled it up and sat with him while he soaked, listening to his stories about Montana's rich past over and over.

This was Becky and Max's last evening together. His shingles were about cleared up and he was being dismissed from her care. He'll be on his own again, Becky thought, and I sure am going to miss him.

## what does a home health care aide do?

"The main purpose of a *home health care aide* is to go in and do personal care," says Becky. "I give them their bath, and if they're too ill to get out of bed, I give them a bed bath. I do for them what I would do for myself. "

Home health care aides make it possible for the elderly and the ill, the disabled and infirm, to stay in their own homes. "My goal is to help them stay independent," says Becky.

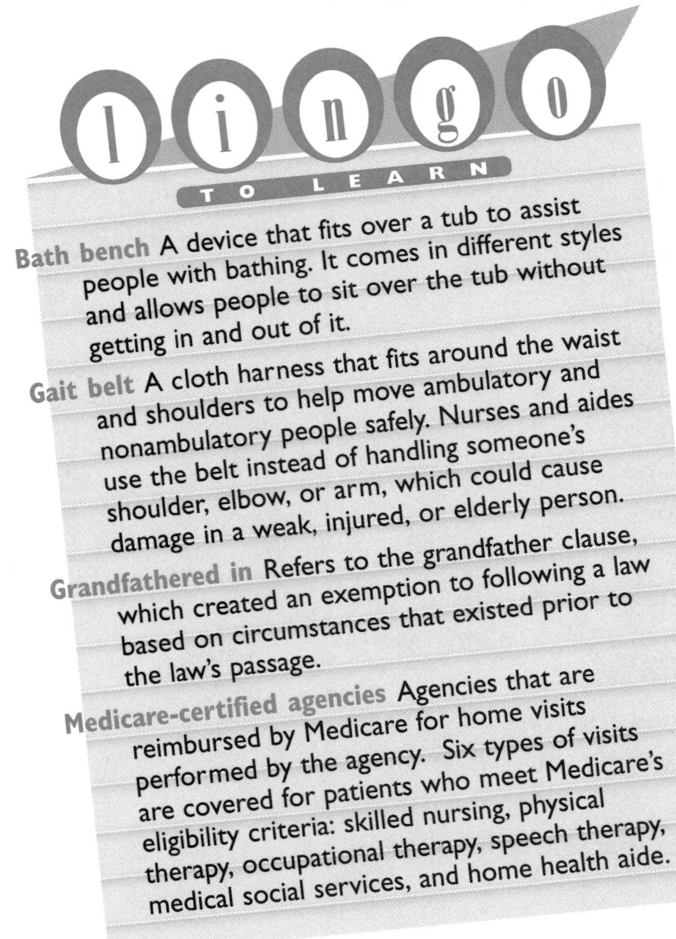

**lingo TO LEARN**

**Bath bench** A device that fits over a tub to assist people with bathing. It comes in different styles and allows people to sit over the tub without getting in and out of it.

**Gait belt** A cloth harness that fits around the waist and shoulders to help move ambulatory and nonambulatory people safely. Nurses and aides use the belt instead of handling someone's shoulder, elbow, or arm, which could cause damage in a weak, injured, or elderly person.

**Grandfathered in** Refers to the grandfather clause, which created an exemption to following a law based on circumstances that existed prior to the law's passage.

**Medicare-certified agencies** Agencies that are reimbursed by Medicare for home visits performed by the agency. Six types of visits are covered for patients who meet Medicare's eligibility criteria: skilled nursing, physical therapy, occupational therapy, speech therapy, medical social services, and home health aide.

Home health care aides work mainly with the elderly. They could, however, work with people of any age who are recovering at home following hospitalization, an accident, or an illness. They also help children whose parents are ill, disabled, or neglected, or give care to people suffering from specific ailments, such as Alzheimer's and AIDS. Becky often has clients with head trauma who were in car accidents. Adults with developmental disabilities who have trouble with routine life skills are also commonly served. So are incapacitated people who have small children at home.

The home health care aide offers assistance in dressing and grooming, including teeth cleaning, nail cleaning, shaving, and fixing hair. Preventing falls is a major part of the job, and aides get their clients up and out of bed and transfer them, usually to a sitting chair or a wheelchair. If the client has a catheter bag, the aide changes it. Becky often cuts hair, but never nails. "We're not allowed to cut nails because of possible injury," says Becky. "Even a small cut can be disastrous to people with diabetes or a blood-clotting disorder." Massages, alcohol rubs, and other therapies may be part of a client's treatment and administered by the aide. Aides often weigh clients.

Aides can also give medications and take their client's vital signs, like blood pressure, pulse, and temperature, but some states restrict aides from doing these tasks. Becky, who works in Montana, does not take vitals. Only the home nurse does that, she says. She is also not allowed to administer any medications, including intravenous medicines like those taken with a hand-held pump, although she can remind clients to take their medicine and assist them in opening the bottle and reading the prescription. Occasionally, aides change nonsterile dressings, use special equipment such as a hydraulic lift, or assist with braces and artificial limbs.

Household chores are another part of an aide's job. Light housekeeping, such as laundry, changing the bed linens, washing the dishes, and fixing light meals is often necessary to help a client. Becky sometimes fixes breakfast or lunch for a client after their bath and runs the vacuum cleaner before she leaves.

Aides keep written records of their visits, the duties performed, and the client's condition and progress. They look for bruises, sores, or anything unusual and report it to their supervisor. It is the aide's responsibility to make sure any changes in a client get noticed and reported. They are the eyes and ears of the nurses and doctors because they see clients on a regular basis.

In home care agencies, aides work under the supervision of a registered nurse, physical therapist, or social worker, who assigns them specific duties. Every client Becky works with comes with a list of duties assigned by the admitting nurse. "I can't even wash a sink full of dirty dishes unless it's on the duty sheet," says Becky. Home health care aides often participate in case reviews, consulting with a team of people who care for the client—nurses, therapists, and health care professionals. The aide, however, goes into the home alone and does her work independently.

Giving personal attention and emotional support to clients is perhaps the most important part of a home health care aide's job. For many people, the aide is the only company they receive on a regular basis. They cannot go out alone, and if they're elderly most of their friends are too old to come and visit. Helping a client get through the day by listening and sharing a part of their life is a crucial part of the job. "I feel clients relive their life through me," says Becky. "You are often their lifeline to the outside world."

**FYI**

The skills used by a certified nursing assistant (CNA) and a home health care aide are similar. The main difference between the two is supervision. CNAs work under supervision in health care facilities. Home health care aides work alone in homes, independent of site supervision.

## what is it like to be a home health care aide?

"I'm the pick-up girl," says Becky. "I usually start with ten clients of my own to see during the week, and by Friday I've picked up another twenty. If someone can't make it into work for whatever reason, I pick up their client load. I'm a widow and my kids are mostly grown, so I don't mind filling in for vacation leaves and holidays. People still need a bath on Christmas."

Becky, who has an associate's degree in health care administration, also helps with the scheduling at Yellowstone County Visiting Nurse Service in Billings, Montana, where she's worked for the past six years. Her days begin around 7:00 in the morning and she tries to end them by 3:00 PM, but it isn't always that easy. "Last night," says Becky, "the phone rings at 10:00 PM. A co-worker calls to tell me she needs to take her son to the ER in the morning and she has two clients that need covering, one at 8:00 AM and one at 9:00. I call everyone else and they're full, so I pick up the 9:00 and reschedule the 8:00. I have clients scheduled every hour till noon, beginning at 7:00 AM.

"The average amount of time we spend in a home is about an hour for a normal bath, an hour and a half if a lady's hair needs fixing," says Becky. "We're allowed two hours, and I need that much time if I'm going to do laundry." Becky sees some clients everyday, others two or three times a week. "A lot depends on the situation. If they're incontinent, they may need a bath everyday. Some clients need to be seen twice a day. If they can't transfer themselves, you need to get them up in the morning, then get them back into bed at night." Before Becky leaves a client at night, she makes sure they have the portable phone beside their bed, the remote control for the TV, a glass of water, and anything else they may need. "This person is here for the night, and they can't get up. You need to put yourself in their place and ask: What would I want to have with me?"

Becky carries a bath bench with her to every home. "Not all tubs have bath chairs in them. My bench hooks onto the far side of the tub and the other half comes over the near side and has legs that sit on the floor. You can sit on it and I can lift your legs over the tub and slide you onto the bench. All aides carry benches." Becky, who is five feet, eleven inches tall, transfers clients using a gait belt, which fits around their waist. "Every aide transfers differently," says Becky, "and transferring clients is a large part of what we do."

Becky's clients know about when she's coming, and they try to be ready for her. "Still," says Becky, "I don't rush them. If they're eating when I arrive, I let them finish and have a cup of coffee with them if I'm invited. I care for my people like they were my own parents or grandparents."

Becky, who's been with Yellowstone County for almost six years, makes her own schedule and enjoys a lot of flexibility. "There's no time clock to punch," says Becky. "You're trusted to use your time wisely. It's professional." The nursing service where Becky works covers a sixty-six-mile radius, much of it rural. Aides use their own cars to travel to clients, and they're reimbursed for gas mileage at the official state rate. "We used to carry pagers, but now we have cell phones," says Becky. "They help tremendously, especially if you're trying to find a farmhouse out in the middle of nowhere."

The majority of Becky's clients are elderly, averaging eighty years old. She also sees some younger people that are quadriplegic and brain injured. "The hardest part of my job is saying good-bye to clients," says Becky, "but I never wish them back. Most of the time they're in pain, and they're ready to leave this world when they go."

## have I got what it takes to be a home health care aide?

Above all, home health care aides are caregivers, and they need to be gentle, doting people. Administering to people in their own home is hard, physically demanding work that requires lifting people into baths, out of bed, and up and down stairs. For this reason, it is important for an aide to be in good physical health. Aides do not have the equipment and facilities that are present in a hospital to help them, and they must be comfortable working alone and independently. Some people's homes are messy and unkempt, which can make the housekeeping aspect of this job an unpleasant one. Clients, too, can be cranky and sore. Aides need to be even tempered and compassionate with a willingness to serve others. Genuine warmth and respect are vital character traits, as are cheerfulness and a sense of humor. Because they work in people's private homes, home health care aides must be discreet and honor confidentiality.

Becky believes being a problem solver is crucial to the job. "You never know what you're going to encounter in someone's home. A toilet could back up while you're on the job. You have to be able to think on your feet. The more you know about life in general, the better you'll be at this job."

Home health care aides are observant. "You're a troubleshooter for the nurse," says Becky. "A suspicious lump could be the start of cancer, and you're the only one who sees it."

Home health care aides must be willing to follow instructions and abide by each client's individual health plan. Aides are considered paraprofessionals, not professional medical personnel, and therefore are limited in what they can and cannot do. They cannot make unilateral decisions and have to be content to stay within their boundaries. For some this may be a hard task. Aides are also required to do a lot of charting and paperwork, a part of the job that Becky finds tedious.

For the true caregiver, though, the field can be inspiring. "The smiles I get from people are my reward," says Becky. "I know I help people. If you like to give to people, there are a lot of people who need that giving. I always wanted to be a missionary. Well, this is my missionary work."

## how do I become a home health care aide?

Most agencies that employ home health care aides require that you have a high school diploma for entry-level positions. Becky always wanted to be a nurse, but she didn't take the math she needed in high school. When she was seventeen and still in high school, she got a job as a nursing assistant in a nursing home and liked it. Between babies and while she raised her family, she worked at nursing homes whenever a job opportunity presented itself.

In 1988, when Becky was working as a nursing assistant in Iowa, federal legislation was passed that required all assistants to be certified. "Because I was working at the time, they grandfathered me in," Becky explains. In 1989 Becky graduated from Iowa Lakes Community College with her associate's degree in health administration. In 1992 she moved to Montana and went to work for Yellowstone County. With her experience and certification, she had no trouble finding work in the home health care field.

## education

### High School

High school courses in consumer and family science that offer cooking, sewing, and meal planning are helpful. Classes in psychology, child development, and family living would also provide you with some good background. You should pay attention in health class, and any training you can get in first aid, hygiene, and the principles of health care are worth having. It's important to have good communication skills, both written and oral, for the large volume of documentation home aides are required to do.

If you're interested in the home health care profession, you shouldn't neglect your studies in math, science, and biology. Any advancement in the home health care aide profession is going to require additional education. If you decide later to become a nurse, therapist, social worker, dietitian, or any of the many careers related to this field, you'll have the required courses to begin your postsecondary training.

### Postsecondary Training

Most home care agencies provide some kind of on-the-job training for their home health care aides. You might receive training in CPR (cardiopulmonary resuscitation), first aid, and how to transfer people. You're also likely to be trained to dress, feed, and bathe clients. The more specific the skill required for certain clients, the more an agency is likely to offer comprehensive training.

Closely related to a home health care aide is a certified nursing assistant (CNA). The work is very similar, but a CNA usually works in a health care or nursing home facility as opposed to a private home. Training programs for CNAs vary from state to state. In Montana, a home health care aide must be a CNA. The same is true in Illinois, but the majority of states don't require this training. CNA training usually consists of a combination of course work and on-the-job instruction in a nursing home facility totaling about eighty hours. A person training to be a CNA learns basic anatomy, how to take vitals and care for specific diseases like Alzheimer's and Parkinson's, and hospice work. CNAs also learn patient transfers, contamination prevention, and the basics of blood-born pathogens like AIDS. Many programs pay CNAs while they train, then issue a pay raise once they've become certified. Certification exams vary from state to state to meet federal requirements, and a CNA must pass a skills competency test before he or she can become certified.

The federal government has issued guidelines for home health care aides who are employed by agencies that receive reimbursement from Medicare, but specific requirements vary from state to state. Federal law requires home health care aides to pass a competency test in twelve areas, including communication skills; observation, reporting, and documentation of patient progress; basic nutrition; reading and recording vital signs; infection control procedures; maintaining a clean and healthy environment; personal hygiene and grooming; and safe transfer techniques. Most of these skills are taught by the employing agency. Federal law suggests that before taking the competency test, home health care aides receive at least seventy-five hours of combined classroom and practical training supervised by a registered nurse. If you go through a CNA training program, all the above requirements will be met.

Some secondary and postsecondary schools, especially vocational training institutes, offer programs for home health care aides and certified nursing assistants. Sometimes, these are dual programs. For more information about these schools, contact the National Association of Health Career Schools (*see* "Sources of Additional Information").

> I care for my people like they were my own parents or grandparents.

## certification or licensing

Certification to be a home health care aide varies from state to state, and most states do not require certification. Some states require more training than the federal guidelines specify, some up to 120 hours. Once you finish your required training, there is no standard certification or license you receive to be a home health care aide.

The National Association for Home Care (NAHC) offers voluntary National Homemaker-Home Health Aide certification that demonstrates an aide's ability to meet industry standards. Their certification standards include seventy-five hours of classroom training, completing a competency checklist of seventeen different skills, and passing a fifty-question multiple choice written exam. The association offers their certification through agencies and schools only, not to individuals. If you complete a formal training program, you may be able to apply, through the school, for NAHC certification. This certification may or may not transfer to other states. To find out an individual state's requirements for home health care aides, contact the health department of the state that interests you, or call the state's association of home health agencies.

## scholarships and grants

Most home care and nursing home facilities offer paid training for aides. To receive financial assistance to attend a school with a home health care training program, contact the school's financial aid office for information about the availability of grants and scholarships.

## internships and volunteerships

Volunteering in a nursing home facility or hospital is a good way to see if the home health care profession is right for you. Many facilities offer volunteer opportunities and welcome the extra help and personal touch that volunteers bring. To explore these opportunities, contact hospitals, nursing and residential care facilities, nonprofit agencies that serve the elderly and disabled, and state and county health and aging services near you for more information about their volunteer programs.

## who will hire me?

Home health care aides work mainly for privately owned home health agencies, visiting nurse associations, residential care facilities such as nursing homes with home visiting departments, hospitals, public health and welfare departments, community volunteer agencies, hospices, and nonprofit agencies that offer respite care. Also called home care organizations, the National Association for Home Care has identified 20,215 home care organizations in the United States and its territories as of December 1996. Of those, 12,181 are Medicare-certified agencies.

You can learn more about possible employment by contacting local agencies and programs that serve the elderly, disabled, and infirm. Visiting your city or county health department and contacting the personnel office may also provide useful information. Often, local organizations like hospices sponsor open houses to introduce their programs and services to the community. This could be a great way to meet the staff of an agency that provides home health care. You may even get to go on a home visit with an aide.

The National Home Care and Hospice Directory contains information on more than 18,500 hospice and home care providers. To obtain a copy of the directory, and for more information about home health care employment, contact the Home Care Aide Association of America (*see* "Sources of Additional Information").

## where can I go from here?

Becky loves her job and feels content working as an aide. At times she thinks about becoming a registered nurse or managing a retirement facility for the elderly. She's not sure she wants to go back to school, however, and either direction would require more education. As Becky readily admits, advancement possibilities as a home health care aide are limited and there are few opportunities in the field without more education. It may be possible to move into some supervisory and management duties, but more formal training is needed for these positions, too. Becky assists her supervisor with scheduling, client-aide matches, and other administrative duties, a position she got mainly because of her associate's degree and twelve years experience in the field. Every supervisor where Becky works is a registered nurse.

Being a private home health care provider and working for one client or family, possibly as a live-in, offers some possibility for higher pay and may appeal to some people more than working for a private or state agency. Even so, says Becky, the job is the same. People may wish to specialize in certain areas, such as working with hospice and the terminally ill, AIDS patients, or quadriplegics. Such work may require additional training, but could bring higher pay and increased responsibility.

Home health care aides who wish to achieve more professional positions and higher paying jobs must be willing to further their education and gain more experience, and it is wise to enter the field with this understanding.

## what are some related jobs?

Home health care aides can move into other caregiving and health-centered fields such as nursing, social work, child care, physical and occupational therapy, and nutrition and diet. Being a home health care aide is a good way to test your ability to be a caregiver and know if it's right for you. The U.S. Department of Labor classifies home health care aides under the headings *Service Occupation* (DOT) and *Humanitarian* (GOE). Also under these headings are nannies, preschool teachers and child care workers, psychiatric technicians, counselors and social workers, people who work in therapy and rehabilitation, and nurses. Under *Humanitarian* is Child and Adult Care, which includes Patient Care. Listed here are emergency medical techni-

## in-depth

In the last century, it wasn't unusual for family photographs to include grandmothers and grandfathers, or an elderly aunt or uncle, posed beside nieces and nephews, sons and daughters, and small children. A typical household often included an elderly parent or injured relative among its members, and parents typically moved in with their children when they got too old to do the strenuous chores that daily living required one hundred years ago. If a person had no family, they usually went to a sanatorium or hospital to live. These institutions were typically unpleasant, unless you were rich, and no one looked forward to spending their old age in one. People with disabilities, contagious diseases, or those in need of constant supervision were also confined to these institutions. Even with a family, the daily business of running a household left little time to tend to the needs of a severely sick or terminally ill family member.

Rural communities started to make visiting nurses available to check on people who lived far from town and didn't have transportation to come for regular doctor's visits.

Eventually, these nurses began to discover that homebound people needed more than just medical care. Patients were grateful for the company, emotional support, and kindness the visiting nurse showed them. Nurses began to spend time with their patients, reading to them, fixing them a cup of tea, and lending them a listening ear. As society grew more fast-paced, and both parents started to work out of the home, the demand for this kind of home attendant grew, and the home health care aide profession was born.

Modern medicine has made it possible for many illnesses to be treated at home, and keeping people in their homes is far less expensive than caring for them in a hospital or nursing home. The medical profession has also discovered that people recover better when they're at home. For all these reasons, home health care continues to grow at a fast pace and is considered a good way to meet people's health and medical needs.

cians, medical assistants, pharmacy technicians, dental assistants, and surgical technicians.

| **Related Jobs** |
| --- |
| Preschool teacher |
| Nannie |
| Nurse |
| Social worker |
| Emergency medical assistant |
| Child care provider |
| Physical therapist |
| Occupational therapist |
| Medical assistant |
| Surgical technician |
| Health care administrator |

## what are the salary ranges?

A home health care aide's salary and benefits depend a lot on where she or he works. According to a 1996 National Association for Home Care survey, the starting average salary for aides working in Medicare-certified agencies ranges from $5.25 to $8.29 an hour. Wages were somewhat higher in the Northeast and West and lower in the Midwest and South. Some aides working for larger agencies may earn as much as $10.00 an hour. Aides are sometimes paid on a per-visit basis, and most aides are not paid for their travel time between jobs. Aides in smaller private companies are regularly hired on a part-time or on-call basis. Larger agencies, however, generally pay better than smaller ones and are more likely to employ aides in full-time positions.

Whether you work for a private visiting nurse facility or a public health service could make a difference. Becky, who works for Yellowstone County, is considered a government employee.

Aides here start at $7.00 an hour and can earn up to $9.00. They receive full insurance benefits including a pension plan, sick and vacation leave, and thirteen paid holidays a year. In addition to gas reimbursement, aides at Yellowstone County are paid a wage for the entire time they work, including their travel time between jobs.

## what is the job outlook?

In 1991 home health care services employed a total of 344,000 people; in 1996, that figure rose to 697,000. The occupation of home health care aide is expected to be one of the fastest growing in the country through the year 2006. Most of the growth will come from the large number of people over seventy expected to need home care in the years to come, and the rapid turnover of employees this business produces. The high turnover is a reflection of the relatively little amount of training the job requires, the low pay, and the high emotional output the work demands. If you're interested in this field, enjoy it, and have even a little bit of experience, you should have no trouble finding a job.

The huge increase in home health care aide jobs in the early 1990s is due mainly to a revision in Medicare that took place in the late 1980s. With an increase in Medicare payments for home health care, the industry continues to grow. In 1963 there were only 1,100 home care agencies; today there are more than 20,000 agencies that offer some kind of home care. The number of Medicare-certified hospices has grown from 31 in 1984 to 2,154 in 1996. Employment for home health care aides has a lot to do with the continuation of Medicare payments. Unfortunately, the Balanced Budget Act of 1997 is designed to reduce the growth in home health care payments, something the industry is bracing for. "A cut in Medicare benefits for home health care means longer hospitals stays and more people in nursing homes," says Becky.

It is unlikely, though, that home health care will be abandoned anytime soon. It is more cost effective, better for recovery, and people want it. Although a cut in Medicare benefits for home care may mean a reduction in certain services for some people, home health care should remain one of the most viable options in health care today.

## how do I learn more?

### sources of additional information

**National Association for Home Care**
228 Seventh Street, SE
Washington, DC 20003
202-547-7424
www.nahc.org

**Home Care Aide Association of America**
228 Seventh Street, SE
Washington, DC 2003
202-547-7424

**Hospice Association of America**
228 Seventh Street, SE
Washington, DC 2003
202-547-7424

**National Association of Health Career Schools**
750 First Street, NE
Suite 940
Washington, DC 20002
NAHCS@aol.com

### bibliography

Following is a sampling of materials relating to the professional concerns and development of home health care aides.

#### Books

Cardoza, Anne. *Opportunities in Homecare Services Careers.* Lincolnwood: VGM Career Horizons, 1994.

Sherman, Margie. *Your Opportunities as a Home Health Aide.* Salem: Energeia, 1995.

#### Periodicals

*AFHHA Insider*. Monthly. Explores home health care industry issues. American Federation of Home Health Agencies, 1320 Fenwick Lane, Suite 100, Silver Spring, MD 20910, 301-588-1454.

*Homecare News.* Monthly. Newspaper. National Association for Home Care. 228 Seventh Street SE, Washington, DC 20003, 202-547-7424.

*Journal of Community Health Nursing*. Quarterly. Focuses on health care issues relevant to all aspects of community practice, including visiting nursing services. Lawrence Erlbaum Associates, Inc., 10 Industrial Way, Mahwah, NJ 07430, 201-236-9500.

# horticultural technician

**Definition**

Horticultural technicians plant, cultivate, and market plants to be used for ornamental purposes. They may also landscape and maintain both private and public grounds.

**Alternative job titles**

Arboriculture technicians

Floriculture technicians

Landscape development technicians

Nursery-operation technicians

Turf grass technicians

**Salary range**

$12,000 to $25,000 to $45,000

**Educational requirements**

High school diploma

**Certification or licensing**

None

**Outlook**

Faster than the average

GOE
02.02.02*

DOT
405*

O*NET
79999C

## High School Subjects

Agriculture
Biology
Chemistry
Earth science
Mathematics

## Personal Interests

Building things
Drawing
The Environment
Plants/Gardening

afternoon in early May, Randy Abell steps carefully between rows of knee-high blue spruce. His head horticulturist, who is kneeling at the base of one of the trees, looks up at his approach.

"Mark, I think these look the best, right here," Randy says, pointing at a nearby row. "Which formula did we use on these?" He and Mark were trying to grow a blue spruce with an even bluer color than is usual, by varying the amounts of nitrogen, phosphorous, potassium, and sulfur in the nutrient formulas.

Mark wipes his hands on his pants, and grins. "I thought those were the best ones, too," he says. "They got the first formula we made up—the one with the least nitrogen."

Randy nods his head. "Well, that's what we expected, then," he says. "Hey, I'll be out the rest of the afternoon. I'm going to meet with the City Planning Commission, and then I'm going out to Brookstone to start working up a design for the new phase of development. Why don't you up-pot those pin oaks?"

Mark stands up and stretches. "Sure will. Sounds like you've got a busy day."

Squinting into the late-afternoon sun, Randy smiles and nods. "Well, it's that time of year."

## what does a horticultural technician do?

The beauty of a thick, green carpet of grass, perfectly sculpted shrubbery, and the bright, splashy colors of a flower bed are pleasures almost everyone has enjoyed, whether at their own homes, at a public park, or simply driving along a highway. Most of us have also seen the other end of the horticultural spectrum—overgrown lots, lawns taken over by weeds, and flowers or houseplants that just won't thrive. Although some may attribute success with plants to a "green thumb," there is actually an art and a science to the placement, growth, and maintenance of trees, plants, and shrubs, as any *horticultural technician* can tell you.

Horticultural technology is typically broken down into the areas of floriculture, nursery operation, turf grass, arboriculture, and landscape development. Technicians may specialize in one of these areas, or may have responsibilities pertaining to several, or even all, areas. In every case, the technician must be well versed in the specific needs of the plants or trees with which he or she is working, including the methods of starting and growing them, soil conditions, temperature, moisture, and light requirements, pest and disease control, and appropriate fertilizers.

Both *floriculture technicians* and *nursery-operation technicians* work in nurseries or greenhouses to raise and sell plants. Floriculture technicians specialize in flowers, while nursery-operations technicians lean more toward shrubs, hedges, and trees. In many cases, however, the two areas may overlap. The duties of these technicians include planting seeds, transplanting seedlings, inspecting crops for problems, culling or pruning plants, and regulating the application of nutrients, herbicides, pesticides, and fungicides. They may also be responsible for planning growing schedules, quantities, and utilization of the growing space to achieve the best yield for the crops.

Technicians in the fields of floriculture and nursery operation sometimes hold a position known as *plant propagator.* Plant propagators initiate new kinds of plant growth, using techniques such as grafting, budding, layering, and rooting. They also develop and revise nutrient formulas for use on the new plants, select growth mediums, and prepare the containers to be used for growing. Other technicians may specialize in determining problems that are preventing plants from growing properly. Called *plant diagnosticians,* these technicians often work in retail nurseries solving problems for, and making recommendations to, customers who are having difficulties with their plants.

*Turf grass technicians* may work in either the private or the public sector. In the private sector, the technician is usually involved in providing lawn care services to homeowners' corporations, or institutions with extensive grounds, like private colleges or country clubs. These services may include mowing, trimming, and irrigation, as well as insect, disease, and weed control, and the proper use of fertilizers for particular types of grass and soil. In the public sector, technicians

**Fungicide** An agent applied to destroy fungi.

**Pesticide** A chemical preparation for destroying pests, such as mites.

**Herbicide** A substance or preparation used for killing plants, especially weeds.

**Mulch** A covering of straw, leaves, or manure spread on the ground around plants to enrich the soil and prevent evaporation and erosion.

**Taproot** A plant's main root, which descends downward.

**Spreader roots** A plant's smaller roots, which grow laterally.

**Weed barrier** A fabric, often used in landscaping, that permits air and moisture to pass through, but prevents weed growth.

**(B&B) "Balled and burlapped"** Refers to the method by which most large trees are transplanted, with their root balls covered in burlap.

**Up-pot** To transfer a plant from one container into a larger container, to allow it more root room and further growth.

**Sod** Ready-made squares of grass used to establish new lawns and renovate old ones.

usually work for local, state, or federal government agencies to plan turf grass areas for parks, playing fields, or grassy areas along public highways.

Some technicians also work in the production of sod, or in producing or selling seeds, fertilizers, insecticides, herbicides, and other equipment and supplies used in the area of horticulture. Finally, some work as *turf grass research and development technicians* for privately owned companies, or local, state, or federal agencies.

*Aboriculture technicians* may also work in either the private sector, owning their own business or working for an established company, or in the public sector, working for a park system, botanical garden, arboretum, or government agency. In both cases, technicians may plant, feed, prune, provide pest control, and diagnose problems or diseases for various types of trees. They may also use their knowledge of different tree types to determine where and how trees should be removed and planted. Some arboriculture technicians, known as *tree movers,* are responsible for the digging up, transportation, and replanting of trees, as well as estimating the cost of the removal. Others use their knowledge to sell trees, tools, fertilizers, peat moss, and other necessary equipment. These technicians also often work with customers to advise them about products and offer suggestions in planting and caring for their trees.

*Landscape development technicians* sketch and develop planting plans, using their knowledge of soil, temperature, and light conditions, and of specific plant requirements. They also compile lists of materials needed, and oversee or carry out the landscaping activities, including purchasing, planting, and caring for the decided-upon trees, shrubs, and plants.

These technicians may work for private companies that provide landscaping services directly to homeowners or businesses, or for a city or county, designing landscaping for parks, parking lots, streets, or municipal buildings. Some technicians in this field may also work for landscape architects, or begin their own businesses.

## what is it like to be a horticultural technician?

Randy Abell owns a small, "full-service" horticultural business. Full-service, as he defines it, means landscaping, groundskeeping, nursery, and garden center divisions. When Randy started Abell Nursery and Landscaping six years ago, he was its sole employee. As a result, he has worked in, and developed a familiarity with, each area of service. Although part of his time is now spent on managerial responsibilities for his ten-employee, $450,000 business, he still works in all phases of the horticultural process. "My typical day depends on which area of the business I'm working in," he says. "If I'm working in the nursery or the garden center, I'm staying in one place, basically. If I'm doing landscaping or lawn care, I'm on-site, wherever that may be."

> "It's like cooking with spices. A dish can turn out 100 different ways. So can a plant, depending on the nutrients."

If he'll be working in the nursery, Randy's day begins in the greenhouse, with the task of trimming and "up-potting" certain plant specimens. By up-potting, or transferring plants from smaller into larger containers, he allows the roots space to expand, thereby producing a larger, more valuable plant.

He must also water all the plants with the appropriate liquid fertilizers. Randy says there are 150 different fertilizers used at his nursery. "Every one has a different formulation that will produce a different result in a plant," he says. "It's like cooking with spices. A dish can turn out one hundred different ways, depending on the spices. So can a plant, depending on the nutrients." Randy works with one of his technicians to determine the most beneficial formulation for each plant. "That may involve a certain amount of research," he says.

On another day, Randy might work in the fields, planting seedlings. To ensure that the seedlings are planted in straight rows for precision in mowing and harvesting, a rope is stretched tightly from one end of a row to the other. Marks on the rope indicate where the seedlings should be planted. "Planting is one of the times when knowing a plant's characteristics is necessary. You need to know how far apart and how deep to plant," Randy says. "Of course, you also need to know what fertilizer type to use, and when to harvest."

Harvesting the plants may be another of the nursery worker's responsibilities, although it requires a more specialized skill. "The tree digger removes the trees, balls, and burlaps them," Randy says. "To do that, you have to be able to look at a tree and know how big a root ball to dig. It's tricky."

Individualized care of various plant specimens is the order of the day when working in the nursery. Individualized lawn care is the business of the groundskeeping crews. "Working on a groundskeeping crew means following a schedule for fertilizing, mowing, trimming, and weed spraying," Randy says. "Every client—every lawn—has a set schedule and a page in the log book. Whenever we do work on that lawn, it gets documented."

Knowing the type, amount, and frequency of fertilizer and weed spray for a particular lawn is a part of the groundskeeping or turf grass technician's job. Another part is knowing when and how to mow. "You go to a job and check the conditions," Randy says. "Sometimes, you'll need to change the mowing height." For example, if the grass is wet, raising the mowing height creates less of a mess. If the ground is very dry, raising the mowing height prevents damage to the grass. If the ground is too wet, it might be wise to postpone mowing till another day. "You have to know what will do the best job," he says.

When working on a landscaping crew, Randy's responsibilities are more varied. Since every job is different, in terms of soil type, lay of the land, proportion of sun to shade, and the client's guidelines and budget, he must be able to plan and coordinate a number of factors to achieve the desired result.

Working from a landscaping design or blueprint, he reads the scale of the design to determine spacing of the plants he is installing. At present, Randy or a design architect are the only ones who develop the landscaping plans for his clients. Randy hopes that won't always be the case. "A technician with a good education and a good experience level could do design," he says. "You have to be aware of which plants will look good and grow well together. You also have to consider soil conditions and drainage and all that."

Once the design has been established, it is basically a matter of laying out the plants accordingly, and planting them. "Once again, you have to know how to plant each type, and what nutrients to add," Randy says. Some soil or drainage conditions require the application of extra materials, such as peat, or the installation of drainage pipes. Some jobs also require landscape construction. "I build retaining walls, borders. Sometimes I contract out for decks or patios," he says.

To finish up a landscaping job, Randy installs a fabric weed barrier on the ground around the plants, cutting a slit or an "X" for each plant to grow through. He then covers the weed barrier with the top dressing chosen by his client, which might be mulch, pine nuggets, or various types of crushed stone. A final duty involves explaining to the client how to care for his or her new plants.

The garden center workers must be the most knowledgeable technicians, according to Randy. "Working in there, you have to know everything."

The garden center is the retail side of the business, where customers come to buy trees, flowers, plants, equipment, and fertilizers. Usually, only one person works in the center at a time, dealing with all the customers who come in. Responsibilities include helping them select their plants and instructing them in planting and care, suggesting fertilizers, pesticides, and fungicides, helping them load up their purchases, and making delivery arrangements. Perhaps the most difficult part of working in the garden center is fielding questions. "I answer questions about why customers' plants aren't doing well, how they can get their flowers to bloom, how tall trees will be at their mature height, what plants look pretty together," Randy says. "I need to either know the answers, or know where to look for the answers to all kinds of questions."

In addition to dealing with the customers, those technicians who work in the garden center often help out in the greenhouse or in the fields.

**To be a successful horticultural technician, you should**

- Enjoy growing and caring for plants and flowers
- Be willing to work outdoors, in varying weather conditions
- Be a good problem solver
- Have patience to perform detailed work
- Work well with your hands
- Be able to distinguish between hundreds, even thousands, of different plants

## have I got what it takes to be a horticultural technician?

Horticultural technicians work very closely with various forms of plant life. They must be able to work effectively, on a basic, hands-on level, performing such activities as planting, watering, and pruning. They must also be able to understand the more theoretical and scientific aspects of horticulture, such as the chemical and biological, in order to determine the growth needs of specific plants.

"This job is anything but boring," Randy says. "There are so many different aspects of it. If you're working in lawn maintenance or landscaping, you're always in different places. If you're in the nursery, you're working with different plants in different stages of growth. If you're in the garden center, you're dealing with different customers and finding solutions to all kinds of problems."

The weather is also an aspect of this job that is variable. "Because of this, it is important for technicians, particularly those in turf grass management and landscaping, to be flexible," he says. "When your work is largely dependent on weather conditions, you have to be flexible. You can't get really rigid about a schedule, because bad weather may come along and just blow that all to pieces." In some areas of the country, horticultural technicians, particularly those in turf grass management and landscaping, may find it difficult to find a job that lasts through the winter months. Conversely, the work load in the spring and summer is often quite heavy, requiring extra long days.

Weather conditions may prove to be a pro or a con of the job, depending upon the season, and the day. "Oh, in the spring, it's great to be in this business," Randy says. "Everyone's cooped up in their offices, and we're outside. But in late July, when it's 100 degrees, not too many people will be jealous of us." Depending upon the climate and temperatures of the region, and upon the area of specialization, horticultural technicians may have to work in rain, mud, or extreme heat or cold. According to Randy, unless the technician plans to work exclusively in a greenhouse or garden center, he or she should enjoy working outdoors, and be able to tolerate varying weather conditions.

It is important for the technician to be able to think on his or her feet, and be a good problem solver. "Sometimes, if you're working in the nursery center," Randy says, "a customer will come in carrying just a little leaf, or a twig, and they'll say 'What is this? I want more of them.'"

FYI

**Tree Trivia**

The life span of a tree in the best urban setting is only about sixty years, while the same species of tree, in a rural forest, may live several times that long.

The technician also gets questions about planting, fertilizing, and caring for various plants. "We grow about fifty varieties, and deal with maybe two hundred," Randy says, "so there's a certain amount of fact-finding and research involved in answering some questions."

The horticultural technician's work may be physically demanding, depending upon the type of duties he or she is performing. Workers specializing in arboriculture, for example, may be required to climb trees. Many tasks include digging, bending, squatting, and lifting.

The hard work, however, is rewarding, Randy says. "In some jobs, you can't see the physical results of your work. This you can see—for years to come. It's gratifying to see your work all over town, to see trees that you started from seedlings." According to him, the business is pleasant, as well as rewarding. "You're always the good guy," he says. "Everyone likes trees and flowers, so you're doing something that makes people happy."

## how do I become a horticultural technician?

Randy started his career with no formal training in horticulture. "I had some experience from working on my dad's tree farm. I'd say maybe I knew five percent of what I needed to know," he says, laughing. As his business developed, so did his knowledge of the field, partly through experience, and partly through training classes and seminars he attended. "The state I live in has a Master Plantsman program through the local chapter of the American Association of Nurserymen," Randy says. "I used that to get a horticultural education, since my teaching degree wasn't going to help me very much in a nursery and landscaping business."

## education

### High School

Entry-level horticultural technicians often begin their careers with only a high school diploma. While in high school, courses in biology and chemistry may be helpful in acquainting the prospective tech with the workings of plant life, and the effects of various nutrients. A background in geometry is also a useful tool, especially in landscape design and maintenance, where area, circumference, and square footage must frequently be determined.

English classes are recommended to increase verbal and written skills, and a background in Latin is very helpful, since those in the industry refer to many plants by their Latin names.

Finally, if classes in botany, horticulture, or agriculture are offered, take them. If these classes are unavailable at your high school, you might consider some other methods of obtaining the same information. Often, local chapters of horticultural associations, area nurseries, or greenhouses offer short classes or workshops in horticultural principles and techniques, which are open to the public for a small fee.

### Postsecondary Training

There are a number of choices for the person considering further training in horticulture. The right path depends upon what type of work you hope to do.

For the horticultural technician starting his or her first job straight out of high school, the majority of training will take place on the job. This is especially common in the areas of turf grass management and landscaping, where entry-level jobs for high school grads are plentiful.

There are also a number of nondegree or certificate programs, which could be taken before starting a career in horticulture, or after, in conjunction with on-the-job training. Many botanical gardens and arboreta offer educational programs as a community service. Many county cooperative extension offices also offer short courses geared toward the general public or the nursery trade, such as the Master Gardener Program. The American Association of Botanical Gardens and Arboreta (AABGA) offers a certificate course, based on a booklet of fundamental topics in practical horticulture. After studying the booklet, candidates may take a written and practical exam. Upon passing the exam, they receive the North American Certificate in Horticulture, which certifies that they have attained the equivalent, in education and experience, of a botanical garden internship, or a two-year degree (*see* "Sources of Additional Information").

A two-year associate's degree is yet another option for post-high school education. An associate's degree in horticultural technology generally requires classes that stress the technical or hands-on aspects of a specialized field. A sampling of possible classes includes plant propagation, lawn and turf management, nursery and garden center management, pest management, plant taxonomy, plant materials, landscape techniques, and fundamentals of horticulture. Many programs may also require some classes in English, computers, math, and science. For the student who wants to enter the horticultural field at a technical or middle management level, an associate's degree is a good choice.

There are also a multitude of programs that lead to bachelor's, master's, and doctorate degrees in various aspects of horticulture. For the student who is motivated to pursue higher education in this field, the sky is the limit.

## scholarships and grants

Many schools that offer undergraduate degrees in horticulture also offer to qualified applicants financial aid in the form of scholarships and grants. The interested student should check with representatives of any given school to see what types of assistance may be available. The Society of American Florists also maintains a list of scholarships available to students specializing in floriculture (*see* "Sources of Additional Information").

Some scholarships or grants may be available through such sources as local civic groups, floriculture, horticulture, or garden organizations, and business groups. Check directly with them for information on educational financial assistance.

## internships and volunteerships

The American Association of Botanical Gardens and Arboreta (AABGA) has a very strong internship program. Offering a directory of internships in more than one hundred public gardens throughout the United States, AABGA is a valuable resource for students who are interested in a practical experience in horticulture.

The length and content of internships vary from garden to garden, as do the wages paid, but most share certain features, such as rotating duties in several areas of the garden, formal and informal classes, field trips, garden programs, and special projects. Specialized internships focus on conservation of rare plants, arid

## in-depth

### Horticultural Therapy

Therapy for plants? No, therapy *with* plants! Horticultural therapists work with the emotionally and mentally handicapped, the blind and disabled, senior citizens, and inner city youth to grow and nurture plants. Through such therapy projects as growing and propagating flowers and houseplants, making flower arrangements, and starting terrariums, their clients receive the physical and emotional benefits of working with living things. Therapists may also be involved in training the mentally retarded, and otherwise disabled persons, for jobs in the field of horticulture. For more information on education, jobs, and developments in the field of horticultural therapy, contact The American Horticultural Therapy Association at 800-634-1603.

### Floral Design

Since ancient times, people have used flowers to express love, congratulations, or sympathy. Today, still, the facet of floriculture known as floral design is thriving. Floral designers create arrangements of live, cut, potted, dried, and artificial flowers and foliage. Their artistic and creative talents are evident in funeral arrangements, table centerpieces, get-well gifts, corsages, and bridal bouquets. Floral designers may learn their skills on the job, through adult education courses, through vocational or commercial schools, or by receiving a degree in floristry or flower shop management from a two-year college. For more information, contact The Society of American Florists at 800-336-4743.

land horticulture, historic garden restoration, plant inventories, children's programs, and zoo horticulture.

In addition, check with your neighbors to see if you can help them with their lawn care and gardening. A summer job mowing lawns is a good introduction into the field.

## who will hire me?

Randy grew up primed for a horticultural career, living on his parents' tree farm until he went to college. Even so, he was certain that he didn't want to join the family business, deciding instead, to get a B.S. in education. It wasn't until he had completed his education and begun his first teaching job that he realized he wasn't going to be happy working inside. He decided to pursue a career in horticulture after all.

"I started out on the lawn maintenance side," he says, "because that was the easiest way to start, and I just expanded from there." Although he started his own business immediately, as opposed to first working for someone else, he now says he would not advise most people to do that. "I think it'd be much easier to work for someone else first, and learn a little bit," he says. "I would especially suggest a job in a nursery or a garden center. That would be great experience."

The area of specialization that the prospective technician chooses to pursue will determine, to some degree, where he or she should look for a job. For those technicians who are interested in floriculture or nursery operation, a logical place to begin is by applying to all greenhouses and nurseries in the area. The job seeker should also remember that many large retail department stores have garden centers, which require knowledgeable staff members. Public botanical gardens and arboreta may also provide good opportunities for employment in these areas of specialization.

Technicians interested in the area of turf grass management might look in the local yellow pages under "lawn care" to get ideas for where to send resumes or applications. They might also check any area golf courses and cemeteries for entry-level openings. Finally, federal, state, and local government agencies may be a good source of possibilities, since they often employ turf grass technicians to maintain public areas.

Prospective arboriculture technicians should check with local park systems, botanical gardens, and arboreta for job openings. Also, like turf grass technicians, arboriculture techs are often hired by government agencies to maintain public areas. If the technician chooses to work in the private sector, he or she might check the "tree service" listings in the local phone book to compile a list of possible contacts. Likewise, landscape development technicians might want to check the phone book to compile a list of area landscape contractors and landscape architects.

Job seekers in all areas of horticulture should regularly check the classified advertising sections of all area newspapers. Entry-level horticultural jobs are frequently listed, especially in the spring and summer months. One particularly good way to break into the field might be for the high school student with an interest in horticulture to take an entry-level summer job in a nursery, public garden, lawn care or landscaping company, or a tree service business. This might provide him or her with an entry into the business on a permanent basis, as well as good job experience.

Technicians who decide to pursue formal schooling in horticulture may get assistance in finding a job through their college placement offices. Finally, there are a number of associations and publications geared toward the horticultural industry, which may be of help to the job seeker (*see* "How Do I Learn More?").

## where can I go from here?

Randy has now achieved his most basic career goal, by owning his own full-service horticultural business. His plans are to expand and refine the existing facets of his company. "What I wanted was to be a full-service company, and now I am," he says. "But some areas of the business are still rudimentary, and I want to develop them much further."

Becoming the owner of their own businesses is a common goal for many workers in certain areas of horticulture, particularly those in turf grass management and landscape development. There are, however, a vast number of other career steps that the horticultural technician could take. Because the field is so wide and diverse, there is no set advancement pattern.

Technicians in the areas of floriculture and nursery operation might move into the position of *horticultural inspector.* Inspectors work for state or federal agencies to inspect plants being transported across state lines or entering the United States for insects, diseases, and legality. With additional experience and further schooling, these techs may also choose to become *horticulturists,* either at a research facility or a large firm.

Turf grass technicians might eventually become *greens superintendents.* Greens superintendents plan for and oversee the construction and maintenance of golf courses, including the supervision of other workers. Other possibilities in the area of turf grass are consulting, sod growing, and research and development positions.

Arboriculture technicians may, with further training, become *tree surgeons,* or *city foresters.* Also, they may decide to start their own businesses, offering all types of tree services, including planting, pruning, spraying, and removal.

The technician who pursues a career in landscape development often advances by becoming the manager or owner of his or her own small landscaping business. They might also move into supervisory positions at public parks, cemeteries, or botanical gardens. By returning to school to obtain a bachelor's degree, the technician could become a *landscape architect.*

Further schooling is also an option for horticultural technicians in other areas of specialization. Bachelor's, master's, or doctoral degrees in horticultural science will open up further possibilities for advancement in the areas of research or teaching.

### Advancement Possibilities

**Horticulturists** conduct experiments and investigations into problems of breeding, production, storage, processing, and transit of fruit, nuts, berries, flowers, and trees; develop new plant varieties; and determine methods of planting, spraying, cultivating, and harvesting.

**City foresters** advise communities on the selection, planting schedules, and proper care of trees. They also plant, feed, spray, and prune trees and may supervise other workers in these activities.

**Turf grass consultants** analyze turf grass problems and recommend solutions. They also determine the best type of turf grass to use for specified areas, mowing schedules, and techniques.

## what are some related jobs?

The U.S. Department of Labor classifies horticultural workers under several headings: *Horticulture Specialty Occupations, Gardening and Groundskeeping Occupations, Plant Life and Related Service Occupations, Tree Farming and Related Occupations,* and *Agriculture, Forestry, and Fishing Managers and Officers* (DOT); *General Supervision, Plants and Animals: Nursery and Gardening, Elemental Work, Plants and Animals: Nursery and Gardening, and Managerial Work, Plants and Animals: Specialty Cropping* (GOE). Included under the Nursery and Groundskeeping heading are individuals who maintain the grounds in cemeteries, and on industrial or commercial sites, who work in seasonal greenery and Christmas trees, and who specialize in growing certain types of flowers, such as roses or orchids. The Plant Specialization category includes workers in the fields of botany, agronomy, mycology, and soil conservation, in addition to horticulture.

| Related Jobs |
| --- |
| Soil conservationists |
| Wood technologists |
| Forest ecologists |
| Tree surgeon helpers |
| Greens superintendents |
| Greenskeepers |
| Flower pickers |
| Rose-grading supervisors |
| Plant breeders |
| Plant pathologists |
| Christmas-tree graders |

## what are the salary ranges?

Since the field of horticulture encompasses so many different types of positions, salary levels for technicians vary significantly. In addition to the technician's chosen area of work, there are several other factors that influence the amount of compensation. When considering potential earnings, the technician should keep in mind that government agencies, colleges, and universities usually pay more than privately owned companies or gardens. Also, salary may vary between regions of the country, with those workers in the Midwest generally earning the highest wages, and those in the southern states earning the lowest. As is the case in most areas of employment, the technician's levels of education and experience are significant factors. Finally, a worker who is able to specialize in a specific function, such as propagating plants, will likely command a higher salary.

Technicians just beginning in either landscape development, turf grass management, or floriculture and nursery operation might expect to earn between $16,000 and $20,000. An overall average wage for technicians in these fields is $25,000, with the average for a managerial position being in the mid-thirties. With experience, a grounds manager might earn as much as $60,000 and nursery operators might earn as much as $45,000.

Beginning arboriculture technicians tend to earn slightly less than workers in the other horticultural fields. Ground workers might expect to make between $15,600 and $29,120 yearly, while those who have to climb the trees might earn as much as $10,000 more a year.

## what is the job outlook?

The job prospects for horticultural technicians are currently favorable. For a number of reasons, their employment is expected to increase faster than average for all occupations through the year 2006.

One major reason for the increase is the continued development and redevelopment of urban areas. With the construction of commercial and industrial buildings, shopping malls, homes, highways, parks, and recreational facilities, there should be an increasing demand for turf grass, arboriculture, and landscaping services. Because this growth will also create a need for more trees, shrubs, and plants, it should translate into an increased demand for nursery operation and floriculture technicians as well.

Another reason for the growth in job opportunities is America's increasing awareness of and commitment to preserving and maintaining the environment. The skills of educated and experienced horticultural technicians will be increasingly necessary to city planners and development committees in their efforts to retain the beauty of nature in the midst of urban expansion. Homeowners and developers are also increasingly using landscaping services to enhance the value and attractiveness of their properties, and many existing buildings and facilities are upgrading their landscaping, as well.

The number of horticultural technicians' jobs may also increase as American homeowners become more and more service oriented. In the past several years, there has been a trend toward homeowners hiring professionals to perform specific tasks, as opposed to performing these tasks themselves. Landscaping, lawn care, and tree services are all examples of this, along with pest control, painting, decorating, repair work, and cleaning services. Obviously, this should create the need for more skilled technicians in these areas.

Finally, because gardening is the number one hobby in the world, there is a continuing, large global market for plants and flowers. Simply the number of trees, plants, shrubs, and flowers purchased yearly—4.5 billion dollars' worth—ensures, to some extent, a favorable job outlook for floriculture and nursery operation technicians.

Employment opportunities in the area of landscaping are tied to local economic conditions. During downturns, the demand for these services often slows, as corporations, governments, and homeowners cut out nonessential spending, which may include landscaping.

# how do I learn more?

## sources of additional information

Following are organizations that provide information on horticultural technician careers, accredited schools and scholarships, and possible employment.

**American Association of Botanical Gardens and Arboreta**
786 Church Road
Wayne, PA 19087
610-688-1120

**Society of American Florists and Ornamental Horticulturists**
1601 Duke Street
Alexandria, VA 22314-3406
703-836-8700

**American Society for Horticultural Science**
600 Cameron Street
Alexandria, VA 22314
703-836-4606

**American Horticultural Therapy Association**
362-A Christopher Avenue
Gaithersburg, MD 20879
800-634-1603

## bibliography

Following is a sampling of materials relating to the professional concerns and development of horticultural technicians.

### Books

Jozwik, Francis X. *How to Find a Good Job Working with Plants, Trees, and Flowers.* Mills, WY: Andmar Press, 1994.

Preece, John E. and Paul E. Read. *The Biology of Horticulture: An Introductory Textbook.* New York: Wiley, 1993.

Rice, Laura Williams and Robert P. Rice, Jr. *Practical Horticulture.* 3rd ed. Englewood Cliffs, NJ: Regents/Prentice Hall, 1996.

### Periodicals

*American Gardener.* Bimonthly. Directed toward both professional and amateur horticulturalists. American Horticultural Society, 7931 East Boulevard Drive, Alexandria, VA 22308, 703-768-5700.

*American Nurseryman.* Biweekly. Discusses the nursery business and topics pertinent to their personnel. American Nurseryman Publishing Co., 77 West Washington Street, Suite 2100, Chicago, IL 60602-2904, 312-782-5505.

*American Society for Horticultural Science Journal.* Bimonthly. Stresses scientific research in the horticultural field. American Society for Horticultural Science, 600 Cameron Street, Alexandria, VA 22314-2562, 703-836-4606.

# job title index